Test Bank for Atkins and Jones's

Chemistry: Molecules, Matter, and Change

Third Edition

Robert J. Balahura
University of Guelph

W. H. Freeman and Company
New York

ISBN: 0-7167-2794-3

Printed in the United States of America

First printing 1997, RRD

Contents

Chapter 1: Matter
Form A

1. The chemical symbol for manganese is
a) Mg b) Md c) Ma d) Mn e) Mo

2. What is the chemical symbol for the element iron?
a) Fe b) I c) Ir d) In e) F

3. The name of the element with symbol Sn is
a) tin b) silicon c) scandium d) selenium e) samarium

4. The number of protons, neutrons, and electrons, respectively, in a neutral atom of ^{19}F (Z = 9) is
a) 9 protons, 10 neutrons, and 9 electrons
b) 9 protons, 19 neutrons, and 9 electrons
c) 19 protons, 10 neutrons, and 19 electrons
d) 10 protons, 9 neutrons, and 10 electrons
e) 9 protons, 9 neutrons, and 9 electrons

5. Which one of the following neutral atoms has 8 protons, 8 neutrons, and 8 electrons?
a) ^{19}F (Z = 9) b) ^{18}O (Z = 8) c) ^{15}N (Z = 7) d) ^{16}O (Z = 8) e) ^{14}C (Z = 6)

6. An isotope of the element uranium has a mass number of 235 and an atomic number of 92. The number of electrons, protons, and neutrons, respectively, in a neutral atom of this isotope is
a) 92, 93, 142 b) 92, 92, and 235 c) 143, 143, and 235 d) 92, 143, and 92 e) 92, 92, and 143

7. Neutral atoms of oxygen-16, oxygen-17, and oxygen-18 all have
a) 16 protons
b) 16 protons and 16 electrons
c) 8 neutrons
d) 8 neutrons and 8 protons
e) 8 electrons and 8 protons

8. The relation between the number of neutrons, N, the mass number, A, and the atomic number, Z, is
a) $N = A + Z$ b) $Z = A + N$ c) $N = A - Z$ d) $N = (A - Z)/2$ e) $A = N - Z$

9. Which of the following has 17 protons, 18 neutrons, and 18 electrons?
a) ^{40}Ar (Z = 40) b) $^{35}Cl^-$ (Z = 17) c) $^{32}S^{2-}$ (Z = 16) d) $^{31}P^{3-}$ (Z = 15) e) ^{28}Si (Z = 14)

10. All of the following have 36 electrons except
a) $^{87}Sr^{2+}$ (Z = 38) b) $^{85}Rb^+$ (Z = 37) c) $^{80}Br^-$ (Z = 35) d) $^{79}Se^{2-}$ (Z = 34) e) $^{84}Kr^{2+}$ (Z = 36)

11. Oxygen has an atomic number of 8 and the isotopes ^{16}O, ^{17}O, and ^{18}O all have
a) 8 neutrons.
b) 8 protons and 10 neutrons.
c) 8 neutrons and 8 electrons.
d) 8 protons and 8 electrons.
e) 8 protons and 8 neutrons.

12. Neon-20 (Z = 10), fluorine-20 (Z = 9), and sodium-20 (Z = 11) all have
a) 11 neutrons. b) 20 neutrons. c) 20 electrons. d) 20 protons. e) a mass number of 20.

13. Fill in the blanks and the correct symbol for "E" in the following table.

Symbol = E	Atomic Number	Number of Electrons	Number of Neutrons
E	26	23	30
^{32}E (Z = 16)	___	___	___
E	19	19	20

14. Fill in the blanks and the correct symbol for "E" (or its ions) in the following table.

Symbol = E	Atomic Number	Number of Electrons	Number of Neutrons
$_{20}E^{2+}$	___	___	20
E^{3+}	___	10	14
E^{1-}	___	10	10

15. Fill in the blanks and the correct symbol for "E" (or its ions) in the following table.

Symbol = E	Atomic Number	Number of Electrons	Number of Neutrons
$_{15}E^{3-}$	___	___	25
E^{-}	___	18	18
E^{+}	___	36	48

16. Which of the following is an alkali metal?
a) K b) Ca c) Sc d) Be e) Ba

17. Which of the following is an alkaline earth metal?
a) Ba b) Cs c) Rb d) Y e) V

18. Which of the following is a halogen?
a) Br b) P c) O d) As e) Rn

19. All of the following are halogens except
a) Cl b) S c) I d) At e) F

20. Which of the following is a transition metal?
a) Mn b) Ca c) Ga d) Al e) Sr

21. All of the following are metalloids except
a) P b) Si c) Ge d) As e) Sb

22. All of the following are group 18 elements except
a) Kr b) Ne c) Xe d) I e) He

23. All of the following elements are solids at room temperature and atmospheric pressure except
a) Hg b) Ba c) I d) Sn e) Cd

24. Which of the following is a noble gas?
a) Xe b) I c) Po d) F e) N

25. Lithium nitride is composed of
a) a group 1 cation and a group 15 anion.
b) a group 1 cation and a group 14 anion.
c) a group 1 cation and a group 16 anion.
d) a group 2 cation and a group 15 anion.
e) a group 2 cation and a group 14 anion.

26. All of the following are atoms except
a) O_2 b) N c) Cl d) Na e) S

27. Which of the following elements are most likely to form a cation?
a) P b) Be c) S d) I e) N

28. Which of the following is likely to form an anion?
a) P b) Ba c) V d) Rb e) Zn

29. Which of the following metals commonly forms a +2 charge?
a) Cs b) Rb c) Li d) Ba e) K

30. Which of the following nonmetals readily forms a –2 charge?
a) P b) N c) S d) Br e) Se

31. What is the charge on the cation in $(NH_4)_3PO_4$?
a) +1 b) +3 c) –3 d) –1 e) +6

32. What is the charge on the cation in $(NH_4)_2CO_3$?
a) +1 b) +2 c) +8 d) –2 e) –6

33. What is the charge on the anion in NaClO?
a) –1 b) –2 c) –3 d) +1 e) –4

34. What is the charge on the anion in $K_2Cr_2O_7$?
a) –1 b) –2 c) –14 d) –6 e) –12

35. An ionic compound has the formula M_2CO_3. What is the charge on M?
a) +1 b) +4 c) +3 d) +2 e) 0

36. Predict the formula of the cation of gallium based on its position on the periodic table.
a) Ga^{3+} b) Ga^{+} c) Ga^{2+} d) Ga^{4+} e) Ga^{3-}

37. Predict the formula of the anion of arsenic based on its position on the periodic table.
a) As^{4-} b) As^{3+} c) As^{3-} d) As^{4+} e) As^{2-}

38. Which of the following would be classified as a mixture?
a) seawater b) boiling water c) ice d) sugar e) baking soda

39. Which of the following is a pure substance?
a) salt b) bourbon whiskey c) tea d) wine e) brass

40. If 1 g of sodium perchlorate is dissolved in water, the sodium perchlorate is referred to as the
a) solute. b) solution. c) solvent. d) precipitate. e) spectator.

41. A homogeneous mixture can be described as
a) one prepared by shaking flour with water.
b) one in which the composition of the mixture is the same throughout the sample.
c) a solution like milk.
d) a substance like a rock.
e) one which is a patchwork of aggregates of different substances.

42. All of the following are heterogeneous mixtures except
a) vodka. b) milk. c) a rock. d) sour cream. e) a precipitate in water.

43. Which of the following is an alloy?
a) A metallic element.
b) Solid copper mixed with molten copper.
c) A homogeneous mixture of two different metallic elements.
d) A homogeneous mixture of two different nonmetals.
e) A metallic coating of one metal on another different metal.

44. Which of the following is a chemical property of methane (natural gas)?
a) Methane-air mixtures burn.
b) Methane is a colorless, odorless gas at room temperature.
c) Methane melts at 89 K.
d) Liquid methane is a poor conductor of electricity.
e) Solid methane has a density of 0.415 g/mL.

45. Which of the following is a chemical property of zinc?
a) Zinc conducts electricity.
b) Zinc is bluish-white in color.
c) Zinc melts at 1180 K.
d) Zinc dissolves in hydrochloric acid with the evolution of hydrogen gas.
e) The density of zinc is 7.14 g/cm^3.

46. Filtration is the process
a) used to separate two gases.
b) used to separate dissolved salts from seawater.
c) used to separate a mixture of sand and gold dust.
d) used to remove dissolved sugar from water.
e) used to separate a solid from a liquid.

47. A solution of two solutes X and Y are passed through a column containing an adsorbent. Solute X was somewhat retained by the adsorbent and the first portion of the solution to emerge from the column contained only solute Y. This type of separation is called
a) chromatography. b) filtration. c) centrifugation. d) recrystallization. e) distillation.

48. Pure water could be obtained from seawater by
a) distillation. b) chromatography. c) precipitation. d) filtration. e) centrifugation.

49. An example of a chemical change is
a) the melting of ice to form water.
b) the evaporation of water on a hot day.
c) the formation of frost when moist air passes over a cold surface.
d) the boiling of water to form steam.
e) the burning of hydrogen in air to form water.

50. An example of a physical property is
a) the reaction of rubidium with water to form rubidium hydroxide.
b) the density of boron.
c) the burning of sulfur to form sulfur dioxide.
d) the reaction of cesium with oxygen to form cesium superoxide.
e) the energy content of liquid sodium.

51. At 25°C, chlorine is a green-yellow poisonous gas with a density of 3×10^{-3} g/cm^3. Chlorine has a melting point of −101°C and a boiling point of −35°C, and the energy required to melt and boil chlorine is 6.4 and 20.4 kJ/mol, respectively. Chlorine burns in hydrogen to form hydrogen chloride. Which of the following is a chemical property of chlorine?
a) Chlorine requires energy to boil.
b) Chlorine liberate energy when it freezes.
c) Chlorine melts at −101°C.
d) Chlorine burns in hydrogen to form hydrogen chloride.
e) Chlorine is green-yellow in color.

52. Potassium is a soft, shiny gray metal which has a density of 0.9 g/cm^3. Potassium melts at 64°C and boils at 774°C, and the energy required for melting and boiling is 2.4 and 79 kJ/mol, respectively. When potassium is added to water, potassium hydroxide is formed along with hydrogen, which ignites during the reaction. Potassium also reacts with oxygen to form KO_2, with chlorine to form KCl, and with sulfur to form K_2S. Which of the following is a physical property of potassium?
a) density
b) Potassium reacts with water.
c) Potassium forms KO_2 when it reacts with oxygen.
d) Potassium reacts with sulfur.
e) Potassium reacts with chlorine.

53. Paper chromatography separations are based on the fact that
a) the components to be separated interact differently with a liquid phase moving along a paper and the paper itself.
b) liquids are adsorbed on calcium carbonate.
c) the components can be distilled.
d) a carrier gas is unreactive.
e) the components to be separated are volatile.

54. Describe how you would separate the main components of a mixture prepared by mixing dirt with an aqueous solution of table salt.

55. Consider the distillation apparatus below. Describe the purpose of parts A, B, and C.

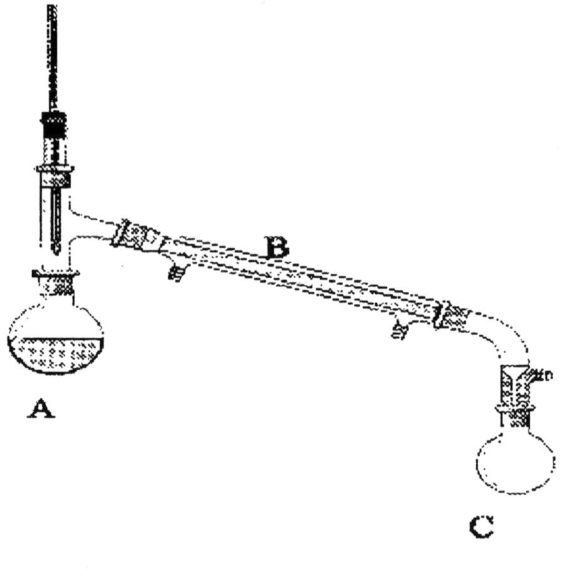

56. What is the modern name for the ferrous ion?
a) ferrous(III) ion b) iron(II) ion c) ferrous(II) ion d) ferric ion e) iron(III) ion

57. What is the modern name for the mercurous ion?
a) mercuric ion b) mercurous(I) ion c) mercury(II) ion d) mercury(I) ion e) mercurous(II) ion

58. A light-green aqueous solution was labeled "iron(III) chlorate". This solution contains the ions
a) Fe^{3+} and ClO^-. b) Fe^{3+} and ClO_2^-. c) Fe^{3+} and ClO_4^-. d) Fe^{2+} and ClO_4^-. e) Fe^{3+} and ClO_3^-.

59. A green aqueous solution was labeled "nickel(II) chloride". This solution contains the ions
a) $2Ni^+$ and $2Cl^-$. b) Ni_2O^{2+} and Cl^-. c) Ni^{4+} and Cl^-. d) NiO^{2+} and Cl^-. e) Ni^{2+} and Cl^-.

60. The name of Br^- is the
a) bromate ion. b) brominium ion. c) bromide ion. d) bromite ion. e) bromine(I) ion.

61. Which of the following is the phosphide ion?
a) PO_3^{3-} b) P^- c) P^{3-} d) PO_4^{3-} e) P^{2-}

62. The name of ClO^- is the
a) chlorite ion. b) oxychloride ion. c) hypochlorous ion. d) perchlorate ion. e) chloric ion.

63. Which of the following is the perchlorate ion?
a) Cl^- b) ClO_4^- c) ClO_3^- d) ClO^- e) ClO_2^-

64. The name of the parent acid of the chlorite ion is
a) hydrogen chloride.
b) hydrochloric acid.
c) hypochlorous acid.
d) chlorous acid.
e) chloric acid.

65. The formula for sulfurous acid is
a) H_2SO_4. b) H_2SO_3. c) H_2S. d) $H_2S_2O_4$. e) HSO.

66. Lead(II) iodide is a bright yellow crystalline compound that is not very soluble in water. What is the formula of this compound?
a) PbI_2 b) PbI c) Pb_2I_4 d) Pb_2I_2 e) PbI_4

67. Which of the following statements is true regarding mass spectrometry?
a) A mass spectrometer is used to separate the components of a homogeneous solution.
b) A mass spectrometer is used to measure the density of a sample.
c) A mass spectrometer is used to weigh very small samples.
d) A mass spectrometer is used to separate the components of a heterogeneous mixture.
e) A mass spectrometer is used to measure the mass and abundance of an isotope.

68. Which of the following ions has a –2 charge?
a) chlorite ion b) phosphate ion c) chlorate ion d) nitrite ion e) sulfite ion

69. Which of the following has a –3 charge?
a) nitride ion b) nitrite ion c) perchlorate ion d) hydrogen phosphate ion e) sulfite ion

70. Which of the following is boron nitride?
a) BN b) B_2N c) B_3N_2 d) BC e) B_3N_2

71. Which of the following is beryllium nitride?
a) Be_2N_2 b) BeN_2 c) Be_3N_2 d) BeN e) Be_2N

72. The binary compound CaS is
a) sulfur calcide. b) calcium sulfite. c) calcous sulfide. d) calcium sulfide. e) calcium sulfate.

73. Which of the following is calcium phosphate?
a) Ca_3PO_4 b) Ca_2PO_4 c) $CaPO_4$ d) $Ca_3(PO_4)_2$ e) $Ca_3(PO_3)_2$

74. The name of the compound Co_2O_3 is
a) dicobalt trioxide.
b) cobalt(III) trioxide.
c) cobalt(II) oxide.
d) cobalt(III) oxide.
e) dicobalt oxide.

75. The name of the compound $CuCl_2{\cdot}2H_2O$ is
a) cuprous(II) chloride dihydrate.
b) copper(III) chloride dihydrate.
c) copper(II) chloride dihydrate.
d) cuprate(II) chloride dihydrate.
e) cupric(III) chloride dihydrate.

76. A faintly-yellow aqueous solution was labeled "$NH_4Fe(SO_4)_2$". Which of the following names corresponds to this label?
a) ammonium iron(II) sulfate
b) ammonium iron(III) sulfate
c) iron(III) hydrogen sulfate
d) ammonia iron(II) hydrogen sulfate
e) iron(II) ammonia sulfate

77. The name of P_2O_5 is
a) diphosphorus oxide.
b) phosphorus pentoxide.
c) phosphorus oxide.
d) diphosphorus pentoxide.
e) diphosphoric oxide.

78. The formula of phosphorus pentachloride is
a) PCl_5 b) P_2Cl_{10} c) PCl_3 d) P_2Cl_5 e) PCl_4

79. Which of the following is diarsenic pentoxide?
a) AsO_5 b) As_2O_5 c) As_3O_5 d) AsO_4 e) As_2O_{10}

80. The formula of hydrazine is
a) N_2H_3 b) NH_3. c) N_2H_4. d) HN_3. e) NO_2.

Chapter 1: Matter
Form B

1. The chemical symbol for potassium is
a) K b) P c) Po d) Pt e) At

2. What is the chemical symbol for the element calcium?
a) Ca b) Cu c) Cs d) C e) Ce

3. The name of the element with symbol Cr is
a) chromium b) copper c) carbon d) cobalt e) cadmium

4. The number of protons, neutrons, and electrons, respectively, in a neutral atom of ^{121}Sb (Z = 51) is
a) 121 protons, 70 neutrons, and 121 electrons
b) 51 protons, 51 neutrons, and 51 electrons
c) 51 protons, 70 neutrons, and 51 electrons
d) 70 protons, 51 neutrons, and 70 electrons
e) 51 protons, 121 neutrons, and 51 electrons

5. Which one of the following neutral atoms has 19 protons, 20 neutrons, and 19 electrons?
a) ^{40}Ca (Z = 20) b) ^{43}K (Z = 19) c) ^{18}Ar (Z = 18) d) ^{19}F (Z = 9) e) ^{39}K (Z = 19)

6. An isotope of the element potassium has a mass number of 39 and an atomic number of 19. The number of electrons, protons, and neutrons, respectively, in a neutral atom of this isotope is
a) 19, 20, and 19 b) 19, 39, and 20 c) 19, 19, and 20 d) 20, 19, and 19 e) 19, 19, and 39

7. Copper has an atomic number of 29. Neutral atoms of copper-63 and copper-65 both have
a) 29 neutrons and 29 protons
b) 36 protons
c) 29 neutrons
d) 34 protons
e) 29 protons and 29 electrons

8. Which of the following has 17 protons, 18 neutrons, and 17 electrons?
a) ^{31}P (Z = 15) b) ^{28}Si (Z = 14) c) ^{35}Cl (Z = 17) d) ^{32}S (Z = 16) e) ^{40}Ar (Z = 18)

9. Which of the following has 10 electrons?
a) $^{27}Al^{3+}$ (Z = 13) b) $^{31}P^{3-}$ (Z = 15) c) ^{24}Mg (Z = 12) d) $^{32}S^{2-}$ (Z = 16) e) ^{28}Si (Z = 14)

10. Which of the following has 20 neutrons?
a) $^{35}Cl^{-}$(Z = 17) b) $^{37}Cl^{-}$(Z = 17) c) ^{40}Ar(Z = 18) d) ^{31}P(Z = 15) e) $^{32}S^{2-}$(Z = 16)

11. Magnesium has an atomic number of 12 and the isotopes ^{24}Mg, ^{25}Mg, and ^{26}Mg all have
a) 12 neutrons.
b) 12 protons and 12 neutrons.
c) 12 protons and 12 electrons.
d) 12 neutrons and 12 electrons.
e) 12 protons and 14 neutrons.

12. Argon-40 (Z = 18), potassium-40 (Z = 19), and calcium-40 (Z = 20) all have
a) 40 electrons. b) a mass number of 40. c) 40 protons. d) 20 neutrons. e) 40 neutrons.

13. Fill in the blanks and the correct symbol for "E" in the following table.

Symbol = E	Atomic Number	Number of Electrons	Number of Neutrons
E	24	24	28
^{75}E (Z = 33)	___	___	___
E	38	38	49

14. Fill in the blanks and the correct symbol for "E" (or its ions) in the following table.

Symbol = E	Atomic Number	Number of Electrons	Number of Neutrons
$_{19}E^{+}$	___	___	20
E^{2-}	___	18	16
E^{2+}	___	10	12

15. Fill in the blanks and the correct symbol for "E" (or its ions) in the following table.

Symbol = E	Atomic Number	Number of Electrons	Number of Neutrons
$_{11}E^{+}$	___	___	12
E^{2-}	___	10	8
E^{-}	___	10	10

16. Which of the following is an alkali metal?
a) Li b) Be c) Ba d) Sc e) Sr

17. Which of the following is an alkaline earth metal?
a) Mg b) K c) Li d) Sc e) Cr

18. All of the following are halogens except
a) Se b) Cl c) F d) Br e) At

19. All of the following are group 15 elements except
a) Si b) P c) As d) Sb e) Bi

20. Which of the following is a metalloid?
a) B b) P c) Se d) C e) As

21. Which of the following is a noble gas element?
a) Ar b) Au c) Cl d) Se e) At

22. All of the following elements are gases at room temperature and atmospheric pressure except
a) Br b) Cl c) O d) N e) Ne

23. Which of the following is a group16 element?
a) Br b) As c) Sb d) Se e) P

24. Which of the following is a group 14 element?
a) Si b) Al c) P d) As e) S

25. Calcium selenide is composed of
a) a group 2 cation and a group 16 anion.
b) a group 2 cation and a group 15 anion.
c) a group 1 cation and a group 16 anion.
d) a group 1 cation and a group 15 anion.
e) a group 2 cation and a group 17 anion.

26. All of the following are atoms except
a) C b) H_2 c) Ne d) Si e) Br

27. Which of the following elements are most likely to form an anion?
a) K b) Cs c) Sr d) Se e) Ba

28. Which of the following is likely to form a cation?
a) Al b) P c) At d) N e) Se

29. All of the following elements commonly form a +1 charge except
a) Be b) Na c) Cu d) Cs e) Rb

30. Which of the following elements readily forms a +3 charge?
a) Al b) Ca c) Na d) Si e) As

31. What is the charge on the cation in $Ca_3(PO_4)_2$?
a) +6 b) +2 c) +3 d) −2 e) −3

32. What is the charge on the cation in $Hg_2(ClO_4)_2$?
a) +2 b) +1 c) +4 d) −1 e) −2

33. What is the charge on the anion in $Pb(NO_3)_2$?
a) −6 b) −2 c) −1 d) −3 e) +2

34. What is the charge on the anion in $Cr_2(SO_3)_3$?
a) −2 b) −6 c) −3 d) −4 e) +3

35. An ionic compound has the formula $M_2(SO_4)_3$. What is the charge on M?
a) +3 b) +2 c) +6 d) +1 e) −2

36. Predict the formula of the cation of silicon based on its position on the periodic table.
a) Si^{2+} b) Si^{4-} c) Si^{3+} d) Si^{6+} e) Si^{4+}

37. Predict the formula of the anion of Se based on its position on the periodic table.
a) Se^{3-} b) Se^{2-} c) Se^{-} d) Se^{4-} e) Se^{2+}

38. Which of the following would be classified as a mixture?
a) air b) graphite c) salt d) methane e) sodium sulfite

39. Which of the following is a pure substance?
a) dry ice (solid carbon dioxide) b) seawater c) blood d) air e) coffee

40. If a solution of silver perchlorate is added to a solution of sodium iodide, solid silver iodide forms. The silver iodide is called a
a) precipitate. b) condensate. c) solute. d) spectator. e) filtrate.

41. All of the following are homogeneous mixtures except
a) coffee with cream. b) vodka. c) sugar dissolved in water. d) household vinegar. e) apple juice.

42. A heterogeneous mixture can be described as
a) one in which the composition of the mixture is the same throughout the sample.
b) a patchwork of aggregates of different substances.
c) being made up of a substance dissolved in water.
d) a solution like that obtained by dissolving table salt in water.
e) a mixture that has the same properties throughout the mixture.

43. Which of the following is a mixture?
a) brass b) ice in water c) sodium dihydrogen phosphate d) C_{60} e) sulfur trioxide gas

44. Which of the following is a chemical property of sulfur?
a) Sulfur is yellow.
b) Sulfur burns in oxygen to form sulfur dioxide.
c) The density of sulfur.
d) Sulfur is tasteless and almost odorless.
e) Sulfur melts at 388 K.

45. Which of the following is a chemical property of ozone, O_3?
a) Ozone is slightly soluble in water.
b) Liquid ozone boils at 161 K.
c) Ozone plus silver produces silver oxide.
d) Ozone gas has a sharp odor.
e) Ozone is a light blue gas at room temperature.

46. Muddy water is placed on a porous surface and clear water allowed to seep through. This process is called
a) filtration. b) recrystallization. c) chromatography. d) centrifugation. e) distillation.

47. A homogeneous mixture contains two pure compounds dissolved in water. The two compounds could be separated by
a) chromatography. b) addition of acid. c) addition of base. d) filtration. e) centrifugation.

48. Distillation can be used to
a) separate a mixture of sand and iron filings.
b) separate nitrogen and oxygen gases.
c) separate carbon dioxide and carbon monoxide gases.
d) separate a volatile component from a solid.
e) purify a precipitate.

49. Which of the following is a chemical property?
a) Sodium metal is soft.
b) Lithium reacts with nitrogen to form lithium nitride.
c) Bromine has a density of 3.1 g/cm^3.
d) Sodium hydrogen carbonate is white.
e) Sulfur melts at 115°C.

50. An example of a physical property is
a) the melting of sodium metal.
b) the reaction of sodium to give sodium hydroxide.
c) the reaction of sodium with chlorine to give sodium chloride.
d) the burning of sodium to give Na_2O_2.
e) the reaction of sodium with sulfur to give sodium sulfide.

51. At 25°C, fluorine is a pale yellow gas with a density of 1.6 x 10^{-3} g/cm^3. The melting point of F_2(g) is –220°C and the boiling point is –183°C, and the energy required to melt and boil fluorine is 0.51 and 6.5 kJ/mol, respectively. Fluorine is the most reactive element, and water actually burns in fluorine to form HF(g). Which of the following is a chemical property of fluorine?
a) Fluorine is pale yellow in color.
b) Fluorine has a density of 1.6 x 10^{-3} g/cm^3.
c) Fluorine boils at –183°C.
d) Fluorine reacts with water to give HF(g).
e) Fluorine liberates energy when it condenses.

52. Potassium is a soft, shiny gray metal which has a density of 0.9 g/cm^3. Potassium melts at 64°C and boils at 774°C, and the energy required for melting and boiling is 2.4 and 79 kJ/mol, respectively. When potassium is added to water, potassium hydroxide is formed along with hydrogen, which ignites during the reaction. Potassium also reacts with oxygen to form KO_2, with chlorine to form KCl, and with sulfur to form K_2S. Which of the following is a chemical property of potassium?
a) Potassium is soft.
b) Potassium liberates energy when it solidifys.
c) Potassium boils at 774°C.
d) Potassium reacts with water.
e) Potassium melts at 64°C.

53. Gas-liquid chromatography can be used to separate
a) lead and cadmium in lakes.
b) iron filings from gold.
c) table salt from dirt on beaches.
d) sodium carbonate from sodium hydrogen carbonate.
e) the components of gasoline.

54. Describe how you would separate a mixture containing methanol, propanol, and butanol.

55. Describe, in detail, how you would apply the scientific method to the following cases.
(a) A fellow student makes the statement "Application of mineral oil to my skin is just as effective as using expensive sun screen preparations."
(b) Your mother sends you off to university with a basket of apples saying "An apple a day keeps the doctor away."
(c) You read in the newspaper that the concentration of aluminum in the local water supply is increasing.

56. What is the modern name of Fe^{3+}?
a) iron(III) ion b) ferrous ion c) ferric(III) ion d) ferric ion e) iron(II) ion

57. What is the modern name of Hg_2^{2+}?
a) mercury(I) ion b) mercurous(I) ion c) mercuric ion d) mercury(II) ion e) mercurous(II) ion

58. A sky-blue aqueous solution was labeled "chromium(II) perchlorate". This solution contains the ions
a) Cr_2O^{2+} and ClO_4^-.
b) Cr_2O^{4+} and Cl^-.
c) Cr^{2+} and Cl^-.
d) Cr^{2+} and ClO_4^-.
e) Cr^{3+} and ClO_4^-.

59. A faintly-green aqueous solution was labeled "ammonium iron(II) sulfate." This solution contains the ions
a) NH_4^+, Fe^{3+}, and SO_4^{2-}.
b) Fe^{2+}, and $NH_3SO_4^{2-}$.
c) Fe^{3+} and SO_4^{2-}.
d) NH_4^+, Fe^{2+}, and SO_4^{2-}.
e) NH_4^+, Fe^{2+}, and SO_3^{2-}.

60. The name of N^{3-} is the
a) hydronitrogen ion. b) nitride ion. c) nitrite ion. d) pernitrate ion. e) nitrate ion.

61. Which of the following is the sulfide ion?
a) SO_4^{2-} b) S^- c) SO_3^{2-} d) S^{2-} e) S^{3-}

62. The name of BrO^- is the
a) bromite ion. b) oxybromide ion. c) hypobromous ion. d) perbromate ion. e) bromic ion.

63. Which of the following is the sulfite ion?
a) HSO_4^- b) SO_4^{2-} c) SO_3^{2-} d) HSO_3^- e) S^{2-}

64. The name of the parent acid of the nitrite ion is
a) hydronitric acid. b) nitrous acid. c) pernitrous acid. d) nitric acid e) pernitrous acid.

65. The formula for perchloric acid is
a) $HClO_4$. b) $HClO$. c) $HClO_3$. d) $HClO_2$. e) $HClO_5$.

66. Silver chromate is a red crystalline compound that is not very soluble in water. What is the formula of this compound?
a) Ag_2CrO_4 b) $AgCrO_4$ c) $Ag_2(CrO_4)_2$ d) Ag_4CrO_4 e) Si_2CrO_4

67. Which of the following statements is true regarding mass spectrometry?
a) A mass spectrum is a plot of the detector signal against the speed of an ion.
b) A mass spectrum is a plot of the detector signal against the magnetic field.
c) A mass spectrum is used to calculate the ionization potential of an atom.
d) A mass spectrum is a plot of mass against ionization potential of an atom.
e) A mass spectrum allows is used to determine the shape of a molecule.

68. Which of the following has a –1 charge?
a) hydrogen phosphate ion b) phosphite ion c) sulfite ion d) nitrite ion e) carbonate

69. Which of the following ions have a –2 charge?
a) sulfide ion b) hydrogen sulfate ion c) nitrate ion d) chlorite ion e) hydrogen carbonate ion

70. Which of the following is gallium sulfide?
a) Ga_2S_3 b) Ga_2S c) GaS d) GaS_3 e) GaS_2

71. Which of the following is gallium arsenide?
a) $GaAs$ b) Ga_2As c) $GaAs_2$ d) $GaAs_4$ e) Ga_2As_3

72. The binary compound GaP is
a) gallium phosphate.
b) gallium phosphide.
c) gallium phosphorus.
d) phosphorus gallide.
e) gallium phosphite.

73. Which of the following is potassium hydrogen phosphate?
a) KH_2PO_4 b) $K_2H_2PO_4$ c) K_2HPO_4 d) $KHPO_4$ e) K_3PO_4

74. The name of the compound Hg_2Cl_2 is
a) mercury(II) chloride.
b) mercury(I) chloride.
c) dimercury dichloride.
d) mercury dichloride.
e) dimercury chloride.

75. The name of the compound $NiCl_2{\cdot}6H_2O$ is
a) nickelate(II) chloride hexahydrate.
b) nickel(II) chloride hexahydrate.
c) nickelic(III) chloride hexahydrate.
d) nickelous(II) chloride hexahydrate.
e) nickel(III) chloride hexahydrate.

76. A solution was prepared by dissolving cobalt(II) nitrate hexahydrate in water. This light pink solution contains the ions

a) Co^{2+} and NO_3^-.
b) Co^+ and NO_3^-.
c) Co^{3+} and NO_3^-.
d) Co_2O^{2+} and NO_3^-.
e) Co^{2+} and NO_2^-.

77. The name of N_2O_4 is

a) dinitrogen dioxygen.
b) dinitrogen tetroxide.
c) dinitrogen tetraoxygen
d) dinitrogen oxide.
e) nitrogen oxide.

78. Which of the following is antimony trichloride?

a) $SbCl_3$ b) A_2Cl_6 c) $AnCl_3$ d) ACl_3 e) Sb_2Cl_6

79. Which of the following is sulfur trioxide?

a) S_3O_3 b) S_2O_3 c) SO_3 d) S_2O_6 e) SO_4

80. The formula of phosphine is

a) P_2H_3. b) P_2H_4. c) HP_3. d) PH_5. e) PH_3.

Chapter 2: Measurements and Moles
Form A

1. The prefix corresponding to the factor 10^{-3} is
a) milli. b) deci. c) nano. d) centi. e) pico.

2. The prefix corresponding to the factor 10^{-2} is
a) centi. b) milli. c) micro. d) pico. e) nano.

3. The prefix "centi" corresponds to the factor
a) 10^{-1}. b) 10^{-2}. c) 10^{-6}. d) 10^{-9}. e) 10^{-3}.

4. A millisecond and a microsecond are
a) 10^{-3} s and 10^{-6} s, respectively.
b) 10^{-6} s and 10^{-9} s, respectively.
c) 10^{-3} s and 10^{-12} s, respectively.
d) 10^{-3} s and 10^{-9} s, respectively.
e) 10^{-2} s and 10^{-6} s, respectively.

5. A bottle of cola purchased in Europe gave the volume as 50 cL. What is the volume in mL?
a) 0.005 L b) 50 mL c) 0.05 L d) 5000 mL e) 500 mL

6. The atmosphere of the earth extends about 60 miles. If 1 km = 0.6214 mi, what distance does this correspond to in km?
a) 97 km b) 9700 km c) 37 km d) 124 km e) 3700 km

7. Which of the following is longest?
a) 2.0×10^{-10} cm b) 2.0 nm c) 2.0×10^{-4} 1m d) 200 pm e) 2.0×10^{-9} dm

8. All of the following are intensive properties of a substance except
a) mass. b) melting point. c) density. d) pressure. e) temperature.

9. A 5-g sample of lead is bluish-white in color and is very soft and malleable. It has a melting point of 600 K and a boiling point of 1893 K. At 20°C its density is 11.35 g/cm^3. Which property of lead is an extensive property?
a) The melting and boiling points of lead.
b) The mass of lead.
c) The density of lead.
d) The soft, malleable qualities of lead.
e) The color of lead.

10. All of the following are intensive properties of a substance except
a) internal energy. b) temperature. c) density. d) solubility. e) color.

11. The density of lead is 11.4 g/mL. Express the density in g/m^3.
a) 1.14×10^{6} g/m^3 b) 1.14×10^{3} g/m^3 c) 1.14×10^{7} g/m^3 d) 1.14×10^{-5} g/m^3 e) 1.14×10^{4} g/m^3

12. A gold coin with a mass of 96.5 g is placed in exactly 100 mL of water in a graduated cylinder. If 5.0 mL of water are displaced from the cylinder, what is the density of the coin?
a) 92 g/mL b) 4.8 g/mL c) 5.2 g/mL d) 38 g/mL e) 19 g/mL

13. Lead is obtained from the mineral galenite. The storage bin for the galenite must be able to hold 50,000 kg of galenite. If the density of galenite is 7.51 g/cm^3, what volume does the storage bin need to be so as to hold the galenite?
a) 6.66 m^3 b) 6.66 x 10^4 m^3 c) 376 m^3 d) 6.66 x 10^{-3} m^3 e) 150 m^3

14. A gold brick measures 5.00 x 10.0 x 20.0 cm. If the density of gold is 19.3 g/cm^3, what is the mass of the brick in pounds? (1 lb = 454 g)
a) 4.26 lb b) 235 lb c) 42.6 lb d) 23.5 lb e) 425 lb

15. Exactly 1 m^3 is equal to
a) 1000 cm^3. b) 1000 L. c) 100 L. d) 10^{-3} L. e) 1 L.

16. The carbon-oxygen bond enthalpy in carbon monoxide is 256 kcal/mol. If 4.184 joules $\neq$ 1 calorie, what is the bond enthalpy in kJ/mol?
a) 61.2 kJ/mol b) 1.07 x 10^6 kJ/mol c) 16.3 kJ/mol d) 6.12 x 10^3 kJ/mol e) 1070 kJ/mol

17. The Br–Br bond length in $Br_2(g)$ is 228 pm. If 1 / = 10^{-10} m, what is the bond length in /?
a) 2.28 / b) 0.228 / c) 1.14 / d) 228 / e) 22.8 /

18. An automobile from Germany had a gasoline comsumption rating of 20 km/L. What is the American rating for this car in mi/gal? (1.6 km = 1 mi and 1 gal = 3.785 L)
a) 47 mi/gal b) 8.5 mi/gal c) 76 mi/gal d) 32 mi/gal e) 120 mi/gal

19. A fast chemistry professor runs a 10 km race at the rate of exactly 6 min/mi. If 1 mi = 1.61 km, how long does it take to run the race?
a) 26.8 min b) 37.3 min c) 36.0 min d) 9.66 min e) 96.6 min

20. What is the Fahrenheit temperature that corresponds to 0 K?
a) –523°F b) –434°F c) –549°F d) –459°F e) –120°F

21. What is the Celsius temperature that corresponds to 98.5°F?
a) 54.7°F b) 36.9°F c) 66.5°F d) 86.7°F e) 22.7°F

22. Which of the measured numbers below has the greatest number of significant figures?
a) 0.3100 b) 0.00310 c) 310. d) 3.1 x 10^{-4} e) 0.000310

23. How many zeros are significant in the measured number 0.001050 miles?
a) 2 b) 1 c) 3 d) 4 e) 5

24. What answer should be reported if 0.57 cm is subtracted from 10.003 cm?
a) 9 cm b) 9.430 cm c) 9.43 cm d) 9.433 cm e) 9.4 cm

25. The number 0.000650 expressed in scientific notation is
a) 6.50×10^{-3} b) 6.5×10^{4} c) 6.5×10^{-4} d) 6.50×10^{-4} e) 6.5×10^{-3}

26. When the calculation

$$4.60 \times 9.315 - 1.0 \times 10^{-2}$$

is carried out, the number of significant figures in the answer is
a) 3. b) 4. c) 2. d) 5. e) 6.

27. When the calculation

$$(32.8070 \times 1.0200) \div 0.000201$$

is carried out, the number of significant figures in the answer is
a) 3. b) 6. c) 5. d) 4. e) 2.

28. Chlorine has two naturally occurring isotopes, Cl-35 and Cl-37, with natural abundance of 75.77% and 24.23%, respectively. The mass of an atom of Cl-35 is 5.807×10^{-23} g and that of an atom of Cl-37 is 6.139×10^{-23} g. What is the average molar mass of chlorine?
a) 34.97 g/mol b) 36.48 g/mol c) 35.97 g/mol d) 36.97 g/mol e) 35.45 g/mol

29. Gallium has two naturally occurring isotopes, Ga-69 and Ga-71, with natural abundance of 60.20% and 39.80%, respectively. The molar mass of Ga-69 is 68.9256 g/mol and the molar mass of Ga-71 is 70.9247 g/mol. What is the average molar mass of gallium?
a) 69.93 g/mol b) 70.13 g/mol c) 69.72 g/mol d) 70.00 g/mol e) 68.93 g/mol

30. If the molar mass of Ni is 58.71 g/mol, what mass of Ni contains 3.022 mol Ni?
a) 19.43 g b) 29.48 g c) 177.4 g d) 51.47 g e) 88.70 g

31. How many moles of nitrogen atoms are contained in 10.62 g of nitrogen gas, N_2?
a) 0.3790 mol b) 2.638 mol c) 0.7580 mol d) 1.319 mol e) 148.8 mol

32. The molar mass of gallium atoms in a natural sample is 69.72 g mol^{-1}. The sample consists of Ga-69 of molar mass 68.9256 g mol^{-1} and Ga-71 of molar mass 70.9247 g mol^{-1}. What are the percentage abundances of these two isotopes?

33. A natural sample of an element consists of three isotopes with the following percentage abundances and molar masses.

% Abundance	Molar Mass, g mol^{-1}
35.39	150.9377
35.25	151.9791
29.36	156.9332

What is the molar mass of the natural sample?

34. How many grams of chlorine dioxide are needed to have 0.250 mol?
a) 22.4 g b) 24.7 g c) 16.9 g d) 67.5 g e) 24.5 g

35. What is the molar mass of nitric acid?
a) 47 g/mol b) 63 g/mol c) 79 g/mol d) 31 g/mol e) 48 g/mol

36. What is the molar mass of ammonium nitrate?
a) 64.05 g/mol b) 160.1 g/mol c) 79.04 g/mol d) 94.06 g/mol e) 80.05 g/mol

37. What is the molar mass of vitamin C, $C_6H_8O_6$?
a) 88.05 g/mol b) 188.1 g/mol c) 176.1 g/mol d) 140.1 g/mol e) 160.1 g/mol

38. How many oxygen atoms are present in 1 millimole of ozone, O_3?
a) 1.81×10^{24} b) 6.02×10^{20} c) 6.02×10^{26} d) 1.81×10^{21} e) 1.81×10^{19}

39. A sample of silver chloride has a mass of 9.000 g. If the sample is found to contain 6.774 g of silver, what is the mass percent silver in the sample?
a) 75.27% b) 24.73% c) 92.69% d) 60.36% e) 49.46%

40. The number of grams of silver in 9.000 g silver chloride is
a) 8.343 g. b) 2.226 g. c) 6.774 g. d) 7.527 g. e) 6.280 g.

41. The mass percent oxygen in copper(II) sulfate pentahydrate is
a) 57.7% b) 36.1% c) 40.1% d) 25.6% e) 90.2%

42. How many grams of oxygen are combined with other elements in 62 g $CuSO_4{,}5H_2O$?
a) 36 g b) 4.0 g c) 16 g d) 25 g e) 21 g

43. How many grams of bromine are required to produce 50.0 g of magnesium bromide?
a) 100. g b) 294 g c) 43.4 g d) 21.7 g e) 3.68 g

44. How many grams of lead are required to produce 75.0 g of lead(IV) oxide?
a) 3.19 g b) 661 g c) 57.3 g d) 86.6 g e) 65.0 g

45. How many grams of copper are required to produce 25.0 g of copper(I) oxide?
a) 5.72 g b) 364 g c) 22.2 g d) 20.0 g e) 11.1 g

46. A red compound composed of lead and oxygen contains 90.66% Pb. What is the empirical formula of the compound?
a) Pb_3O_4 b) PbO c) PbO_2 d) PbO_9 e) Pb_9O

47. The mass percent composition of hydrogen peroxide is 94.1% oxygen and 5.90% hydrogen. What is the empirical formula of hydrogen peroxide?
a) HO b) H_3O c) H_2O_2 d) HO_2 e) H_2O_3

48. Calcium metal of mass 1.284 g reacts with phosphorus to form 1.945 g of calcium phosphide. What is the empirical formula of calcium phosphide?
a) CaP_2 b) Ca_3P_4 c) CaP_3 d) Ca_2P e) Ca_3P_2

49. Iron of mass 2.00 g reacts with sulfur to form 4.296 g of "fool's gold". What is the empirical formula of "fool's gold"?
a) FeS_2 b) FeS c) FeS_3 d) Fe_2S_3 e) Fe_2S

50. Xenon of mass 5.08 g reacts with fluorine to form 9.49 g of a xenon fluoride. What is the empirical formula of this compound?
a) XeF_2 b) XeF_4 c) XeF_6 d) XeF e) Xe_2F

51. Which of the following formulas must be an empirical formula?
a) As_4O_6 b) Na_2O_2 c) P_4O_{10} d) $Na_2S_2O_4$ e) $Al_2(SO_4)_3$

52. The molar mass of hydrazine is 32 g mol^{-1} and its empirical formula is NH_2. What is its molecular formula?
a) N_2H_4 b) HH_3 c) NH_2 d) N_4H_8 e) N_2H_2

53. The molar mass of aspirin is 180 g mol^{-1} and its empirical formula is $C_9H_8O_4$. What is its molecular formula?
a) $C_{18}H_{16}O_8$ b) $C_{12}H_{11}O_7$ c) $C_{27}H_{24}O_{12}$ d) $C_9H_8O_4$ e) $C_9H_{12}O_8$

54. The molar mass of fructose, the main sweetener in jams, is 180 g mol^{-1} and its empirical formula is CH_2O. What is its molecular formula?
a) $C_8H_{16}O_8$ b) $C_6H_{12}O_6$ c) $C_6H_{10}O_4$ d) $C_4H_8O_4$ e) $C_3H_6O_3$

55. Ketoprofen is an anti-inflammatory drug which is 75.57% C, 5.55% H, and 18.88% O. If the molecular mass of ketoprofen is 254 g mol^{-1}, what is its molecular formula?

56. Ascorbic acid, commonly called vitamin C, is 41.39% C, 3.47% H, and 55.14% O. If the molecular mass of vitamin C is 174 g mol^{-1}, what is its molecular formula?

57. Calculate the molarity of a solution prepared by dissolving 0.584 g NaCl in enough water to prepare 25.00 mL of solution.
a) 9.99×10^{-3} M b) 1.36 M c) 23.4 M d) 0.399 M e) 0.0234 M

58. Calculate the molarity of a solution prepared by dissolving 2.073 g $AgClO_4$ in enough water to prepare 75.00 mL of solution.
a) 1.333×10^{-4} M b) 0.1333 M c) 9.999×10^{-3} M d) 0.02764 M e) 0.7499 M

59. Calculate the molarity of a solution prepared by dissolving 4.328 g Na_2SO_4 in enough water to prepare 1250 mL of solution.
a) 0.3278 M b) 3.278×10^{-2} M c) 2.098×10^{-4} M d) 0.5714 M e) 0.02098 M

60. Calculate the molarity of a solution prepared by dissolving 20.0 g of glucose, $C_6H_{12}O_6$, in enough water to prepare 2500 mL of solution.
a) 0.0444 M b) 0.111 M c) 4.44×10^{-5} M d) 0.278 M e) 1.44 M

61. Calculate the molarity of a solution prepared by dissolving 16.834 g $LiClO_4$ in enough water to prepare 1.500 L of solution.
a) 0.1055 M b) 4.213 M c) 0.1582 M d) 1.055×10^{-4} M e) 11.22 M

62. Calculate the mass of sodium chloride required to make 75.00 mL of a 0.538 M NaCl(aq) solution.
a) 6.90 g b) 8.16 g c) 2.36 g d) 0.690 g e) 23.6 g

63. Calculate the mass of sodium sulfate required to make 750.0 mL of a 0.152 M Na_2SO_4(aq) solution.
a) 4.54 g b) 0.691 g c) 29.9 g d) 18.8 g e) 0.114 g

64. What volume of 0.202 M NaCl solution should you transfer so as to have a sample of solution that contains 1.00 millimole NaCl?
a) 0.202 mL b) 4.95 L c) 4.95 mL d) 2.02 L e) 202 mL

65. What volume of 0.463 M urea, $(NH_2)_2CO$(aq), should you transfer so as to have a sample of solution that contains 1.00 mol urea?
a) 21.6 mL b) 2.16 L c) 0.463 L d) 0.463 mL e) 2.16 mL

66. If you have available 200 mL of 0.500 M NaCl(aq), what volume of this solution should you transfer so as to have a sample of solution that contains 0.150 mol NaCl?
a) 100 mL b) Not enough 0.500 M NaCl(aq) is available. c) 75 mL d) 30 mL e) 15 mL

67. A mass of 0.839 mg of the analgesic acetaminophen, $C_8H_9NO_2$, is required in an experiment. The analgesic is only sightly soluble in water but is available as a 5.00×10^{-4} M soution. What volume of solution must be used to deliver this mass?
a) 3.94 mL b) 5.55×10^{-3} mL c) 11.1 mL d) 5.55 mL e) 90.1 mL

68. A solution was prepared by dissolving 5.00 g sodium chloride in 100.0 mL of 1.49 M NaCl(aq). What is the molarity of NaCl in the final solution?
a) 6.50 M b) 0.00570 M c) 0.856 M d) 2.35 M e) 2.20 M

69. A solution is prepared by dissolving 26.8 g of KOH in approximately 100 mL of water and then adding water to a final volume of 250.0 mL. What is the molarity of KOH(aq) in this solution?
a) 3.18 M b) 0.478 M c) 4.78 M d) 0.191 M e) 1.91 M

70. Acetic acid, CH_3COOH, can be purchased as "glacial acetic acid," which is 100% CH_3COOH and has a density of 1.05 g mL^{-1}. Calculate the molarity of glacial acetic acid.

71. Calculate the volume of 0.202 M NaCl(aq) that should be used to make 125 mL of 0.00320 M NaCl(aq).
a) 1.98 mL b) 25.3 mL c) 7.89 mL d) 0.400 mL e) 0.0808 mL

72. Calculate the volume of a solution of 2.38 M glucose, $C_6H_{12}O_6$(aq), that should be used to make 3.00 L of 0.0333 M glucose(aq).
a) 238 mL b) 26.4 mL c) 42.0 mL d) 99.9 mL e) 37.9 mL

73. Calculate the volume of 1.83 M sodium thiosulfate solution, $Na_2S_2O_3$(aq), that should be used to make 1.75 L of 0.0292 M $Na_2S_2O_3$.
a) 35.8 mL b) 27.9 mL c) 32.7 mL d) 110 mL e) 93.5 mL

74. How many mL of water should be added to 50.00 mL of 1.25 M HCl(aq) to prepare 0.800 M HCl(aq)?
a) 78.1 mL b) 32.0 mL c) 28.1 mL d) 50.0 mL e) 42.3 mL

75. What volume of 0.799 M HCl(aq) should be diluted to 750 mL with water to reduce its concentration to 0.0340 M HCl(aq)?
a) 20.4 mL b) 319 mL c) 31.9 mL d) 204 mL e) 1.76 L

76. It is desired to add 2.95 g of gold(III) chloride to a bath used in gold plating. The gold(III) chloride is available as a 2.00 M aqueous solution. What volume of this solution must be added?
a) 16.9 mL b) 4.86 mL c) 19.4 mL d) 1.69 mL e) 9.72 mL

77. A solution is prepared by mixing equal volumes of 0.20 M Na_2SO_4(aq), 0.30 M $MgSO_4$(aq), and 0.40 M $NaNO_3$(aq). What is the molarity of the sodium ion in the final solution? Assume that the volumes are additive.
a) 0.27 M b) 0.30 M c) 0.60 M d) 0.90 M e) 0.80 M

78. A solution was prepared by mixing 35.0 mL of 0.200 M NaI(aq) with 28.5 mL of 0.350 M NaI(aq). Assuming that the volumes are additive, what is the molarity of NaI(aq) in the final solution?
a) 0.596 M b) 0.267 M c) 0.485 M d) 0.00866 M e) 0.0170 M

79. A solution was prepared by mixing 557 mL of 0.100 M $AgNO_3$(aq) with 12.8 mL of 1.89 M $AgNO_3$(aq) and then diluting to a final volume of 800.0 mL. What is the molarity of Ag^+ in this solution?
a) 0.726 M b) 0.100 M c) 0.110 M d) 1.99 M e) 1.89 M

80. A solution was prepared by weighing 0.269 g of ammonium iron(II) sulfate hexahydrate, $(NH_4)_2Fe(SO_4)_2{,}6H_2O$, and diluting to 500.0 mL in a volumetric flask. A 1.00-mL sample of this solution was transferred to 250-mL volumetric flask and diluted to the mark with water. What is the final concentration of iron in solution?
a) 5.49×10^{-6} M b) 6.86×10^{-4} M c) 1.37×10^{-6} M d) 8.57×10^{-5} M e) 5.49×10^{-9} M

Chapter 2: Measurements and Moles
Form B

1. The prefix corresponding to the factor 10^{-9} is
a) nano. b) pico. c) micro. d) kilo. e) milli.

2. The prefix corresponding to the factor 10^{-12} is
a) pico. b) micro. c) nano. d) kilo. e) centi.

3. The prefix "micro" corresponds to the factor
a) 10^{-12}. b) 10^{-2}. c) 10^{-3}. d) 10^{-6}. e) 10^{-9}.

4. A centiliter and a microliter are
a) 10^{-1} L and 10^{-9} L, respectively.
b) 10^{-1} L and 10^{-6} L, respectively.
c) 10^{-2} L and 10^{-6} L, respectively.
d) 10^{-6} L and 10^{-9} L, respectively.
e) 10^{-2} L and 10^{-9} L, respectivey.

5. One dm^3 is equivalent to
a) 100 L. b) 100 mL. c) 10 L. d) 10 cm^3. e) 1 L.

6. How many picometers would be equivalent to 5 cm?
a) 5×10^{14} pm b) 5×10^{8} pm c) 5×10^{6} pm d) 5×10^{12} pm e) 5×10^{10} pm

7. Which of the following is the largest volume?
a) 500 mL b) 50 cm^3 c) 5×10^{6} 1L d) 5 cL e) 0.5 dm^3

8. All of the following are extensive properties of a substance except
a) color. b) length. c) mass. d) volume. e) internal energy.

9. A 1-g sample of white phosphorus is a white, transparent, waxy crystalline solid. It has a melting point of 44°C and a boiling point of 280°C and a density of 1.8 g/cm^3. The energy required to melt white P is 2.5 kJ/mol. White P is extremely reactive and ignites spontaneously in air. Which property of white P is an extensive property?
a) The energy required to melt white P.
b) The color of white P.
c) The melting point of white P.
d) The density of white P.
e) The boiling point of white P.

10. All of the following are extensive properties of a substance except
a) hardness. b) mass. c) volume. d) length. e) energy content.

11. The density of platinum is 21.4 g/cm^3. Express the density in g/m^3.
a) 2.14×10^{6} g/m^3 b) 2.14×10^{3} g/m^3 c) 2.14×10^{7} g/m^3 d) 2.14×10^{-5} g/m^3 e) 2.14×10^{4} g/m^3

12. A lead ball with a mass of 36.0 g is placed in exactly 100 mL of water in a graduated cylinder. If 3.2 mL of water are displaced from the cylinder, what is the density of the ball?
 a) 32.8 g/mL b) 41.5 g/mL c) 8.9 g/mL d) 1.2 g/mL e) 11 g/mL

13. Bismuth has a density of 9.80g/cm^3. What is the volume of a chunk of bismuth that has a mass of 59.5 g?
 a) 0.165 cm^3 b) 583 cm^3 c) 9.80 cm^3 d) 6.07 cm^3 e) 59.5 cm^3

14. A gold brick measures 2.00 x 4.00 x 8.00 in. If the density of gold is 19.3 g/cm^3, what is the mass of the brick in kg? (1 in = 2.54 cm)
 a) 54.3 kg b) 20.2 kg c) 12.4 kg d) 1235 kg e) 3.14 kg

15. Exactly 1 L is equal to
 a) 10^{-3} m^3. b) 10^{-6} m^3. c) 10^3 m^3. d) 10 m^3. e) 10^{-3} cm^3.

16. The nitrogen-nitrogen bond enthalpy in nitrogen gas is 944 kJ/mol. If 4.184 joules ≠ 1 calorie, what is the bond enthalpy in kcal/mol?
 a) 39.50 kcal/mol b) 3950 kcal/mol c) 226 kcal/mol d) 226 x 10^3 kcal/mol e) 22.6 kcal/mol

17. The F–F bond length in F_2(g) is 1.42 /. If 1 / = 10^{-10} m, what is the bond length in pm?
 a) 14.2 pm b) 71 pm c) 1420 pm d) 142 pm e) 0.142 pm

18. An American automobile had a "mileage" rating of 36 mi/gal. What is the rating in km/L? (1.6 km = 1 mi and 1 gal = 3.785 L)
 a) 15 km/L b) 23 km/L c) 9.5 km/L d) 85 km/L e) 58 km/L

19. If 1 mi = 1.61 km, what is a speed of 1.0 km/min in mi/hr?
 a) 17 mi/hr b) 60 mi/hr c) 37 mi/hr d) 97 mi/hr e) 1.6 mi/hr

20. What is the Fahrenheit temperature that corresponds to 298 K?
 a) 32°F b) 45°F c) 43°F d) 77°F e) 57°F

21. What is the Celsius temperature that corresponds to 15°F?
 a) −17°C b) −5.0°C c) 17°C d) −9.4°C e) −24°C

22. Which of the measured numbers below has the greatest number of significant figures?
 a) 6.02350 b) 0.0060235 c) 0.00060235 d) 6.0235 x 10^3 e) 60235.

23. How many zeros are significant in the measured number 0.070090 g?
 a) 4 b) 1 c) 2 d) 3 e) 5

24. What answer should be reported if 4.560 is added to 2.6 x 10^{-3}?
 a) 4.563 b) 4.56 c) 4.6 d) 4.5626 e) 7.2

25. The number 65000.0 expressed in scientific notation is
 a) 6.5 x 10^4 b) 6.5000 x 10^{-4} c) 6.5 x 10^{-4} d) 6.50000 x 10^4 e) 6.5000 x 10^4

26. When the calculation

$$2.090 \times 6.230 - 2.20 \times 10^{-2}$$

is carried out, the number of significant figures in the answer is

a) 4. b) 6 c) 5. d) 3 e) 2.

27. When the calculation

$$(6.02300 \times 10^{23} \div 2.1200 \times 10^{-9}) \times 70.000$$

is carried out, the number of significant figures in the answer is

a) 5. b) 6. c) 2. d) 3. e) 4.

28. Lithium has two naturally occurring isotopes, Li-6 and Li-7, with natural abundance of 7.42% and 92.58%, respectively. The mass of an atom of Li-6 is 9.988×10^{-24} g and the mass of an atom of Li-7 is 1.165×10^{-23} g. What is the average molar mass of lithium?

a) 5.982 g/mol b) 6.941 g/mol c) 6.515 g/mol d) 6.015 g/mol e) 7.016 g/mol

29. Europium has two naturally occurring isotopes, Eu-151 and Eu-153, with natural abundance of 47.8% and 52.2%, respectively. The molar mass of Eu-151 is 150.9199 g/mol and the molar mass of Eu-153 is 152.9212 g/mol. What is the average molar mass of europium?

a) 151.92 g/mol b) 152.00 g/mol c) 151.89 g/mol d) 151.96 g/mol e) 152.92 g/mol

30. If the molar mass of chlorine gas, Cl_2, is 70.9 g/mol, what mass of chlorine gas contains 0.652 moles of chlorine gas?

a) 46.2 g b) 23.1 g c) 109 g d) 9.20 g e) 92.4 g

31. How many moles of fluorine atoms are contained in 4.139 g of fluorine gas, F_2?

a) 4.590 mol b) 78.64 mol c) 0.1089 mol d) 9.181 mol e) 0.2178 mol

32. The molar mass of magnesium in a natural sample is 24.31 g mol^{-1}. The sample consists of Mg-24 of molar mass 23.9850 g mol^{-1}, Mg-25 of molar mass 24.9858 g mol^{-1}, and Mg-26 of molar mass 25.9826 g mol^{-1}. If the percentage abundance of Mg-25 is 10.00%, calculate the percentage abundances of the other two isotopes of magnesium.

33. A natural sample of europium consists of two isotopes, Eu-151 and Eu-153 of molar masses 150.9199 g mol^{-1} and 152.9212 g mol^{-1}, respectively. If the molar mass of a natural sample is 151.96 g mol^{-1}, what are the percentage abundances of the isotopes?

34. How many grams of titanium dioxide are needed to have 0.850 mol?

a) 95.1 g b) 67.9 g c) 94.0 g d) 54.3 g e) 10.6 g

35. What is the molar mass of dinitrogen pentoxide?

a) 150 g/mol b) 108 g/mol c) 44 g/mol d) 30 g/mol e) 94 g/mol

36. What is the molar mass of sodium hydrogen carbonate?

a) 107.0 g/mol b) 83.00 g/mol c) 68.01 g/mol d) 106.0 g/mol e) 84.01 g/mol

37. What is the molar mass of rubbing alcohol, C_3H_8O?
a) 60.09 g/mol b) 120.2 g/mol c) 30.05 g/mol d) 48.08 g/mol e) 59.09 g/mol

38. How many nitrogen atoms are present in 3 millimoles of nitrogen gas, N_2?
a) 3.61×10^{19} b) 3.61×10^{21} c) 1.81×10^{21} d) 1.81×10^{19} e) 1.20×10^{21}

39. A sample of water has a mass of 50.0 g. If the sample is found to contain 5.60 g of hydrogen, what is the mass percent hydrogen in the sample?
a) 11.2% b) 5.60% c) 89.0% d) 4.40% e) 64.2%

40. The number of grams of hydrogen in 50.0 g water is
a) 4.03 g. b) 5.93 g. c) 5.60 g. d) 3.60 g. e) 36.0 g

41. The mass percent sulfur in potassium sulfate is
a) 18.4%. b) 23.7%. c) 20.3% d) 36.8% e) 26.9%

42. How many grams of hydrogen are combined with other elements in 62 g $CuSO_4{,}5H_2O$?
a) 5.0 g b) 4.0 g c) 0.50 g d) 1.3 g e) 2.5 g

43. How many grams of fluorine are required to produce 50.0 g of barium fluoride?
a) 66.6 g b) 3.51 g c) 5.42 g d) 100. g e) 10.8 g

44. How many grams of chromium are required to produce 75.0 g of chromium(VI) oxide?
a) 144 g b) 69.3 g c) 1.33 g d) 39.0 g e) 26.4 g

45. How many grams of nickel are required to produce 25.0 g of nickel boride, Ni_2B?
a) 11.4 g b) 22.8 g c) 301 g d) 2.11 g e) 5.13 g

46. A green compound composed of chromium and oxygen contains 68.42% Cr. What is the empirical formula of the compound?
a) Cr_3O b) Cr_2O_3 c) CrO_3 d) Cr_2O e) Cr_2O_2

47. The mass percent composition of hydrazine is 87.4% nitrogen and 12.6% hydrogen. What is the empirical formula of hydrazine?
a) NH_2 b) NH_3 c) N_2H_3 d) N_2H_6 e) N_2H

48. Sodium metal of mass 0.584 g reacts with phosphorus to form 0.846 g of sodium phosphide. What is the empirical formula of sodium phosphide?
a) Na_3P b) NaP c) Na_2P d) Na_3P_3 e) Na_4P

49. A sample of phosphorus of mass 3.654 g reacts with chlorine to form 16.20 g of a phosphorus chloride. What is the empirical formula of this compound?
a) P_2Cl_3 b) PCl_4 c) PCl_3 d) PCl_2 e) P_3Cl

50. Krypton of mass 3.90 g reacts with fluorine to form 5.67 g of a krypton fluoride. What is the empirical formula of this compound?
a) KrF_2 b) KrF c) KrF_4 d) KrF_6 e) Kr F_8

51. Which of the following formulas must be an empirical formula?
a) $C_2H_8N_2$ b) $Ca_{10}(OH)_2(PO_4)_6$ c) P_4 d) $Pb_3(AsO_4)_2$ e) N_2H_4

52. The molar mass of terephthalic acid is 166 g mol^{-1} and its empirical formula is $C_4H_3O_2$. What is its molecular formula?
a) $C_4H_3O_2$ b) $C_8H_6O_4$ c) $C_{10}H_5O_6$ d) $C_{12}H_9O_6$ e) $C_8H_9O_8$

53. The molar mass of the amino acid histidine is 155 g mol^{-1} and its empirical formula is $C_6H_9N_3O_2$. What is its molecular formula?
a) $C_{18}H_{27}N_9O_6$ b) $C_9H_{12}N_3O_2$ c) $C_{12}H_{18}N_6O_4$ d) $C_6H_9N_3O_2$ e) $C_6H_{12}N_3O_4$

54. The molar mass of cyclohexane is 84 g mol^{-1} and its empirical formula is CH_2. What is its molecular formula?
a) C_6H_{12} b) C_7H_{14} c) C_4H_8 d) C_5H_{10} e) C_8H_{16}

55. Ibuprofen is an anti-inflammatory drug which is 75.69% C, 8.80% H, and 15.51% O. If the molecular mass of ibuprofen is 206 g mol^{-1}, what is its molecular formula?

56. Caffeine is 49.47% C, 5.19% H, 28.86% N, and 16.48% O. If the molecular mass of caffeine is 194 g mol^{-1}, what is its molecular formula?

57. Calculate the molarity of a solution prepared by dissolving 0.584 g NaCl in enough water to prepare 750 mL of solution.
a) 0.0455 M b) 0.0133 M c) 1.33×10^{-5} M d) 0.0400 M e) 7.79×10^{-4} M

58. Calculate the molarity of a solution prepared by dissolving 2.073 g $AgClO_4$ in enough water to prepare 1250 mL of solution.
a) 7.999×10^{-3} M b) 9.999×10^{-3} M c) 0.3438 M d) 0.08000 M e) 0.1659 M

59. Calculate the molarity of a solution prepared by dissolving 4.329 g Na_2SO_4 in enough water to prepare 50.00 mL of solution.
a) 5.246×10^{-4} M b) 0.7624 M c) 0.5246 M d) 1.312 M e) 1.906 M

60. Calculate the molarity of a solution prepared by dissolving 28.0 g glucose, $C_6H_{12}O_6$, in enough water to prepare 375 mL of solution.
a) 2.41 M b) 0.0172 M c) 0.583 M d) 0.199 M e) 0.414 M

61. Calculate the molarity of a solution prepared by dissolving 7.771 g $LiClO_4$ in enough water to prepare 25.00 mL of solution.
a) 0.2423 M b) 1.826 M c) 0.07304 M d) 2.922 M e) 0.5476 M

62. Calculate the mass of sodium chloride required to make 125.0 mL of a 0.470 M NaCl(aq) solution.
a) 0.0588 g b) 1.01 g c) 4.55 g d) 3.76 g e) 3.43 g

63. Calculate the mass of sodium sulfate required to make 2.50 L of a 1.09 M Na_2SO_4(aq) solution.
a) 450 g b) 2.73 g c) 180 g d) 379 g e) 72.0 g

64. What volume of 0.202 M NaCl solution should you transfer so as to have a sample of solution that contains 0.354 mol NaCl?
a) 1.75 L b) 5.71 mL c) 71.5 mL d) 1.75 mL e) 0.571 L

65. What volume of 2.00 M urea, $(NH_2)_2CO$(aq), should you transfer so as to have a sample of solution that contains 1.19 mol urea?
a) 168 mL b) 2.38 L c) 595 mL d) 59.5 mL e) 1.68 L

66. If you have available 500 mL of 0.250 M NaCl(aq), what volume of this solution should you transfer so as to have a sample of solution that contains 0.130 mol NaCl(aq)?
a) Not enough 0.250 M NaCl(aq) is available. b) 65 mL c) 125 mL d) 32.5 mL e) 3.85 mL

67. A mass of 5.00 1g of the analgesic acetaminophen, $C_8H_9NO_2$, is required in an experiment. The analgesic is only slightly soluble in water but is available as a 3.12×10^{-5} M solution. What volume of solution must be used to deliver this mass?
a) 3.31×10^{-5} mL b) 1.06 mL c) 3.31×10^{-2} mL d) 0.242 mL e) 0.756 mL

68. A solution was prepared by dissolving 14.88 g of sodium hydroxide in 300.0 mL of 0.4780 M NaOH(aq). What is the molarity of NaOH in the final solution?
a) 1.593 M b) 0.515 M c) 0.850 M d) 1.848 M e) 1.718 M

69. A solution is prepared by dissolving 55.1 g of KCl in approximately 75 mL of water and then adding water to a final volume of 125 mL. What is the molarity of KCl(aq) in this solution?
a) 14.8 M b) 3.70 M c) 9.85 M d) 0.739 M e) 5.91 M

70. A bottle of sulfuric acid purchased for laboratory use is labeled "96.7% H_2SO_4." This means that for every 100 g of this solution, 96.7% is H_2SO_4 and 3.3% is water. If the density of this solution is 1.845 g mL^{-1}, calculate the molarity of the solution.

71. Calculate the volume of 0.202 M NaCl(aq) that should be used to make 750 mL of 0.321 M NaCl(aq).
a) 838 mL b) 152 mL c) 48.6 mL d) 1.19 L e) 11.6 L

72. Calculate the volume of a solution of 0.204 M glucose, $C_6H_{12}O_6(aq)$, that should be used to make 25.00 mL of 8.16×10^{-3} M glucose(aq).
a) 0.625 mL b) 3.06 mL c) 1.00 mL d) 2.04 mL e) 5.10 mL

73. Calculate the volume of 0.755 M sodium thiosulfate solution, $Na_2S_2O_3(aq)$, that should be used to make 15.0 L of 0.319 M $Na_2S_2O_3$.
a) 11.3 L b) 158 mL c) 2.36 L d) 3.61 L e) 6.34 L

74. How many mL of water should be added to 75.00 mL of 0.889 M HCl(aq) to prepare 0.800 M HCl(aq)?
a) 7.5 mL b) 67.5 mL c) 8.3 mL d) 83.3 mL e) 21.7 mL

75. What volume of 0.136 M HCl(aq) should be diluted to 250 mL with water to reduce its concentration to 0.00500 M HCl(aq)?
a) 9.19 mL b) 6.90 mL c) 17.0 mL d) 1.25 mL e) 3.40 mL

76. It is desired to add 8.37 g of gold(III) chloride to a bath used in gold plating. The gold(III) chloride is available as a 1.98 M aqueous solution. What volume of this solution must be added?
a) 13.9 mL b) 5.46 mL c) 27.6 mL d) 71.8 mL e) 54.6 mL

77. A solution is prepared by mixing equal volumes of 0.20 M $Na_2SO_4(aq)$, 0.30 M $MgSO_4(aq)$, and 0.40 M $NaNO_3(aq)$. What is the molarity of the sulfate ion in the final solution? Assume that the volumes are additive.
a) 0.90 M b) 0.17 M c) 0.23 M d) 0.50 M e) 0.40 M

78. A solution was prepared by mixing 50.0 mL of 0.150 M $Na_2SO_4(aq)$ with 125 mL of 0.262 M $Na_2SO_4(aq)$. Assuming that the volumes are additive, what is the molarity of $Na_2SO_4(aq)$ in the final solution?
a) 0.230 M b) 0.0403 M c) 0.187 M d) 0.0429 M e) 0.460 M

79. A solution was prepared by mixing 250 mL of 0.547 M NaOH(aq) with 50.0 mL of 1.62 M NaOH(aq) and then diluting to a final volume of 1.50 L. What is the molarity of Na^+ in this solution?
a) 0.726 M b) 0.0912 M c) 0.145 M d) 0.217 M e) 0.0540 M

80. A solution was prepared by weighing 0.269 g of ammonium iron(II) sulfate hexahydrate, $(NH_4)_2Fe(SO_4)_2{,}6H_2O$, and diluting to 500.0 mL in a volumetric flask. A 5.00-mL sample of this solution was transferred to a 250-mL volumetric flask and diluted to the mark with water. What is the concentration of sulfate ions in solution?
a) 0.0274 M b) 5.49×10^{-5} M c) 2.74×10^{-5} M d) 0.0233 M e) 2.74×10^{-8} M

Chapter 3: Chemical Reactions: Modifying Matter
Form A

1. Calcium reacts with water to form calcium hydroxide and hydrogen. What are the physical states of calcium, calcium hydroxide, and hydrogen, respectively?
a) s, s, aq b) l, s, g c) l, aq, g d) s, aq, g e) s, s, g

2. The combustion of butane, $C_4H_{10}(g)$, produces carbon dioxide and water vapor. In the balanced equation for this reaction, what are the coefficients of butane, oxygen, carbon dioxide, and water, respectively?
a) 1, 6, 8, 10 b) 1, 6, 4, 5 c) 2, 13, 4, 10 d) 2, 13, 8, 10 e) 1, 13, 4, 5

3. When aluminum metal is dissolved in perchloric acid, aluminum(III) perchlorate and hydrogen gas are formed. In the balanced equation for this reaction, what is the coefficient of hydrogen gas?
a) 3 b) 2 c) 1 d) 4 e) 5

4. Calcium reacts with water to form calcium hydroxide and hydrogen. In the balanced equation for this reaction, what is the coefficient of hydrogen?
a) 1 b) 2 c) 3 d) 4 e) ½

5. The coefficients of calcium hydroxide and hydrogen, respectively, in the balanced equation for the reaction of calcium with water are
a) 2 and 2. b) 1 and 1. c) 1 and 2. d) 1 and 3. e) 2 and 1.

6. The following equation is unbalanced.
$$F_2 + H_2O \rightarrow O_3 + HF$$
In the balanced equation, the coefficient of HF is
a) 6 b) 3 c) 2 d) 4 e) 8

7. The following equation is unbalanced.
$$KH + H_2O \rightarrow KOH + H_2$$
In the balanced equation, the coefficient of H_2 is
a) 1 b) 2 c) 3 d) 4 e) 5

8. Which of the following give(s) an aqueous solution that conducts electricity?
a) O_2
b) sucrose, $C_{12}H_{22}O_{12}$
c) ethylene glycol, $HOCH_2CH_2OH$ (antifreeze)
d) ethanol, CH_3CH_2OH
e) $Pb(NO_3)_2$

9. Which of the following is a weak electrolyte when dissolved in water?
a) ammonia b) hydrogen chloride c) sodium hydroxide d) perchloric acid e) sodium oxide

10. When aqueous solutions of barium nitrate and ammonium sulfate are mixed, what are the "spectator ions" in the reaction?
a) Ba^{2+} and SO_4^{2-} b) NH_4^+ and NO_3^- c) Ba^{2+} and NO_3^- d) SO_4^{2-} and NO_3^- e) NH_4^+ and SO_4^{2-}

11. When aqueous solutions of cadmium nitrate and sodium sulfide are mixed, what are the "spectator ions" in the reaction?
a) Cd^{2+} and S^{2-} b) Na^+ and S^{2-} c) Cd^{2+} and NO_3^- d) Na^+ and NO_3^- e) S^{2-} and NO_3^-

12. Which of the following is soluble in water?
a) lead(II) chloride b) lead(II) sulfide c) lead(II) carbonate d) lead(II) sulfate e) lead(II) acetate

13. Which of the following is soluble in water?
a) mercury(I) nitrate
b) lead (II) sulfate
c) barium carbonate
d) mercury(I) chloride
e) lead(II) iodide

14. Which of the following is soluble in water?
a) $HgCl_2$ b) Hg_2Cl_2 c) $PbCl_2$ d) AgCl e) PbI_2

15. Which of the following is insoluble in water?
a) Na_2SO_4 b) $MgCl_2$ c) $CaCO_3$ d) $(NH_4)_2CO_3$ e) $Pb(ClO_4)_2$

16. Which of the following is insoluble in water?
a) $CaCl_2$ b) $HgCl_2$ c) $PbCl_2$ d) $BaCl_2$ e) RbCl

17. Which of the following pairs of solutions will give a precipitate when mixed?
a) $Pb(NO_3)_2$(aq) and KI(aq)
b) $AgNO_3$(aq) and $LiClO_4$(aq)
c) K_2SO_4(aq) and $Cu(NO_3)_2$(aq)
d) $Ca(CH_3COO)_2$(aq) and NH_4Cl(aq)
e) $NaCH_3COO$(aq) and $CaCl_2$(aq)

18. Which of the following pairs of solutions will give a precipitate when mixed?
a) $AgNO_3$(aq) and $NaCH_3COO$(aq)
b) Na_2SO_4(aq) and $CuCl_2$(aq)
c) Na_2CO_3(aq) and $Ca(ClO_4)_2$(aq)
d) $Hg(NO_3)_2$(aq) and NaCl(aq)
e) NH_4Cl(aq) and $Ca(CH_3COO)_2$(aq)

19. Consider the possible reactions below:
1. $Hg_2(ClO_4)_2(aq) + NH_4I(aq) \rightarrow$
2. $Hg(ClO_4)_2(aq) + NaI(aq) \rightarrow$
3. $HgI_2(aq) + KI(aq) \rightarrow$
4. $Hg_2(NO_3)_2(aq) + KI(aq) \rightarrow$

Which of the possible reactions above gives the net ionic equation, $Hg_2^{2+}(aq) + 2I^-(aq) \rightarrow Hg_2I_2(s)$?
a) 1, 2, 3, and 4 b) 1 and 2 c) 1 and 4 d) 2 and 3 e) 1 only

20. Write the balanced net ionic equation for the reaction that occurs when solid calcium carbonate is added to an aqueous solution of perchloric acid.

21. Write the balanced net ionic equation for the reaction which occurs when aqueous solutions of ammonium chromate and lead(II) nitrate are mixed. Identify the spectator ions.

22. Complete and balance the equations for the reactions which occur, if any, for the following situations. Write the net ionic equation where appropriate.
a) Aqueous solutions of ammonium carbonate and calcium perchlorate are mixed.
b) $K_2SO_4(aq) + MgCl_2(aq) \rightarrow$
c) $Na_2CO_3(aq) + HCl(aq) \rightarrow$

23. All of the following are acids except
a) CH_3CH_2COOH b) $NaCH_3COO$ c) HI d) $HClO_4$ e) HBr

24. When CaO(s) dissolves in water,
a) $O^{2-}(aq)$ reacts with water to give hydroxide ions.
b) oxygen gas is produced.
c) the resulting solution is neutral.
d) hydrogen gas is produced.
e) the resulting solution is acidic.

25. What type of reagent is required to convert NO_2^- to HNO_2?
a) acid b) base c) reducing agent d) oxidizing agent e) neutralization reagent

26. What type of reagent is required to convert SO_3^{2-} to HSO_3^-?
a) acid b) base c) reducing agent d) oxidizing agent e) neutralization agent

27. Which of the following aqueous reagents, when mixed, give both a neutralization reaction and a precipitation reaction?
a) $CaCO_3$ and HCl
b) $Ba(OH)_2$ and H_2SO_4
c) KOH and H_2SO_4
d) H_2SO_4 and Na_2CO_3
e) $Mg(OH)_2$ and HNO_3

28. Which of the following oxides gives a basic solution when dissolved in water?
a) calcium oxide b) dinitrogen pentoxide c) carbon dioxide d) sulfur dioxide e) selenium trioxide

29. Which of the following oxides gives an acidic solution when dissolved in water?
a) P_4O_{10} b) CaO c) Na_2O d) SrO e) K_2O

30. Which of the following oxides gives an acidic solution when dissolved in water?
a) barium oxide b) calcium oxide c) potassium oxide d) sulfur trioxide e) sodium oxide

31. All of the following elements give amphoteric oxides except
a) Sr b) Al c) Ge d) Sb e) Pb

32. Which of the following is an amphoteric oxide?
a) Ga_2O_3 b) SiO_2 c) P_4O_6 d) Fe_2O_3 e) Na_2O

33. Which of the following is correct with respect to the reactions of Ga_2O_3?
a) Ga_2O_3 dissolves in water to give an acidic solution.
b) Ga_2O_3 reacts with water to give hydrogen and oxygen gases.
c) Ga_2O_3 is insoluble in NaOH.
d) Ga_2O_3 reacts with base to give $Ga(OH)_4^-$.
e) Ga_2O_3 does not react with acid or base.

34. What is the net ionic equation for the reaction that occurs when aqueous nitric acid is added to aqueous sodium hydroxide?
a) $HNO_2(aq) + OH^-(aq) \rightarrow H_2O(l) + NO_2^-(aq)$
b) $H_3O^+(aq) + NaOH(aq) \rightarrow 2H_2O(l) + Na^+(aq)$
c) $H_3O^+(aq) + OH^-(aq) \rightarrow 2H_2O(l)$
d) $2H_3O^+(aq) + Na(OH)_2(aq) \rightarrow 3H_2O(l) + Na^+(aq)$
e) $HNO_3(aq) + NaOH(aq) \rightarrow Na^+(aq) + NO_3^-(aq) + H_2O(l)$

35. What salt is produced by the reaction between aqueous solutions of calcium hydroxide and sulfuric acid?
a) $CaSO_4$ b) CaO c) $Ca(HSO_4)_2$ d) $Ca(HSO_4)(OH)$ e) H_2O

36. When lithium carbonate is added to aqueous perchloric acid, a gas is evolved. What are the products of this reaction?
a) $CO(g)$ and $LiClO_4(aq)$
b) $CO_2(g)$, $H_2O(l)$, and $LiClO_4(aq)$
c) $Cl_2(g)$ and $LiClO_4(aq)$
d) $Cl_2(g)$, $O_2(g)$, and $LiClO_4(aq)$
e) $O_2(g)$ and $LiClO_4(aq)$

37. What is the oxidation number of chlorine in HOCl?
a) +1 b) −1 c) +2 d) +3 e) 0

38. What is the oxidation number of aluminum in $Na[Al(OH)_4]$?
a) +3 b) +1 c) +2 d) 0 e) +4

39. What is the oxidation number of chromium in Na_2CrO_4?
a) +6 b) +3 c) +4 d) +7 e) 0

40. What is the oxidation number of sulfur in $Na_2S_2O_3$?
a) +2 b) +4 c) +6 d) +8 e) 0

41. Predict the strongest reducing agent for each pair and explain your reasoning.
   a) N_2 or N_2H_4
   b) NH_3 or N_2H_4
   c) P_4 or H_3PO_4

42. In a redox reaction, the reducing agent
   a) loses electrons and is reduced.
   b) loses electrons and is oxidized.
   c) contains an element that undergoes a decrease in oxidation number.
   d) gains electrons and is oxidized.
   e) gains electrons and is reduced.

43. Consider the following reaction:

$$2MnO_4^-(aq) + 5HOOCCOOH(aq) + 6H^+(aq) \rightarrow 2Mn^{2+}(aq) + 8H_2O(l) + 10CO_2(aq)$$

   In this reaction,
   a) electrons are transferred from MnO_4^- to HOOCOOH.
   b) HOOCCOOH is reduced.
   c) no electrons are transferred since the oxidation number of Mn does not change.
   d) electrons are transferred from HOOCCOOH to MnO_4^-.
   e) MnO_4^- is oxidized.

44. Consider the following reaction:

$$Cd(s) + 2AgCl(s) \rightarrow Cd^{2+}(aq) + 2Ag(s) + 2Cl^-(aq)$$

   In this reaction,
   a) electrons are transferred from silver in AgCl(s) to Cd(s).
   b) electrons are transferred from Cd(s) to silver in AgCl(s).
   c) no electrons are transferred since the oxidation number of chlorine atoms does not change.
   d) the oxidizing agent is Cd(s).
   e) the reducing agent is AgCl(s).

45. Consider the following reaction:

$$2Fe^{3+}(aq) + 2Hg(l) + 2Cl^-(aq) \rightarrow 2Fe^{2+}(aq) + Hg_2Cl_2(s)$$

   In this reaction,
   a) Hg(l) is the oxidizing agent.
   b) $Fe^{3+}(aq)$ is oxidized.
   c) Hg(l) is reduced.
   d) Hg(l) is oxidized.
   e) $Fe^{3+}(aq)$ is the reducing agent.

46. Consider the following reaction:

$$IO_3^-(aq) + 2Tl^+(aq) + 2Cl^-(aq) + 6H^+(aq) \rightarrow ICl_2^-(aq) + 2Tl^{3+}(aq) + 3H_2O(l)$$

   In this reaction,
   a) $IO_3^-(aq)$ undergoes oxidation.
   b) $Tl^+(aq)$ is the oxidizing agent.
   c) $IO_3^-(aq)$ undergoes reduction.
   d) $Cl^-(aq)$ undergoes oxidation.
   e) $Tl^+(aq)$ gains electrons.

47. Consider the following reaction:

$$CH_2(COOH)_2(aq) + 6Ce^{4+}(aq) + 2H_2O(l) \rightarrow 2CO_2(aq) + HCOOH(aq) + 6Ce^{3+}(aq) + 6H^+(aq)$$

In this reaction,

a) $Ce^{4+}(aq)$ transfers electrons to $CH_2(COOH)_2(aq)$.
b) $CH_2(COOH)_2(aq)$ is the oxidizing agent.
c) $CH_2(COOH)_2(aq)$ transfers electrons to $Ce^{4+}(aq)$.
d) $Ce^{4+}(aq)$ is the reducing agent.
e) hydrogen atoms in $H_2O(l)$ lose electrons.

48. Consider the following reaction:

$$2Mn(OH)_3(s) + 6H^+(aq) + 6I^-(aq) \rightarrow 2Mn^{2+}(aq) + 2I_3^-(aq) + 6H_2O(l)$$

In this reaction,

a) $I^-(aq)$ transfers electrons to manganese in $Mn(OH)_3(s)$.
b) $Mn(OH)_3(s)$ is the reducing agent.
c) $H^+(aq)$ is the oxidizing agent.
d) $I^-(aq)$ is the oxidizing agent.
e) $I^-(aq)$ gains electrons from $Mn(OH)_3(s)$.

49. When an acidic solution of potassium permanganate is mixed with a sodium chloride solution, Mn^{2+} ions and chlorine gas are formed. Which species is oxidized?

a) H_2O b) MnO_4^- c) Cl_2 d) Cl^- e) H^+

50. When an acidic solution of potassium permanganate is mixed with a sodium chloride solution, Mn^{2+} ions and chlorine gas are formed. Which species is the oxidizing agent?

a) H_2O b) MnO_4^- c) Cl^- d) H^+ e) Mn^{2+}

51. Charcoal "filters" are used to remove dissolved chlorine gas from drinking water. In acidic solution, $Cl_2(aq)$, reacts with carbon to form carbon dioxide and chloride ions. Which species is oxidized?

a) carbon b) Cl_2 c) CO_2 d) H_2O e) H^+

52. Charcoal "filters" are used to remove dissolved chlorine gas from drinking water. In acidic solution, $Cl_2(aq)$, reacts with carbon to form carbon dioxide and chloride ions. Which species is the oxidizing agent?

a) Cl_2 b) carbon c) CO_2 d) H_2O e) Cl^-

53. When aqueous sodium hydroxide is mixed with aluminum shavings, hydrogen gas and the aluminate ion, $Al(OH)_4^-$, are formed. Which species is oxidized?

a) Al b) OH^- c) H_2O d) $Al(OH)_4^-$ e) This is not a redox reaction and thus no species is oxidized.

54. When aqueous sodium hydroxide is mixed with aluminum shavings, hydrogen gas and the aluminate ion, $Al(OH)_4^-$, are formed. Which species is the oxidizing agent?
a) OH^- b) Al c) H_2O d) H_2 e) This is not a redox reaction and thus no species is oxidized.

55. If an aqueous solution of sulfuric acid is added to a pool of mercury, SO_2(aq) and mercury(II) sulfate are formed. Which species is oxidized?
a) Hg(l) b) H_2SO_4(aq) c) H_2O(l) d) SO_2(aq) e) $HgSO_4$(aq)

56. If an aqueous solution of sulfuric acid is added to a pool of mercury, SO_2(aq) and mercury(II) sulfate are formed. Which species is the oxidizing agent?
a) H_2SO_4(aq) b) Hg(l) c) H_2O(l) d) SO_2(aq) e) $HgSO_4$(aq)

57. When concentrated sulfuric acid is added to P_4O_{10}(s), sulfur trioxide gas and concentrated phosphoric acid are formed. Which of the following is correct?
a) The species that is reduced is H_3PO_4(l).
b) This is not a redox reaction.
c) The species that is oxidized is SO_3(g).
d) The reducing agent is P_4O_{10}(s).
e) The oxidizing agent is H_2SO_4(l)

58. If zinc metal is added to a basic aqueous solution containing nitrate ions, ammonia and $Zn(OH)_4^{2-}$(aq) are formed. Which species is reduced?
a) NO_3^-(aq) b) Zn(s) c) OH^-(aq) d) H_2O(l) e) ammonia

59. If zinc metal is added to a basic aqueous solution containing nitrate ions, ammonia and $Zn(OH)_4^{2-}$(aq) are formed. Which species is the reducing agent?
a) Zn(s) b) NO_3^-(aq) c) OH^-(aq) d) H_2O(l) e) $Zn(OH)_4^-$

60. If an acidic solution of potassium dichromate, $K_2Cr_2O_7$(aq), is added to aqueous ethanol, CH_3CH_2OH(aq), acetic acid, CH_3COOH(aq) and Cr^{3+}(aq) are produced. Which species is reduced?
a) $Cr_2O_7^{2-}$(aq) b) Cr^{3+}(aq) c) CH_3CH_2OH(aq) d) H^+(aq) e) H_2O(l)

61. If an acidic solution of potassium dichromate, $K_2Cr_2O_7$(aq), is added to aqueous ethanol, CH_3CH_2OH(aq), acetic acid, CH_3COOH(aq) and Cr^{3+}(aq) are produced. Which species is the reducing agent?
a) CH_3CH_2OH(aq) b) H^+(aq) c) H_2O(l) d) CH_3COOH(aq) e) $Cr_2O_7^{2-}$(aq)

62. When an aqueous basic solution of sodium hypochlorite, NaOCl(aq), is added to $Fe(OH)_3$(s), the ferrate ion, FeO_4^{2-}(aq), and chloride ion are produced. Which species is the oxidizing agent?
a) H_2O(l) b) OH^-(aq) c) Cl^-(aq) d) OCl^-(aq) e) $Fe(OH)_3$(s)

63. When an aqueous basic solution of sodium hypochlorite, NaOCl(aq), is added to $Fe(OH)_3$(s), the ferrate ion, FeO_4^{2-}(aq), and chloride ion are produced. Which species undergoes oxidation?
a) $Fe(OH)_3$(s) b) OH^-(aq) c) H_2O(l) d) FeO_4^{2-}(aq) e) OCl^-(aq)

64. Ammonia gas and NO(g) react to form nitrogen gas and H_2O(g). Which species is the oxidizing agent?
a) NO(g)
b) NH_3(g)
c) H_2O(g)
d) N_2(g)
e) Since this is not a redox reaction, there is no oxidizing agent.

65. Ammonia gas and NO(g) react to form nitrogen gas and H_2O(g). Which species is the reducing agent?
a) NH_3(g)
b) NO(g)
c) H_2O(g)
d) N_2(g)
e) Since this is not a redox reaction, there is no reducing agent.

66. Ammonia gas and NO(g) react to form nitrogen gas and H_2O(g). Which of the following is correct?
a) This is not a redox reaction.
b) Reduction produces H_2O(g).
c) Both oxidation and reduction produce N_2(g).
d) NO(g) is the reducing agent.
e) Oxidation produces H_2O(g).

67. Dinitrogen tetroxide, N_2O_4(l) and hydrazine, N_2H_4(l), react to form nitrogen gas and H_2O(g). Which of the following is correct?
a) N_2O_4(l) is the reducing agent.
b) Oxidation produces H_2O(g).
c) This is not a redox reaction.
d) Reduction produces H_2O(g).
e) Both oxidation and reduction produce N_2(g).

68. What type of reagent is required to convert NO_2^- to N_2?
a) oxidizing agent b) reducing agent c) acid d) base e) precipitating agent

69. What type of reagent is required to convert N_2O_4 to N_2?
a) reducing agent b) oxidizing agent c) acid d) base e) neutralization reagent

70. In a redox reaction, the reducing agent
a) is the species that is reduced.
b) gains electrons.
c) contains an element that undergoes a decrease in oxidation number.
d) contains an element that undergoes an increase in oxidation number.
e) reacts with hydrogen gas.

71. Consider the reaction below:

$$Cu(s) + 2H_2SO_4(aq, conc.) \rightarrow Cu^{2+}(aq) + SO_4^{2-}(aq) + SO_2(g) + 2H_2O(l)$$

Identify the oxidizing agent.

a) H_2SO_4 b) Cu c) Cu^{2+} d) SO_4^{2-} e) SO_2

72. Consider the reaction below:

$$4Fe^{2+}(aq) + 4H^{+}(aq) + O_2(g) \rightarrow 4Fe^{3+}(aq) + 2H_2O(l)$$

Identify the oxidizing agent.

a) Fe^{3+} b) H^+ c) Fe^{2+} d) O_2 e) H_2O

73. Consider the following reaction:

$$2Cr(OH)_3(s) + 3ClO^-(aq) + 4OH^-(aq) \rightarrow 2CrO_4^{2-}(aq) + 3Cl^-(aq) + 5H_2O(l)$$

Identify the oxidizing agent.

a) OH^- b) ClO^- c) Cl^- d) CrO_4^{2-} e) $Cr(OH)_3$

74. Consider the following reaction:

$$NH_4^+(aq) + NO_3^-(aq) \rightarrow N_2O(g) + 2H_2O(l)$$

Identify the oxidizing agent.

a) NO_3^- b) H_2O c) N_2O d) NH_4^+ e) H^+

75. Which of the following reactions is a redox reaction?

a) $2Fe(s) + 3Cl_2(g) \rightarrow 2FeCl_3(s)$

b) $Na_3P(s) + 3HCl(aq) \rightarrow 3NaCl(aq) + PH_3(g)$

c) $FeS(s) + 2HCl(aq) \rightarrow FeCl_2(aq) + H_2S(g)$

d) $NH_3(g) + HCl(g) \rightarrow NH_4Cl(s)$

e) $MgO(s) + SO_3(g) \rightarrow MgSO_4(s)$

76. All of the following are redox reactions except

a) $Zn(s) + 2HCl(aq) \rightarrow ZnCl_2(aq) + H_2(g)$

b) $NH_3(g) + HCl(g) \rightarrow NH_4Cl(s)$

c) $2KClO_3(s) \rightarrow 2KCl(s) + 3O_2(g)$

d) $2Mg(s) + O_2(g) \rightarrow 2MgO(s)$

e) $2Na(s) + S(s) \rightarrow Na_2S(s)$

77. What type of reagent is required to convert N_2H_4 to N_2?

a) reducing agent b) oxidizing agent c) acid d) base e) neutralization agent

78. How many electrons are transferred when NO_3^- is converted to ammonia?

a) 8 b) 7 c) 6 d) 3 e) 2

79. All of the following can act as reducing agents except

a) NO_3^-. b) NO_2^-. c) NO. d) NH_3. e) N_2H_4.

80. Write the balanced equation for the combustion of ethanol, CH_3CH_2OH, to carbon dioxide and water vapor. Identify the oxidizing agent and the reducing agent.

Chapter 3: Chemical Reactions: Modifying Matter
Form B

1. Sodium reacts with water to form sodium hydroxide and hydrogen. What are the physical states of sodium, sodium hydroxide, and hydrogen, respectively?
a) l, aq, g b) s, aq, g c) l, s, g d) s, s, g e) s, s, aq

2. When acetylene, $C_2H_2(g)$, is burned, the products are carbon dioxide and water vapor. In the balanced equation for this reaction, what are the coefficients of acetylene, oxygen, carbon dioxide, and water, respectively?
a) 2, 5, 4, 2 b) 1, 2, 2, 1 c) 1, 5, 2, 2 d) 2, 2, 4, 2 e) 1, 1, 1, 1

3. When aluminum metal is dissolved in perchloric acid, aluminum(III) perchlorate and hydrogen gas are formed. In the balanced equation for this reaction, what is the coefficient of perchloric acid?
a) 6 b) 3 c) 2 d) 1 e) 4

4. Calcium reacts with water to form calcium hydroxide and hydrogen. In the balanced equation for this reaction, what is the coefficient of water?
a) 2 b) 1 c) 3 d) 4 e) ½

5. The coefficients of sodium hydroxide and hydrogen, respectively, in the balanced equation for the reaction of sodium with water are
a) 2 and 1. b) 2 and 3. c) 1 and 1. d) 1 and 2. e) 2 and 2.

6. The following equation is unbalanced.

$$P_4 + Cl_2 \rightarrow PCl_3$$

In the balanced equation, the coefficient of Cl_2 is
a) 6 b) 3 c) 9 d) 12 e) 2

7. The following equation is unbalanced.

$$Cs + H_2O \rightarrow CsOH + H_2$$

In the balanced equation, the coefficient of CsOH is
a) 2 b) 1 c) 3 d) 4 e) 5

8. Which of the following give(s) an aqueous solution that does not conduct electricity?
a) methanol, CH_3OH b) NaOH c) KCl d) Na_2SO_4 e) $LiClO_4$

9. Which of the following is a weak electrolyte when dissolved in water?
a) $HClO_4$ b) acetic acid, CH_3COOH c) LiOH d) H_2SO_4 e) HCl

10. When aqueous solutions of calcium chloride and ammonium carbonate are mixed, what are the "spectator ions" in the reaction?
a) Ca^{2+} and $CO_3{}^{2-}$ b) $NH_4{}^+$ and $CO_3{}^{2-}$ c) Cl^- and $CO_3{}^{2-}$ d) Ca^{2+} and Cl^- e) $NH_4{}^+$ and Cl^-

11. When aqueous solutions of cadmium chloride and ammonium sulfide are mixed, what are the "spectator ions" in the reaction?
a) Cd^{2+} and S^{2-} b) NH_4^+ and Cl^- c) Cd^{2+} and Cl^- d) NH_4^+ and S^{2-} e) Cl^- and S^{2-}

12. Which of the following is soluble in water?
a) lead(II) sulfate
b) lead(II) nitrate
c) lead(II) phosphate
d) lead(II) chromate
e) lead(II) bromide

13. Which of the following is soluble in water?
a) barium sulfate
b) mercury(I) nitrate
c) lead(II) chromate
d) barium perchlorate
e) lead(II) acetate

14. Which of the following is soluble in water?
a) Na_3PO_4 b) $MgCO_3$ c) $Ca_3(PO_4)_2$ d) $BaSO_4$ e) CuS

15. Which of the following is insoluble in water?
a) $Ba(OH)_2$ b) $(NH_4)_2S$ c) CaC_2O_4 d) CaO e) $NaNO_3$

16. Which of the following is insoluble in water?
a) K_2SO_4 b) $MgSO_4$ c) $CaSO_4$ d) $BaSO_4$ e) $(NH_4)_2SO_4$

17. Which of the following pairs of solutions will give a precipitate when mixed?
a) $Hg(CH_3COO)_2(aq)$ and $NH_4Cl(aq)$
b) $Hg_2(CH_3COO)_2(aq)$ and NaI(aq)
c) $K_2SO_4(aq)$ and $Cu(NO_3)_2(aq)$
d) $NaCH_3COO(aq)$ and $Hg(ClO_4)_2(aq)$
e) $AgNO_3(aq)$ and $LiClO_4(aq)$

18. Which of the following pairs of solutions will give a precipitate when mixed?
a) $Hg(NO_3)_2(aq)$ and NaCl(aq)
b) $Na_3PO_4(aq)$ and $Ca(ClO_4)_2(aq)$
c) $NH_4Cl(aq)$ and $Ca(CH_3COO)_2(aq)$
d) $AgNO_3(aq)$ and $NaCH_3COO(aq)$
e) $Na_2SO_4(aq)$ and $CuCl_2(aq)$

19. Consider the possible reactions below:
1. $Hg_2(CH_3COO)_2(aq) + NH_4I(aq) \rightarrow$
2. $Hg(NO_3)_2(aq) + NaI(aq) \rightarrow$
3. $HgCl_2(aq) + KI(aq) \rightarrow$
4. $Hg_2(NO_3)_2(aq) + KI(aq) \rightarrow$

Which of the possible reactions above gives the net ionic equation, $Hg_2^{2+}(aq) + 2I^-(aq) \rightarrow Hg_2I_2(s)$?
a) 2 and 3 b) 2 and 4 c) 1 only d) 1 and 4 e) 1, 2, 3, and 4

20. Write the balanced net ionic equation for the reaction that occurs, if any, when aqueous solutions of sodium sulfate and sulfuric acid are mixed.

21. Write the balanced net ionic equation for the reaction which occurs when aqueous solutions of magnesium perchlorate and sodium hydroxide are mixed.

22. Complete and balance the equations for any reactions which will occur for the following situations. Write the net ionic equation where appropriate.
    a) $BaSO_4(s) + NaCl(aq) \rightarrow$
    b) $Pb(NO_3)_2(aq) + Na_2CrO_4(aq) \rightarrow$
    c) $Ca(OH)_2(aq) + H_2SO_4(aq) \rightarrow$

23. All of the following are bases except
  a) $Sr(OH)_2$ b) CH_3COOH c) KOH d) NH_3 e) CH_3NH_2

24. When BaO(s) dissolves in water,
  a) the resulting solution is acidic.
  b) hydrogen gas is produced.
  c) O^{2-}(aq) reacts with water to give hydroxide ions.
  d) the resulting solution is neutral.
  e) oxygen gas is produced.

25. What type of reagent is required to convert HNO_2 to NO_2^-?
  a) base b) acid c) reducing agent d) oxidizing agent e) neutralization reagent

26. What type of reagent is required to convert HSO_3^- to SO_3^{2-}?
  a) base b) acid c) reducing agent d) oxidizing agent e) neutralization agent

27. Which of the following aqueous reagents, when mixed, give both a neutralization reaction and a precipitation reaction?
  a) KOH and H_2CO_3
  b) $CaCO_3$ and HCl
  c) $CaCl_2$ and H_2CO_3
  d) H_2SO_4 and Na_2CO_3
  e) $Ca(OH)_2$ and H_2CO_3

28. Which of the following oxides gives a basic solution when dissolved in water?
  a) dinitrogen pentoxide
  b) selenium dioxide
  c) carbon dioxide
  d) magnesium oxide
  e) sulfur trioxide

29. Which of the following oxides gives an acidic solution when dissolved in water?
  a) SO_2 b) CaO c) Na_2O d) BaO e) K_2O

30. Which of the following oxides gives a basic solution when dissolved in water?
 a) K_2O b) N_2O_5 c) SeO_3 d) P_4O_{10} e) SO_2

31. All of the following elements give amphoteric oxides except
 a) Hg b) Al c) Sn d) Bi e) Be

32. Which of the following is an amphoteric oxide?
 a) SnO b) CaO c) BaO d) K_2O e) MgO

33. Which of the following is correct with respect to the reactions of PbO?
 a) PbO does not react with acid or base.
 b) PbO dissolves in water to give hydroxide ions.
 c) PbO reacts with base to give $Pb(OH)_4^-$.
 d) PbO reacts with water to give hydrogen gas.
 e) PbO is insoluble in base.

34. What is the net ionic equation for the reaction that occurs when aqueous calcium hydroxide is added to aqueous nitric acid?
 a) $2H_3O^+(aq) + Ca(OH)_2(aq) \rightarrow 4H_2O(l) + Ca^{2+}(aq)$
 b) $H_3O^+(aq) + NO_3^-(aq) + Ca^+(aq) + OH^-(aq) \rightarrow CaNO_3(aq) + 2H_2O(l)$
 c) $H_3O^+(aq) + Ca(OH)(aq) \rightarrow 2H_2O(l) + Ca^{2+}(aq)$
 d) $H_3O^+(aq) + OH^-(aq) \rightarrow 2H_2O(l)$
 e) $HNO_3(aq) + OH^-(aq) \rightarrow H_2O(l) + NO_3^-(aq)$

35. What salt is produced by the reaction between aqueous solutions of potassium hydroxide and phosphoric acid?
 a) K_3PO_4 b) KH_2PO_4 c) KO_2 d) K_2HPO_4 e) H_2O

36. When calcium carbonate is added to aqueous sulfuric acid, a gas is evolved. What are the products of this reaction?
 a) $H_2(g)$, CO(g), and $CaSO_4(s)$
 b) $CO_2(g)$, $H_2O(l)$, and $CaSO_4(s)$
 c) CO(g) and $CaSO_4(s)$
 d) $O_2(g)$, CO(g), and $CaSO_4(s)$
 e) CO(g), $H_2O(g)$, and $CaSO_4(s)$

37. What is the oxidation number of each of the hydrogen atoms in $NaHCO_3$?
 a) +1 b) −1 c) −2 d) +2 e) 0

38. What is the oxidation number of carbon in CaC_2?
 a) −1 b) −2 c) +4 d) −4 e) 0

39. What is the oxidation number of manganese in $(NH_4)_2MnO_4$?
 a) +6 b) +7 c) +2 d) +4 e) 0

40. What is the oxidation number of carbon in $NaHCO_3$?

a) +4 b) +3 c) +6 d) –4 e) 0

41. Predict the strongest oxidizing agent for each pair and explain your reasoning.

a) NO_3^- or NO

b) H_2S or S_8

c) ClO^- or ClO_4^-

42. In a redox reaction, the oxidizing agent

a) loses electrons and is oxidized.
b) loses electrons and is reduced.
c) gains electrons and is reduced.
d) gains electrons and is oxidized.
e) contains an element that undergoes an increase in oxidation number.

43. Consider the following reaction:

$$2MnO_4^-(aq) + 5HOOCCOOH(aq) + 6H^+(aq) \rightarrow 2Mn^{2+}(aq) + 8H_2O(l) + 10CO_2(aq)$$

In this reaction,

a) HOOCCOOH is reduced.
b) MnO_4^- is oxidized.
c) no electrons are transferred since the oxidation number of Mn does not change.
d) the oxidizing agent is MnO_4^-.
e) electrons are transferred from MnO_4^- to HOOCOOH.

44. Consider the following reaction:

$$Cd(s) + 2AgCl(s) \rightarrow Cd^{2+}(aq) + 2Ag(s) + 2Cl^-(aq)$$

In this reaction,

a) the reducing agent is AgCl(s).
b) no electrons are transferred since the oxidation number of chlorine atoms does not change.
c) the reducing agent is Cd(s).
d) the oxidizing agent is Cd(s).
e) electrons are transferred from silver in AgCl(s) to Cd(s).

45. Consider the following reaction:

$$2Fe^{3+}(aq) + 2Hg(l) + 2Cl^-(aq) \rightarrow 2Fe^{2+}(aq) + Hg_2Cl_2(s)$$

In this reaction,

a) Fe^{3+}(aq) is the reducing agent
b) Fe^{3+}(aq) loses electrons.
c) Hg(l) loses electrons.
d) Hg(l) is reduced.
e) Hg(l) is the oxidizing agent.

46. Consider the following reaction:

$$IO_3^-(aq) + 2Tl^+(aq) + 2Cl^-(aq) + 6H^+(aq) \rightarrow ICl_2^-(aq) + 2Tl^{3+}(aq) + 3H_2O(l)$$

In this reaction,

a) $Tl^+(aq)$ is the oxidizing agent.
b) $Cl^-(aq)$ undergoes oxidation.
c) $IO_3^-(aq)$ undergoes oxidation.
d) $Tl^+(aq)$ gains electrons.
e) iodine atoms in $IO_3^-(aq)$ are reduced.

47. Consider the following reaction:

$$CH_2(COOH)_2(aq) + 6Ce^{4+}(aq) + 2H_2O(l) \rightarrow 2CO_2(aq) + HCOOH(aq) + 6Ce^{3+}(aq) + 6H^+(aq)$$

In this reaction,

a) $Ce^{4+}(aq)$ gains electrons from $CH_2(COOH)_2(aq)$.
b) $Ce^{4+}(aq)$ transfers electrons to $CH_2(COOH)_2(aq)$.
c) hydrogen atoms in $H_2O(l)$ lose electrons.
d) $CH_2(COOH)_2(aq)$ is the oxidizing agent.
e) $Ce^{4+}(aq)$ is the reducing agent.

48. Consider the following reaction:

$$2Mn(OH)_3(s) + 6H^+(aq) + 6I^-(aq) \rightarrow 2Mn^{2+}(aq) + 2I_3^-(aq) + 6H_2O(l)$$

In this reaction,

a) $I^-(aq)$ gains electrons from $Mn(OH)_3(s)$.
b) $I^-(aq)$ is the oxidizing agent.
c) $H^+(aq)$ is the oxidizing agent.
d) $Mn(OH)_3(s)$ gains electrons from $I^-(aq)$.
e) $Mn(OH)_3(s)$ is the reducing agent.

49. When an acidic solution of potassium permanganate is mixed with a sodium chloride solution, Mn^{2+} ions and chlorine gas are formed. Which species is reduced?

a) H_2O b) Mn^{2+} c) Cl^- d) H^+ e) MnO_4^-

50. When an acidic solution of potassium permanganate is mixed with a sodium chloride solution, Mn^{2+} ions and chlorine gas are formed. Which species is the reducing agent?

a) MnO_4^- b) H^+ c) Cl_2 d) H_2O e) Cl^-

51. Charcoal "filters" are used to remove dissolved chlorine gas from drinking water. In acidic solution, $Cl_2(aq)$, reacts with carbon to form carbon dioxide and chloride ions. Which species is reduced?

a) Cl_2 b) carbon c) Cl^- d) H_2O e) H^+

52. Charcoal "filters" are used to remove dissolved chlorine gas from drinking water. In acidic solution, $Cl_2(aq)$, reacts with carbon to form carbon dioxide and chloride ions. Which species is the reducing agent?

a) carbon b) Cl_2 c) CO_2 d) H_2O e) H^+

53. When aqueous sodium hydroxide is mixed with aluminum shavings, hydrogen gas and the aluminate ion, $Al(OH)_4^-$, are formed. Which species is reduced?
 a) OH^- b) Al c) H_2O d) H_2 e) This is not a redox reaction and thus no species is oxidized.

54. When aqueous sodium hydroxide is mixed with aluminum shavings, hydrogen gas and the aluminate ion, $Al(OH)_4^-$, are formed. Which species is the reducing agent?
 a) Al b) OH^- c) H_2O d) $Al(OH)_4^-$ e) This is not a redox reaction and thus no species is oxidized.

55. If an aqueous solution of sulfuric acid is added to a pool of mercury, $SO_2(aq)$ and mercury(II) sulfate are formed. Which species is reduced?
 a) $H_2SO_4(aq)$ b) Hg(l) c) $H_2O(l)$ d) $SO_2(aq)$ e) $HgSO_4(aq)$

56. If an aqueous solution of sulfuric acid is added to a pool of mercury, $SO_2(aq)$ and mercury(II) sulfate are formed. Which species is the reducing agent?
 a) Hg(l) b) $H_2SO_4(aq)$ c) $H_2O(l)$ d) $SO_2(aq)$ e) $HgSO_4(aq)$

57. When $P_4O_{10}(s)$ is added to water, phosphoric acid is produced. Which of the following is correct?
 a) The reducing agent is P_4O_{10}.
 b) The reducing agent is water.
 c) The species that undergoes reduction is water.
 d) This is not a redox reaction.
 e) The species that undergoes oxidation is P_4O_{10}.

58. If zinc metal is added to a basic aqueous solution containing nitrate ions, ammonia and $Zn(OH)_4^{2-}(aq)$ are formed. Which species is oxidized?
 a) Zn(s) b) $NO_3^-(aq)$ c) $OH^-(aq)$ d) $H_2O(l)$ e) $Zn(OH)_4^-$

59. If zinc metal is added to a basic aqueous solution containing nitrate ions, ammonia and $Zn(OH)_4^{2-}(aq)$ are formed. Which species is the oxidizing agent?
 a) $NO_3^-(aq)$ b) Zn(s) c) $OH^-(aq)$ d) $H_2O(l)$ e) ammonia

60. If an acidic solution of potassium dichromate, $K_2Cr_2O_7(aq)$, is added to aqueous ethanol, $CH_3CH_2OH(aq)$, acetic acid, $CH_3COOH(aq)$ and $Cr^{3+}(aq)$ are produced. Which species is the oxidizing agent?
 a) $Cr_2O_7^{2-}(aq)$ b) $CH_3CH_2OH(aq)$ c) $H_2O(l)$ d) $Cr^{3+}(aq)$ e) $H^+(aq)$

61. If an acidic solution of potassium dichromate, $K_2Cr_2O_7(aq)$, is added to aqueous ethanol, $CH_3CH_2OH(aq)$, acetic acid, $CH_3COOH(aq)$ and $Cr^{3+}(aq)$ are produced. Which species is oxidized?
 a) $CH_3CH_2OH(aq)$ b) $CH_3COOH(aq)$ c) $H_2O(l)$ d) $Cr_2O_7^{2-}(aq)$ e) $H^+(aq)$

62. When an aqueous basic solution of sodium hypochlorite, NaOCl(aq), is added to $Fe(OH)_3(s)$, the ferrate ion, $FeO_4^{2-}(aq)$, and chloride ion are produced. Which species undergoes reduction?
 a) $Fe(OH)_3(s)$ b) $OCl^-(aq)$ c) $OH^-(aq)$ d) $H_2O(l)$ e) $Cl^-(aq)$

63. When an aqueous basic solution of sodium hypochlorite, NaOCl(aq), is added to $Fe(OH)_3$(s), the ferrate ion, FeO_4^{2-}(aq), and chloride ion are produced. Which species is the reducing agent?
a) FeO_4^{2-}(aq) b) $Fe(OH)_3$(s) c) OH^-(aq) d) H_2O(l) e) OCl^-(aq)

64. Ammonia gas and NO(g) react to form nitrogen gas and H_2O(g). Which species undergoes reduction?
a) NO(g)
b) NH_3(g)
c) H_2O(g)
d) N_2(g)
e) Since this is not a redox reaction, no species undergoes reduction.

65. Ammonia gas and NO(g) react to form nitrogen gas and H_2O(g). Which species undergoes oxidation?
a) NH_3(g)
b) NO(g)
c) H_2O(g)
d) N_2(g)
e) Since this is not a redox reaction, no species undergoes oxidation.

66. Ammonia gas and NO(g) react to form nitrogen gas and H_2O(g). Which of the following is correct?
a) This is not a redox reaction.
b) NO(g) is the reducing agent.
c) Oxidation produces H_2O(g).
d) Reduction produces H_2O(g).
e) Electrons are transferred from ammonia to NO(g).

67. Dinitrogen tetroxide, N_2O_4(l) and hydrazine, N_2H_4(l), react to form nitrogen gas and H_2O(g). Which of the following is correct?
a) N_2O_4(l) is the reducing agent.
b) Reduction produces H_2O(g).
c) This is not a redox reaction.
d) Oxidation produces H_2O(g).
e) Electrons are transferred from hydrazine to dinitrogen tetroxide.

68. What type of reagent is required to convert N_2 to NO_2^-?
a) oxidizing agent b) reducing agent c) acid d) base e) precipitating agent

69. What type of reagent is required to convert N_2 to N_2O_4?
a) reducing agent b) oxidizing agent c) acid d) base e) neutralization agent

70. In a redox reaction, the oxidizing agent
a) is the species that is oxidized.
b) contains an element that undergoes a decrease in oxidation number.
c) contains an element that undergoes an increase in oxidation number.
d) reacts with oxygen gas.
e) loses electrons.

71. Consider the reaction below:

$$Cu(s) + 2H_2SO_4(aq, conc.) \rightarrow Cu^{2+}(aq) + SO_4^{2-}(aq) + SO_2(g) + 2H_2O(l)$$

Identify the reducing agent.
a) Cu b) H_2SO_4 c) Cu^{2+} d) SO_4^{2-} e) SO_2

72. Consider the reaction below:

$$4Fe^{2+}(aq) + 4H^+(aq) + O_2(g) \rightarrow 4Fe^{3+}(aq) + 2H_2O(l)$$

Identify the reducing agent.
a) O_2 b) Fe^{3+} c) Fe^{2+} d) H^+ e) H_2O

73. Consider the following reaction:

$$2Cr(OH)_3(s) + 3ClO^-(aq) + 4OH^-(aq) \rightarrow 2CrO_4^{2-}(aq) + 3Cl^-(aq) + 5H_2O(l)$$

Identify the reducing agent.
a) $Cr(OH)_3$ b) CrO_4^{2-} c) OH^- d) ClO^- e) Cl^-

74. Consider the following reaction:

$$NH_4^+(aq) + NO_3^-(aq) \rightarrow N_2O(g) + 2H_2O(l)$$

Identify the reducing agent.
a) H^+ b) N_2O c) H_2O d) NO_3^- e) NH_4^+

75. Which of the following is a redox reaction?
a) $2NaCl(s) + H_2SO_4 \rightarrow Na_2SO_4(s) + 2HCl(g)$
b) $CaC_2(s) + 2H_2O(l) \rightarrow Ca(OH)_2(aq) + C_2H_2(g)$
c) $MgSO_4(s) \rightarrow MgO(s) + SO_3(g)$
d) $NaCl(aq) + AgNO_3(aq) \rightarrow AgCl(s) + NaNO_3(aq)$
e) $2Mg(s) + O_2(g) \rightarrow 2MgO(s)$

76. All of the following are redox reactions except
a) $CaCO_3(s) + H_2SO_4(aq) \rightarrow CaSO_4(s) + H_2O(l) + CO_2(g)$
b) $2Mg(s) + O_2(g) \rightarrow 2MgO(s)$
c) $2Na(s) + S(s) \rightarrow Na_2S(s)$
d) $Zn(s) + 2HCl(aq) \rightarrow ZnCl_2(aq) + H_2(g)$
e) $2KClO_3(s) \rightarrow 2KCl(s) + 3O_2(g)$

77. What type of reagent is required to convert N_2 to N_2H_4?
a) oxidizing agent b) reducing agent c) acid d) base e) neutralization agent

78. How many electrons are transferred when sulfite, SO_3^{2-}, is converted to H_2S?
a) 6 b) 8 c) 4 d) 7 e) 2

79. All of the following can act as oxidizing agents except
a) $S_2O_3^{2-}$. b) SO_4^{2-}. c) S_2Cl_2. d) SO_3^{2-}. e) H_2S.

80. Write the balanced equation for the combustion of hydrazine, $N_2H_4(l)$, to nitrogen gas and water. Identify the oxidizing agent and the reducing agent.

Chapter 4: Reaction Stoichiometry: Chemistry's Accounting Form A

1. How many moles of beryllium atoms can be extracted from 12 moles of beryl, $Be_3Al_2Si_6O_{18}$?

a) 12 mol b) 1.8×10^{24} mol c) 36 mol d) 2.2×10^{25} mol e) 4 mol

2. How many moles of boron atoms can be extracted from 2.5 moles of borax, $Na_2B_4O_7 \cdot 10H_2O$?

a) 100 mol b) 2.4×10^{24} mol c) 6.0×10^{24} mol d) 10 mol e) 2.5 mol

3. How many moles of silver chloride can be precipitated when excess sodium chloride is added to an aqueous solution that contains 0.50 moles of silver nitrate?

a) 1.5 mol b) 0.50 mol c) 1.0 mol d) 0.13 mol e) 0.25 mol

4. How many moles of sodium chloride must be added to an aqueous solution that contains 2.0 moles of silver nitrate in order to precipitate 0.50 moles of silver chloride?

a) 0.25 mol b) 1.0 mol c) 1.5 mol d) 2.0 mol e) 0.50 mol

5. How many moles of sodium chloride must be added to an aqueous solution that contains 2.0 moles of silver nitrate in order to precipitate 3.0 moles of silver chloride?

a) 3.0 mol
b) Not enough silver nitrate is available to precipitate 3.0 moles of silver chloride.
c) 2.0 mol
d) 1.0 mol
e) 1.5 mol

6. How many moles of sodium chloride must be added to an aqueous solution that contains 2.0 moles of silver nitrate in order to precipitate 2.0 moles of silver chloride?

a) 3.0 mol
b) 1.5 mol
c) Not enough silver nitrate is available to precipitate 2.0 moles of silver chloride.
d) 1.0 mol
e) 2.0 mol

7. How many moles of carbon are required to react completely with 0.422 moles of $SnO_2(s)$ according to the equation below?

$$SnO_2(s) + 2C(s) \rightarrow Sn(s) + 2CO(g)$$

a) 0.844 mol b) 0.633 mol c) 0.211 mol d) 0.422 mol e) 1.27 mol

8. How many moles of carbon monoxide gas are produced when 0.422 moles of $SnO_2(s)$ react with excess carbon according to the equation below?

$$SnO_2(s) + 2C(s) \rightarrow Sn(s) + 2CO(g)$$

a) 0.844 mol b) 0.211 mol c) 0.422 mol d) 1.27 mol e) 0.633 mol

9. How many moles of carbon monoxide gas are produced from the reaction of 2.88 moles of $SnO_2(s)$ with 2.88 moles of carbon according to the equation below?

$$SnO_2(s) + 2C(s) \rightarrow Sn(s) + 2CO(g)$$

a) 1.44 mol b) 2.88 mol c) 4.32 mol d) 5.76 mol e) 8.64 mol

10. Consider the following reaction:

$$3NO_2(g) + H_2O(l) \rightarrow 2HNO_3(l) + NO(g)$$

How many moles of nitric acid are produced starting from 5.00 moles of $NO_2(g)$ and excess water?

a) 0.300 mol b) 10.0 mol c) 7.50 mol d) 1.67 mol e) 3.33 mol

11. Consider the following reaction:

$$3NO_2(g) + H_2O(l) \rightarrow 2HNO_3(l) + NO(g)$$

How many moles of $NO_2(g)$ are required to react with 2.00 moles of water to produce 3.00 moles of nitric acid?

a) 4.00 mol b) 4.50 mol c) 1.50 mol d) 3.00 mol e) 9.00 mol

12. Consider the following reaction:

$$3NO_2(g) + H_2O(l) \rightarrow 2HNO_3(l) + NO(g)$$

How many moles of $NO_2(g)$ are required to react with 1.00 mole of water to produce 3.00 moles of nitric acid?

a) 6.00 mol
b) 4.50 mol
c) 2.00 mol
d) 1.50 mol
e) Not enough water is available to produce 3.00 moles of nitric acid.

13. Copper(II) nitrate decomposes upon heating to form copper(II) oxide, nitrogen dioxide gas, and oxygen gas. If 1.0 mole of copper(II) nitrate decomposes, how many moles of nitrogen dioxide would be formed?

a) 4.0 mol b) 6.0 mol c) 0.50 mol d) 2.0 mol e) 1.0 mol

14. Copper(II) nitrate decomposes upon heating to form copper(II) oxide, nitrogen dioxide gas, and oxygen gas. How many moles of copper(II) nitrate are required to produce *at least* 1.0 mole of either nitrogen dioxide and oxygen gas?

a) 0.50 mol b) 4.0 mol c) 2.0 mol d) 0.25 mol e) 1.0 mol

15. Ammonium nitrate decomposes to "laughing gas," $N_2O(g)$, and water. If a sample of ammonium nitrate gives 0.50 moles of water, how many moles of $N_2O(g)$ are also formed?

a) 0.13 mol b) 2.0 mol c) 0.25 mol d) 1.0 mol e) 0.50 mol

16. Consider the following reaction:

$$5NaN_3(s) + NaNO_3(aq) \rightarrow 3Na_2O(s) + 8N_2(g)$$

If 2.0 moles of $N_2(g)$ are produced, how many moles of $Na_2O(s)$ are produced?

a) 0.75 mol b) 8.0 mol c) 5.3 mol d) 0.63 mol e) 0.38 mol

17. Calculate the volume of 0.121 M KOH(aq) needed to react completely with 25.0 mL of 0.528 M $HNO_3(aq)$.

a) 54.5 mL b) 109 mL c) 5.73 mL d) 175 mL e) 218 mL

18. Calculate the volume of 1.00 M $Ca(OH)_2(aq)$ needed to react completely with 50.0 mL of 2.24 M $HNO_3(aq)$.

a) 224 mL b) 168 mL c) 56.0 mL d) 22.4 mL e) 112 mL

19. Calculate the number of moles of NaOH(aq) needed to react completely with 125 mL of 6.00 M $H_2SO_4(aq)$.

a) 0.0416 mol b) 1.50 mol c) 0.375 mol d) 3.00 mol e) 0.750 mol

20. Calculate the number of moles of $Ca(OH)_2(aq)$ needed to react completely with 125 mL of 6.00 M HCl(aq).

a) 3.00 mol b) 0.0416 mol c) 1.50 mol d) 0.750 mol e) 0.375 mol

21. Rust, $Fe_2O_3(s)$, can be removed from porcelain according to the reaction below.

$$Fe_2O_3(s) + 6Na_2C_2O_4(aq) + 6HCl(aq) \rightarrow 2Na_3Fe(C_2O_4)_3(aq) + 6NaCl(aq) + 3H_2O(l)$$

How many moles of rust can be removed by 375 mL of 2.00 M $Na_2C_2O_4(aq)$?

a) 0.750 mol b) 0.125 mol c) 0.375 mol d) 0.250 mol e) 4.50 mol

22. Rust, $Fe_2O_3(s)$, can be removed from porcelain according to the reaction below.

$$Fe_2O_3(s) + 6Na_2C_2O_4(aq) + 6HCl(aq) \rightarrow 2Na_3Fe(C_2O_4)_3(aq) + 6NaCl(aq) + 3H_2O(l)$$

How many grams of rust can be removed by 375 mL of 2.00 M $Na_2C_2O_4(aq)$?

a) 59.9 g b) 719 g c) 120 g d) 39.9 g e) 20.0 g

23. Consider the following reactions.

$$ZrC(s) + 4Cl_2(g) \rightarrow ZrCl_4(g) + CCl_4(g)$$

$$ZrCl_4(g) + 2Mg(s) \rightarrow Zr(s) + 2MgCl_2(l)$$

How many moles of Zr(s) are produced from 5.00 moles of chlorine gas?

a) 2.00 mol b) 5.00 mol c) 1.25 mol d) 0.800 mol e) 1.00 mol

24. Consider the following reactions.

$$ZrC(s) + 4Cl_2(g) \rightarrow ZrCl_4(g) + CCl_4(g)$$

$$ZrCl_4(g) + 2Mg(s) \rightarrow Zr(s) + 2MgCl_2(l)$$

How many moles of chlorine gas are required to produce 1.50 moles of $MgCl_2(l)$?

a) 0.750 mol b) 12.0 mol c) 1.50 mol d) 6.00 mol e) 3.00 mol

25. Potassium perchlorate can be produced from the reactions below.

$$Cl_2(g) + 2KOH(aq) \rightarrow KCl(aq) + KClO(aq) + H_2O(l)$$
$$3KClO(aq) \rightarrow 2KCl(aq) + KClO_3(aq)$$
$$4KClO_3(aq) \rightarrow 3KClO_4(s) + KCl(aq)$$

How many moles of chlorine gas are required to produce 10.0 moles of $KClO_4(s)$?

a) 40.0 mol b) 3.00 mol c) 3.33 mol d) 30.0 mol e) 10.3 mol

26. Potassium perchlorate can be produced from the reactions below.

$$Cl_2(g) + 2KOH(aq) \rightarrow KCl(aq) + KClO(aq) + H_2O(l)$$
$$3KClO(aq) \rightarrow 2KCl(aq) + KClO_3(aq)$$
$$4KClO_3(aq) \rightarrow 3KClO_4(s) + KCl(aq)$$

How many moles of $KClO_4(s)$ are produced from 5.00 mol of KOH(aq)?

a) 0.833 mol b) 0.625 mol c) 7.50 mol d) 5.00 mol e) 2.50 mol

27. Consider the following reactions.

$$4Bi(s) + 3O_2(g) \rightarrow 2Bi_2O_3(s)$$
$$Bi_2O_3(s) + 6HCl(aq) \rightarrow 2BiCl_3(aq) + 3H_2O(l)$$

If 5.00 L of 0.200 M $BiCl_3(aq)$ are produced, how many moles of Bi(s) are used?

a) 1.00 mol b) 2.00 mol c) 4.00 mol d) 0.250 mol e) 0.200 mol

28. Consider the following reactions for the determination of dissolved oxygen in water.

$$2MnSO_4(aq) + 4NaOH(aq) + O_2(aq) \rightarrow 2MnO_2(s) + 2Na_2SO_4(aq) + 2H_2O(l)$$
$$MnO_2(s) + 2H_2SO_4(aq) + 2NaI(aq) \rightarrow MnSO_4(aq) + I_2(aq) + Na_2SO_4(aq) + 2H_2O(l)$$
$$I_2(aq) + 2Na_2S_2O_3(aq) \rightarrow Na_2S_4O_6(aq) + 2NaI(aq)$$

If 9.00 mL of 0.0240 M $Na_2S_2O_3(aq)$ are used in the analysis, how many moles of dissolved oxygen were determined?

a) 2.16×10^{-4} mol b) 2.70×10^{-5} mol c) 1.08×10^{-4} mol d) 5.40×10^{-5} mol e) 4.32×10^{-4} mol

29. Consider the following reaction:

$$5NaN_3(s) + NaNO_3(aq) \rightarrow 3Na_2O(s) + 8N_2(g)$$

If 1.50 g of $N_2(g)$ are produced, how many grams of $Na_2O(s)$ are also produced?

a) 2.05 g b) 1.24 g c) 3.32 g d) 2.22 g e) 8.85 g

30. Consider the following reaction:

$$4FeS_2(s) + 11O_2(g) \rightarrow 2Fe_2O_3(s) + 8SO_2(g)$$

How many grams of sulfur dioxide are produced from the reaction of 1.000 kg of $FeS_2(s)$?

a) 534.0 g b) 267.0 g c) 2136 g d) 1068 g e) 4272 g

31. A 25.00-mL sample of $HClO_4(aq)$ was titrated to the stoichiometric point with 36.80 mL of 0.1011 M NaOH(aq). What is the molarity of the perchloric acid solution?

a) 0.06020 M b) 0.09301 M c) 0.1488 M d) 0.003720 M e) 0.06868 M

32. A 50.00-mL sample of $HClO_4$(aq) was titrated to the stoichiometric point with 26.80 mL of 1.02 M NaOH(aq). What is the mass of perchloric acid in the solution?
a) 1.09 g b) 0.0273 g c) 54.9 g d) 2.75 g e) 1.93 g

33. A 50.0-mL sample of sulfuric acid from a lake near a mine was titrated to the stoichiometric point with 17.52 mL of 0.0120 M NaOH(aq). What is the molarity of sulfuric acid in the sample?
a) 0.00841 M b) 0.000105 M c) 0.00210 M d) 0.00420 M e) 0.00526 M

34. A 25.0 mL sample of KOH(aq) was titrated to the stoichiometric point with 37.5 mL of 0.400 M phosphoric acid. What is the concentration of KOH(aq)?
a) 0.555 M b) 0.0450 M c) 0.200 M d) 1.80 M e) 0.600 M

35. In a titration, a 1.00-g sample of an acid requires 42.0 mL of 0.600 M KOH(aq) for complete reaction. If the acid has 2 acidic hydrogens per molecule, calculate the molar mass of the acid.
a) 79.4 g/mol b) 159 g/mol c) 19.8 g/mol d) 39.7 g/mol e) 126 g/mol

36. In a titration, a 1.00-g sample of an acid, HX, requires 20.3 mL of 0.300 M NaOH(aq) for complete reaction. Calculate the molar mass of the acid.
a) 328 g/mol b) 82.1 g/mol c) 164 g/mol d) 244 g/mol e) 41.1 g/mol

37. How many mL of 0.200 M H_2SO_4(aq) must be added to 25.0 mL of 0.0888 M $BaCl_2$(aq) to precipitate all the barium as barium sulfate?
a) 22.2 mL b) 5.55 mL c) 25.0 mL d) 11.1 mL e) 9.01 mL

38. How many mL of 0.100 M HCl(aq) are neutralized by 1.00 g of $Mg(OH)_2$(s)?
a) 343 mL b) 85.7 mL c) 686 mL d) 171 mL e) 5.85 mL

39. A 25.0-mL sample of oxalic acid, $H_2C_2O_4$(with two acidic protons), was titrated to the stoichiometric point with 28.3 mL of 0.101 M NaOH(aq). Calculate the molarity of $H_2C_2O_4$(aq).
a) 0.0572 M b) 0.114 M c) 0.0286 M d) 0.229 M e) 0.0715 M

40. Consider the following net ionic equation.

$$6Fe^{2+}(aq) + Cr_2O_7^{2-}(aq) + 14H^+(aq) \rightarrow 6Fe^{3+}(aq) + 2Cr^{3+}(aq) + 7H_2O(l)$$

When a 25.0-mL sample of an iron(II) solution was titrated, 28.7 mL of 0.200 M $K_2Cr_2O_7$(aq) was required to reach the stoichiometric point. What is the molar concentration of iron(II)?
a) 0.689 M b) 0.230 M c) 1.38 M d) 0.459 M e) 0.0383 M

41. Manganese in an ore can be determined by treating the ore with a measured, excess quantity of sodium oxalate to reduce $MnO_2(s)$ to $MnCl_2(aq)$ followed by determination of the unreacted sodium oxalate by titration with potassium permanganate.

$$MnO_2(s) + Na_2C_2O_4(aq) + 4HCl(aq) \rightarrow 2MnCl_2(aq) + 2CO_2(g) + 2H_2O(l) + 2NaCl(aq)$$

$$2KMnO_4(aq) + 5Na_2C_2O_4(aq) + 16HCl(aq) \rightarrow 2MnCl_2(aq) + 10CO_2(g) + 8H_2O(l) + 10NaCl(aq)$$

If a sample is treated with 50.0 mL of 0.275 M $Na_2C_2O_4(aq)$ and the unreacted $Na_2C_2O_4(aq)$ requires 18.28 mL of 0.1232 M $KMnO_4(aq)$, calculate the number of grams of manganese in the sample.

42. Consider the following reaction.

$$PCl_3(l) + Cl_2(g) \rightarrow PCl_5(s)$$

Calculate the number of grams of $PCl_5(s)$ produced from 55.8 g of $PCl_3(l)$ if the percent yield is 78%.

43. In a combustion analysis of a 3.00-mg sample of a compound containing only carbon and hydrogen, it was found that 6.00 mg of carbon dioxide were produced. What is the percent hydrogen in the sample?

a) 45.3% b) 50.0% c) 27.3% d) 54.7% e) 25.0%

44. Consider the following reaction.

$$2Al_2O_3(l) \rightarrow 4Al(l) + 3O_2(g)$$

If 500 g of Al(l) are produced from 1500 g of $Al_2O_3(l)$, what is the percent yield?

a) 63.0% b) 79.0% c) 37.0% d) 33.3% e) 53.0%

45. Consider the following reaction.

$$5NaN_3(s) + NaNO_3(aq) \rightarrow 3Na_2O(s) + 8N_2(g)$$

If 2.50 g of $NaN_3(s)$ react with excess $NaNO_3(aq)$, 0.995 g of $N_2(g)$ is recovered. What is the percent yield?

a) 57.7% b) 72.9% c) 25.0% d) 81.4% e) 41.7%

46. Consider the following reaction.

$$6Na(l) + Al_2O_3(s) \rightarrow 2Al(l) + 3Na_2O(s)$$

When 5.52 g of sodium react with excess $Al_2O_3(s)$, 1.00 g of Al(l) is produced. What is the percent yield?

a) 18.1% b) 15.4% c) 46.3% d) 39.1% e) 11.1% 7

47. Consider the following reaction.

$$3H_2(g) + WO_3(s) \rightarrow 3H_2O(g) + W(s)$$

When 500.0 g of $WO_3(s)$ react with excess hydrogen, 375 g of W(s) are obtained. What is the percent yield?

a) 75.0% b) 94.6% c) 79.2% d) 31.5% e) 78.4% 7

48. A sample of an unknown compound of mass 0.141 g reacts with chlorine gas and all the carbon in the compound is converted to CCl_4. If the mass of CCl_4 is 0.282 g, calculate the percent by mass of carbon in the unknown compound.

a) 15.6% b) 50.0% c) 7.80% d) 1.28% e) 2.20%

49. Consider the following reaction.

$$3H_2(g) + WO_3(s) \rightarrow 3H_2O(g) + W(s)$$

When 500.0 g of $WO_3(s)$ react with excess hydrogen, 375 g of W(s) are obtained. What is the percent yield?

a) 94.6% b) 75.0% c) 79.2% d) 31.5% e) 78.4%

50. A sample of an unknown compound of mass 0.141 g reacts with chlorine gas and all the carbon in the compound is converted to CCl_4. If the mass of CCl_4 is 0.282 g, calculate the percent by mass of carbon in the unknown compound.

a) 2.20% b) 50.0% c) 7.80% d) 1.28% e) 15.6%

51. How many moles of carbon monoxide gas are produced from the reaction of 5.00 moles of $SnO_2(s)$ with 9.00 moles of carbon according to the equation below?

$$SnO_2(s) + 2C(s) \rightarrow Sn(s) + 2CO(g)$$

a) 9.00 mol b) 7.50 mol c) 4.50 mol d) 18.0 mol e) 5.00 mol

52. Consider the following reaction:

$$3NO_2(g) + H_2O(l) \rightarrow 2HNO_3(l) + NO(g)$$

How many moles of nitric acid are produced starting from 5.00 moles of $NO_2(g)$ and 2.00 moles of water?

a) 10.0 mol b) 7.50 mol c) 4.00 mol d) 1.67 mol e) 3.33 mol

53. Ammonium nitrate decomposes to "laughing gas," $N_2O(g)$, and water. How many moles of ammonium nitrate are required to produce *at least* 1.0 mole of $N_2O(g)$ and 1.0 mole of water?

a) 0.50 mol b) 0.25 mol c) 1.0 mol d) 2.0 mol e) 4.0 mol

54. Consider the following reaction:

$$5NaN_3(s) + NaNO_3(aq) \rightarrow 3Na_2O(s) + 8N_2(g)$$

If 3.0 moles of NaN_3 and 3.0 moles of $NaNO_3$ react, how many moles of N_2 are produced?

a) 2.7 mol b) 4.8 mol c) 24 mol d) 1.9 mol e) 0.38 mol

55. Consider the following reaction:

$$5NaN_3(s) + NaNO_3(aq) \rightarrow 3Na_2O(s) + 8N_2(g)$$

If 1.0 g of NaN_3 reacts with 25 mL of 0.10 M $NaNO_3(aq)$, how many moles of $N_2(g)$ are produced?

a) 0.0040 mol b) 0.0094 mol c) 0.024 mol d) 0.020 mol e) 0.0031 mol

56. Consider the reaction below:

$$6Li(s) + N_2(g) \rightarrow 2Li_3N(s)$$

The following represents a mixture of Li(s) and $N_2(g)$ just before reaction occurs.

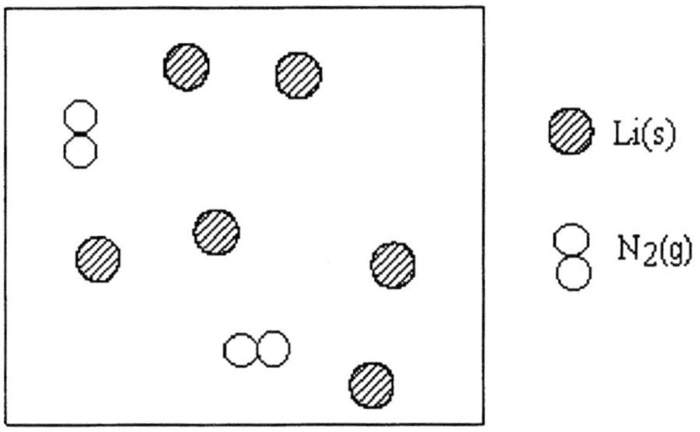

If each symbol represents 1 mole of Li(s) and 1 mole of $N_2(g)$, what is the limiting reactant and how many moles of the excess reactant remain after the reaction is complete?

a) $N_2(g)$, 2 mol Li(s)
b) Li(s), 1 mol $N_{2(g)}$
c) Li(s), 0.5 mol $N_2(g)$
d) $N_2(g)$, 4 mol Li(s)
e) $N_2(g)$, 3 mol Li(s)

57. Consider the following reaction:

$$3NO_2(g) + H_2O(l) \rightarrow 2HNO_3(l) + NO(g)$$

How many moles of the excess reactant remain if 2.00 moles of $H_2O(l)$ and 5.00 moles of $NO_2(g)$ are used?

a) 3.00 mol $NO_2(g)$
b) 0.33 mol $H_2O(l)$
c) 1.00 mol $NO_2(g)$
d) 4.00 mol $NO_2(g)$
e) 1.67 mol $H_2O(l)$

58. Consider the following reaction:

$$6Na(l) + Al_2O_3(s) \rightarrow 2Al(l) + 3Na_2O(s)$$

How many moles of Al(l) are produced from 1.00 mole of each reactant?

a) 3.00 mol b) 1.00 mol c) 0.333 mol d) 2.00 mol e) 0.167 mol

59. Consider the following reaction:

$$6Na(l) + Al_2O_3(s) \rightarrow 2Al(l) + 3Na_2O(s)$$

How grams of $Na_2O(s)$ are produced from 6.00 moles of each reactant?

a) 31.0 g b) 558 g c) 0.290 g d) 372 g e) 186 g

60. Consider the following reaction:

$$6Na(l) + Al_2O_3(s) \rightarrow 2Al(l) + 3Na_2O(s)$$

How moles of $Na_2O(s)$ are produced from 5.00 g of each reactant?

a) 0.147 mol b) 0.217 mol c) 0.0490 mol d) 0.434 mol e) 0.109 mol

61. Consider the following reaction:

$$6Na(l) + Al_2O_3(s) \rightarrow 2Al(l) + 3Na_2O(s)$$

How grams of Al(l) are produced from 5.00 g of each reactant?

a) 1.76 g b) 5.86 g c) 2.65 g d) 1.96 g e) 1.32 g

62. A manufacturer of children's wagons has in stock 500 wagon bodies, 2200 wheels, and 750 axles. If each wagon produced has 1 body, 4 wheels, and 2 axles, how many wagons can be assembled?

a) 375 b) 625 c) 550 d) 500 e) 250

63. A manufacturer of children's wagons has in stock 500 wagon bodies, 2200 wheels, and 750 axles. If each wagon produced has 1 body, 4 wheels, and 2 axles, how many components remain after the maximum number of wagons is assembled?

a) 250 bodies and 250 axles
b) 125 bodies and 500 wheels
c) 125 bodies and 1825 wheels
d) 200 wheels
e) 125 bodies and 700 wheels

64. Consider the following reaction:

$$4FeS_2(s) + 11O_2(g) \rightarrow 2Fe_2O_3(s) + 8SO_2(g)$$

How many grams of $Fe_2O_3(s)$ are produced from the reaction of 1.000 kg of $FeS_2(s)$ and 1.000 kg of oxygen gas?

a) 751.2 g b) 1331 g c) 2662 g d) 907.4 g e) 665.6 g

65. Consider the following reaction:

$$4FeS_2(s) + 11O_2(g) \rightarrow 2Fe_2O_3(s) + 8SO_2(g)$$

How many grams of $Fe_2O_3(s)$ are produced from 500.0 g of $FeS_2(s)$ and 400.0 g of oxygen gas?

a) 353.0 g b) 332.8 g c) 1996 g d) 266.2 g e) 1331 g

66. The reaction of antimony, Sb(s), with chlorine, $Cl_2(g)$, is shown pictorially below.

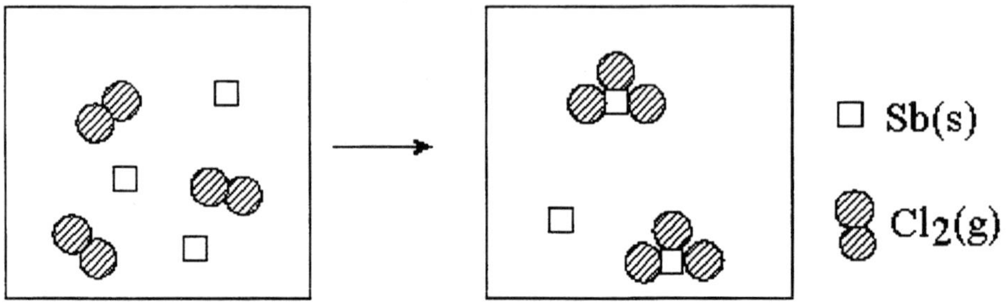

Which of the following gives the stoichiometric reaction?

a) $2Sb(s) + 3Cl_2(g) \rightarrow Sb_2Cl_6(s)$
b) $Sb(s) + 3Cl_2(g) \rightarrow SbCl_6(s)$
c) $3Sb(s) + 3Cl_2(g) \rightarrow 3SbCl_3(s)$
d) $2Sb(s) + 3Cl_2(g) \rightarrow 2SbCl_3(s)$
e) $3Sb(s) + 3Cl_2(g) \rightarrow Sb_3Cl_6(s)$

67. Consider the reaction below.

$$2Sb(s) + 3Cl_2(g) \rightarrow 2SbCl_3(s)$$

Determine the limiting reactant and the number of moles of the other reactant remaining when 6 moles of antimony and 6 moles of chlorine gas react.

a) Sb(s) and 2 moles of $Cl_2(g)$ remaining
b) Sb(s) and 4 moles of $Cl_2(g)$ remaining
c) $Cl_2(g)$ and 3 moles of Sb(s) remaining
d) $Cl_2(g)$ and 4 moles of Sb(s) remaining
e) $Cl_2(g)$ and 2 moles of Sb(s) remaining

68. The reaction of an element, A(s), with an element B(s), is shown pictorially below.

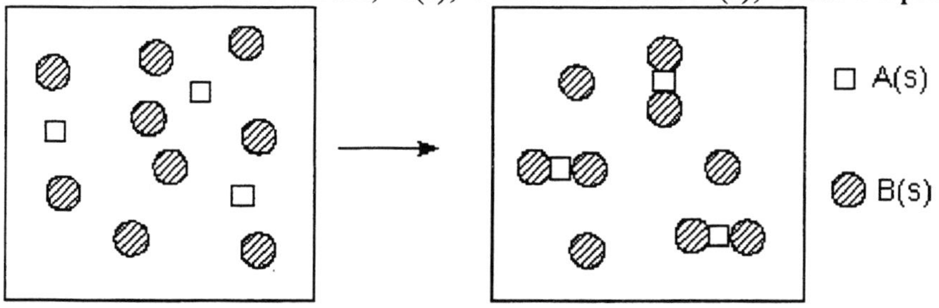

Which of the following gives the stoichiometric reaction?

a) $A(s) + 2B(s) \rightarrow AB_2(s)$
b) $A(s) + 5B(s) \rightarrow AB_2(s)$
c) $3A(s) + 9B(s) \rightarrow 3AB_3(s)$
d) $3A(s) + 2B(s) \rightarrow A_3B_2(s)$
e) $3A(s) + 9B(s) \rightarrow A_3B_9(s)$

69. Consider the following reaction.

$$2Mg(s) + O_2(g) \rightarrow 2MgO(s)$$

If the reactants are in the proportions shown below,

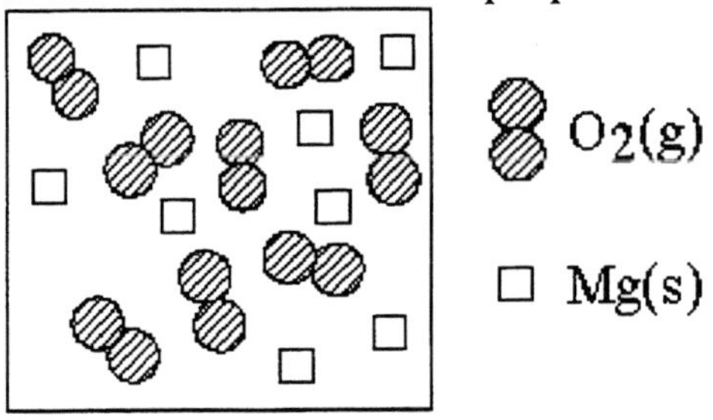

which of the following represents the product mixture?

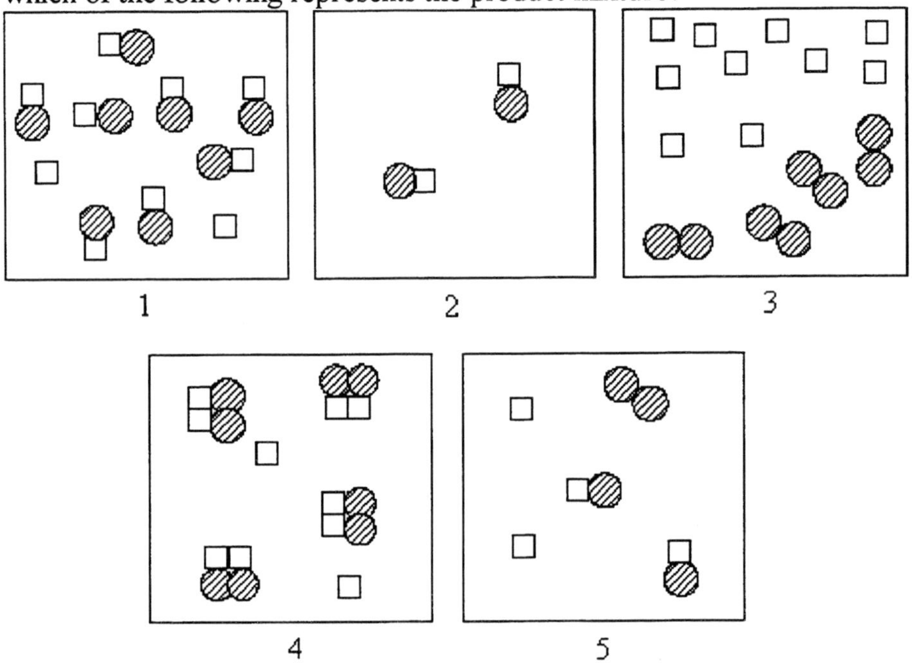

a) 1 b) 2 c) 3 d) 4 e) 5

70. Consider the following reaction.

$$2Mg(s) + O_2(g) \rightarrow 2MgO(s)$$

If 10 moles of magnesium react with 4 moles of oxygen gas, which is the limiting reactant and how many moles of the excess reactant remain?

a) magnesium and 1 mole of oxygen gas remains
b) oxygen gas and 2 moles of magnesium remain
c) magnesium and no oxygen gas remains
d) oxygen gas and 1 mole of magnesium remains
e) oxygen gas and no magnesium remains

71. How many moles of OH^- ions are contained in 125 mL of 0.625 M $Ba(OH)_2(aq)$?

a) 0.0781 mol b) 0.0391 mol c) 0.624 mol d) 0.156 mol e) 0.312 mol

72. How many moles of H^+ ions are contained in 125 mL of an aqueous solution which is 0.625 M oxalic acid, $H_2C_2O_4$(with two acidic protons)?
a) 0.624 mol b) 0.156 mol c) 0.312 mol d) 0.0391 mol e) 0.0781 mol

73. Consider the following reaction.

$$Cl_2(g) + 2NaOH(aq) \rightarrow NaCl(aq) + NaOCl(aq) + H_2O(l)$$

How many grams of sodium hypochlorite can be produced from 50.0 g $Cl_2(g)$ and 500.0 mL of 2.00 M NaOH(aq)?
a) 26.3 g b) 37.2 g c) 74.5 g d) 149 g e) 52.5 g

74. If 16.0 mL of 1.80 M $AgNO_3$(aq) are added to 64.00 mL of 0.200 M HCl(aq), how many moles of AgCl(s) are produced?
a) 1.80 mol b) 0.200 mol c) 0.0256 mol d) 0.0288 mol e) 0.0128 mol

75. Consider the reaction below.

$$2Fe(s) + 3Cl_2(g) \rightarrow 2FeCl_3(s)$$

The diagram below shows a mixture of Fe(s) and $Cl_2(g)$ where each symbol is 1 mole of the species.

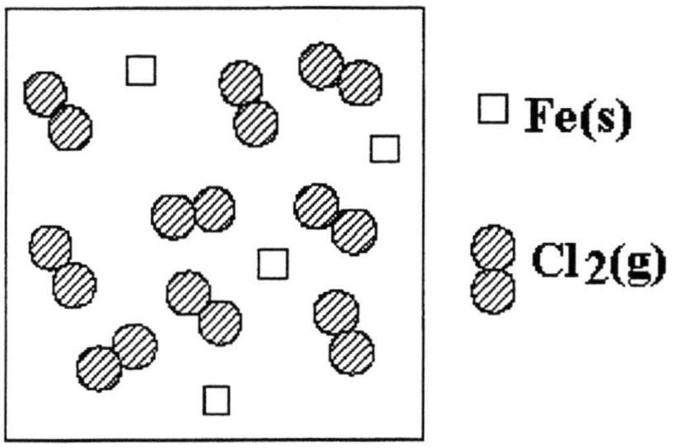

(a) What is the limiting reactant?
(b) What is the excess reactant?
(c) How many moles of excess reactant remain after the reaction is complete?
(d) Draw the product mixture using the same symbols as above.

76. Combustion analysis of a sample of a compound containing only carbon and hydrogen gives 0.500 g of carbon dioxide and 0.102 g of water. What is the empirical formula of the compound?
a) CH b) C_2H c) CH_3 d) CH_2 e) C_5H_4

77. A sample of phosphorus of mass 0.808 g reacts with oxygen to form 1.852 g of a phosphorus oxide. Determine the empirical formula of the oxide.
a) P_4O_{10} b) P_4O_5 c) P_2O_5 d) PO_3 e) P_2O

78. An analysis of 15.0 mg of acetaminophen showed that it contained 9.534 mg C, 0.900 mg H, 1.391 mg N, and the remainder oxygen. What is the empirical formula of acetaminophen?
a) $C_6H_7NO_2$ b) $C_8H_9NO_2$ c) C_2H_4NO d) C_4H_8NO e) C_4H_4NO

79. Combustion analysis of a 15.0-mg sample of acetaminophen produced 8.10 mg of water. What is the percent hydrogen in acetaminophen?
a) 6.00% b) 12.0% c) 3.00% d) 24.7% e) 27.0%

80. Combustion analysis of a 15.0-mg sample of a common headache remedy which contains carbon, hydrogen, nitrogen, and oxygen gave 35.0 mg of carbon dioxide and 8.10 mg of water. In a separate analysis, a 30.0-mg sample gave 3.378 mg of ammonia. The molar mass was also determined to be 151 g/mol. What is the molecular formula of this compound?

Chapter 4: Reaction Stoichiometry: Chemistry's Accounting Form B

1. How many moles of vanadium atoms can be extracted from 4 moles of vanadinite, $(PbO)_9(V_2O_5)_3PbCl_2$.
a) 3.6×10^{24} mol b) 24 mol c) 8 mol d) 1.8×10^{24} mol e) 12 mol

2. How many moles of cobalt atoms can be extracted from 6.3 moles of linnaetite, Co_3S_4?
a) 1.1×10^{25} mol b) 19 mol c) 6.3 mol d) 3.0 mol e) 3.8×10^{24} mol

3. How many moles of barium sulfate can be precipitated when excess sodium sulfate is added to an aqueous solution that contains 0.50 moles of barium chloride?
a) 0.13 mol b) 1.0 mol c) 0.50 mol d) 1.5 mol e) 0.25 mol

4. How many moles of sodium sulfate must be added to an aqueous solution that contains 2.0 moles of barium chloride in order to precipitate 0.50 moles of barium sulfate?
a) 0.25 mol b) 1.0 mol c) 0.50 mol d) 2.0 mol e) 1.5 mol

5. How many moles of sodium sulfate must be added to an aqueous solution that contains 2.0 moles of barium chloride in order to precipitate 3.0 moles of barium sulfate?
a) 3.0 mol
b) 1.5 mol
c) 2.0 mol
d) Not enough barium chloride is available to precipitate 3.0 moles of barium sulfate.
e) 1.0 mol

6. How many moles of sodium sulfate must be added to an aqueous solution that contains 2.0 moles of barium chloride in order to precipitate 2.0 moles of barium sulfate?
a) 1.0 mol
b) 2.0 mol
c) Not enough barium chloride is available to precipitate 2.0 moles of barium sulfate.
d) 3.0 mol
e) 1.5 mol

7. How many moles of carbon are required to react completely with 1.88 moles of $SnO_2(s)$ according to the equation below?
$$SnO_2(s) + 2C(s) \rightarrow Sn(s) + 2CO(g)$$
a) 0.47 mol b) 0.94 mol c) 7.52 mol d) 3.76 mol e) 1.88 mol

8. How many moles of carbon monoxide gas are produced when 2.88 moles of $SnO_2(s)$ react with excess carbon according to the equation below?
$$SnO_2(s) + 2C(s) \rightarrow Sn(s) + 2CO(g)$$
a) 1.4.32 mol b) 1.44 mol c) 5.76 mol d) 2.88 mol e) 11.5 mol

9. How many moles of carbon monoxide gas are produced from the reaction of 5.00 moles of $SnO_2(s)$ with 5.00 moles of carbon according to the equation below?

$$SnO_2(s) + 2C(s) \rightarrow Sn(s) + 2CO(g)$$

a) 5.00 mol b) 1.25 mol c) 2.50 mol d) 7.50 mol e) 10.0 mol

10. Consider the following reaction:

$$3NO_2(g) + H_2O(l) \rightarrow 2HNO_3(l) + NO(g)$$

How many moles of nitric acid are produced starting from 0.250 moles of $NO_2(g)$ and excess water?

a) 0.167 mol b) 0.375 mol c) 2.67 mol d) 0.0833 mol e) 0.500 mol

11. Consider the following reaction:

$$3NO_2(g) + H_2O(l) \rightarrow 2HNO_3(l) + NO(g)$$

How many moles of $NO_2(g)$ are required to react with 3.00 moles of water to produce 3.00 moles of nitric acid?

a) 1.50 mol b) 4.50 mol c) 3.00 mol d) 6.00 mol e) 9.00 mol

12. Consider the following reaction:

$$3NO_2(g) + H_2O(l) \rightarrow 2HNO_3(l) + NO(g)$$

How many moles of $NO_2(g)$ are required to react with 1.50 moles of water to produce 4.00 moles of nitric acid?

a) Not enough water is available to produce 4.00 moles of nitric acid.
b) 8.00 mol
c) 3.00 mol
d) 6.00 mol
e) 1.33 mol

13. Copper(II) nitrate decomposes upon heating to form copper(II) oxide, nitrogen dioxide gas, and oxygen gas. If 1.0 mole of copper(II) nitrate decomposes, how many moles of oxygen gas would be formed?

a) 1.5 mol b) 0.25 mol c) 2.0 mol d) 0.50 mol e) 1.0 mol

14. Copper(II) nitrate decomposes upon heating to form copper(II) oxide, nitrogen dioxide gas, and oxygen gas. How many moles of copper(II) nitrate are required to produce *at least* 2.0 mole of either nitrogen dioxide and oxygen gas?

a) 2.0 mol b) 1.0 mol c) 6.0 mol d) 8.0 mol e) 4.0 mol

15. Ammonium nitrate decomposes to "laughing gas," $N_2O(g)$, and water. If a sample of ammonium nitrate gives 4.0 moles of water, how many moles of $N_2O(g)$ are also formed?

a) 0.25 mol b) 0.50 mol c) 4.0 mol d) 2.0 mol e) 1.0 mol

16. Consider the following reaction:

$$5NaN_3(s) + NaNO_3(aq) \rightarrow 3Na_2O(s) + 8N_2(g)$$

If 3.0 moles of $N_2(g)$ are produced, how many moles of $NaN_3(s)$ were used?

a) 1.6 mol b) 24 mol c) 4.8 mol d) 15 mol e) 1.9 mol

17. Calculate the volume of 0.121 M KOH(aq) needed to react completely with 50.0 mL of 0.528 M HNO_3(aq).
a) 26.4 mL b) 4.58 mL c) 109 mL d) 218 mL e) 436 mL

18. Calculate the volume of 1.00 M HNO_3(aq) needed to react completely with 25.0 mL of 1.84 M $Ca(OH)_2$(aq).
a) 21.7 mL b) 184 mL c) 46.0 mL d) 368 mL e) 92.0 mL

19. Calculate the number of moles of KOH(aq) needed to react completely with 375 mL of 6.00 M H_2SO_4(aq).
a) 0.563 mol b) 9.00 mol c) 4.50 mol d) 1.13 mol e) 2.25 mol

20. Calculate the number of moles of $Ba(OH)_2$(aq) needed to react completely with 25.0 mL of 2.00 M $HClO_4$(aq).
a) 0.0500 mol b) 0.0125 mol c) 0.0250 mol d) 0.400 mol e) 0.100 mol

21. Rust, Fe_2O_3(s), can be removed from porcelain according to the reaction below.

$$Fe_2O_3(s) + 6Na_2C_2O_4(aq) + 6HCl(aq) \rightarrow 2Na_3Fe(C_2O_4)_3(aq) + 6NaCl(aq) + 3H_2O(l)$$

How many moles of rust can be removed by 500.0 mL of 2.84 M $Na_2C_2O_4$(aq)?
a) 0.236 mol b) 1.42 mol c) 0.500 mol d) 0.118 mol e) 8.52 mol

22. Rust, Fe_2O_3(s), can be removed from porcelain according to the reaction below.

$$Fe_2O_3(s) + 6Na_2C_2O_4(aq) + 6HCl(aq) \rightarrow 2Na_3Fe(C_2O_4)_3(aq) + 6NaCl(aq) + 3H_2O(l)$$

How many grams of rust can be removed by 500.0 mL of 2.84 M $Na_2C_2O_4$(aq)?
a) 1360 g b) 18.8 g c) 37.7 g d) 227 g e) 79.9 g

23. Consider the following reactions.

$$ZrC(s) + 4Cl_2(g) \rightarrow ZrCl_4(g) + CCl_4(g)$$

$$ZrCl_4(g) + 2Mg(s) \rightarrow Zr(s) + 2MgCl_2(l)$$

How many moles of Zr(s) are produced from 9.00 moles of chlorine gas?
a) 2.25 mol b) 1.00 mol c) 0.444 mol d) 9.00 mol e) 2.00 mol

24. Consider the following reactions.

$$ZrC(s) + 4Cl_2(g) \rightarrow ZrCl_4(g) + CCl_4(g)$$

$$ZrCl_4(g) + 2Mg(s) \rightarrow Zr(s) + 2MgCl_2(l)$$

How many moles of chlorine gas are required to produce 2.25 moles of $MgCl_2$(l)?
a) 4.50 mol b) 1.13 mol c) 9.00 mol d) 18.0 mol e) 2.25 mol

25. Potassium perchlorate can be produced from the reactions below.

$$Cl_2(g) + 2KOH(aq) \rightarrow KCl(aq) + KClO(aq) + H_2O(l)$$
$$3KClO(aq) \rightarrow 2KCl(aq) + KClO_3(aq)$$
$$4KClO_3(aq) \rightarrow 3KClO_4(s) + KCl(aq)$$

How many moles of chlorine gas are required to produce 5.00 moles of $KClO_4(s)$?

a) 1.67 mol b) 15.0 mol c) 20.0 mol d) 1.50 mol e) 5.13 mol

26. Potassium perchlorate can be produced from the reactions below.

$$Cl_2(g) + 2KOH(aq) \rightarrow KCl(aq) + KClO(aq) + H_2O(l)$$
$$3KClO(aq) \rightarrow 2KCl(aq) + KClO_3(aq)$$
$$4KClO_3(aq) \rightarrow 3KClO_4(s) + KCl(aq)$$

How many moles of $KClO_4(s)$ are produced from 10.0 mol of $KOH(aq)$?

a) 15.0 mol b) 1.25 mol c) 1.67 mol d) 5.00 mol e) 10.0 mol

27. Consider the following reactions.

$$4Bi(s) + 3O_2(g) \rightarrow 2Bi_2O_3(s)$$
$$Bi_2O_3(s) + 6HCl(aq) \rightarrow 2BiCl_3(aq) + 3H_2O(l)$$

If 2.00 L of 0.200 M $BiCl_3(aq)$ are produced, how many moles of Bi(s) are used?

a) 0.800 mol b) 1.60 mol c) 4.00 mol d) 0.400 mol e) 0.100 mol

28. Consider the following reactions for the determination of dissolved oxygen in water.

$$2MnSO_4(aq) + 4NaOH(aq) + O_2(aq) \rightarrow 2MnO_2(s) + 2Na_2SO_4(aq) + 2H_2O(l)$$
$$MnO_2(s) + 2H_2SO_4(aq) + 2NaI(aq) \rightarrow MnSO_4(aq) + I_2(aq) + Na_2SO_4(aq) + 2H_2O(l)$$
$$I_2(aq) + 2Na_2S_2O_3(aq) \rightarrow Na_2S_4O_6(aq) + 2NaI(aq)$$

If 8.00 mL of 0.0240 M $Na_2S_2O_3(aq)$ are used in the analysis, how many moles of dissolved oxygen were determined?

a) 4.80×10^{-5} mol b) 2.40×10^{-5} mol c) 3.84×10^{-4} mol d) 9.60×10^{-5} mol e) 1.92×10^{-4} mol

29. Consider the following reaction:

$$5NaN_3(s) + NaNO_3(aq) \rightarrow 3Na_2O(s) + 8N_2(g)$$

If 2.00 g of $N_2(g)$ are produced, how many grams of $Na_2O(s)$ are also produced?

a) 4.42 g b) 1.66 g c) 3.32 g d) 0.750 g e) 7.08 g

30. Consider the following reaction:

$$4FeS_2(s) + 11O_2(g) \rightarrow 2Fe_2O_3(s) + 8SO_2(g)$$

How many grams of $Fe_2O_3(s)$ are produced from the reaction of 1.000 kg of $FeS_2(s)$?

a) 665.6 g b) 2662 g c) 5325 g d) 1331 g e) 751.2 g

31. A 50.00-mL sample of $HClO_4(aq)$ was titrated to the stoichiometric point with 26.80 mL of 1.022 M NaOH(aq). What is the molarity of the perchloric acid solution?

a) 0.5478 M b) 0.02739 M c) 0.1.570 M d) 0.2739 M e) 0.001369 M

32. A 25.00-mL sample of $HClO_4$(aq) was titrated to the stoichiometric point with 42.32 mL of 1.02 M NaOH(aq). What is the mass of perchloric acid in the solution?
a) 1.73 g b) 2.56 g c) 4.34 g d) 0.108 g e) 0.0432 g

33. A 20.0-mL sample of sulfuric acid from a lake near a mine was titrated to the stoichiometric point with 9.92 mL of 0.0120 M NaOH(aq). What is the molarity of sulfuric acid in the sample?
a) 0.00238 M b) 0.00398 M c) 0.00595 M d) 0.00298 M e) 0.0119 M

34. A 25.0 mL sample of KOH(aq) was titrated to the stoichiometric point with 14.5 mL of 0.400 M phosphoric acid. What is the concentration of KOH(aq)?
a) 0.0773 M b) 0.232 M c) 0.696 M d) 4.31 M e) 0.145 M

35. In a titration, a 0.500-g sample of an acid requires 21.0 mL of 0.600 M KOH(aq) for complete reaction. If the acid has 2 acidic hydrogens per molecule, calculate the molar mass of the acid.
a) 79.4 g/mol b) 159 g/mol c) 39.7 g/mol d) 19.8 g/mol e) 126 g/mol

36. In a titration, a 2.00-g sample of an acid, HX, requires 40.6 mL of 0.300 M NaOH(aq) for complete reaction. Calculate the molar mass of the acid.
a) 244 g/mol b) 41.1 g/mol c) 164 g/mol d) 328 g/mol e) 82.1 g/mol

37. How many mL of 0.500 M H_2SO_4(aq) must be added to 50.0 mL of 0.490 M $BaCl_2$(aq) to precipitate all the barium as barium sulfate?
a) 24.5 mL b) 40.8 mL c) 50.0 mL d) 49.0 mL e) 98.0 mL

38. How many mL of 0.200 M HCl(aq) are neutralized by 5.00 g of $Mg(OH)_2$(s)?
a) 21.4 mL b) 2330 mL c) 857 mL d) 429 mL e) 1710 mL

39. A 25.0-mL sample of oxalic acid, $H_2C_2O_4$(with two acidic protons), was titrated to the stoichiometric point with 10.3 mL of 0.202 M NaOH(aq). Calculate the molarity of $H_2C_2O_4$(aq).
a) 0.0832 M b) 0.0208 M c) 0.0416 M d) 0.166 M e) 0.0520 M

40. Consider the following net ionic equation.

$$6Fe^{2+}(aq) + Cr_2O_7^{2-}(aq) + 14H^+(aq) \rightarrow 6Fe^{3+}(aq) + 2Cr^{3+}(aq) + 7H_2O(l)$$

When a 50.0-mL sample of an iron(II) solution was titrated, 28.7 mL of 0.200 M $K_2Cr_2O_7$(aq) was required to reach the stoichiometric point. What is the molar concentration of iron(II)?
a) 0.0191 M b) 0.344 M c) 0.115 M d) 0.689 M e) 0.0344 M

41. Manganese in an ore can be determined by treating the ore with a measured, excess quantity of sodium oxalate to reduce $MnO_2(s)$ to $MnCl_2(aq)$ followed by determination of the unreacted sodium oxalate by titration with potassium permanganate.

$$MnO_2(s) + Na_2C_2O_4(aq) + 4HCl(aq) \rightarrow 2MnCl_2(aq) + 2CO_2(g) + 2H_2O(l) + 2NaCl(aq)$$

$$2KMnO_4(aq) + 5Na_2C_2O_4(aq) + 16HCl(aq) \rightarrow 2MnCl_2(aq) + 10CO_2(g) + 8H_2O(l) + 10NaCl(aq)$$

If a sample is treated with 100.0 mL of 0.200 M $Na_2C_2O_4(aq)$ and the unreacted $Na_2C_2O_4(aq)$ requires 32.44 mL of 0.1000 M $KMnO_4(aq)$, calculate the number of grams of manganese in the sample.

42. Consider the following reaction.

$$2SbF_3(s) + 3CCl_4(l) \rightarrow 3CCl_2F_2(g) + 2SbCl_3(s)$$

If the actual yield is 92%, how many grams of $CCl_4(l)$ are required to produce 36.8 g of $CCl_2F_2(g)$?

43. In a combustion analysis of a 2.50-mg sample of a compound containing only carbon and hydrogen, it was found that 9.00 mg of water were produced. What is the percent carbon in the sample?

a) 60.0% b) 20.0% c) 27.8% d) 40.0% e) 30.0%

44. Consider the following reaction.

$$2Al_2O_3(l) \rightarrow 4Al(l) + 3O_2(g)$$

If 750 g of Al(l) are produced from 1500 g of $Al_2O_3(l)$, what is the percent yield?

a) 94.5% b) 50.0% c) 52.9% d) 26.5% e) 100%

45. Consider the following reaction.

$$5NaN_3(s) + NaNO_3(aq) \rightarrow 3Na_2O(s) + 8N_2(g)$$

If 5.00 g of $NaN_3(s)$ react with excess $NaNO_3(aq)$, 0.995 g of $N_2(g)$ is recovered. What is the percent yield?

a) 46.2% b) 69.0% c) 19.9% d) 73.9% e) 28.9%

46. Consider the following reaction.

$$6Na(l) + Al_2O_3(s) \rightarrow 2Al(l) + 3Na_2O(s)$$

When 5.52 g of sodium react with excess $Al_2O_3(s)$, 1.50 g of Al(l) is produced. What is the percent yield?

a) 69.4% b) 27.2% c) 81.5% d) 34.7% e) 23.1%

47. Consider the following reaction.

$$3H_2(g) + WO_3(s) \rightarrow 3H_2O(g) + W(s)$$

When 10.0 kg of $WO_3(s)$ react with excess hydrogen, 6.55 kg of W(s) are obtained. What is the percent yield?

a) 40.3% b) 65.5% c) 82.6% d) 79.3% e) 26.3% 7

48. A sample of an unknown compound of mass 0.141 g reacts with chlorine gas and all the carbon in the compound is converted to CCl_4. If the mass of CCl_4 is 0.564 g, calculate the percent by mass of carbon in the unknown compound.
a) 31.2% b) 25.0% c) 7.80% d) 20.8% e) 33.3%

49. Consider the following reaction.
$3H_2(g) + WO_3(s) \rightarrow 3H_2O(g) + W(s)$
When 10.0 kg of $WO_3(s)$ react with excess hydrogen, 6.55 kg of W(s) are obtained. What is the percent yield?
a) 82.6% b) 65.5% c) 40.3% d) 79.3% e) 26.3%

50. A sample of an unknown compound of mass 0.141 g reacts with chlorine gas and all the carbon in the compound is converted to CCl_4. If the mass of CCl_4 is 0.564 g, calculate the percent by mass of carbon in the unknown compound.
a) 31.2% b) 25.0% c) 7.80% d) 20.8% e) 33.3%

51. How many moles of carbon monoxide gas are produced from the reaction of 5.00 moles of $SnO_2(s)$ with 8.00 moles of carbon according to the equation below?
$SnO_2(s) + 2C(s) \rightarrow Sn(s) + 2CO(g)$
a) 10.0 mol b) 16.0 mol c) 5.00 mol d) 4.00 mol e) 8.00 mol

52. Consider the following reaction:
$3NO_2(g) + H_2O(l) \rightarrow 2HNO_3(l) + NO(g)$
How many moles of nitric acid are produced starting from 8.00 moles of $NO_2(g)$ and 3.00 moles of water?
a) 16.0 mol b) 12.0 mol c) 2.67 mol d) 3.00 mol e) 5.33 mol

53. Ammonium nitrate decomposes to "laughing gas," $N_2O(g)$, and water. How many moles of ammonium nitrate are required to produce *at least* 3.0 moles of $N_2O(g)$ and 3.0 moles of water?
a) 3.0 mol b) 2.0 mol c) 0.75 mol d) 1.5 mol e) 6.0 mol

54. Consider the following reaction:
$5NaN_3(s) + NaNO_3(aq) \rightarrow 3Na_2O(s) + 8N_2(g)$
If 15 moles of NaN_3 and 4.0 moles of $NaNO_3$ react, how many moles of N_2 are produced?
a) 24 mol b) 9.4 mol c) 2.4 mol d) 32 mol e) 0.50 mol

55. Consider the following reaction:
$5NaN_3(s) + NaNO_3(aq) \rightarrow 3Na_2O(s) + 8N_2(g)$
If 1.0 g of NaN_3 reacts with 25 mL of 0.20 M $NaNO_3(aq)$, how many moles of $N_2(g)$ are produced?
a) 0.040 mol b) 0.0094 mol c) 0.0031 mol d) 0.024 mol e) 0.12 mol

56. Consider the reaction below:

$6Li(s) + N_2(g) \rightarrow 2Li_3N(s)$

The following represents a mixture of Li(s) and $N_2(g)$ just before reaction occurs.

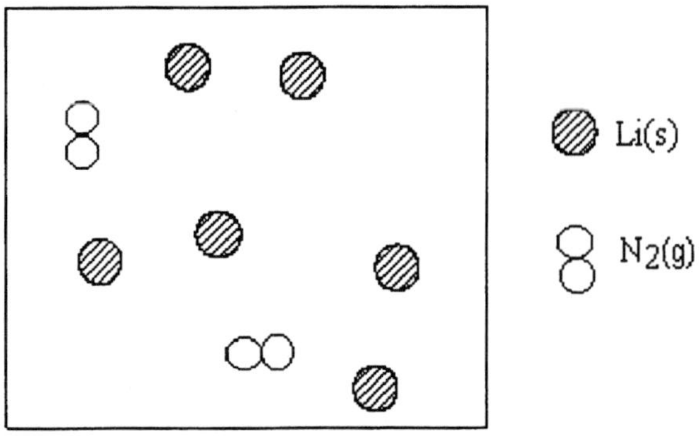

If each symbol represents 1 mole of Li(s) and 1 mole of $N_2(g)$, how many moles of $Li_3N(s)$ are produced and how many moles of excess reactant remain after the reaction is complete?

a) 1 mol of $Li_3N(s)$ and 1 mol of $N_2(g)$
b) 2 mol of $Li_3N(s)$ and 2 mol of $N_2(g)$
c) 2 mol $Li_3N(s)$ and no excess reactant
d) 3 mol of $Li_3N(s)$ and 1 mol of $N_2(g)$
e) 2 mol of $Li_3N(s)$ and 1 mol of $N_2(g)$

57. Consider the following reaction:

$3NO_2(g) + H_2O(l) \rightarrow 2HNO_3(l) + NO(g)$

How many moles of the excess reactant remain if 4.00 moles of $H_2O(l)$ and 10.0 moles of $NO_2(g)$ are used?

a) 3.70 mol $H_2O(l)$
b) 2.00 mol $NO_2(g)$
c) 6.00 mol $NO_2(g)$
d) 0.67 mol $H_2O(l)$
e) 8.00 mol $NO_2(g)$

58. Consider the following reaction:

$6Na(l) + Al_2O_3(s) \rightarrow 2Al(l) + 3Na_2O(s)$

How many moles of Al(l) are produced from 3.00 mole of each reactant?

a) 0.500 mol b) 6.00 mol c) 1.00 mol d) 9.00 mol e) 3.00 mol

59. Consider the following reaction:

$6Na(l) + Al_2O_3(s) \rightarrow 2Al(l) + 3Na_2O(s)$

How grams of $Na_2O(s)$ are produced from 3.00 moles of each reactant?

a) 186 g b) 93.0 g c) 0.145 g d) 15.5 g e) 279 g

60. Consider the following reaction:

$6Na(l) + Al_2O_3(s) \rightarrow 2Al(l) + 3Na_2O(s)$

How moles of Al(l) are produced from 5.00 g of each reactant?

a) 0.0490 mol b) 0.217 mol c) 0.0980 mol d) 0.0652 mol e) 0.0725 mol

61. Consider the following reaction:

$$6Na(l) + Al_2O_3(s) \rightarrow 2Al(l) + 3Na_2O(s)$$

How grams of $Na_2O(s)$ are produced from 5.00 g of each reactant?

a) 9.11 g b) 3.04 g c) 6.76 g d) 13.5 g e) 26.9 g

62. A manufacturer of children's wagons has in stock 500 wagon bodies, 1360 wheels, and 750 axles. If each wagon produced has 1 body, 4 wheels, and 2 axles, how many wagons can be assembled?

a) 340 b) 500 c) 375 d) 250 e) 680

63. A manufacturer of children's wagons has in stock 500 wagon bodies, 1360 wheels, and 750 axles. If each wagon produced has 1 body, 4 wheels, and 2 axles, how many components remain after the maximum number of wagons is assembled?

a) 160 bodies and 35 axles
b) 320 bodies and 80 wheels
c) 160 bodies and 70 axles
d) 160 bodies and 410 axles
e) 160 wheels

64. Consider the following reaction:

$$4FeS_2(s) + 11O_2(g) \rightarrow 2Fe_2O_3(s) + 8SO_2(g)$$

How many grams of $Fe_2O_3(s)$ are produced from 500.0 g of $FeS_2(s)$ and 500.0 g of oxygen gas?

a) 1331 g b) 665.5 g c) 453.7 g d) 375.6 g e) 332.8 g

65. Consider the following reaction:

$$4FeS_2(s) + 11O_2(g) \rightarrow 2Fe_2O_3(s) + 8SO_2(g)$$

How many grams of $Fe_2O_3(s)$ are produced from 500.0 g of $FeS_2(s)$ and 350.0 g of oxygen gas?

a) 332.8 g b) 399.3 g c) 317.6 g d) 1331 g e) 9607 g

66. The reaction of antimony, Sb(s), with chlorine, $Cl_2(g)$, is shown pictorially below.

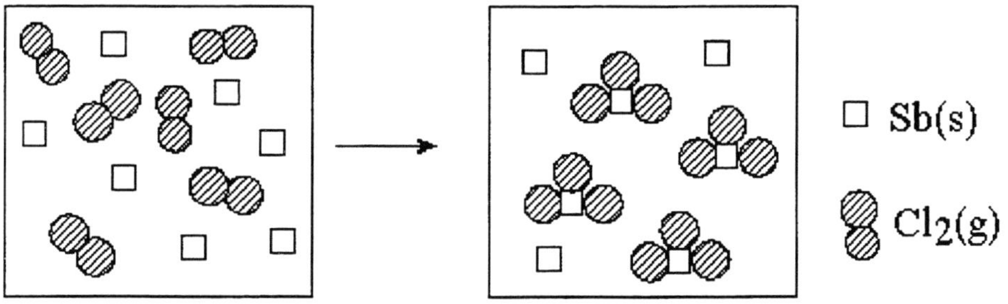

Which of the following gives the stoichiometric reaction?

a) $2Sb(s) + 3Cl_2(g) \rightarrow Sb_2Cl_6(s)$
b) $2Sb(s) + 3Cl_2(g) \rightarrow 2SbCl_3(s)$
c) $6Sb(s) + 6Cl_2(g) \rightarrow 2Sb_3Cl_6(s)$
d) $3Sb(s) + 3Cl_2(g) \rightarrow Sb_3Cl_6(s)$
e) $4Sb(s) + 6Cl_2(g) \rightarrow 4SbCl_3(s)$

67. Consider the reaction below.

$$6Li(s) + N_2(g) \rightarrow 2Li_3N(s)$$

Determine the limiting reactant and the number of moles of the other reactant remaining when 3 moles of lithium and 3 moles of nitrogen gas react.

a) $N_2(g)$ and 0.5 moles of Li(s) remaining
b) Li(s) and 2.5 moles of $N_2(g)$ remaining
c) $N_2(g)$ and 3 moles of Li(s) remaining
d) Li(s) and 2 moles of $N_2(g)$ remaining
e) Li(s) and 0.5 moles of $N_2(g)$ remaining

68. The reaction of an element, A(s), with an element, B(s), is shown pictorially below.

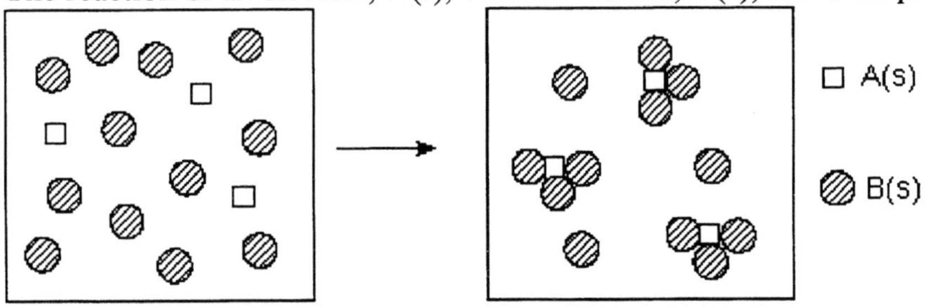

Which of the following gives the stoichiometric reaction?

a) $3A(s) + 12\,B(s) \rightarrow 3AB_3(s) + 3B(s)$
b) $3A(s) + 12B(s) \rightarrow A_3B_{12}(s)$
c) $3A(s) + 9B(s) \rightarrow A_3B_9(s)$
d) $3A(s) + 3B(s) \rightarrow A_3B_3(s)$
e) $A(s) + 3B(s) \rightarrow AB_3(s)$

69. Consider the following reaction.

$$2Mg(s) + O_2(g) \rightarrow 2MgO(s)$$

If the reactants are in the proportions shown below,

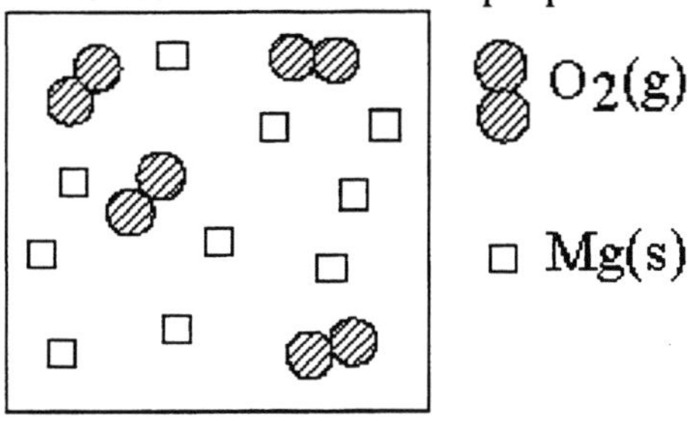

which of the following represents the product mixture?

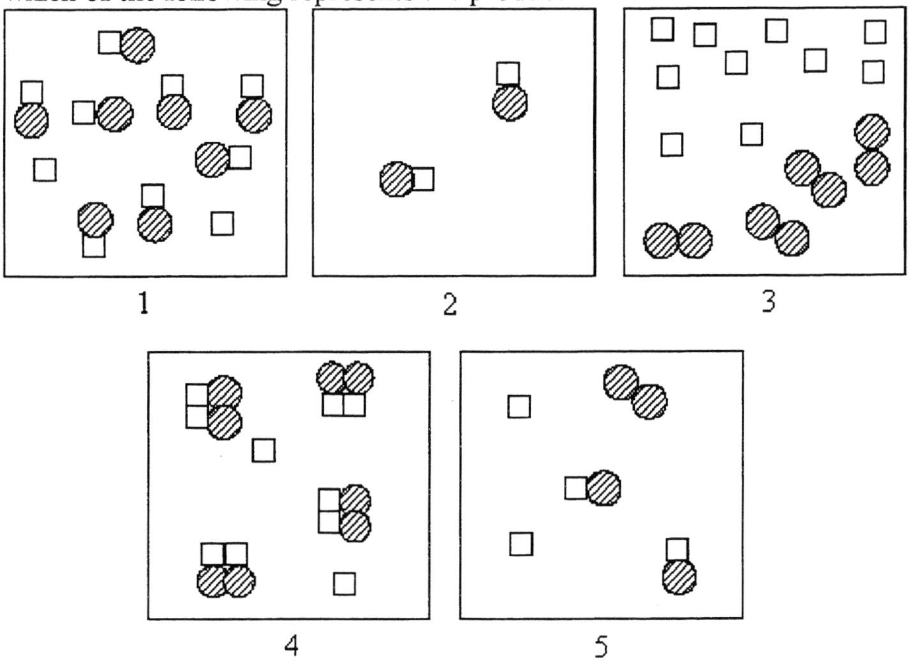

a) 1 b) 2 c) 3 d) 4 e) 5

70. Consider the following reaction.

$$2Mg(s) + O_2(g) \rightarrow 2MgO(s)$$

If 8 moles of magnesium react with 8 moles of oxygen gas, which is the limiting reactant and how many moles of the excess reactant remain?

a) magnesium and no oxygen gas remains
b) magnesium and 4 moles of oxygen gas remain
c) magnesium and 2 moles of oxygen gas remain
d) oxygen gas and no magnesium remains
e) oxygen gas and 6 moles of magnesium remain

71. How many moles of OH^- ions are contained in 375 mL of 0.200 M $Ba(OH)_2(aq)$?

a) 0.300 mol b) 0.0375 mol c) 0.075 mol d) 0.150 mol e) 0.600 mol

72. How many moles of H^+ ions are contained in 375 mL of an aqueous solution which is 0.200 M oxalic acid, $H_2C_2O_4$(with two acidic protons)?
a) 0.0375 mol b) 0.300 mol c) 0.150 mol d) 0.600 mol e) 0.0750 mol

73. Consider the following reaction.

$$Cl_2(g) + 2NaOH(aq) \rightarrow NaCl(aq) + NaOCl(aq) + H_2O(l)$$

How many grams of sodium hypochlorite can be produced from 25.0 g $Cl_2(g)$ and 500.0 mL of 2.00 M NaOH(aq)?
a) 13.1 g b) 52.5 g c) 26.2 g d) 37.2 g e) 20.0 g

74. If 10.0 mL of 1.20 M $AgNO_3$(aq) are added to 64.00 mL of 0.200 M HCl(aq), how many moles of AgCl(s) are produced?
a) 0.0128 mol b) 0.200 mol c) 1.20 mol d) 0.0120 mol e) 0.0240 mol

75. Consider the reaction below.

$$2Fe(s) + 3Cl_2(g) \rightarrow 2FeCl_3(s)$$

The diagram below shows a mixture of Fe(s) and $Cl_2(g)$ where each symbol is 1 mole of the species.

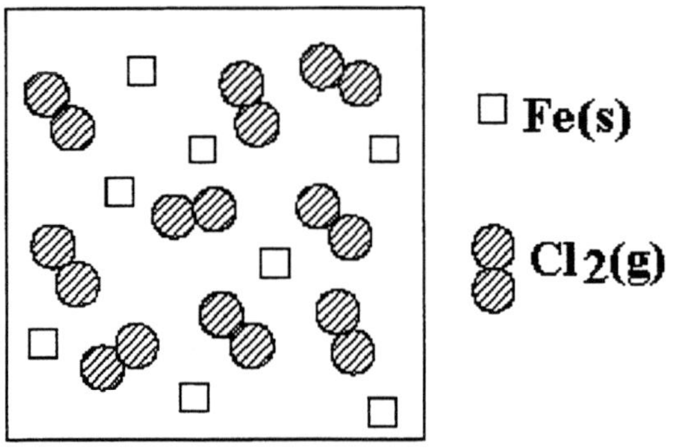

(a) What is the limiting reactant?
(b) What is the excess reactant?
(c) How many moles of excess reactant remain after the reaction is complete?
(d) Draw the product mixture using the same symbols as above.

76. Combustion analysis of a sample of a compound containing only carbon and hydrogen gives 1.00 g of carbon dioxide and 0.102 g of water. What is the empirical formula of the compound?
a) C_2H b) CH_3 c) CH d) CH_2 e) C_5H_4

77. A sample of phosphorus of mass 1.616 g reacts with chlorine to form 7.165 g of a phosphorus chloride. Determine the empirical formula of the chloride.
a) PCl_3 b) P_2Cl_6 c) P_2Cl_3 d) P_9Cl_4 e) P_3Cl

78. An analysis of 12.0 mg of ibuprofen showed that it contained 9.083 mg C, 1.056 mg H, and the remainder oxygen. What is the empirical formula of ibuprofen?
a) C_7H_9O b) $C_{14}H_{17}O_2$ c) $C_{13}H_{18}O_2$ d) C_6H_9O e) $C_6H_9O_2$

79. Combustion analysis of a 12.0-mg sample of ibuprofen produced 9.50 mg of water. What is the percent hydrogen in ibuprofen?
a) 8.80% b) 17.6% c) 35.2% d) 4.40% e) 39.6%

80. Combustion analysis of a 18.0-mg sample of a common headache remedy which contains carbon, hydrogen, nitrogen, and oxygen gave 42.0 mg of carbon dioxide and 9.72 mg of water. In a separate analysis, a 35.0-mg sample gave 3.941 mg of ammonia. The molar mass was also determined to be 151 g/mol. What is the molecular formula of this compound?

Chapter 5: The Properties of Gases
Form A

1. The atmospheric pressure at the top of a ski run was 615 Torr. What is this pressure in kPa?
a) 125 kPa b) 0.809 kPa c) 82.0 kPa d) 1.24 kPa e) 101 kPa

2. The gauge pressure of an automobile tire is typically 32 to 34 lb in^{-2}. Which of the following pressures corresponds most closely to this pressure range?
a) 1500 kPa b) 1500 Torr c) 230 kPa d) 1500 mmHg e) 45,000 Pa

3. A gas sample has a volume of 9.8 L at 720 Torr. If the temperature is constant, what volume does the gas sample have at 1.5 atm?
a) 9.3 L b) 6.2 L c) 5.7 L d) 6.9 L e) 0.16 L

4. The pressure of a gas in a rigid container at 18°C is 135 kPa. What is the pressure of the gas at 85°C?
a) 101 kPa b) 772 kPa c) 638 kPa d) 110 kPa e) 166 kPa

5. A 2.50-L sample of neon has a pressure of 1.59 atm at 45.0°C. What volume would the sample have if the pressure is 725 Torr and the temperature 22.0°C?
a) 3.87 L b) 2.04 L c) 3.07 L d) 0.733 L e) 1.62 L

6. Ammonia burns in air as follows:

$$4NH_3(g) + 3O_2(g) \rightarrow 2N_2(g) + 6H_2O(g)$$

How many liters of $N_2(g)$ are formed by burning 2.50 L of ammonia in excess oxygen with all species at the same temperature and pressure?
a) 2.00 L b) 2.50 L c) 5.00 L d) 0.625 L e) 1.25 L

7. Ammonia burns in air as follows:

$$4NH_3(g) + 3O_2(g) \rightarrow 2N_2(g) + 6H_2O(g)$$

How many liters of $N_2(g)$ are formed by burning 4.50 L of ammonia and 4.50 L of oxygen with all species at the same temperature and pressure?
a) 2.00 L b) 3.00 L c) 4.50 L d) 2.25 L e) 9.00 L

8. The value of the gas constant in units of L kPa K^{-1} mol^{-1} is
a) 8.314 b) 0.08206 c) 62.359 d) 8.314×10^3 e) 8.206×10^2

9. Which of the following gases are at the same temperature and pressure? All the containers are non-rigid and have the same volume.

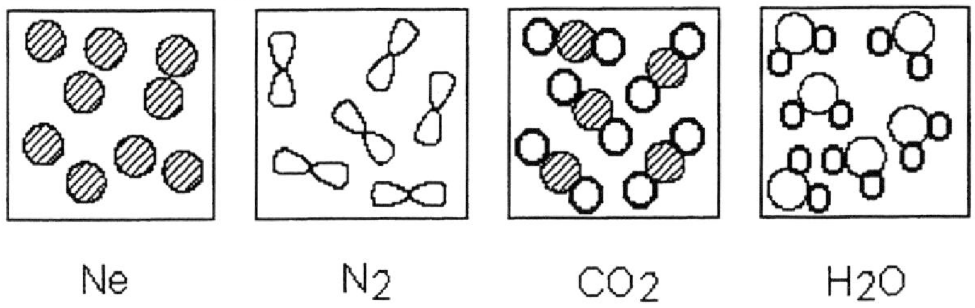

a) Ne and CO_2
b) N_2 and H_2O
c) CO_2 and H_2O
d) N_2 and CO_2
e) N_2, CO_2, and H_2O
S 5.9

10. A 5.00-L sample of xenon gas has a pressure of 155 Torr at –90°C. What mass of Xe is present?
a) 18.1 g b) 4.53 g c) 6.78×10^3 g d) 8.92 g e) 0.0880 g

11. A 2.50-L sample of sulfur dioxide gas has a pressure of 345 Torr at –95°C. What mass of SO_2 is present?
a) 3.78×10^3 g b) 0.0491 g c) 4.98 g d) 9.33 g e) 2.41 g

12. What is the pressure of 10.0 g of nitrogen gas in a container of volume 3.00 L at a temperature of –88.0°C?
a) 366 kPa b) 174 kPa c) 87.1 kPa d) 183 kPa e) 1.81 kPa

13. What is the volume of 10.0 g of nitrogen gas at –88°C and 183 kPa?
a) 6.00 L b) 5.86 L c) 22.4 L d) 1.43 L e) 3.00 L

14. What is the pressure of 20.0 g of oxygen gas in a container of volume 7.50 L at 39°C?
a) 1.60 atm b) 0.534 atm c) 4.27 atm d) 0.267 atm e) 2.13 atm

15. What is the volume of 20.0 g of oxygen gas at 39°C and 2.13 atm?
a) 5.63 atm b) 1.88 atm c) 0.939 atm d) 15.0 L e) 7.51 L

16. What mass of carbon monoxide will exert the same pressure as 10.0 g of chlorine gas in the same container under the same conditions?
a) 7.90 g b) 3.95 g c) 88.5 g d) 10.0 g e) 2.80 g

17. A 1.00-L sample of a gas at 705 Torr and 47°C has a mass of 0.988 g. The gas is
a) CO b) CO_2 c) CH_4 d) CF_4 e) Cl_2

18. A 1.00-L sample of a gas at 115 kPa and 36°C has a mass of 2.87 g. The gas is
a) SO_2 b) NO_2 c) CF_4 d) SO_3 e) Cl_2

19. What is the molar volume of nitrogen at 300°C and 0.500 atm?
a) 94.0 L b) 22.4 L c) 188 L d) 49.2 L e) 47.0 L

20. What is the molar volume of chlorine at 25°C and 98.0 kPa?
a) 2.12 L b) 12.6 L c) 22.4 L d) 25.3 L e) 50.6 L

21. What mass of sulfur dioxide will exert the same pressure as 50.0 mg of carbon dioxide in the same container under the same conditions?
a) 36.4 mg b) 24.3 mg c) 1.14 mg d) 50.0 mg e) 72.8 mg

22. What mass of carbon monoxide will exert the same pressure as 10.0 g chlorine in the same container under the same conditions?
a) 0.141 g b) 25.3 g c) 10.0 g d) 7.90 g e) 3.95 g

23. How many molecules of N_2 (g) occupy 1.50μL at 25°C and 15.0 Torr?
a) 1.21×10^{-9} molecules
b) 4.98×10^{32} molecules
c) 4.03×10^{16} molecules
d) 6.02×10^{23} molecules
e) 7.29×10^{14} molecules

24. How many molecules of carbon dioxide gas occupy 1.00 mL at 0°C and 1.00×10^{-3} Torr?
a) 3.54×10^{13} molecules
b) 4.46×10^{-8} molecules
c) 2.69×10^{16} molecules
d) 5.87×10^{-11} molecules
e) 6.02×10^{23} molecules

25. What volume of CO_2(g) at 25°C and 760 Torr is produced when 1.00 kg of calcium carbonate is used to neutralize a sulfuric acid spill? The equation for the reaction is

$$CaCO_3(s) + H_2SO_4(aq) \rightarrow CaSO_4(s) + CO_2(g) + H_2O(l)$$

a) 244 L b) 122 L c) 22.4 L d) 321 L e) 488 L

26. A possible reaction for the decomposition of sodium azide in automobile air bags is

$$6NaN_3(s) + 2Fe_2O_3(s) \rightarrow 3Na_2O_2(s) + 4Fe(s) + 9N_2(g)$$

What mass of NaN_3 is required to produce 40.0 L of N_2(g) at 20°C and 800 Torr?

27. What volume of carbon dioxide measured at 40°C and 760 Torr would be produced by the combustion of 454 g of glucose, $C_6H_{12}O_6$(s)?

28. The number of acidic hydrogens, n, in a molecule can be determined using the reaction below.

$$H_nA(s) + nCH_3MgCl(solution) \rightarrow nCH_4(g) + (MgCl)_nA(s)$$

If 359 mg of citric acid, $C_6H_8O_7$, react with excess CH_3MgCl, 145 mL of methane at 25°C and 720 Torr are produced. Calculate the number of acidic hydrogens in citric acid.

29. Sulfur dioxide in air can be determined by titration with potassium permanganate.

$$5SO_2(aq) + 2MnO_4^-(aq) + H_3O^+(aq) + H_2O(l) \rightarrow 2Mn^{2+}(aq) + 5HSO_4^-(aq)$$

Exactly 200 L of air at 25°C are drawn through an aqueous solution to absorb the SO_2. This solution required 21.34 mL of 0.005000 M $KMnO_4(aq)$ to reach the stoichiometric point. Calculate the pressure of $SO_2(g)$ in the air.

30. Standard temperature and pressure, STP, refer to
a) 298 K and 760 Torr b) 0°C and 1 atm c) 0°C and 1 Pa d) 0°C and 202 kPa e) 25°C and 1 atm

31. Calculate the volume occupied by 10.0 moles of chlorine gas at STP.
a) 22.4 L b) 112 L c) 448 L d) 245 L e) 224 L

32. Calculate the volume occupied by 2.88 moles of hydrogen at STP.
a) 32.3 L b) 64.5 L c) 22.4 L d) 129 L e) 70.6 L

33. Calculate the volume occupied by 10.0 g of nitrogen gas at STP.
a) 8.00 L b) 22.4 L c) 8.75 L d) 16.0 L e) 4.00 L

34. Calculate the volume occupied by 5.21 g of fluorine gas at STP.
a) 1.54 L b) 3.07 L c) 22.4 L d) 6.14 L e) 3.36 L

35. Under what set of conditions would carbon dioxide have the greatest molar volume?
a) high temperature and low pressure
b) The molar volume is always 22.4 L.
c) low temperature and high pressure
d) high temperature and high pressure
e) low temperature and low pressure

36. The volume of 1.10 g of an unknown diatomic gas is 625 mL at STP. What is the molar mass of the gas?
a) 15.4 g/mol b) 39.4 g/mol c) 35.8 g/mol d) 30.7 g/mol e) 25.4 g/mol

37. The volume of 2.81 g of an unknown gas is 2.25 L at STP. What is the molar mass of the gas?
a) 17.9 g/mol b) 100 g/mol c) 50.4 g/mol d) 28.0 g/mol e) 142 g/mol

38. Calculate the volume of carbon dioxide produced at STP by the combustion of 1.25 moles of propane, $C_3H_8(g)$.
a) 22.4 L b) 28.0 L c) 67.2 L d) 84.0 L e) 168 L

39. Calculate the volume of carbon dioxide produced at STP by the combustion of 454 grams of propane, $C_3H_8(g)$.
a) 692 L b) 1.84 L c) 923 L d) 231 L e) 1.38 L

40. Consider the following reaction.

$$4NH_4ClO_4(s) \rightarrow 2Cl_2(g) + 8H_2O(g) + 2N_2O(g) + 3O_2(g)$$

Calculate the total volume of all gases produced at STP by the decomposition of two moles of ammonium perchlorate.

a) 78.4 L b) 168 L c) 336 L d) 7.50 L e) 15.0 L

41. Consider the following reaction.

$$4NH_4ClO_4(s) \rightarrow 2Cl_2(g) + 8H_2O(g) + 2N_2O(g) + 3O_2(g)$$

Calculate the total volume of all gases produced at STP by the decomposition of 1.00 kg of ammonium perchlorate.

a) 2860 L b) 715 L c) 383 L d) 128 L e) 31.9 L

42. Consider the following reaction.

$$4NH_4ClO_4(s) \rightarrow 2Cl_2(g) + 8H_2O(g) + 2N_2O(g) + 3O_2(g)$$

How many moles of ammonium perchlorate are needed to produce a total volume of 1000 L of product gases at STP?

a) 44.6 mol b) 267 mol c) 670 mol d) 11.9 mol e) 179 mol

43. Consider the following reaction.

$$CS_2(g) + 3O_2(g) \rightarrow CO_2(g) + 2SO_2(g)$$

If 5.00 moles of $CS_2(g)$ are burned, what volume of $SO_2(g)$ is produced at STP?

a) 56.0 L b) 224 L c) 112 L d) 10.0 L e) 22.4 L

44. Consider the following reaction.

$$CS_2(g) + 3O_2(g) \rightarrow CO_2(g) + 2SO_2(g)$$

If 5.00 grams of $CS_2(g)$ are burned, what volume of $SO_2(g)$ is produced at STP?

a) 8.41 L b) 2.94 L c) 2.00 L d) 1.47 L e) 0.735 L

45. Consider the following reaction.

$$CS_2(g) + 3O_2(g) \rightarrow CO_2(g) + 2SO_2(g)$$

If 8.83 L of $SO_2(g)$ are produced at STP, how many grams of $CS_2(g)$ are required?

a) 98.9 g b) 12.6 g c) 15.0 g d) 25.3 g e) 30.0 g

46. Consider the following reaction.

$$CS_2(g) + 3O_2(g) \rightarrow CO_2(g) + 2SO_2(g)$$

If the volumes of all products and reactants are measured at 75°C and 1 atm, what is the total volume of products if 1.00 L of each reactant is used?

a) 1.00 L b) 1.67 L c) 1.33 L d) 1.50 L e) 3.00 L

47. Consider the combustion of butane at 0°C and 1 atm:

$$2C_4H_{10}(g) + 13O_2(g) \rightarrow 8CO_2(g) + 10H_2O(l)$$

What mass of $O_2(g)$ is needed to react with 5.00 mL of butane?

a) 0.0929 g b) 5.49×10^{-4} g c) 2.75×10^{-4} g d) 0.0464 g e) 0.0325 g

48. Consider the combustion of butane at 35°C and 95.0 kPa:

$$2C_4H_{10}(g) + 13O_2(g) \rightarrow 8CO_2(g) + 10H_2O(l)$$

What mass of $O_2(g)$ is needed to react with 25.0 mL of butane?

a) 6.03 g b) 0.386 g c) 1.70 g d) 3.40 g e) 0.193 g

49. Which of the following gases has the lowest density at STP?

a) methane, CH_4 b) argon c) fluorine d) nitrogen e) ozone, O_3

50. Which of the following gases has the highest density at STP?

a) sulfur trioxide, SO_3 b) carbon dioxide c) NO_2 d) argon e) nitrogen

51. Which of the following gases has the lowest density at STP?

a) SO_3 b) NO_2 c) acetylene, C_2H_2 d) chlorine e) argon

52. Under what set of conditions would the density of argon be smallest?

a) the density of argon is independent of P and T
b) high pressure and high temperature
c) low pressure and low temperature
d) low pressure and high temperature
e) high pressure and low temperature

53. A sample of gaseous hydrogen has a volume of 15.8 L at 1.78 atm and 95°C. Calculate the density of hydrogen under these conditions.

a) 0.236 g/L b) 0.457 g/L c) 0.118 g/L d) 0.0589 g/L e) 0.00662 g/L

54. Calculate the density of nitrogen at 695 Torr and –15°C.

a) 2.42 g/L b) 1.08 g/L c) 1.21 g/L d) 0.605 g/L e) 9.07 g/L

55. Calculate the density of chlorine at 675 Torr and –15°C.

a) 2.97 g/L b) 2.66 g/L c) 1.49 g/L d) 1.33 g/L e) 5.95 g/L

56. Calculate the density of xenon at 2.00 atm and –25°C.

a) 6.45 g/L b) 25.8 g/L c) 0.127 g/L d) 12.9 g/L e) 10.7 g/L

57. An unknown sample of a gas has a density of 2.45 g/L at 1.50 atm and 25°C. Which of the following is the gas?

a) Ar b) HCl c) CO_2 d) HF e) N_2

58. An unknown sample of a gas has a density of 3.09 g/L at 245 kPa and 75°C. Which of the following is the gas?

a) HCl b) HF c) BF_3 d) NO_2 e) SO_2

59. The density of the vapor of a volatile organic compound is 0.960 g/L at 185 Torr and 225°C. What is the molar mass of this compound?

a) 175 g/mol b) 161 g/mol c) 21.5 g/mol d) 168 g/mol e) 72.8 g/mol

60. The density of the vapor of a volatile organic compound is 0.371 g/L at 18.5 kPa and 300°C. What is the molar mass of the gas?
a) 95.5 g/mol b) 50.0 g/mol c) 943 g/mol d) 694 g/mol e) 135 g/mol

61. Consider the following reaction.
$CS_2(g) + 3O_2(g) \rightarrow CO_2(g) + 2SO_2(g)$
If the volumes of all products and reactants are measured at 75°C and 1 atm, what is the total volume of products if 2.50 L of CS_2 and 9.00 L of O_2 are used?
a) 9.00 L b) 7.50 L c) 2.50 L d) 5.00 L e) 10.0 L

62. Consider the following reaction.
$CS_2(g) + 3O_2(g) \rightarrow CO_2(g) + 2SO_2(g)$
If 7.00 L of products are formed at 100°C and 1 atm, what volume of $SO_2(g)$ was produced?
a) 4.67 L b) 2.00 L c) 3.50 L d) 7.00 L e) 2.33 L

63. Consider the following reaction.
$CS_2(g) + 3O_2(g) \rightarrow CO_2(g) + 2SO_2(g)$
If 14.0 L of products are formed at 100°C and 1 atm, what volume of $CO_2(g)$ was produced?
a) 2.00 L b) 9.33 L c) 3.50 L d) 4.67 L e) 7.00 L

64. A gas mixture contains 0.0500 moles of hydrogen, 0.0400 moles of carbon dioxide, and 0.0325 moles of nitrogen in a 2.00 L flask. If the total pressure in the flask is 1000 Torr, what is the partial pressure of carbon dioxide?
a) 0.215 atm b) 0.161 atm c) 1.32 atm d) 0.327 atm e) 0.430 atm

65. A 750-mL sample of nitrogen was collected by displacement of water from a container at 30°C and an atmospheric pressure of 742 Torr. If the vapor pressure of water at 30°C is 31.8 Torr, calculate the number of moles of nitrogen gas produced.
a) 0.221 mol b) 0.0294 mol c) 0.285 mol d) 0.0282 mol e) 0.211 mol

66. A 500-mL sample of dry air (measured at sea-level pressure, 760 Torr) contains 0.2314 g of oxygen at 25°C. Calculate the partial pressure of oxygen in the sample.
a) 0.707 atm b) 0.0297 atm c) 0.354 atm d) 0.0358 atm e) 0.231 atm

67. Consider the following reaction.
$Xe(g) + 2F_2(g) \rightarrow XeF_4(g)$
If 1.50 L of xenon at 0.750 atm are mixed with 3.00 L of fluorine at 1.00 atm in a 5.00 L flask at 350°C, what is the partial pressure of $XeF_4(g)$, assuming 100% yield?
a) 0.225 atm b) 0.300 atm c) 0.750 atm d) 0.113 atm e) 0.450 atm

68. Consider the following reaction.

$$Xe(g) + 2F_2(g) \rightarrow XeF_4(g)$$

If 1.50 L of xenon at 0.750 atm are mixed with 2.00 L of fluorine at 1.00 atm in a 5.00 L flask at 350°C, what is the partial pressure of the reactant remaining, assuming 100% yield?

a) 0.0174 atm Xe(g)
b) 0.122 atm $F_2(g)$
c) 0.0249 atm Xe(g)
d) 0.175 atm $F_2(g)$
e) 0.0498 atm Xe(g)

69. The interhalogen compound ClF_3 can be prepared by direct combination of the elements:

$$Cl_2(g) + 3F_2 \rightarrow 2ClF_3(g)$$

If 2.00 L of chlorine at 1.20 atm are mixed with 2.00 L of fluorine at 3.10 atm in a 5.00 L flask at 240°C, what is the partial pressure of $ClF_3(g)$, assuming 100% yield?

a) 0.827 atm b) 1.24 atm c) 0.480 atm d) 0.960 atm e) 0.414 atm

70. The interhalogen compound ClF_3 can be prepared by direct combination of the elements:

$$Cl_2(g) + 3F_2 \rightarrow 2ClF_3(g)$$

If 1.50 L of chlorine at 1.20 atm are mixed with 2.00 L of fluorine at 3.10 atm in a 5.00 L flask at 240°C, what is the partial pressure of $ClF_3(g)$, assuming 100% yield?

a) 0.827 atm b) 0.720 atm c) 1.08 atm d) 0.360 atm e) 0.414 atm

71. The interhalogen compound ClF_3 can be prepared by direct combination of the elements:

$$Cl_2(g) + 3F_2 \rightarrow 2ClF_3(g)$$

If 2.00 L of chlorine at 1.20 atm are mixed with 2.00 L of fluorine at 3.10 atm in a 5.00 L flask at 240°C, what is the partial pressure of the reactant remaining, assuming 100% yield?

a) 0.827 atm $Cl_2(g)$
b) 0.0665 atm $Cl_2(g)$
c) 0.758 atm $F_2(g)$
d) 0.0333 atm $Cl_2(g)$
e) 0.253 atm $F_2(g)$

72. When ammonium nitrate is heated it decomposes to form $N_2O(g)$ and water vapor. A sample of ammonium nitrate was decomposed in an evacuated reaction vessel and the measured gas pressure was 900 Torr. What is the partial pressure of the water vapor?

a) 600 Torr b) 675 Torr c) 450 Torr d) 300 Torr e) 900 Torr

73. Which gaseous molecules (choose one species) effuse slowest?

a) $SO_2(g)$ b) Ar(g) c) NO(g) d) Ne(g) e) $CO_2(g)$

74. The time required for 0.010 moles of $SO_3(g)$ to effuse through an opening was 22 s. Under the same conditions, an unknown gas required 30 s. The unknown gas is probably

a) $SF_4(g)$ b) $SiF_4(g)$ c) $PF_5(g)$ d) $SF_6(g)$ e) $CF_4(g)$

75. The rate of diffusion of HTe(g) at 25°C would be what factor times the rate of diffusion of gaseous ammonia at 25°C?
a) 0.362 b) 0.761 c) 0.131 d) 2.76 e) 0.181

76. Consider the following gases at constant temperature: O_3, NO, SO_2, and Ar. Rank the gases with respect to their average molecular speed, the fastest moving on the left.
a) $SO_2>O_3>Ar>NO$
b) $O_3>SO_2>NO>Ar$
c) $SO_2>O_3>NO>Ar$
d) $NO>Ar>O_3>SO_2$
e) $Ar>NO>O_3>SO_2$

77. In air at 25°C, which gas has the largest average speed?
a) $O_2(g)$ b) $H_2(g)$ c) $N_2(g)$ d) Ne(g) e) $Cl_2(g)$

78. In air at 25°C, which gas has the lowest average speed?
a) $H_2O(g)$ b) $CO_2(g)$ c) CO(g) d) Ne(g) e) $CH_4(g)$

79. At 25°C the molecules of an unknown gas travel four times slower on the average as helium atoms at the same temperature. What is the molar mass of the unknown gas?
a) 32 g/mol b) 16 g/mol c) 144 g/mol d) 256 g/mol e) 64 g/mol

80. Consider the gas ethylene, $C_2H_4(g)$ at 48 atm and 25°C. Calculate the volume of ethylene using the van der Waals equation and the ideal gas equation. The van der Waals constants for ethylene are $a = 4.47\ L^2\ atm\ mol^{-2}$ and $b = 0.057\ L\ mol^{-1}$. Explain the results.

Chapter 5: The Properties of Gases
Form B

1. The atmospheric pressure at the top of a ski hill was 675 Torr. What is this pressure in kPa?
a) 0.888 kPa b) 1.13 kPa c) 114 kPa d) 101 kPa e) 90.0 kPa

2. The gauge pressure of a nearly "flat" automobile tire is about 12 lb in^{-2}. Which of the following pressures corresponds most closely to this pressure?
a) 93000 Pa b) 930 kPa c) 83 kPa d) 930 mmHg e) 930 Torr

3. A gas sample has a volume of 12.4 L at 650 Torr. If the temperature is constant, what volume does the gas have at 3.2 atm.
a) 0.30 L b) 3.3 L c) 46 L d) 4.5 L e) 26 L

4. The pressure of a gas in a rigid container at 18°C is 745 Torr. What is the pressure of the gas at –75°C?
a) 285 Torr b) 1100 Torr c) 760 Torr d) 77.3 Torr e) 507 Torr

5. A 3.00-L sample of argon has a pressure of 345 kPa at 55.0°C. What volume would the sample have if the pressure is 0.839 atm and the temperature exactly 0°C?
a) 1030 L b) 60.4 L c) 22.4 L d) 0.615 L e) 10.1 L

6. Ammonia burns in air as follows:

$$4NH_3(g) + 3O_2(g) \rightarrow 2N_2(g) + 6H_2O(g)$$

How many liters of $N_2(g)$ are formed by burning 4.50 L of ammonia in excess oxygen with all species at the same temperature and pressure?
a) 1.13 L b) 9.00 L c) 4.50 L d) 2.25 L e) 2.00 L

7. Ammonia burns in air as follows:

$$4NH_3(g) + 3O_2(g) \rightarrow 2N_2(g) + 6H_2O(g)$$

How many liters of $N_2(g)$ are formed by burning 4.00 L of ammonia and 2.90 L of oxygen with all species at the same temperature and pressure?
a) 1.93 L b) 2.90 L c) 1.45 L d) 2.00 L e) 8.00 L

8. The value of the gas constant in units of L atm K^{-1} mol^{-1} is
a) 0.08206 b) 8.314 c) 62.359 d) 8.314×10^3 e) 8.206×10^2

9. Which of the following gases are at the same temperature and pressure? All the containers are non-rigid and have the same volume.

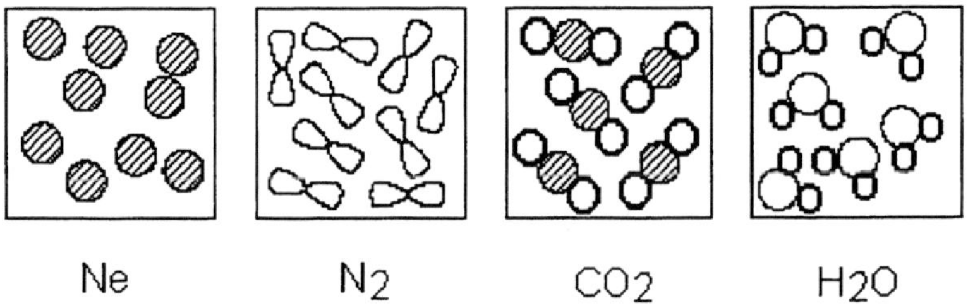

a) Ne and N_2 b) N_2 and CO_2 c) CO_2 and H_2O d) CO_2 and Ne e) H_2O and N_2

10. A 7.50-L sample of xenon gas has a pressure of 155 Torr at –110°C. What mass of Xe is present?
a) 0.148 g b) 15.0 g c) 6.39 g d) 1.14×10^4 g e) 22.2 g

11. A 12.0-L sample of sulfur dioxide gas has a pressure of 285 Torr at –110°C. What mass of SO_2 is present?
a) 9.17 g b) 0.213 g c) 81.3 g d) 1.64×10^4 g e) 21.6 g

12. What is the pressure of 10.0 g of nitrogen gas in a container of volume 3.00 L at a temperature of –88.0°C?
a) 183 atm b) 1.81 atm c) 3.61 atm d) 0.860 atm e) 1.72 atm

13. What is the volume of 10.0 g of nitrogen gas at 88°C and 183 kPa?
a) 6.00 L b) 11.7 L c) 3.00 L d) 1.43 L e) 5.86 L

14. What is the pressure of 20.0 g of oxygen gas in a container of volume 2.20 L at –101°C?
a) 8.02 atm b) 4.01 atm c) 2.35 atm d) 4.71 atm e) 8.72 atm

15. What is the volume of 20.0 g of oxygen gas at –77°C and 0.680 atm?
a) 29.6 L b) 14.8 L c) 22.4 L d) 26.4 L e) 5.81 L

16. What mass of carbon monoxide will exert the same pressure as 4.60 g of chlorine gas in the same container under the same conditions?
a) 1.82 g b) 4.60 g c) 2.80 g d) 40.7 g e) 3.63 g

17. A 1.00-L sample of a gas at 705 Torr and 47°C has a mass of 2.51 g. The gas is
a) CF_4 b) O_2 c) Cl_2 d) SO_3 e) CO_2

18. A 1.00-L sample of a gas at 115 kPa and 36°C has a mass of 0.761 g. The gas is
a) NH_3 b) NO c) CO d) acetylene, C_2H_2 e) N_2

19. What is the molar volume of nitrogen at 400°C and 1.25 atm?
a) 44.2 L b) 88.4 L c) 52.5 L d) 22.4 L e) 26.3 L

20. What is the molar volume of chlorine at –112°C and 92 kPa?
a) 22.4 L b) 29.1 L c) 34.8 L d) 7.27 L e) 14.5 L

21. What mass of sulfur dioxide will exert the same pressure as 75.0 mg of carbon dioxide in the same container under the same conditions?
a) 36.4 mg b) 1.70 mg c) 51.5 mg d) 109 mg e) 75.0 mg

22. What mass of carbon monoxide will exert the same pressure as 18.0 g chlorine in the same container under the same conditions?
a) 0.254 g b) 10.0 g c) 7.11 g d) 45.6 g e) 14.2 g

23. How many molecules of N_2(g) occupy 1.50 L at –65°C and 2.00 Torr?
a) 2.31×10^{-4} molecules
b) 1.39×10^{20} molecules
c) 6.02×10^{23} molecules
d) 1.06×10^{23} molecules
e) 3.84×10^{-28} molecules

24. How many molecules of carbon dioxide gas occupy 1.00 μL at 0°C and 1.00×10^{-3} Torr?
a) 4.46×10^{-11} molecules
b) 6.02×10^{23} molecules
c) 3.54×10^{10} molecules
d) 5.87×10^{-14} molecules
e) 2.69×10^{13} molecules

25. What volume of CO_2(g) at 25°C and 760 Torr is produced when 500 g of calcium carbonate is used to neutralize a sulfuric acid spill? The equation for the reaction is

$$CaCO_3(s) + H_2SO_4(aq) \rightarrow CaSO_4(s) + CO_2(g) + H_2O(l)$$

a) 244 L b) 22.4 L c) 161 L d) 61.0 L e) 122 L

26. A possible reaction for the decomposition of sodium azide in automobile air bags is

$$6NaN_3(s) + 2Fe_2O_3(s) \rightarrow 3Na_2O_2(s) + 4Fe(s) + 9N_2(g)$$

What mass of NaN_3 is required to produce 40.0 L of N_2(g) at 10°C and 900 Torr?

27. What volume of carbon dioxide measured at 25°C and 820 Torr would be produced by the combustion of 1.00 kg of glucose, $C_6H_{12}O_6$(s)?

28. The number of acidic hydrogens, n, in a molecule can be determined using the reaction below.

$$H_nA(s) + nCH_3MgCl(solution) \rightarrow nCH_4(g) + (MgCl)_nA(s)$$

If 200 mg of pyrogallol, $C_6H_6O_3$, react with excess CH_3MgCl, 120 mL of methane at 25°C and 740 Torr are produced. Calculate the number of acidic hydrogens in citric acid.

29. Sulfur dioxide in air can be determined by titration with potassium permanganate.

$$5SO_2(aq) + 2MnO_4^-(aq) + H_3O^+(aq) + H_2O(l) \rightarrow 2Mn^{2+}(aq) + 5HSO_4^-(aq)$$

Exactly 120 L of air at 25°C are drawn through an aqueous solution to absorb the SO_2. This solution required 31.34 mL of 0.006000 M $KMnO_4(aq)$ to reach the stoichiometric point. Calculate the pressure of $SO_2(g)$ in the air.

30. The molar volume of nitrogen gas at STP is about
a) 49.0 L b) 24.5 L c) 11.2 L d) 22.4 L e) 44.8 L

31. Calculate the volume occupied by 5.75 moles of chlorine gas at STP.
a) 258 L b) 141 L c) 129 L d) 64.4 L e) 22.4 L

32. Calculate the volume occupied by 0.138 moles of hydrogen at STP.
a) 22.4 L b) 6.18 L c) 3.38 L d) 1.55 L e) 3.09 L

33. Calculate the volume occupied by 12.4 g of nitrogen gas at STP.
a) 4.96 L b) 10.9 L c) 22.4 L d) 19.8 L e) 9.92 L

34. Calculate the volume occupied by 20.0 g of fluorine gas at STP.
a) 23.6 L b) 5.89 L c) 12.9 L d) 22.4 L e) 11.8 L

35. All of the gases below are at the same temperature and pressure. Which gas has the greatest density?
a) CO_2 b) Cl_2 c) H_2O d) SO_2 e) NO_2

36. The volume of 3.30 g of an unknown diatomic gas is 625 mL at STP. What is the molar mass of the gas?
a) 118 g/mol b) 46.2 g/mol c) 76.2 g/mol d) 92.1 g/mol e) 107 g/mol

37. The volume of 6.46 g of an unknown gas is 2.25 L at STP. What is the molar mass of the gas?
a) 327 g/mol b) 41.2 g/mol c) 230 g/mol d) 116 g/mol e) 64.4 g/mol

38. Calculate the volume of carbon dioxide produced aat STP by the combustion of 0.750 moles of propane, $C_3H_8(g)$.
a) 5.60 L b) 50.4 L c) 22.4 L d) 16.8 L e) 67.2 L

39. Calculate the volume of carbon dioxide produced at STP by the combustion of 227 grams of propane, $C_3H_8(g)$.
a) 462 L b) 0.690 L c) 0.920 L d) 116 L e) 346 L

40. Consider the following reaction.

$$4NH_4ClO_4(s) \rightarrow 2Cl_2(g) + 8H_2O(g) + 2N_2O(g) + 3O_2(g)$$

Calculate the total volume of all gases produced at STP by the decomposition of 5.50 moles of ammonium perchlorate.
a) 20.6 L b) 14.7 L c) 123 L d) 462 L e) 8.21 L

41. Consider the following reaction.

$$4NH_4ClO_4(s) \rightarrow 2Cl_2(g) + 8H_2O(g) + 2N_2O(g) + 3O_2(g)$$

Calculate the total volume of all gases produced at STP by the decomposition of 2.00 kg of ammonium perchlorate.

a) 1430 L b) 256 L c) 63.8 L d) 5720 L e) 766 L

42. Consider the following reaction.

$$4NH_4ClO_4(s) \rightarrow 2Cl_2(g) + 8H_2O(g) + 2N_2O(g) + 3O_2(g)$$

How many moles of ammonium perchlorate are needed to produce a total volume of 500 L of product gases at STP?

a) 134 mol b) 89.5 mol c) 335 mol d) 5.95 mol e) 22.3 mol

43. Consider the following reaction.

$$CS_2(g) + 3O_2(g) \rightarrow CO_2(g) + 2SO_2(g)$$

If 2.50 moles of $CS_2(g)$ are burned, what volume of $SO_2(g)$ is produced at STP?

a) 112 L b) 56.0 L c) 5.00 L d) 22.4 L e) 28.0 L

44. Consider the following reaction.

$$CS_2(g) + 3O_2(g) \rightarrow CO_2(g) + 2SO_2(g)$$

If 15.0 grams of $CS_2(g)$ are burned, what volume of $SO_2(g)$ is produced at STP?

a) 8.83 L b) 2.21 L c) 2.00 L d) 25.2 L e) 4.41 L

45. Consider the following reaction.

$$CS_2(g) + 3O_2(g) \rightarrow CO_2(g) + 2SO_2(g)$$

If 26.5 L of $SO_2(g)$ are produced at STP, how many grams of $CS_2(g)$ are required?

a) 297 g b) 37.8 g c) 90.0 g d) 45.0 g e) 75.9 g

46. Consider the following reaction.

$$CS_2(g) + 3O_2(g) \rightarrow CO_2(g) + 2SO_2(g)$$

If the volumes of all products and reactants are measured at 75°C and 1 atm, what is the total volume of products if 3.00 L of each reactant are used?

a) 3.00 L b) 4.50 L c) 9.00 L d) 4.00 L e) 5.00 L

47. Consider the combustion of butane at 0°C and 1 atm:

$$2C_4H_{10}(g) + 13O_2(g) \rightarrow 8CO_2(g) + 10H_2O(l)$$

What mass of $O_2(g)$ is needed to react with 25.0 mL of butane?

a) 2.75×10^{-3} g b) 0.465 g c) 0.163 g d) 1.38×10^{-3} g e) 0.232 g

48. Consider the combustion of butane at 35°C and 95.0 kPa:

$$2C_4H_{10}(g) + 13O_2(g) \rightarrow 8CO_2(g) + 10H_2O(l)$$

What mass of $O_2(g)$ is needed to react with 5.00 mL of butane?

a) 17.0 g b) 30.2 g c) 1.93 g d) 0.0386 g e) 8.50 g

49. Which of the following gases has the lowest density at STP?
a) carbon dioxide b) chlorine c) sulfur dioxide, SO_2 d) ozone, O_3 e) xenon, Xe

50. Which of the following gases has the highest density at STP?
a) carbon monoxide b) neon c) methane, CH_4 d) He e) hydrogen

51. Which of the following gases has the lowest density at STP?
a) acetylene, C_2H_2 b) ammonia, NH_3 c) neon d) carbon monoxide e) nitrogen

52. Under what conditions would the density of chlorine be largest?
a) the density of chlorine is independent of P and T
b) high pressure and high temperature
c) low pressure and low temperature
d) low pressure and high temperature
e) high pressure and low temperature

53. A sample of gaseous hydrogen has a volume of 14.0 L at 3.89 atm and 88°C. Calculate the density of hydrogen under these conditions.
a) 0.263 g/L b) 1.08 g/L c) 0.525 g/L d) 0.131 g/L e) 2.59×10^{-3} g/L

54. Calculate the density of nitrogen at 655 Torr and 39°C.
a) 7.07 g/L b) 1.94 g/L c) 0.484 g/L d) 0.968 g/L e) 7.54 g/L

55. Calculate the density of chlorine at 885 Torr and 75°C.
a) 5.78 g/L b) 2.89 g/L c) 13.4 g/L d) 6.70 g/L e) 1.45 g/L

56. Calculate the density of xenon at 0.500 atm and –25°C.
a) 6.45 g/L b) 3.23 g/L c) 2.68 g/L d) 1.61 g/L e) 32.0 g/L

57. An unknown sample of a gas has a density of 5.54 g/L at 2.20 atm and 37°C. Which of the following is the gas?
a) SO_3 b) CO_2 c) SO_2 d) Cl_2 e) Kr

58. An unknown sample of a gas has a density of 1.10 g/L at 91.0 kPa and 25°C. Which of the following is the gas?
a) ethane, C_2H_6 b) SF_6 c) butane, C_4H_{10} d) SO_3 e) Xe

59. The density of the vapor of a volatile organic compound is 0.542 g/L at 135 Torr and 285°C. What is the molar mass of this compound?
a) 140 g/mol b) 184 g/mol c) 71.4 g/mol d) 238 g/mol e) 476 g/mol

60. The density of the vapor of a volatile organic compound is 0.827 g/L at 35.6 kPa and 275°C. What is the molar mass of the gas?
a) 53.1 g/mol b) 794 g/mol c) 106 g/mol d) 398 g/mol e) 155 g/mol

61. Consider the following reaction.

$$CS_2(g) + 3O_2(g) \rightarrow CO_2(g) + 2SO_2(g)$$

If the volumes of all products and reactants are measured at 75°C and 1 atm, what is the total volume of products if 5.00 L of CS_2 and 18.0 L of O_2 are used?

a) 5.00 L b) 10.0 L c) 18.0 L d) 20.0 L e) 15.0 L

62. Consider the following reaction.

$$CS_2(g) + 3O_2(g) \rightarrow CO_2(g) + 2SO_2(g)$$

If 14.0 L of products are formed at 100°C and 1 atm, what volume of $SO_2(g)$ was produced?

a) 9.33 L b) 4.66 L c) 4.00 L d) 7.00 L e) 14.0 L

63. Consider the following reaction.

$$CS_2(g) + 3O_2(g) \rightarrow CO_2(g) + 2SO_2(g)$$

If 7.00 L of products are formed at 100°C and 1 atm, what volume of $CO_2(g)$ was produced?

a) 3.50 L b) 1.75 L c) 4.67 L d) 2.33 L e) 1.00 L

64. A gas mixture contains 0.0500 moles of hydrogen, 0.0400 moles of carbon dioxide, and 0.0325 moles of nitrogen in a 2.00 L flask. If the total pressure in the flask is 1000 Torr, what is the partial pressure of hydrogen?

a) 0.537 atm b) 0.161 atm c) 0.0658 atm d) 0.408 atm e) 0.269 atm

65. A 350-mL sample of nitrogen was collected by displacement of water from a container at 30°C and an atmospheric pressure of 736 Torr. If the vapor pressure of water at 30°C is 31.8 Torr, calculate the number of moles of nitrogen gas produced.

a) 0.0130 mol b) 0.0136 mol c) 0.0978 mol d) 0.132 mol e) 0.102 mol

66. A 500-mL sample of dry air (measured at sea-level pressure, 760 Torr) contains 0.0129 g of argon at 25°C. Calculate the partial pressure of argon in the sample.

a) 0.0158 atm b) 0.000323 atm c) 0.00160 atm d) 0.00132 atm e) 0.00930 atm

67. Consider the following reaction.

$$Xe(g) + 2F_2(g) \rightarrow XeF_4(g)$$

If 1.50 L of xenon at 0.750 atm are mixed with 2.00 L of fluorine at 1.00 atm in a 5.00 L flask at 350°C, what is the partial pressure of $XeF_4(g)$, assuming 100% yield?

a) 0.750 atm b) 0.400 atm c) 0.225 atm d) 0.200 atm e) 0.100 atm

68. Consider the following reaction.

$$Xe(g) + 2F_2(g) \rightarrow XeF_4(g)$$

If 1.50 L of xenon at 0.750 atm are mixed with 3.00 L of fluorine at 1.00 atm in a 5.00 L flask at 350°C, what is the partial pressure of the reactant remaining, assuming 100% yield?

a) 0.0750 atm $F_2(g)$
b) 0.225 atm $Xe(g)$
c) 0.250 atm $Xe(g)$
d) 0.300 atm $F_2(g)$
e) 0.150 atm $F_2(g)$

69. The interhalogen compound ClF_3 can be prepared by direct combination of the elements:

$$Cl_2(g) + 3F_2 \rightarrow 2ClF_3(g)$$

If 1.80 L of chlorine at 1.20 atm are mixed with 2.00 L of fluorine at 3.10 atm in a 5.00 L flask at 240°C, what is the partial pressure of $ClF_3(g)$, assuming 100% yield?

a) 0.864 atm b) 0.432 atm c) 0.827 atm d) 1.24 atm e) 0.414 atm

70. The interhalogen compound ClF_3 can be prepared by direct combination of the elements:

$$Cl_2(g) + 3F_2 \rightarrow 2ClF_3(g)$$

If 1.60 L of chlorine at 1.20 atm are mixed with 2.00 L of fluorine at 3.10 atm in a 5.00 L flask at 240°C, what is the partial pressure of $ClF_3(g)$, assuming 100% yield?

a) 0.827 atm b) 0.384 atm c) 1.15 atm d) 0.768 atm e) 0.414 atm

71. The interhalogen compound ClF_3 can be prepared by direct combination of the elements:

$$Cl_2(g) + 3F_2 \rightarrow 2ClF_3(g)$$

If 2.00 L of chlorine at 1.00 atm are mixed with 2.00 L of fluorine at 3.10 atm in a 5.00 L flask at 240°C, what is the partial pressure of the reactant remaining, assuming 100% yield?

a) 0.0210 atm $F_2(g)$
b) 0.0105 atm $F_2(g)$
c) 0.0333 atm $Cl_2(g)$
d) 0.0420 atm $F_2(g)$
e) 0.0665 atm $Cl_2(g)$

72. When ammonium nitrate is heated it decomposes to form $N_2O(g)$ and water vapor. A sample of ammonium nitrate was decomposed in an evacuated reaction vessel and the measured gas pressure was 1260 Torr. What is the partial pressure of the water vapor?

a) 630 Torr b) 840 Torr c) 210 Torr d) 420 Torr e) 1260 Torr

73. Which gaseous molecules (choose one species) effuse fastest?

a) $H_2O(g)$ b) Ar(g) c) $CO_2(g)$ d) NO(g) e) Ne(g)

74. The time required for 0.010 moles of $SO_3(g)$ to effuse through an opening was 22 s. Under the same conditions, an unknown gas required 28 s. The unknown gas is

a) Xe(g) b) $UF_6(g)$ c) $SiF_4(g)$ d) $CF_4(g)$ e) $SF_6(g)$

75. The rate of diffusion of HTe(g) at 25°C would be what factor times the rate of diffusion of argon at 25°C?

a) 0.555 b) 1.80 c) 0.308 d) 3.25 e) 0.278

76. Consider the following gases at constant temperature: O_3, NO, O_2, and Ar. Rank the gases with respect to their average molecular speed, the fastest moving on the left.

a) $NO > O_2 > O_3 > SO_2$
b) $O_2 > NO > O_3 > SO_2$
c) $SO_2 > O_3 > NO > O_2$
d) $O_3 > SO_2 > NO > O_2$
e) $SO_2 > O_3 > O_2 > NO$

77. In air at 25°C, which gas has the largest average speed?
a) He(g) b) Ne(g) c) O_2(g) d) N_2(g) e) Cl_2(g)

78. In air at 25°C, which gas has the lowest average speed?
a) SO_3(g) b) NO(g) c) CO(g) d) CO_2(g) e) Ne(g)

79. At 25°C the molecules of an unknown gas travel six times slower on the average as helium atoms at the same temperature. What is the molar mass of the unknown gas?
a) 144 g/mol b) 16 g/mol c) 32 g/mol d) 64 g/mol e) 256 g/mol

80. If 3.00 moles of sulfur dioxide occupy 4.00 L at 120°C, calculate the pressure using both the van der Waals and ideal gas equations. The van der Waals constants for sulfur dioxide are a = 6.87 L^2 atm mol^{-2} and b = 0.0568 L mol^{-1}. Explain the results.

Chapter 6: Thermochemistry: The Fire Within
Form A

1. When a reaction takes place where the products have a lower energy than the reactants,
a) the energy of the system increases.
b) energy is absorbed.
c) energy is released.
d) the thermal motion of the atoms in the surroundings decreases.
e) the reaction is endothermic.

2. A piece of aluminum of mass 12.0 g at 95.0°C was placed in a calorimeter that contained 100.0 g of water at 15.0°C. If the temperature of the water rose to 17.0°C, what is the specific heat capacity of aluminum? The specific heat capacity of water is 4.184 $J \cdot (°C)^{-1} \cdot g^{-1}$.
a) 0.73 $J \cdot (°C)^{-1} \cdot g^{-1}$
b) 4.6 $J \cdot (°C)^{-1} \cdot g^{-1}$
c) 7.6 $J \cdot (°C)^{-1} \cdot g^{-1}$
d) 4.1 $J \cdot (°C)^{-1} \cdot g^{-1}$
e) 0.89 $J \cdot (°C)^{-1} \cdot g^{-1}$

3. A piece of stainless steel of mass 25.0 g at 88.0°C was placed in a calorimeter that contained 150.0 g of water at 20.0°C. If the temperature of the water rose to 21.4°C, what is the specific heat capacity of the stainless steel? The specific heat capacity of water is 4.184 $J \cdot (°C)^{-1} \cdot g^{-1}$.
a) 1.8 $J \cdot (°C)^{-1} \cdot g^{-1}$
b) 0.53 $J \cdot (°C)^{-1} \cdot g^{-1}$
c) 0.40 $J \cdot (°C)^{-1} \cdot g^{-1}$
d) 1.6 $J \cdot (°C)^{-1} \cdot g^{-1}$
e) 8.1 $J \cdot (°C)^{-1} \cdot g^{-1}$

4. A piece of graphite of mass 75.0 g at 92.0°C was placed in a calorimeter that contained 200.0 g of water at 12.0°C. If the temperature of the water rose to 16.8°C, what is the specific heat capacity of graphite? The specific heat capacity of water is 4.184 $J \cdot (°C)^{-1} \cdot g^{-1}$.
a) 0.58 $J \cdot (°C)^{-1} \cdot g^{-1}$
b) 3.2 $J \cdot (°C)^{-1} \cdot g^{-1}$
c) 4.5 $J \cdot (°C)^{-1} \cdot g^{-1}$
d) 2.5 $J \cdot (°C)^{-1} \cdot g^{-1}$
e) 0.71 $J \cdot (°C)^{-1} \cdot g^{-1}$

5. Calculate the heat required to raise the temperature of 50.0 g of water from 25.0°C to 39.0°C. The specific heat capacity of water is 4.184 $J \cdot (°C)^{-1} \cdot g^{-1}$.
a) 8.16 kJ b) 4.18 kJ c) 5.23 kJ d) 0.163 kJ e) 2.93 kJ

6. How much heat is absorbed by a 25.0-g sample of gold at 25.0°C when it is immersed in boiling water? The specific heat capacity of gold is 0.128 $J \cdot (°C)^{-1} \cdot g^{-1}$ and the specific heat capacity of water is 4.184 $J \cdot (°C)^{-1} \cdot g^{-1}$.
a) 0.0800 kJ b) 0.320 kJ c) 1.00 kJ d) 0.240 kJ e) 4.18 kJ

7. How much heat is absorbed by a 75.0-g sample of aluminum at 25.0°C when it is immersed in boiling water? The specific heat capacity of aluminum is 0.900 $J \cdot (°C)^{-1} \cdot g^{-1}$ and the specific heat capacity of water is 4.184 $J \cdot (°C)^{-1} \cdot g^{-1}$.
a) 21.2 kJ b) 6.75 kJ c) 5.06 kJ d) 4.18 kJ e) 1.69 kJ

8. A reaction occurs in a calorimeter that contains 120.0 g of water at 25.00°C, and 3.60 kJ of heat is evolved. If the specific heat capacity of water is 4.184 J · (°C)$^{-1}$ · g^{-1}, what is the final temperature of the water?
a) 26.81 °C b) 17.83 °C c) 29.18 °C d) 32.17 °C e) 25.13 °C

9. A calorimeter containing 200 mL of water was calibrated by carrying out a reaction which released 15.6 kJ of heat. If the temperature of the calorimeter rose 3.25°C, what is the heat capacity of this calorimeter?
a) 8.37 kJ (°C)$^{-1}$ b) 2.58 kJ (°C)$^{-1}$ c) 4.80 kJ (°C)$^{-1}$ d) 24.0 kJ (°C)$^{-1}$ L^{-1} e) 0.960 kJ L (°C)$^{-1}$

10. The reaction of 50 mL of hydrochloric acid with 50 mL of sodium hydroxide resulted in a 1.15°C increase in temperature in a calorimeter with a heat capacity of 4.81 kJ (°C)$^{-1}$. What is the heat output of the neutralization reaction?
a) 4.18 kJ b) 0.418 kJ c) 1.32 kJ d) 5.53 kJ e) 0.553 kJ

11. The units of heat capacity are
a) J · K^{-1} · g^{-1} b) J · K^{-1} c) J d) J · K e) J · g^{-1}

12. When ethanol evaporates at constant pressure, the sign of the enthalpy change for the process
a) depends on the container volume.
b) is negative.
c) depends on the temperature.
d) is positive.
e) cannot be determined.

13. When 25.0 g of water vapor condenses, 56.5 kJ of heat is released. How much heat is required to vaporize 50.0 g of water?
a) 113 kJ b) 1410 kJ c) 56.5 kJ d) 2830 kJ e) 28.3 kJ

14. A compound has an enthalpy of sublimation of 96.0 kJ · mol^{-1} and an enthalpy of melting of 6.2 kJ/mol. Calculate the enthalpy of vaporization of this compound.

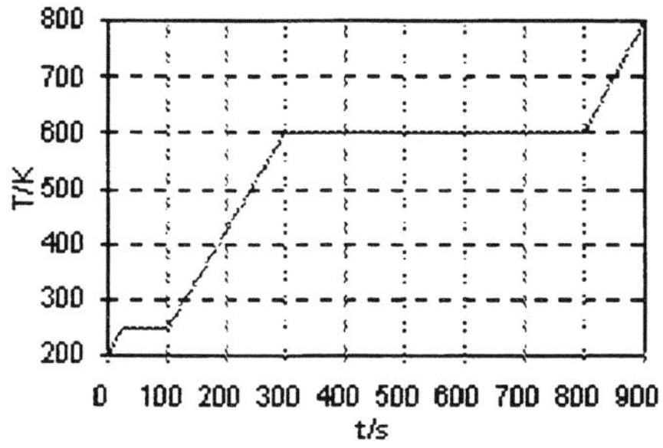

a) +6.2 kJ · mol^{-1} b) +102.2 kJ · mol^{-1}1 c) +89.8 kJ · mol^{-1} d) +96.0 kJ · mol^{-1} e) –89.8 kJ · mol^{-1}

15. The enthalpy of sublimation of sulfur dioxide is 32.3 kJ · mol^{-1} and its enthalpy of vaporization is 24.9 kJ/mol. Calculate the enthalpy of freezing.
a) –7.4 kJ · mol^{-1} b) +57.2 kJ · mol^{-1} c) –57.2 kJ · mol^{-1} d) +7.4 kJ · mol^{-1} e) –32.3 kJ · mol^{-1}

16. At 0°C the enthalpy of melting of water is 6.0 kJ · mol^{-1}. The enthalpy change when 36 g of water freezes is
a) –12 kJ b) +6.0 kJ c) +12 kJ d) –6.0 kJ e) –216 kJ

17. At the freezing point of ethanol, C_2H_5OH, –115°C, the enthalpy of melting is 4.60 kJ · mol^{-1}. The enthalpy change when 23.0 g of ethanol freezes is
a) +4.60 kJ b) –106 kJ c) +2.30 kJ d) –2.30 kJ e) –4.60 kJ

18. The diagram below is a heating curve for 15.0 g of mercury with a heating rate of 400 J · min^{-1}.

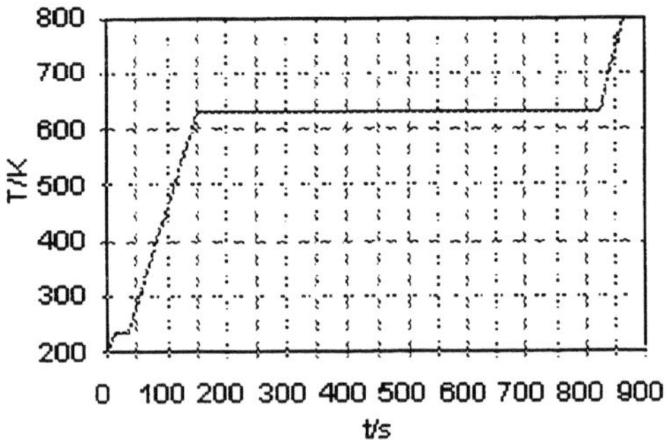

a) +60.2 kJ · mol^{-1} b) +3.61 x 10^3 kJ c) +903 kJ d) +67.5 kJ e) +4.5 kJ · mol^{-1}

19. Consider the heating curve for one mole of a substance that melts at –23°C and boils at 327°C.

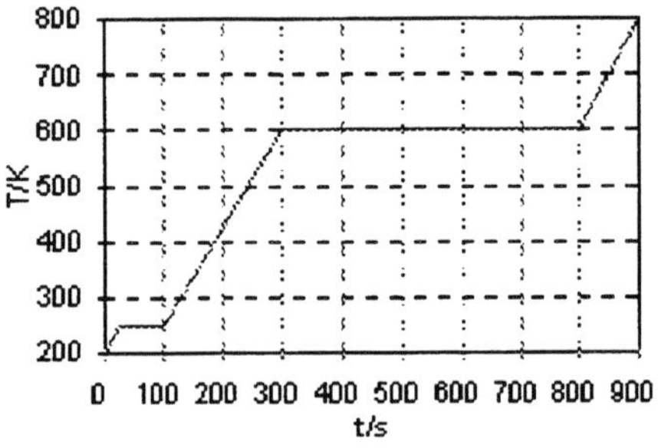

The horizontal portion of the heating curve at 250 K represents the
a) enthalpy of vaporization of the substance.
b) enthalpy of sublimation of the substance.
c) heat required to bring the substance to its melting point.
d) heat required to bring the substance to its boiling point.
e) enthalpy of fusion of the substance.

20. When 9.00 g of methane, $CH_4(g)$, is burned in a calorimeter with a heat capacity of 66.0 kJ · K^{-1}, the temperature of the calorimeter assembly rises by 7.59 K. What is the standard enthalpy of combustion of methane?
a) –1252 kJ · mol^{-1} b) –31.3 kJ · mol^{-1} c) –281 kJ · mol^{-1} d) –501 kJ · mol^{-1} e) –891 kJ · mol^{-1}

21. When 4.00 g of nitric acid are dissolved in 1.00 L of water in a calorimeter with a heat capacity of 1.22 $kJ \cdot K^{-1}$, the temperature increases 1.61 K. What is the molar enthalpy of solution of nitric acid?
a) $-125\ kJ \cdot mol^{-1}$ b) $-20.8\ kJ \cdot mol^{-1}$ c) $-8.02\ kJ \cdot mol^{-1}$ d) $-11.9 \cdot mol^{-1}$ e) $-30.9\ kJ \cdot mol^{-1}$

22. When 5.00 g of ammonium nitrate are dissolved in 1.00 L of water in a calorimeter with a heat capacity of 2.30 $kJ \cdot K^{-1}$, the temperature drops 0.760 K. What is the molar enthalpy of solution of ammonium nitrate?
a) $+109\ kJ \cdot mol^{-1}$ b) $+28.0\ kJ \cdot mol^{-1}$ c) $+5.29\ kJ \cdot mol^{-1}$ d) $+9.15\ kJ \cdot mol^{-1}$ e) $+48.4\ kJ \cdot mol^{-1}$

23. If 100 mL of 1.00 M barium chloride are added to 100 mL of 1.00 M potassium sulfate in a calorimeter with a heat capacity of 1.52 $kJ \cdot K^{-1}$, the temperature rises by 1.73 K. Calculate the enthalpy change for

$$Ba^{2+}(aq) + SO_4^{2-}(aq) \rightarrow BaSO_4(s)$$

a) −11.4 kJ b) −8.79 kJ c) −13.2 kJ d) −26.3 kJ e) −17.6 kJ

24. When sulfur dioxide is formed by burning monoclinic sulfur, the reaction enthalpy is −297.16 kJ, and when rhombic sulfur is burned to form $SO_2(g)$, the reaction enthalpy is −296.83 kJ. Calculate the reaction enthalpy for

$$S(monoclinic) \rightarrow S(rhombic)$$

a) +0.33 kJ b) +593.99 kJ c) −0.33 kJ d) −593.99 kJ e) 0

25. Consider the following reaction:

$$Fe_2O_3(s) + 3CO(g) \rightarrow 2Fe(l) + 3CO_2(g) \qquad \Delta H^o = 14.0\ kJ$$

Calculate the enthalpy change when 1.00 kg of $Fe_2O_3(s)$ reacts with 1.00 kg of CO(g).

26. The standard reaction enthalpy for

$$2HN_3(g) \rightarrow H_2(g) + 3N_2(g)$$

a) $-2\ \Delta H_f^o[HN_3(g)]$
b) $-\Delta H_f^o[HN_3(g)]$
c) $+\Delta H_f^o[H_2(g)] + 3\ \Delta H_f^o[N_2(g)]$
d) $+2\Delta H_f^o[HN_3(g)]$
e) $+\Delta H_f^o[HN_3(g)]$

27. The standard reaction enthalpy for

$$2CO(g) \rightarrow 2C(s) + O_2(g)$$

a) $+\Delta H_f^o[CO(g)]$
b) $-\Delta H_f^o[CO(g)]$
c) $-2\ \Delta H_f^o[CO(g)]$
d) $+2\ \Delta H_f^o[C(s)] + \Delta H_f^o[O_2(g)]$
e) $+2\ \Delta H_f^o[CO(g)]$

28. Calculate the reaction enthalpy for

$$H_2O(l) \rightarrow H_2(g) + 1/2O_2(g)$$

from the data

$$2H_2(g) + O_2(g) \rightarrow 2H_2O(l) \qquad \Delta H^o = -571.6\ kJ$$

a) +285.8 kJ b) −1143 kJ c) −285.8 kJ d) +571.6 kJ e) −571.6 kJ

29. Calculate the reaction enthalpy for

$CH_4(g) + 4S(s) \rightarrow CS_2(l) + 2H_2S(g)$

from the data

$C(s) + 2H_2(g) \rightarrow CH_4(g)$ $\Delta H^o = -74.8$ kJ

$C(s) + 2S(s) \rightarrow CS_2(l)$ $\Delta H^o = +87.9$ kJ

$S(s) + H_2(g) \rightarrow H_2S(g)$ $\Delta H^o = -20.6$ kJ

a) +121.5 kJ b) +203.9 kJ c) −28.1 kJ d) −7.5 kJ e) +142.1 kJ

30. Calculate the reaction enthalpy for

$C_6H_6(l) \rightarrow 3C_2H_2(g)$

from the data

$2C_2H_2(g) + 5O_2(g) \rightarrow 4CO_2(g) + 2H_2O(l)$ $\Delta H^o = -2600$ kJ

$2C_6H_6(l) + 15O_2(g) \rightarrow 12CO_2(g) + 6H_2O(l)$ $\Delta H^o = -6536$ kJ

a) −3936 kJ b) −668 kJ c) −2636 kJ d) +632 kJ e) +7168 kJ

31. Calculate the reaction enthalpy for

$2PbO(s) + O_2(g) \rightarrow 2PbO_2(s)$

from the data

$2Pb(s) + O_2(g) \rightarrow 2PbO(s)$ $\Delta H^o = -435.8$ kJ

$Pb(s) + O_2(g) \rightarrow PbO_2(s)$ $\Delta H^o = -276.6$ kJ

a) +159.2 kJ b) −159.2 kJ c) −117.4 kJ d) −595.0 kJ e) +117.4 kJ

32. Calculate the reaction enthalpy for the combustion of methane, $CH_4(g)$,

$CH_4(g) + 2O_2(g) \rightarrow CO_2(g) + 2H_2O(l)$

from the data

$C(s) + O_2(g) \rightarrow CO_2(g)$ $\Delta H^o = -393.5$ kJ

$H_2(g) + 1/2O_2(g) \rightarrow H_2O(l)$ $\Delta H^o = -285.8$ kJ

$C(s) + 2H_2(g) \rightarrow CH_4(g)$ $\Delta H^o = -74.8$ kJ

a) −890.3 kJ b) −178.1 kJ c) −252.9 kJ d) −965.1 kJ e) −679.3 kJ

33. Calculate the reaction enthalpy for the combustion of 1 mole of methanol, $CH_3OH(l)$, from the data

$C(s) + O_2(g) \rightarrow CO_2(g)$ $\Delta H^o = -393.5$ kJ

$H_2(g) + 1/2O_2(g) \rightarrow H_2O(l)$ $\Delta H^o = -285.8$ kJ

$C(s) + 2H_2(g) + 1/2O_2(g) \rightarrow CH_3OH(l)$ $\Delta H^o = -238.9$ kJ

a) −178.1 kJ b) −762.2 kJ c) −965.1 kJ d) −215.4 kJ e) −417.0 kJ

34. The combustion of ethanol can result in the formation of $H_2O(l)$ or $H_2O(g)$ along with carbon dioxide. Which reaction, the production of $H_2O(l)$ or $H_2O(g)$, gives off more heat? Verify your explanation with the appropriate calculations.

35. If the standard enthalpy of combustion of cyclohexane, $C_6H_{12}(g)$, is –3953 kJ · mol^{-1}, how many liters of cyclohexane measured at 25°C and 107.2 kPa must be burned to liberate 1000.0 kJ of heat?

36. The combustion of nitromethane, $CH_3NO_2(l)$, produces nitrogen, carbon dioxide, and water vapor. Calculate the standard enthalpy of combustion of nitromethane from the following data.

$C(s) + 3/2H_2(g) + 1/2N_2(g) + O_2(g) \rightarrow CH_3NO_2(l)$ $\Delta H° = -113.1$ kJ

$C(s) + O_2(g) \rightarrow CO_2(g)$ $\Delta H° = -393.51$ kJ

$H_2(g) + 1/2O_2(g) \rightarrow H_2O(g)$ $\Delta H° = -241.82$ kJ

37. If the standard enthalpy of combustion of ethanol, $C_2H_5OH(l)$, is –1368 kJ · mol^{-1}, calculate the heat output if 1.00 kg of ethanol is burned.
a) 63.0 x 10^3 kJ b) 33.7 x 10^3 kJ c) 1.37 x 10^3 kJ d) 29.7 x 10^3 kJ e) 33.7 kJ

38. If the standard enthalpy of combustion of glucose, $C_6H_{12}O_6(l)$, is –2808 kJ · mol^{-1}, calculate the heat output if 500 g of glucose is burned.
a) 2.81 x 10^2 kJ b) 15.6 x 10^3 kJ c) 7.79 x 10^3 kJ d) 1.01 x 10^3 kJ e) 32.1 kJ

39. If the standard enthalpy of combustion of glucose, $C_6H_{12}O_6(l)$, is –2808 kJ · mol^{-1}, calculate the mass of glucose needed to supply 550 kJ of heat.
a) 920 g b) 0.196 g c) 35.3 g d) 5.88 g e) 70.6 g

40. If the standard enthalpy of combustion of coal is –394 kJ · mol^{-1}, calculate the mass of coal needed to supply 550 kJ of heat.
a) 16.8 g b) 0.716 g c) 8.60 g d) 18.2 g e) 32.8 g

41. The standard enthalpy of combustion of benzene, $C_6H_6(l)$, is –3268 kJ · mol^{-1} at 25°C. Which thermochemical equation gives this value?
a) $C_6H_6(l) \rightarrow 6C(s) + 3H_2(g)$
b) $C_6H_6(l) + 9O(g) \rightarrow 6CO(g) + 3H_2O(l)$
c) $C_6H_6(l) + 9/2O_2(g) \rightarrow 6CO(g) + 3H_2O(l)$
d) $C_6H_6(l) + 15/2O_2(g) \rightarrow 6CO_2(g) + 3H_2O(l)$
e) $2C_6H_6(l) + 15O_2(g) \rightarrow 12CO_2(g) + 6H_2O(l)$

42. The standard enthalpy of combustion of carbon is –394 kJ · mol^{-1} at 25°C. Which thermochemical equation gives this value?
a) $C(s) \rightarrow C(g)$
b) $C(g) + O_2(g) \rightarrow CO_2(g)$
c) $C(s) + O_2(g) \rightarrow CO_2(g)$
d) $C(s) + 2O(g) \rightarrow CO_2(g)$
e) $C(s) + 1/2O_2(g) \rightarrow CO(g)$

43. Calculate the standard enthalpy of combustion of hydrazine, $N_2H_4(l)$, given

$H_2O(l) \rightarrow H_2(g) + 1/2O_2(g)$ $\Delta H° = +285.83$ kJ

$N_2(g) + 2H_2(g) \rightarrow N_2H_4(l)$ $\Delta H° = +50.63$ kJ

a) -622.29 kJ · mol^{-1}
b) -235.20 kJ · mol^{-1}
c) -50.63 kJ · mol^{-1}
d) -336.46 kJ · mol^{-1}
e) -521.03 kJ · mol^{-1}

44. If the standard enthalpy of combustion of pentane, $C_5H_{12}(g)$, is -3537 kJ · mol^{-1}, calculate the enthalpy change when 2.00 g of pentane are burned.

a) -196 kJ b) -1769 kJ c) -3537 kJ d) -98.0 kJ e) -49.0 kJ

45. Combustion reactions

a) are endothermic.
b) are exothermic.
c) have $\Delta H° = 0$
d) result in an increase in energy of the system.
e) take in heat from the surroundings.

46. Which thermochemical equation gives the value of the standard enthalpy of formation for HI(g)?

a) $H_2(g) + I_2(s) \rightarrow 2HI(g)$
b) $1/2H_2(g) + 1/2I_2(s) \rightarrow HI(g)$
c) $H(g) + 1/2I_2(s) \rightarrow HI(g)$
d) $1/2H_2(g) + 1/2I_2(g) \rightarrow HI(g)$
e) $H(g) + I(g) \rightarrow HI(g)$

47. Which thermochemical equation gives the value of the standard enthalpy of formation for $PbBr_2(s)$?

a) $Pb(s) + Br_2(s) \rightarrow PbBr_2(s)$
b) $Pb(s) + Br_2(g) \rightarrow PbBr_2(s)$
c) $Pb(s) + Br_2(l) \rightarrow PbBr_2(s)$
d) $Pb(l) + Br_2(l) \rightarrow PbBr_2(s)$
e) $Pb(s) + Br_2(s) \rightarrow PbBr_2(s)$

48. Which thermochemical equation gives the value of the standard enthalpy of formation for $CaCO_3(s)$?

a) $Ca(s) + C(s) + 3/2O_2(g) \rightarrow CaCO_3(s)$
b) $Ca(l) + C(g) + 3/2O_2(g) \rightarrow CaCO_3(s)$
c) $2Ca(s) + 2C(s) + 3O_2(g) \rightarrow 2CaCO_3(s)$
d) $Ca(s) + C(g) + 3/2O_2(g) \rightarrow CaCO_3(s)$
e) $2Ca(s) + 2C(g) + 3O_2(g) \rightarrow 2CaCO_3(s)$

49. Which thermochemical equation gives the value of the standard enthalpy of formation for $NaN_3(s)$?

a) $2Na(s) + 3N_2(g) \rightarrow 2NaN_3(s)$
b) $Na(s) + 3N(g) \rightarrow NaN_3(s)$
c) $Na(g) + 3N(g) \rightarrow NaN_3(s)$
d) $Na(l) + 3/2N_2(g) \rightarrow NaN_3(s)$
e) $Na(s) + 3/2N_2(g) \rightarrow NaN_3(s)$

50. Which of the following has a standard enthalpy of formation equal to zero?

a) Hg(l) b) Na(l) c) $I_2(g)$ d) C(g) e) S(g)

51. Which of the following has a standard enthalpy of formation equal to zero?
a) Ne(g) b) Ne_2(g) c) C(diamond) d) C_{60}(s) e) C(g)

52. All of the following have a standard enthalpy of formation equal to zero except
a) C(diamond) b) Fe(s) c) Hg(l) d) Ne(g) e) Cl_2(g)

53. All of the following have a standard enthalpy of formation equal to zero except
a) Na(l) b) Ca(s) c) Hg(l) d) Zn(s) e) Xe(g)

54. The standard enthalpy of formation of H_2O_2(l) is –187.8 kJ/mol and of H_2O(l) is –285.8 kJ · mol^{-1}. Calculate the reaction enthalpy for

$$2H_2O_2(l) \rightarrow 2H_2O(l) + O_2(g)$$

a) –98.0 kJ b) –196.0 kJ c) –142.9 kJ d) +98.0 kJ e) +196.0 kJ

55. The standard enthalpy of formation of NO(g) is +90.3 kJ · mol^{-1} and of NO_2(g) is +33.2 kJ · mol^{-1}. Calculate the reaction enthalpy for

$$2NO(g) + O_2(g) \rightarrow 2NO_2(g)$$

a) –114.2 kJ b) –57.1 kJ c) +123.5 kJ d) –23.9 kJ e) –47.8 kJ

56. The standard enthalpy of formation of benzene, C_6H_6(l), is +49.0 kJ · mol^{-1} and of acetylene, C_2H_2(g), is +226.7 kJ · mol^{-1}. What is the reaction enthalpy for formation of 1 mole of benzene from acetylene?
a) +147.0 kJ b) –631.1 kJ c) +631.1 kJ d) +177.7 kJ e) –177.7 kJ

57. The reaction enthalpy for

$$3PbO_2(s) \rightarrow Pb_3O_4(s) + O_2(g)$$

is +22.8 kJ. The standard enthalpy of formation of Pb_3O_4(s) is –175.6 kJ · mol^{-1}. What is the standard enthalpy of formation of PbO_2(s)?
a) –50.9 kJ/mol b) –198.4 kJ/mol c) –66.1 kJ/mol d) –152.8 kJ/mol e) –595.2 kJ/mol

58. The reaction enthalpy for

$$2SO_3(g) \rightarrow 2SO_2(g) + O_2(g)$$

is +198.4 kJ. The standard enthalpy of formation of SO_2(g) is –296.8 kJ · mol^{-1}. What is the standard enthalpy of formation of SO_3(g)?
a) –495.2 kJ · mol^{-1}
b) –792.0 kJ · mol^{-1}
c) –396.0 kJ · mol^{-1}
d) –247.6 kJ · mol^{-1}
e) –197.6 kJ · mol^{-1}

59. Ethanol, C_2H_5OH(l), can be made by reaction of ethylene, C_2H_4(g), with H_2O(g). The standard enthalpies of formation of ethanol, ethylene, and H_2O(g) are –277.69, +52.26, and –241.82 kJ · mol^{-1}, respectively. What is the reaction enthalpy for the synthesis of one mole of ethanol?
a) –16.39 kJ b) +16.39 kJ c) –88.13 kJ d) –571.77 kJ e) –467.25 kJ

60. The standard enthalpy of combustion of ethanol, $C_2H_5OH(l)$, is –1368 kJ · mol^{-1}. The standard enthalpies of formation of $H_2O(l)$ and $CO_2(g)$ are –285.83 and –393.51 kJ · mol^{-1}, respectively. What is the standard enthalpy of formation of ethanol?
a) –277 kJ · mol^{-1} b) –689 kJ · mol^{-1} c) +277 kJ · mol^{-1} d) –456 kJ · mol^{-1} e) +689 kJ · mol^{-1}

61. The standard enthalpy of combustion of butane, $C_4H_{10}(g)$, is –2878 kJ · mol^{-1}. The standard enthalpies of formation of $H_2O(l)$ and $CO_2(g)$ are –285.83 and –393.51 kJ · mol^{-1}, respectively. What is the standard enthalpy of formation of butane?
a) –443 kJ · mol^{-1} b) +125 kJ · mol^{-1} c) +689 kJ · mol^{-1} d) –125 kJ · mol^{-1} e) –689 kJ · mol^{-1}

62. The standard enthalpies of combustion of $C_2H_2(g)$, C(s), and $H_2(g)$ are –1300, –394, and –286 kJ · mol^{-1}, respectively. What is the standard enthalpy of formation of $C_2H_2(g)$?
a) –226 kJ · mol^{-1} b) –620 kJ · mol^{-1} c) +620 kJ · mol^{-1} d) +226 kJ · mol^{-1} e) –520 kJ · mol^{-1}

63. The standard enthalpies of combustion of $CH_3OH(g)$, C(s), and hydrogen gas are –764, –394, and –286 kJ · mol^{-1}, respectively. What is the standard enthalpy of formation of $CH_3OH(g)$?
a) +620 kJ · mol^{-1} b) –620 kJ · mol^{-1} c) –202 kJ · mol^{-1} d) –509 kJ · mol^{-1} e) +202 kJ · mol^{-1}

64. If the standard enthalpy of formation of $SO_3(g)$ is –395.7 kJ · mol^{-1}, calculate the reaction enthalpy for

$$2SO_3(g) \rightarrow 2S(s) + 3O_2(g)$$

a) –395.7 kJ b) +395.7 kJ c) +791.4 kJ d) +197.9 kJ e) –791.4 kJ

65. If the standard enthalpy of formation of $PCl_3(g)$ is –287.0 kJ · mol^{-1}, calculate the reaction enthalpy for

$$2PCl_3(g) \rightarrow 2P(s) + 3Cl_2(g)$$

a) –143.5 kJ b) +287.0 kJ c) +574.0 kJ d) –287.0 kJ e) –574.0 kJ

66. If the standard enthalpies of formation of carbon dioxide, $CO_2(g)$, $H_2O(l)$, and toluene, $C_7H_8(l)$, are –393.5, –285.8, and +12.0 kJ · mol^{-1}, respectively, calculate the standard enthalpy of combustion of toluene.
a) –3886 kJ · mol^{-1}
b) –667.3 kJ · mol^{-1}
c) –3910 kJ · mol^{-1}
d) –5029 kJ · mol^{-1}
e) –691.3 kJ · mol^{-1}

67. If the standard enthalpies of formation of carbon dioxide, $CO_2(g)$, $H_2O(l)$, and sucrose, $C_{12}H_{22}O_{11}(s)$, are –393.5, –285.8, and –2222 kJ · mol^{-1}, respectively, calculate the standard enthalpy of combustion of sucrose.
a) –2901 kJ · mol^{-1}
b) –5644 kJ · mol^{-1}
c) –10088 kJ · mol^{-1}
d) –3283 kJ · mol^{-1}
e) –1543 kJ · mol^{-1}

68. If the standard enthalpy of formation of Hg(g) is +61.32 kJ · mol^{-1} at 25°C, calculate the reaction enthalpy for the vaporization of 1 mole of mercury at 25°C and 1 atm.
a) +122.6 kJ b) +61.32 kJ c) –61.32 kJ d) 0 e) +30.66 kJ

69. The standard enthalpies of formation of SO_2(g) and SO_3(g) are –296.83 and –395.72 kJ · mol^{-1}, respectively. Calculate the reaction enthalpy for

$$2SO_3(g) \rightarrow 2SO_2(g) + O_2(g)$$

a) –197.78 kJ b) +197.78 kJ c) +98.89 kJ d) –1385 kJ e) –98.89 kJ

70. The standard enthalpies of formation of C(s, diamond) and CO_2(g) are +1.895 and –393.51 kJ · mol^{-1}, respectively. Calculate the reaction enthalpy for the combustion of 1 mole of diamond.
a) –393.51 kJ b) –395.41 kJ c) –391.61 kJ d) –397.31 kJ e) –1.895 kJ

71. Ozone, O_3(g), is depleted by the reaction

$$O_3(g) + O(g) \rightarrow 2O_2(g) \qquad \Delta H^\circ = -391.87 \text{ kJ}$$

The standard enthalpy of formation of oxygen atoms is +249.17 kJ · mol^{-1}. Calculate the standard enthalpy of formation of ozone.
a) –142.70 kJ · mol^{-1}
b) +142.70 kJ · mol^{-1}
c) –106.47 kJ · mol^{-1}
d) +641.04 kJ · mol^{-1}
e) –641.04 kJ · mol^{-1}

72. The standard enthalpies of formation of ozone, O_3(g), and oxygen atoms are +142.70 and +249.17 kJ · mol^{-1}. Calculate the reaction enthalpy for the reaction of ozone with oxygen atoms to produce oxygen.
a) +106.47 kJ b) –391.87 kJ c) +391.87 kJ d) +641.04 kJ e) –641.04 kJ

73. The standard enthalpies of formation of ozone, O_3(g), and oxygen atoms are +142.70 and +249.17 kJ · mol^{-1}. Calculate the reaction enthalpy for the reaction of oxygen atoms with oxygen to produce ozone.
a) –391.87 kJ b) +106.47 kJ c) +641.04 kJ d) +391.87 kJ e) –106.47 kJ

74. Consider the reaction below:

$$2H_2S(g) + 3O_2(g) \rightarrow 2SO_2(g) + 2H_2O(g) \qquad \Delta H^\circ = -1036.04 \text{ kJ}$$

The standard enthalpies of formation of SO_2(g) and H_2O(g) are –296.83 and –241.82 kJ · mol^{-1}. Calculate the standard enthalpy of formation of H_2S(g).
a) –497.39 kJ · mol^{-1}
b) –41.26 kJ · mol^{-1}
c) +20.63 kJ · mol^{-1}
d) –20.63 kJ · mol^{-1}
e) +41.26 kJ · mol^{-1}

75. The standard enthalpies of formation of PbS(s), SO_2(g), and PbO(s) are –98.3, –296.8, and –217.3 kJ · mol^{-1}, respectively. Calculate the reaction enthalpy for the roasting of PbS(s).

$$PbS(s) + 3O_2(g) \rightarrow 2SO_2(g) + 2PbO(s)$$

a) –415.8 kJ b) –1126.5 kJ c) –612.4 kJ d) –929.9 kJ e) –98.3 kJ

76. The final step of lead production is the reduction of $PbO(s)$ with carbon.

$$PbO(s) + C(s) \rightarrow Pb(s) + CO(g)$$

The standard enthalpies of formation of $PbO(s)$ and $CO(g)$ are –217.3 and –110.5 $kJ \cdot mol^{-1}$, respectively. Calculate the reaction enthalpy for the reduction of $PbO(s)$.

a) +327.8 kJ b) +106.8 kJ c) –327.8 kJ d) +110.5 kJ e) –106.8 kJ

77. The standard enthalpies of formation of $MnO(s)$ and $MnO_2(s)$ are –385.2 and –520.0 $kJ \cdot mol^{-1}$, respectively. Calculate the reaction enthalpy for

$$2MnO(s) + O_2(g) \rightarrow 2MnO_2(s)$$

a) –905.2 kJ b) –1040.0 kJ c) –269.6 kJ d) –1810.4 kJ e) –134.8 kJ

78. Calculate the enthalpy change for the combustion of 10.0 g of $Mg(s)$ given the standard enthalpy of formation of $MgO(s)$, –601.70 $kJ \cdot mol^{-1}$.

a) – 60.17 kJ b) –247.5 kJ c) –495.0 kJ d) –601.7 kJ e) –123.8 kJ

79. At 25°C the enthalpy of vaporization of mercury is +61.32 $kJ \cdot mol^{-1}$ and the standard enthalpy of formation of $HgO(s)$ is –90.83 $kJ \cdot mol^{-1}$. Calculate the reaction enthalpy for

$$2Hg(g) + O_2(g) \rightarrow 2HgO(s)$$

a) –181.66 kJ b) –29.51 kJ c) –304.30 kJ d) –59.02 kJ e) –152.15 kJ

80. Write thermochemical equations to define

(a) the standard enthalpy of formation of ammonium perchlorate.

(b) the standard enthalpy of combustion of sucrose, $C_{12}H_{22}O_{11}(s)$.

(c) the standard enthalpy of sublimation of graphite.

Chapter 6: Thermochemistry: The Fire Within
Form B

1. When a reaction takes place where the products have a higher energy than the reactants,
a) energy is released..
b) the reaction is exothermic.
c) the thermal motion of the atoms in the surroundings increases.
d) energy is absorbed.
e) the energy of the system decreases.

2. A piece of brass of mass 24.0 g at 95.0°C was placed in a calorimeter that contained 100.0 g of water at 15.0°C. If the temperature of the water rose to 16.7°C, what is the specific heat capacity of the brass? The specific heat capacity of water is 4.184 J · (°C)$^{-1}$ · g^{-1}.
a) 1.8 J · (°C)$^{-1}$ · g^{-1}
b) 0.31 J · (°C)$^{-1}$ · g^{-1}
c) 3.7 J · (°C)$^{-1}$ · g^{-1}
d) 2.0 J · (°C)$^{-1}$ · g^{-1}
e) 0.37 J · (°C)$^{-1}$ · g^{-1}

3. A piece of gold of mass 25.0 g at 88.0°C was placed in a calorimeter that contained 75.0 g of water at 25.00°C. If the temperature of the water rose to 25.70°C, what is the specific heat capacity of gold? The specific heat capacity of water is 4.184 J · (°C)$^{-1}$ · g^{-1}.
a) 0.10 J · (°C)$^{-1}$ · g^{-1}
b) 0.35 J · (°C)$^{-1}$ · g^{-1}
c) 0.14 J · (°C)$^{-1}$ · g^{-1}
d) 4.2 J · (°C)$^{-1}$ · g^{-1}
e) 5.2 J · (°C)$^{-1}$ · g^{-1}

4. A piece of marble of mass 45.0 g at 92.0°C was placed in a calorimeter that contained 100.0 g of water at 12.0°C. If the temperature of the water rose to 18.6°C, what is the specific heat capacity of the marble? The specific heat capacity of water is 4.184 J g^{-1} (°C)$^{-1}$.
a) 0.84 J · (°C)$^{-1}$ · g^{-1}
b) 5.1 J · (°C)$^{-1}$ · g^{-1}
c) 3.3 J · (°C)$^{-1}$ · g^{-1}
d) 2.4 J · (°C)$^{-1}$ · g^{-1}
e) 0.67 J · (°C)$^{-1}$ · g^{-1}

5. Calculate the heat required to raise the temperature of 50.0 g of water from 20.0°C to 75.0°C. The specific heat capacity of water is 4.184 J · (°C)$^{-1}$ · g^{-1}.
a) 4.18 kJ b) 11.5 kJ c) 0.232 kJ d) 15.7 kJ e) 0.639 kJ

6. How much heat is absorbed by a 75.0-g sample of gold at 25.0°C when it is immersed in boiling water? The specific heat capacity of gold is 0.128 J · (°C)$^{-1}$ · g^{-1} and the specific heat capacity of water is 4.184 J · (°C)$^{-1}$ · g^{-1}.
a) 0.960 kJ b) 0.720 kJ c) 3.00 kJ d) 4.18 kJ e) 0.240 kJ

7. How much heat is absorbed by a 250.0-g sample of aluminum at 25.0°C when it is immersed in boiling water? The specific heat capacity of aluminum is 0.900 J · (°C)$^{-1}$ · g^{-1} and the specific heat capacity of water is 4.184 J · (°C)$^{-1}$ · g^{-1}.
a) 22.5 kJ b) 70.6 kJ c) 16.9 kJ d) 5.63 kJ e) 4.18 kJ

8. A reaction occurs in a calorimeter that contains 240.0 g of water at 25.00°C, and 5.60 kJ of heat is evolved. If the specific heat capacity of water is 4.184 J · (°C)$^{-1}$ · g^{-1}, what is the final temperature of the water?
a) 30.58 °C b) 29.18 °C c) 19.42 °C d) 25.31 °C e) 35.24 °C

9. A calorimeter containing 200 mL of water was calibrated by carrying out a reaction which released 31.2 kJ of heat. If the temperature of the calorimeter rose 3.25°C, what is the heat capacity of this calorimeter?
a) 5.16 kJ (°C)$^{-1}$ b) 48.0 kJ (°C)$^{-1}$ L^{-1} c) 16.7 kJ (°C)$^{-1}$ d) 9.60 kJ (°C)$^{-1}$ e) 1.92 kJ L (°C)$^{-1}$

10. The reaction of 50 mL of hydrochloric acid with 50 mL of sodium hydroxide resulted in a 1.08°C increase in temperature in a calorimeter with a heat capacity of 6.62 kJ (°C)$^{-1}$. What is the heat output of the neutralization reaction?
a) 0.613 kJ b) 6.13 kJ c) 0.715 kJ d) 7.15 kJ e) 1.71 kJ

11. A joule is defined as
a) 1 kg · m^{-2} · s^{-2} b) 1 L · atm c) 1 kg · m^{2} · s^{-2} d) 1 kg · s^{-2} e) 1 L · atm^{-2} s^{-2}

12. When benzene vapor condenses at constant pressure, the sign of the enthalpy change for the process
a) depends on the container volume.
b) depends on the temperature.
c) is positive.
d) cannot be determined.
e) is negative.

13. When 25.0 g of ethanol vapor condenses, 23.6 kJ of heat is released. How much heat is required to vaporize 50.0 g of ethanol?
a) 590 kJ b) 94.4 kJ c) 47.2 kJ d) 1180 kJ e) 23.6 kJ

14. A compound has an enthalpy of sublimation of 86.0 kJ · mol^{-1} and an enthalpy of melting of 7.2 kJ · mol^{-1}. Calculate the enthalpy of vaporization of this compound.
a) +93.2 kJ · mol^{-1} b) +86.0 kJ · mol^{-1} c) +78.8 kJ · mol^{-1} d) +7.2 kJ · mol^{-1} e) –78.8 kJ · mol^{-1}

15. The enthalpy of sublimation of mercury is 61.6 kJ · mol^{-1} and its enthalpy of vaporization is 59.3 kJ · mol^{-1}. Calculate the enthalpy of fusion.
a) –2.3 kJ · mol^{-1} b) +2.3 kJ · mol^{-1} c) –120.9 kJ · mol^{-1} d) +120.9 kJ · mol^{-1} e) –61.6 kJ · mol^{-1}

16. At 100°C the enthalpy of vaporization of water is 40.7 kJ · mol^{-1}. The enthalpy change when 36.0 g of water vapor condenses is
a) –40.7 kJ b) –1.47 x 10^{3} kJ c) –81.4 kJ d) +40.7 kJ e) +81.4 kJ

17. At the boiling point of ethanol, C_2H_5OH, 78°C, the enthalpy of vaporization is 43.5 kJ · mol^{-1}. The enthalpy change when 23.0 g of ethanol vapor condenses is
a) +43.5 kJ b) +87.0 kJ c) +21.8 kJ d) –43.5 kJ e) –21.8 kJ

18. The diagram below is a heating curve for 15.0 g of mercury with a heating rate of 400 · min^{-1}.

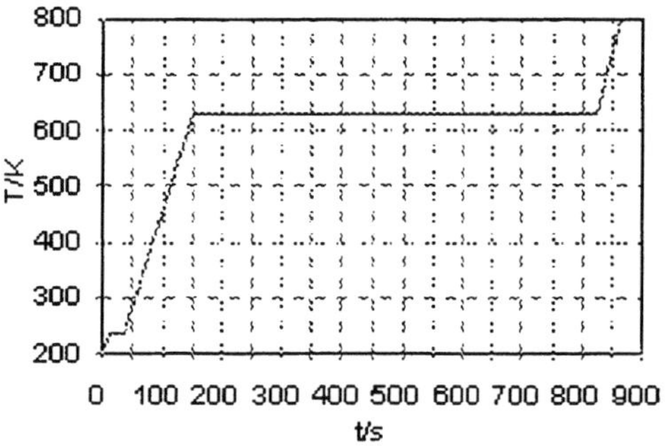

Estimate the enthalpy of fusion of mercury based on

a) +33.4 kJ b) +0.167 kJ · mol^{-1} c) +2.2 kJ · mol^{-1} d) +60.2 kJ e) +2.14 x 10^3 kJ

19. Consider the heating curve for one mole of a substance that melts at –23°C and boils at 327 °C.

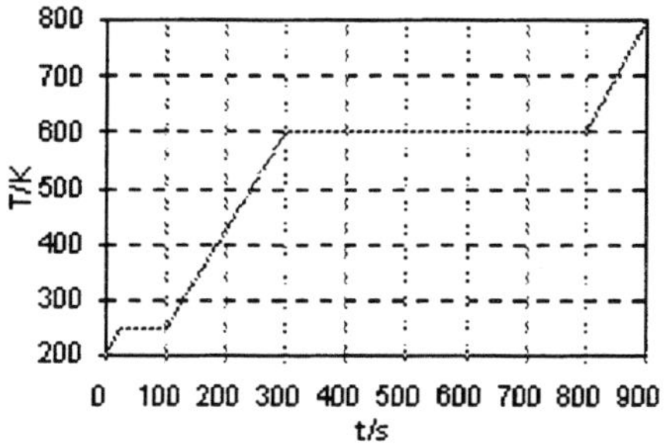

If the heat is added at a constant rate of 180 J/min, the molar enthalpy of vaporization of the substance is

a) +1.5 kJ · mol^{-1} b) +0.24 kJ · mol^{-1} c) +0.60 kJ · mol^{-1} d) +2.7 · mol^{-1} e) +3.0 kJ · mol^{-1}

20. When 5.00 g of methane, $CH_4(g)$, is burned in a calorimeter with a heat capacity of 66.0 kJ · K^{-1}, the temperature of the calorimeter assembly rises by 4.21 K. What is the standard enthalpy of combustion of methane?

a) –889 kJ · mol^{-1} b) –1254 kJ · mol^{-1} c) –86.8 kJ · mol^{-1} d) –278 kJ · mol^{-1} e) –1389 kJ · mol^{-1}

21. When 3.00 g of nitric acid are dissolved in 1.00 L of water in a calorimeter with a heat capacity of 1.22 kJ · K^{-1}, the temperature increases 1.21 K. What is the molar enthalpy of solution of nitric acid?

a) –125 kJ · mol^{-1} b) –11.9 kJ· mol^{-1} c) –8.02 kJ · mol^{-1} d) –20.8 kJ · mol^{-1} e) –31.0 kJ · mol^{-1}

22. When 10.0 g of ammonium nitrate are dissolved in 1.00 L of water in a calorimeter with a heat capacity of 2.30 kJ · K^{-1}, the temperature drops 1.52 K. What is the molar enthalpy of solution of ammonium nitrate?

a) +5.29 kJ · mol^{-1} b) +28.0 kJ · mol^{-1} c) +48.4 kJ · mol^{-1} d) +9.15 kJ · mol^{-1} e) +109 kJ · mol^{-1}

23. If 100 mL of 1.00 M silver nitrate are added to 100 mL of 1.00 M potassium iodide in a calorimeter with a heat capacity of 2.30 kJ/K, the temperature rises 4.87 K. Calculate the enthalpy change for

$$Ag^+(aq) + I^-(aq) \rightarrow AgI(s)$$

a) –9.45 kJ b) –4.72 kJ c) –21.2 kJ d) –56.0 kJ e) –112 kJ

24. When SnO_2(s) is formed from the combustion of gray tin, the reaction enthalpy is –578.6 kJ, and when white tin is burned to form SnO_2(s), the reaction enthalpy is –580.7 kJ. Calculate the reaction enthalpy for

$$Sn(gray) \rightarrow Sn(white)$$

a) +2.1 kJ b) –2.1 kJ c) –1159.3 kJ d) +1159.3 kJ e) 0

25. Consider the following reaction:

$$2SO_2(g) + O_2(g) \rightarrow 2SO_3(g) \qquad \Delta H^\circ = -198 \text{ kJ}$$

Calculate the enthalpy change when 200 L of SO_3(g) are produced at 25°C and 4.22 atm.

26. The standard reaction enthalpy for

$$2PCl_3(g) \rightarrow 2P(s) + 3Cl_2(g)$$

a) $+2\ \Delta H_f^\circ[PCl_3(g)]$
b) $+\Delta H_f^\circ[PCl_3(g)]$
c) $-2\ \Delta H_f^\circ[PCl_3(g)]$
d) $+2\ \Delta H_f^\circ[P(s)] + 3\ \Delta H_f^\circ[Cl_2(g)]$
e) $-\Delta H_f^\circ[PCl_3(g)]$

27. The standard enthalpy for

$$2H_2O(l) \rightarrow 2H_2(g) + O_2(g)$$

a) $-2\ \Delta H_f^\circ[H_2O(l)]$
b) $-\Delta H_f^\circ[H_2O(l)]$
c) $+2\ \Delta H_f^\circ[H_2(g)] + \Delta H_f^\circ[O_2(g)]$
d) $+\Delta H_f^\circ[H_2O(l)]$
e) $+2\ \Delta H_f^\circ[H_2O(l)]$

28. Calculate the reaction enthalpy for

$$2S(s) + 3O_2(g) \rightarrow 2SO_3(g)$$

from the data

$$SO_3(g) \rightarrow S(s) + 3/2O_2(g) \qquad \Delta H^\circ = +395.7 \text{ kJ}$$

a) +791.4 kJ b) –395.7 kJ c) –197.9 kJ d) +395.7 kJ e) –791.4 kJ

29. Calculate the reaction enthalpy for the decomposition of hydrogen peroxide,

$$2H_2O_2(l) \rightarrow 2H_2O(l) + O_2(g)$$

from the data

$$H_2(g) + O_2(g) \rightarrow H_2O_2(l) \qquad \Delta H^\circ = -187.8 \text{ kJ}$$

$$2H_2(g) + O_2(g) \rightarrow 2H_2O(l) \qquad \Delta H^\circ = -571.6 \text{ kJ}$$

a) –89.8 kJ b) +473.8 kJ c) –947.2 kJ d) –98.0 kJ e) –196.0 kJ

30. Calculate the reaction enthalpy for the combustion of propane, $C_3H_8(g)$,

$$C_3H_8(g) + 5O_2(g) \rightarrow 3CO_2(g) + 4H_2O(l)$$

from the data

$2H_2(g) + O_2(g) \rightarrow 2H_2O(l)$ $\Delta H^o = -572$ kJ

$C(s) + O_2(g) \rightarrow CO_2(g)$ $\Delta H^o = -394$ kJ

$3C(s) + 4H_2(g) \rightarrow C_3H_8(g)$ $\Delta H^o = -106$ kJ

a) −106 kJ b) −1072 kJ c) −2326 kJ d) −2220 kJ e) −2432 kJ

31. Calculate the reaction enthalpy for

$$PCl_3(l) + Cl_2(g) \rightarrow PCl_5(s)$$

from the data

$2P(s) + 3Cl_2(g) \rightarrow 2PCl_3(l)$ $\Delta H^o = -639$ kJ

$2P(s) + 5Cl_2(g) \rightarrow 2PCl_5(s)$ $\Delta H^o = -887$ kJ

a) −1526 kJ b) −320 kJ c) −124 kJ d) −444 kJ e) −248 kJ

32. Calculate the reaction enthalpy for the combustion of ethylene, $C_2H_4(g)$,

$$C_2H_4(g) + 3O_2(g) \rightarrow 2CO_2(g) + 2H_2O(l)$$

from the data

$C(s) + O_2(g) \rightarrow CO_2(g)$ $\Delta H^o = -393.5$ kJ

$H_2(g) + 1/2O_2(g) \rightarrow H_2O(l)$ $\Delta H^o = -285.8$ kJ

$2C(s) + 2H_2(g) \rightarrow C_2H_4(g)$ $\Delta H^o = +52.3$ kJ

a) −1410.9 kJ b) −215.4 kJ c) −731.6 kJ d) −1358.6 kJ e) −267.7 kJ

33. Calculate the reaction enthalpy for the combustion of 1 mole of ethanol, $C_2H_5OH(l)$, from the data

$C(s) + O_2(g) \rightarrow CO_2(g)$ $\Delta H^o = -393.5$ kJ

$H_2(g) + 1/2O_2(g) \rightarrow H_2O(l)$ $\Delta H^o = -285.8$ kJ

$2C(s) + 3H_2(g) + 1/2O_2(g) \rightarrow C_2H_5OH(l)$ $\Delta H^o = -277.7$ kJ

a) −1366.7 kJ b) −348.1 kJ c) −70.4 kJ d) −1644.4 kJ e) −1922.1 kJ

34. If the standard enthalpy of combustion of ethylene, $C_2H_4(g)$, is −1411 kJ · mol^{-1}, how many liters of ethylene measured at 25°C and 98.6 kPa must be burned to liberate 8.92 x 10^3 kJ of heat?

35. If the standard enthalpy of combustion of acetylene, $C_2H_2(g)$, is −1300 kJ · mol^{-1}, how many liters of acetylene measured at 25°C and 2.62 atm must be burned to liberate 10.0 x 10^6 kJ of heat?

36. Calculate the reaction enthalpy for

$C_6H_4(OH)_2(l) \rightarrow C_6H_4O_2(l) + H_2(g)$

from the following data.

$C_6H_4(OH)_2(l) + H_2O_2(l) \rightarrow C_6H_4O_2(l) + 2H_2O(l)$	$\Delta H° = -206.4$ kJ
$H_2(g) + O_2(g) \rightarrow H_2O_2(l)$	$\Delta H° = -187.8$ kJ
$H_2(g) + 1/2O_2(g) \rightarrow H_2O(l)$	$\Delta H° = -285.8$ kJ

37. If the standard enthalpy of combustion of ethanol, $C_2H_5OH(l)$, is –1368 kJ · mol^{-1}, calculate the heat output if 500 g of ethanol is burned.

a) 14.8×10^3 kJ b) 6.85×10^2 kJ c) 16.9 kJ d) 16.9×10^3 kJ e) 31.5×10^3 kJ

38. If the standard enthalpy of combustion of glucose, $C_6H_{12}O_6(l)$, is –2808 kJ · mol^{-1}, calculate the heat output if 250 g of glucose is burned.

a) 5.05×10^2 kJ b) 3.90×10^3 kJ c) 7.80×10^3 kJ d) 1.42×10^2 kJ e) 16.1 kJ

39. If the standard enthalpy of combustion of glucose, $C_6H_{12}O_6(l)$, is –2808 kJ · mol^{-1}, calculate the mass of glucose needed to supply 1650 kJ of heat.

a) 0.588 g b) 106 g c) 2.76 kg d) 212 g e) 17.6 g

40. If the standard enthalpy of combustion of coal is –394 kJ · mol^{-1}, calculate the mass of coal needed to supply 1650 kJ of heat.

a) 50.3 g b) 25.8 g c) 98.4 g d) 54.6 g e) 2.15 g

41. The standard enthalpy of combustion of methanol, $CH_3OH(l)$, is –726 kJ · mol^{-1} at 25°C. Which thermochemical equation gives this value?

a) $2CH_3OH(l) + 3O_2(g) \rightarrow 2CO_2(g) + 4H_2O(l)$
b) $CH_3OH(l) + 3/2O_2(g) \rightarrow CO_2(g) + 2H_2O(l)$
c) $CH_3OH(l) + 3O(g) \rightarrow CO_2(g) + 2H_2O(l)$
d) $CH_3OH(l) + O_2(g) \rightarrow CO(g) + 2H_2O(l)$
e) $CH_3OH(l) + 2O(g) \rightarrow CO(g) + 2H_2O(l)$

42. The standard enthalpy of combustion of hydrogen is –286 kJ · mol^{-1} at 25°C. Which thermochemical equation gives this value?

a) $H_2(g) + O(g) \rightarrow H_2O(l)$
b) $2H_2(g) + O_2(g) \rightarrow 2H_2O(l)$
c) $H_2(g) + 1/2O_2(g) \rightarrow H_2O(l)$
d) $H_2(g) \rightarrow 2H(g)$
e) $2H(g) + O(g) \rightarrow H_2O(l)$

43. Calculate the standard enthalpy of combustion of gaseous hydrazine, $N_2H_4(g)$, given

$H_2O(l) \rightarrow H_2(g) + 1/2O_2(g)$	$\Delta H° = +285.83$ kJ
$N_2(g) + 2H_2(g) \rightarrow N_2H_4(g)$	$\Delta H° = +95.40$ kJ

a) –381.23 kJ · mol^{-1}
b) –190.43 kJ · mol^{-1}
c) –476.26 kJ · mol^{-1}
d) –667.06 kJ · mol^{-1}
e) –95.40 kJ · mol^{-1}

44. If the standard enthalpy of combustion of octane, $C_8H_{18}(l)$, is $-5471\ kJ \cdot mol^{-1}$, calculate the enthalpy change when 1.00 kg of octane are burned.
a) –2736 kJ b) -23.9×10^3 kJ c) –5471 kJ d) -47.9×10^3 kJ e) -95.8×10^3 kJ

45. All of the following reactions or processes are endothermic except
a) combustion. b) vaporization. c) sublimation. d) melting. e) $I_2(s) \rightarrow I_2(g)$

46. Which thermochemical equation gives the value of the standard enthalpy of formation for HCl(g)?
a) $H(g) + 1/2Cl_2(l) \rightarrow HCl(g)$
b) $H(g) + Cl(g) \rightarrow HCl(g)$
c) $1/2H_2(g) + 1/2Cl_2(l) \rightarrow HCl(g)$
d) $1/2H_2(g) + 1/2Cl_2(g) \rightarrow HCl(g)$
e) $H_2(g) + Cl_2(g) \rightarrow 2HCl(g)$

47. Which thermochemical equation gives the value of the standard enthalpy of formation for $Hg_2Cl_2(s)$?
a) $2Hg(l) + Cl_2(l) \rightarrow Hg_2Cl_2(s)$
b) $2Hg(s) + Cl_2(g) \rightarrow Hg_2Cl_2(s)$
c) $2Hg(l) + Cl_2(g) \rightarrow Hg_2Cl_2(s)$
d) $2Hg(s) + Cl_2(l) \rightarrow Hg_2Cl_2(s)$
e) $Hg(l) + Cl(g) \rightarrow HgCl(s)$

48. Which thermochemical equation gives the value of the standard enthalpy of formation for $NH_4ClO_4(s)$?
a) $N(g) + 2H_2(g) + Cl_2(l) + 2O_2(g) \rightarrow NH_4ClO_4(s)$
b) $N(g) + 2H_2(g) + Cl(g) + 2O_2(g) \rightarrow NH_4ClO_4(s)$
c) $N_2(g) + 4H_2(g) + Cl_2(g) + 4O_2(g) \rightarrow 2NH_4ClO_4(s)$
d) $N(g) + 4H(g) + Cl(g) + 4O(g) \rightarrow NH_4ClO_4(s)$
e) $1/2N_2(g) + 2H_2(g) + 1/2Cl_2(g) + 2O_2(g) \rightarrow NH_4ClO_4(s)$

49. Which thermochemical equation gives the value of the standard enthalpy of formation for $XeO_3(s)$?
a) $Xe_2(g) + 3O_2(g) \rightarrow 2XeO_3$
b) $1/2Xe_2(g) + 3O(g) \rightarrow XeO_3(s)$
c) $1/2Xe_2(g) + 3/2O_2(g) \rightarrow XeO_3(s)$
d) $Xe(g) + 3O(g) \rightarrow XeO_3(s)$
e) $Xe(g) + 3/2O_2(g) \rightarrow XeO_3(s)$

50. Which of the following has a standard enthalpy of formation equal to zero?
a) $F_2(g)$ b) $Br_2(g)$ c) $I_2(g)$ d) C(g) e) S(g)

51. Which of the following has a standard enthalpy of formation equal to zero?
a) $I_2(s)$ b) $Br_2(s)$ c) $Br_2(g)$ d) $I_2(g)$ e) $I_2(l)$

52. All of the following have a standard enthalpy of formation equal to zero except
a) $Cl_2(g)$ b) $I_2(s)$ c) $O_2(g)$ d) $F_2(g)$ e) $Br_2(g)$

53. All of the following have a standard enthalpy of formation equal to zero except
a) $Br_2(l)$ b) $Cl_2(g)$ c) $I_2(g)$ d) Kr(g) e) Al(s)

54. The standard enthalpy of formation of H_2O_2(l) is –187.8 kJ · mol^{-1} and of H_2O(l) is –285.8 kJ · mol^{-1}. Calculate the reaction enthalpy for

$H_2O(l) + 1/2O_2(g) \rightarrow H_2O_2(l)$

a) –187.8 kJ b) –473.6 kJ c) –98.0 kJ d) +187.8 kJ e) +98.0 kJ

55. The standard enthalpy of formation of NO(g) is +90.3 kJ · mol^{-1} and of NO_2(g) is +33.2 kJ · mol^{-1}. Calculate the reaction enthalpy for

$NO_2(g) \rightarrow NO(g) + 1/2O_2(g)$

a) +123.5 kJ b) –57.1 kJ c) –123.5 kJ d) +90.3 kJ e) +57.1 kJ

56. The standard enthalpy of formation of cyclopropane, C_3H_6(g), is +53.3 kJ · mol^{-1} and of cyclohexane, C_6H_{12}(l), is –156.4 kJ · mol^{-1}. What is the reaction enthalpy for formation of 1 mole of cyclohexane from cyclopropane?

a) –263.0 kJ b) –209.7 kJ c) +156.4 kJ d) +106.6 kJ e) +263.0 kJ

57. The reaction enthalpy for

$2SnO_2(s) \rightarrow 2SnO(s) + O_2(g)$

is +589.8 kJ. The standard enthalpy of formation of SnO(s) is –285.8 kJ · mol^{-1}. What is the standard enthalpy of formation of SnO_2(s)?

a) –1161.4 kJ · mol^{-1}
b) –580.7 kJ · mol^{-1}
c) –304.0 kJ · mol^{-1}
d) +580.7 kJ · mol^{-1}
e) –608.0 kJ · mol^{-1}

58. The reaction enthalpy for

$2CO_2(g) \rightarrow 2CO(g) + O_2(g)$

is +566.0 kJ. The standard enthalpy of formation of CO(g) is –110.5 kJ · mol^{-1}. What is the standard enthalpy of formation of CO_2(g)?

a) –172.5 kJ · mol^{-1}
b) –676.5 kJ · mol^{-1}
c) –338.3 kJ · mol^{-1}
d) –787.0 kJ · mol^{-1}
e) –393.5 kJ · mol^{-1}

59. Ethane, C_2H_6(g), can be made by reaction of hydrogen gas with acetylene, C_2H_2(g). The standard enthalpies of formation of ethane and acetylene are –84.68 and +226.73 kJ · mol^{-1}, respectively. What is the reaction enthalpy for the production of one mole of ethane?

a) –142.05 kJ b) –311.41 kJ c) +142.05 kJ d) +84.68 kJ e) +311.41 kJ

60. The standard enthalpy of combustion of benzene, C_6H_6(l), is –3268 kJ · mol^{-1}. The standard enthalpies of formation of H_2O(l) and CO_2(g) are –285.83 and –393.51 · mol^{-1}, respectively. What is the standard enthalpy of formation of benzene?

a) –689 kJ · mol^{-1} b) +689 kJ · mol^{-1} c) –49 kJ · mol^{-1} d) +49 kJ · mol^{-1} e) –436 kJ · mol^{-1}

61. The standard enthalpy of combustion of octane, $C_8H_{18}(l)$, is $-5471\ kJ \cdot mol^{-1}$. The standard enthalpies of formation of $H_2O(l)$ and $CO_2(g)$ are -285.83 and $-393.51\ kJ \cdot mol^{-1}$, respectively. What is the standard enthalpy of formation of octane?
a) $-689\ kJ \cdot mol^{-1}$ b) $+250\ kJ \cdot mol^{-1}$ c) $-438\ kJ \cdot mol^{-1}$ d) $+689\ kJ \cdot mol^{-1}$ e) $-250\ kJ \cdot mol^{-1}$

62. The standard enthalpies of combustion of $CH_3OH(l)$, C(s), and hydrogen gas are -726, -394, and $-286\ kJ \cdot mol^{-1}$, respectively. What is the standard enthalpy of formation of $CH_3OH(l)$?
a) $-620\ kJ \cdot mol^{-1}$ b) $-240\ kJ \cdot mol^{-1}$ c) $-484\ kJ \cdot mol^{-1}$ d) $+620\ kJ \cdot mol^{-1}$ e) $-240\ kJ \cdot mol^{-1}$

63. The standard enthalpies of combustion of $CH_4(g)$, C(s), and hydrogen gas are -890, -394, and $-286\ kJ \cdot mol^{-1}$, respectively. What is the standard enthalpy of formation of $CH_4(g)$?
a) $-445\ kJ \cdot mol^{-1}$ b) $-76\ kJ \cdot mol^{-1}$ c) $+76\ kJ \cdot mol^{-1}$ d) $-620\ kJ \cdot mol^{-1}$ e) $+620\ kJ \cdot mol^{-1}$

64. If the standard enthalpy of formation of $PH_3(g)$ is $+5.4\ kJ \cdot mol^{-1}$, calculate the reaction enthalpy for

$$2PH_3(g) \rightarrow 2P(s) + 3H_2(g)$$

a) +5.4 kJ b) +10.8 kJ c) −10.8 kJ d) −2.7 kJ e) −5.4 kJ

65. If the standard enthalpy of formation of $N_2O(g)$ is $+82.05\ kJ \cdot mol^{-1}$, calculate the reaction enthalpy for

$$2N_2O(g) \rightarrow 2N_2(g) + O_2(g)$$

a) +82.05 kJ b) −82.05 kJ c) −41.03 kJ d) −164.10 kJ e) +164.10 kJ

66. If the standard enthalpies of formation of carbon dioxide, $CO_2(g)$, $H_2O(l)$, and glucose, $C_6H_{12}O_6(s)$, are -393.5, -285.8, and $-1268\ kJ \cdot mol^{-1}$, respectively, calculate the standard enthalpy of combustion of glucose.
a) $-5344\ kJ \cdot mol^{-1}$
b) $-2808\ kJ \cdot mol^{-1}$
c) $-588.7\ kJ \cdot mol^{-1}$
d) $-1947\ kJ \cdot mol^{-1}$
e) $-1950\ kJ \cdot mol^{-1}$

67. If the standard enthalpies of formation of carbon dioxide, $CO_2(g)$, $H_2O(l)$, and benzoic acid, $C_6H_5COOH(s)$, are -393.5, -285.8, and $-385.1\ kJ \cdot mol^{-1}$, respectively, calculate the standard enthalpy of combustion of benzoic acid.
a) $-6454\ kJ \cdot mol^{-1}$
b) $-1064\ kJ \cdot mol^{-1}$
c) $-294.2\ kJ \cdot mol^{-1}$
d) $-3227\ kJ \cdot mol^{-1}$
e) $-3997\ kJ \cdot mol^{-1}$

68. If the standard enthalpy of formation of $Br_2(g)$ is +30.91 kJ at 25°C, calculate the reaction enthalpy for the vaporization of 1 mole of bromine at 25°C.
a) +15.46 kJ b) +30.91 kJ c) +61.82 kJ d) 0 e) −30.91 kJ

69. The standard enthalpies of formation of CO(g) and CO_2(g) are –110.53 and –393.51 kJ · mol^{-1}, respectively. Calculate the reaction enthalpy for

$$2CO_2(g) \rightarrow 2CO(g) + O_2(g)$$

a) +282.98 kJ b) +565.96 kJ c) –282.98 kJ d) –565.96 kJ e) –1008 kJ

70. The standard enthalpies of combustion of diamond and graphite are –395.41 and –393.51 kJ · mol^{-1}, respectively. What is the standard enthalpy of formation of diamond?

a) +788.92 kJ · mol^{-1}
b) –788.92 kJ · mol^{-1}
c) –1.90 kJ · mol^{-1}
d) –395.41 kJ · mol^{-1}
e) +1.90 kJ · mol^{-1}

71. Ozone, O_3(g), is depleted by the reaction

$$O_3(g) + O(g) \rightarrow 2O_2(g) \qquad \Delta H^{o} = -391.87\ \text{kJ}$$

The standard enthalpy of formation of ozone is +142.70 kJ · mol^{-1}. Calculate the standard enthalpy of formation of oxygen atoms.

a) +498.34 kJ · mol^{-1}
b) +249.17 kJ · mol^{-1}
c) –534.57 kJ · mol^{-1}
d) +534.57 kJ · mol^{-1}
e) +1069.1 kJ · mol^{-1}

72. The standard enthalpies of formation of ozone, O_3(g), and oxygen atoms are +142.70 and +249.17 kJ · mol^{-1}. Calculate the reaction enthalpy for

$$2O_2(g) \rightarrow O_3(g) + O(g)$$

a) –641.04 kJ b) +106.47 kJ c) –391.87 kJ d) +641.04 kJ e) +391.87 kJ

73. Given the following data

$$O_3(g) + O(g) \rightarrow 2O_2(g) \qquad \Delta H^{o} = -391.87\ \text{kJ}$$
$$O_2(g) + O(g) \rightarrow O_3(g) \qquad \Delta H^{o} = -106.47\ \text{kJ}$$

calculate the reaction enthalpy for

$$2O_3(g) \rightarrow 3O_2(g)$$

a) –285.40 kJ b) –498.34 kJ c) –677.27 kJ d) +498.34 kJ e) +285.40 kJ

74. Consider the reaction below:

$$2H_2S(g) + 3O_2(g) \rightarrow 2SO_2(g) + 2H_2O(g) \qquad \Delta H^{o} = -1036.04\ \text{kJ}$$

The standard enthalpies of formation of H_2S(g) and H_2O(g) are –20.63 and –241.82 kJ · mol^{-1}. Calculate the standard enthalpy of formation of SO_2(g).

a) –814.85 kJ · mol^{-1}
b) +296.83 kJ · mol^{-1}
c) –593.66 kJ · mol^{-1}
d) +593.66 kJ · mol^{-1}
e) –296.83 kJ · mol^{-1}

75. Calculate the reaction enthalpy for

$$PbS(s) + 3O_2(g) \rightarrow 2SO_2(g) + 2PbO(s)$$

from the data

$S(s) + Pb(s) \rightarrow PbS(s)$ $\Delta H^\circ = -98.3$ kJ

$S(s) + O_2(g) \rightarrow SO_2(g)$ $\Delta H^\circ = -296.8$ kJ

$Pb(s) + 1/2O_2 \rightarrow PbO(s)$ $\Delta H^\circ = -217.3$ kJ

a) −1126.5 kJ b) −929.9 kJ c) −612.4 kJ d) −415.8 kJ e) −98.3 kJ

76. The final step of lead production is the reduction of PbO(s) with carbon.

$PbO(s) + C(s) \rightarrow Pb(s) + CO(g)$ $\Delta H^\circ = +106.8$ kJ

The standard enthalpy of formation of carbon monoxide is $-110.5\ \text{kJ} \cdot \text{mol}^{-1}$. Calculate the standard enthalpy of formation of PbO(s).

a) $-217.3\ \text{kJ} \cdot \text{mol}^{-1}$ b) $-3.7\ \text{kJ} \cdot \text{mol}^{-1}$ c) $+3.7\ \text{kJ} \cdot \text{mol}^{-1}$ d) $+217.3\ \text{kJ} \cdot \text{mol}^{-1}$ e) $+106.8\ \text{kJ} \cdot \text{mol}^{-1}$

77. Consider the reaction

$2MnO(s) + O_2(g) \rightarrow 2MnO_2(g)$ $\Delta H^\circ = -269.6$ kJ

If the standard enthalpy of formation of $MnO_2(s)$ is $-520.0\ \text{kJ} \cdot \text{mol}^{-1}$, calculate the standard enthalpy of formation of MnO(s).

a) $-770.4\ \text{kJ} \cdot \text{mol}^{-1}$
b) $-1309.6\ \text{kJ} \cdot \text{mol}^{-1}$
c) $-654.8\ \text{kJ} \cdot \text{mol}^{-1}$
d) $-385.2\ \text{kJ} \cdot \text{mol}^{-1}$
e) $-250.4\ \text{kJ} \cdot \text{mol}^{-1}$

78. Calculate the enthalpy change for the combustion of 5.00 g of Mg(s) given the standard enthalpy of formation of MgO(s), $-601.70\ \text{kJ} \cdot \text{mol}^{-1}$.

a) −61.88 kJ b) −123.8 kJ c) −247.5 kJ d) − 120.3 kJ e) −601.7 kJ

79. At 25°C the enthalpy of vaporization of bromine is $+30.91\ \text{kJ} \cdot \text{mol}^{-1}$ and the standard enthalpy of formation of HBr(g) is $-36.40\ \text{kJ} \cdot \text{mol}^{-1}$. Calculate the reaction enthalpy for

$$Br_2(g) + H_2(g) \rightarrow 2HBr(g)$$

a) −51.86 kJ b) −5.49 kJ c) −103.71 kJ d) −41.89 kJ e) −67.31 kJ

80. Write thermochemical equations to define

(a) the standard enthalpy of formation of acetic acid.

(b) the standard enthalpy of combustion of pentane, $C_5H_{12}(g)$.

(c) the standard enthalpy of formation of chlorine atoms, Cl(g).

Chapter 7: Inside the Atom
Form A

1. Calculate the wavelength of blue light of frequency 6.40×10^{14} Hz.
a) 214 nm b) 311 nm c) 640 nm d) 936 nm e) 468 nm

2. Calculate the wavelength of yellow light of frequency 5.20×10^{14} Hz.
a) 576 nm b) 173 nm c) 382 nm d) 520 nm e) 1150 nm

3. Calculate the wavelength of infrared radiation of frequency 3.00×10^{14} Hz.
a) 1000 nm b) 300 nm c) 663 nm d) 2000 nm e) 100 nm

4. Calculate the wavelength of radio waves corresponding to the FM station with frequency "92.1 on your dial" where the frequency is in MHz.
a) 2158 nm b) 6.52 m c) 30700 nm d) 3260 nm e) 3.26 m

5. Calculate the wavelength of radio waves corresponding to the FM station with frequency "101.5 on your dial" where the frequency is in MHz.
a) 3383 nm b) 2.96 m c) 5.92 m d) 1958 nm e) 2960 nm

6. Calculate the wavelength of x-rays of frequency 5.00×10^{17} Hz.
a) 1.20 nm b) 0.398 nm c) 0.600 m d) 1.20 m e) 0.600 nm

7. Calculate the frequency of blue light of wavelength 470 nm.
a) 6.38×10^{14} Hz b) 4.70×10^{14} Hz c) 1.57×10^{15} Hz d) 1.57×10^{-15} Hz e) 6.38×10^{16} Hz

8. Calculate the frequency of yellow light of wavelength 580 nm.
a) 5.80×10^{14} Hz b) 1.93×10^{-15} Hz c) 5.17×10^{14} Hz d) 5.17×10^{16} Hz e) 1.93×10^{15} Hz

9. Calculate the frequency of infrared radiation of wavelength 1000 nm.
a) 3.00×10^{16} Hz b) 3.33×10^{15} Hz c) 3.33×10^{-15} Hz d) 3.00×10^{14} Hz e) 1.00×10^{14} Hz

10. Calculate the frequency of radio waves of wavelength 3.50 m.
a) 1.17×10^{8} Hz b) 8.57×10^{6} MHz c) 857 MHz d) 8.57×10^{7} Hz e) 1.17×10^{-8} Hz

11. Calculate the frequency of x-rays of wavelength of 6.00×10^{-10} m.
a) 2.00×10^{-18} Hz b) 5.00×10^{8} Hz c) 2.00×10^{18} Hz d) 6.00×10^{17} Hz e) 5.00×10^{17} Hz

12. Calculate the energy per photon of microwaves of frequency 3.00×10^{11} Hz. Planck's constant is 6.63×10^{-34} J · s.
a) 2.21×10^{-45} J b) 1.99×10^{-22} J c) 5.97×10^{-14} J d) 120 J e) 4.52×10^{44} J

13. Calculate the energy per photon of ultraviolet radiation of frequency 3.00×10^{15} Hz. Planck's constant is 6.63×10^{-34} J · s.
a) 1200 kJ b) 2.21×10^{-49} J c) 5.97×10^{-10} J d) 4.52×10^{48} J e) 1.99×10^{-18} J

14. Calculate the energy per photon of radio waves of frequency 107.3 MHz. Planck's constant is 6.63×10^{-34} J · s.
a) 7.11×10^{-26} J b) 1.62×10^{41} J c) 2.13×10^{-17} J d) 6.18×10^{-42} J e) 0.0428 J

15. Calculate the energy per photon of blue light of frequency 6.40×10^{14} Hz. Planck's constant is 6.63×10^{-34} J · s.
a) 1.27×10^{-10} J b) 256 kJ c) 9.65×10^{47} J d) 1.04×10^{-48} J e) 4.24×10^{-19} J

16. Calculate the energy per mole of photons of microwaves of frequency 3.00×10^{11} Hz. Planck's constant is 6.63×10^{-34} J · s.
a) 2.21×10^{-45} J b) 1.99×10^{-22} J c) 4.52×10^{44} J d) 120 J e) 5.97×10^{-14} J

17. Calculate the energy per mole of photons of ultraviolet radiation of frequency 3.00×10^{15} Hz. Planck's constant is 6.63×10^{-34} J · s.
a) 4.52×10^{48} J b) 1.99×10^{-18} J c) 5.97×10^{-10} J d) 1200 kJ e) 2.21×10^{-49} J

18. Calculate the energy per mole of photons of radio waves of frequency 107.3 MHz. Planck's constant is 6.63×10^{-34} J · s.
a) 2.13×10^{-17} J b) 6.18×10^{-42} J c) 0.0428 J d) 7.11×10^{-26} J e) 1.62×10^{41} J

19. Calculate the energy per mole of photons of blue light of frequency 6.40×10^{14} Hz. Planck's constant is 6.63×10^{-34} J · s.
a) 4.24×10^{-19} J b) 1.27×10^{-10} J c) 1.04×10^{-48} J d) 9.65×10^{47} J e) 256 kJ

20. Which of the following has the greatest frequency?
a) x–rays b) blue light c) red light d) microwaves e) radiowaves

21. Which of the following has the lowest frequency?
a) microwaves b) x-rays c) yellow light d) ultraviolet radiation e) violet light

22. Which of the following has the longest wavelength?
a) microwaves b) infrared radiation c) ultraviolet radiation d) x-rays e) violet light

23. Which of the following has the shortest wavelength?
a) infrared radiation b) ultraviolet radiation c) γ-rays d) microwaves e) green light

24. The wavelike character of ordinary objects like a thrown baseball cannot be detected because
a) of the uncertainty principle.
b) their kinetic energy is too small.
c) their frequencies are so small.
d) their wavelengths are so short.
e) their wavelengths are so long.

25. The absorbance of a dye at 660 nm in a 1 cm cell is 0.761. If the concentration of the dye is 5.40×10^{-4} M, calculate the molar absorption coefficient of the dye. Describe the color and intensity of the solution of the dye.

26. A CO_2 laser produces radiation of 10,600 nm. If the output is 0.150 J/pulse, how many pulses are required to produce 4.00×10^{20} photons?

27. A pointing device uses a ruby laser with an output of 0.500 milliwatts. How many photons are produced if the pointing device is left on for 30 s?

28. Calculate the de Broglie wavelength of a 155 g ball travelling at 85.0 miles per hour.

29. Which of the following is true?
a) An electron in an *s*-orbital has a nonzero probability of being found at the nucleus.
b) An *s*-orbital becomes more dense as the distance from the nucleus increases.
c) A *p*-orbital has a spherical boundary surface.
d) An electron in a *p*-orbital has a nonzero probability of being found at the nucleus.
e) An *s*-orbital has two lobes on opposite sides of the nucleus.

30. Calculate the frequency of a photon emitted by a hydrogen atom when an electron makes a transition from the third to the second principal quantum level.
[$\mathfrak{R} = 3.29 \times 10^{15}$ Hz; $h = 6.63 \times 10^{-34}$ J · s]
a) 4.57×10^{14} Hz b) 1.31×10^{3} Hz c) 2.99×10^{3} Hz d) 1.83×10^{8} Hz e) 2.18×10^{-18} Hz

31. Calculate the frequency of a photon emitted by a hydrogen atom when an electron makes a transition from the state with n = 5 to n = 2.
[$\mathfrak{R} = 3.29 \times 10^{15}$ Hz; $h = 6.63 \times 10^{-34}$ J · s]
a) 2.18×10^{-18} Hz b) 6.91×10^{14} Hz c) 2.99×10^{3} Hz d) 1.31×10^{3} Hz e) 1.83×10^{8} Hz

32. How many orbitals are there in the shell with $n = 3$?
a) 9 b) 18 c) 4 d) 18 e) 3

33. An electron in a hydrogen atom has the quantum numbers $n = 4$, $l = 1$, $m_\ell = 0$. In what type of orbital is the electron located?
a) 3*d* b) 4*s* c) 4*d* d) 4*p* e) 3*p*

34. An electron in a hydrogen atom has the quantum numbers $n = 4$, $l = 3$, $m_\ell = 0$. In what type of orbital is the electron located?
a) 3*d* b) 4*p* c) 4*d* d) 4*s* e) 4*f*

35. How many orbitals are there with $l = 2$?
a) 5 b) 4 c) 7 d) 2 e) 1

36. An electron in an atom has a magnetic quantum number of –2. What is the lowest possible value for the principal quantum number of this electron?
a) 3 b) 4 c) 2 d) 1 e) –2

37. Which of the following is a possible set of quantum numbers for a 4*d*-electron?
a) $n = 4, l = 3, m_\ell = 2, m_s = ½$
b) $n = 4, l = 1, m_\ell = 0, m_s = ½$
c) $n = 4, l = 4, m_\ell = -2, m_s = ½$
d) $n = 4, l = 2, m_\ell = 0, m_s = -½$
e) $n = 4, l = 1, m_\ell = -1, m_s = -½$

38. Which of the following is a possible set of quantum numbers for a 4*p*-electron?
a) $n = 4, l = 3, m_\ell = 2, m_s = ½$
b) $n = 4, l = 1, m_\ell = 0, m_s = -½$
c) $n = 4, l = 2, m_\ell = -1, m_s = -½$
d) $n = 4, l = 2, m_\ell = 0, m_s = ½$
e) $n = 4, l = 4, m_\ell = -2, m_s = ½$

39. Which of the following is a possible set of quantum numbers for a 3*d*-electron?
a) $n = 3, l = 1, m_\ell = 0, m_s = ½$
b) $n = 3, l = 0, m_\ell = 0, m_s = ½$
c) $n = 3, l = 1, m_\ell = 1, m_s = ½$
d) $n = 3, l = 1, m_\ell = -1, m_s = -½$
e) $n = 3, l = 2, m_\ell = 0, m_s = -½$

40. Which of the following represents a possible set of quantum numbers for an electron in an atom?
a) $n = 3, l = 2, m_\ell = 0, m_s = 0$
b) $n = 3, l = 3, m_\ell = -1, m_s = -½$
c) $n = 3, l = 0, m_\ell = 1, m_s = -½$
d) $n = 3, l = 0, m_\ell = 1, m_s = ½$
e) $n = 3, l = 2, m_\ell = -2, m_s = -½$

41. Which subshell can hold the greatest number of electrons?
a) 3*d* b) 4*d* c) 5*d* d) 6*p* e) 4*f*

42. For a 6*p* subshell, what is the most positive value of m_ℓ?
a) +1 b) –1 c) 0 d) +5 e) +6

43. The atomic orbital for an electron in an atom with the quantum numbers $n = 4, l = 3, m_\ell = 0, m_s = -½$ is
a) 4*s* b) 4*p* c) 3*d* d) 4*f* e) 4*d*

44. If the azimuthal quantum number is 3, the type and number of orbitals is
a) 4*f*, 7 b) 3*p*, 3 c) 3*d*, 5 d) 4*d*, 5 e) 4*s*, 1

45. A 4*s*-electron in the potassium atom is lower in energy than a 3*d*-electron due to
a) the relative sizes of the 4*s*- and 3*d*-orbitals.
b) the shapes of the 3*d*-orbitals.
c) a low ionization energy of potassium.
d) the fact that there are five 3*d*-orbitals.
e) penetration and shielding.

46. What is the ground-state electron configuration of an aluminum atom?
a) $[Ne]3s^23p^3$ b) $[Ne]3s^23p^2$ c) $[Ne]3s^13p^2$ d) $[Ne]3s^2$ e) $[Ne]3s^23p^1$

47. What is the ground-state electron configuration of a bromine atom?
a) $[Ar]4s^24p^4$ b) $[Ar]4s^24p^5$ c) $[Ar]4s^24p^6$ d) $[Ar]4s^24p^2$ e) $[Ar]4s^24p^3$

48. What is the ground-state electron configuration of a cobalt atom?
a) $[Ar]3d^84s^2$ b) $[Ar]3d^74s^2$ c) $[Ar]3d^54s^2$ d) $[Ar]3d^54s^1$ e) $[Ar]3d^64s^2$

49. What is the ground-state electron configuration of a chromium atom?
a) $[Ar]3d^54s^1$ b) $[Ar]3d^44s^2$ c) $[Ar]3d^44s^1$ d) $[Ar]3d^54s^2$ e) $[Ar]3d^64s^1$

50. The highest-energy electron in an atom has quantum numbers $n = 3$, $l = 1$, $m_\ell = 0$, $m_s = ½$. This atom could be
a) sulfur b) potassium c) calcium d) bromine e) arsenic

51. Which atom contains an electron with quantum numbers $n = 3$, $l = 2$, $m_\ell = 2$, $m_s = ½$?
a) iron b) calcium c) sulfur d) argon e) sodium

52. How many unpaired electrons are there in the ground state of the ion Cu^+?
a) 0 b) 1 c) 3 d) 5 e) 6

53. What is the electron configuration of the Fe^{2+} ion?
a) $[Ar]3d^54s^1$ b) $[Ar]3d^5$ c) $[Ar]3d^54s^2$ d) $[Ar]3d^44s^2$ e) $[Ar]3d^6$

54. What is the electron configuration of the Se^{2-} ion?
a) [Kr] b) $[Kr]5s^1$ c) $[Ar]3d^{10}4s^24p^4$ d) $[Ar]3d^{10}4s^2$ e) $[Kr]5s^2$

55. What is the electron configuration of the Ni^{4+} ion?
a) $[Ar]3d^54s^1$ b) $[Ar]3d^64s^2$ c) $[Ar]3d^6$ d) $[Ar]3d^44s^2$ e) $[Ar]3d^8$

56. A certain ion has the ground-state electron configuration $[Ar]3d^{10}$. The ion is likely
a) Ga^+ b) Cu^{2+} c) Ni^{2+} d) Cl^- e) Zn^{2+}

57. What main group element forms a positive ion having the ground-state electron configuration [Ar]?
a) potassium b) sulfur c) argon d) scandium e) barium

58. What is the valence-electron configuration of the atoms of the group that contains gallium?
a) np^3 b) ns^2np^2 c) ns^2np^1 d) $nd^{10}ns^2$ e) $nd^{10}np^3$

59. Which of the following atoms has the greatest number of valence-shell electrons?
a) chlorine b) arsenic c) sulfur d) lead e) barium

60. What is the valence-electron configuration of the halogens?
a) ns^2np^6 b) ns^2np^4 c) ns^2np^5 d) ns^2 e) ns^1

61. Which of the following atoms would has the smallest radius?
a) oxygen b) sulfur c) carbon d) silicon e) lithium

62. Which of the following species has the smallest radius?
a) Ca^{2+} b) S^{2-} c) K d) K^+ e) Cl^-

63. Which of the following species has the largest radius?
a) Ca^{2+} b) S^{2-} c) S d) K^+ e) Cl^-

64. Arrange Cl^-, Al^{3+}, and Ca^{2+} in order of increasing ionic radii.
a) $Ca^{2+} < Al^{3+} < Cl^-$
b) $Cl^- < Ca^{2+} < Al^{3+}$
c) $Ca^{2+} < Cl^- < Al^{3+}$
d) $Ca^{2+} < Cl^- < Al^{3+}$
e) $Al^{3+} < Ca^{2+} < Cl^-$

65. Which of the following has the highest first ionization energy?
a) P b) Mg c) S d) Al e) Si

66. Which of the following represents the second ionization energy of the element E?
a) $E(g) \rightarrow E^{2+}(g) + 2e^-(g)$
b) $E(s) \rightarrow E^+(g) + e^-(g)$
c) $E(g) \rightarrow E^+(g) + e^-(g)$
d) $E(s) \rightarrow E^{2+}(g) + 2e^-(g)$
e) $E^+(g) \rightarrow E^{2+}(g) + e^-(g)$

67. For the elements Be, B, N, and O, the first ionization energy increases in the order
a) Be < O < B < N
b) Be < B < N < O
c) B < Be < O < N
d) B < Be < N < O
e) Be < B < O < N

68. Which of the following has the highest first ionization energy?
a) oxygen b) sulfur c) iodine d) cesium e) boron

69. Which of the following has the highest first ionization energy?
a) nitrogen b) beryllium c) carbon d) boron e) oxygen

70. Which of the following has the most metallic character?
a) Ba b) K c) Ca d) Sr e) Sc

71. Which of the following has the lowest second ionization energy?
a) Ca b) Ne c) K d) Ar e) Na

72. Metallic elements
a) have high ionization energies.
b) do not react with acids.
c) have low ionization energies.
d) form covalent halides.
e) have electron affinities higher than fluorine.

73. Which of the following processes, with all species in the gas phase, would require the most energy?
a) $Mg^{2+} \rightarrow Mg^{3+} + e^-$
b) $K \rightarrow K^+ + e^-$
c) $Mg^+ \rightarrow Mg^{2+} + e^-$
d) $Mg \rightarrow Mg^{2+} + 2e^-$
e) $Mg \rightarrow Mg^+ + e^-$

74. Which of the following ions would least likely form?
a) Be^{2+} b) Se^{2-} c) Si^{2+} d) F^{2-} e) Fe^{2+}

75. Aluminum has similar properties to
a) beryllium. b) litium. c) silicon. d) calcium. e) sodium.

76. Which of the following has the largest electron affinity?
a) F b) O c) O^{-} d) N e) F^{-}

77. All of the following are metalloids except
a) Al b) Si c) As d) Sb e) Ge

78. Which of the following is a *p*-block element?
a) arsenic b) calcium c) potassium d) vanadium e) copper

Chapter 7: Inside the Atom
Form B

1. Calculate the wavelength of green light of frequency 5.70×10^{14} Hz.
a) 190 nm b) 1050 nm c) 349 nm d) 570 nm e) 526 nm

2. Calculate the wavelength of red light of frequency 4.30×10^{14} Hz.
a) 1400 nm b) 697 nm c) 462 nm d) 430 nm e) 143 nm

3. Calculate the wavelength of microwaves of frequency 3.00×10^{11} Hz.
a) 663 nm b) 0.200 cm c) 1000 nm d) 300 nm e) 0.100 cm

4. Calculate the wavelength of radio waves corresponding to the FM station with frequency "107.3 on your dial" where the frequency is in MHz.
a) 2.80 m b) 5.62 m c) 2800 nm d) 1853 nm e) 35767 nm

5. Calculate the wavelength of radio waves corresponding to the FM station with frequency "94.1 on your dial" where the frequency is in MHz.
a) 3137 nm b) 6.38 m c) 3.19 m d) 2112 nm e) 3190 nm

6. Calculate the wavelength of ultraviolet radiation of frequency 2.00×10^{15} Hz.
a) 99.4 nm b) 0.300 cm c) 150 nm d) 300 nm e) 200 nm

7. Calculate the frequency of green light of wavelength 530 nm.
a) 1.77×10^{-15} Hz b) 5.30×10^{14} Hz c) 5.66×10^{16} Hz d) 5.66×10^{14} Hz e) 1.77×10^{15} Hz

8. Calculate the frequency of red light of wavelength 700 nm.
a) 4.29×10^{14} Hz b) 2.33×10^{15} Hz c) 4.29×10^{16} Hz d) 2.33×10^{-15} Hz e) 7.00×10^{14} Hz

9. Calculate the frequency of microwaves of wavelength 10 mm.
a) 3.00×10^{16} Hz b) 1.00×10^{16} Hz c) 3.00×10^{18} Hz d) 3.33×10^{17} Hz e) 3.33×10^{-17} Hz

10. Calculate the frequency of radio waves of wavelength 4.50 m.
a) 667 MHz b) 6.67×10^{6} MHz c) 1.50×10^{8} Hz d) 1.50×10^{-8} Hz e) 6.67×10^{7} Hz

11. Calculate the frequency of ultraviolet radiation of wavelength 300 nm.
a) 3.00×10^{15} Hz
b) 1.00×10^{-17} Hz
c) 1.00×10^{-15} Hz
d) 1.00×10^{17} Hz
e) 1.00×10^{15} Hz

12. Calculate the energy per photon of infrared radiation of frequency 3.00×10^{14} Hz. Planck's constant is 6.63×10^{-34} J · s.
a) 1.99×10^{-19} J b) 2.21×10^{-48} J c) 120 kJ d) 5.97×10^{-11} J e) 4.52×10^{47} J

13. Calculate the energy per photon of radio waves of frequency 92.1 MHz. Planck's constant is 6.63×10^{-34} J · s.
a) 0.0368 J b) 6.11×10^{-26} J c) 7.20×10^{-42} J d) 1.83×10^{-17} J e) 1.39×10^{41} J

14. Calculate the energy per photon of radio waves of frequency 94.1 MHz. Planck's constant is 6.63×10^{-34} J · s.
a) 0.0376 J b) 7.05×10^{-42} J c) 6.24×10^{-26} J d) 1.87×10^{-17} J e) 1.42×10^{41} J

15. Calculate the energy per photon of red light of frequency 4.30×10^{14} Hz. Planck's constant is 6.63×10^{-34} J · s.
a) 172 kJ b) 8.55×10^{-11} J c) 2.85×10^{-19} J d) 1.54×10^{-48} J e) 6.49×10^{47} J

16. Calculate the energy per mole of photons of infrared radiation of frequency 3.00×10^{14} Hz. Planck's constant is 6.63×10^{-34} J · s.
a) 4.52×10^{47} J b) 2.21×10^{-48} J c) 1.99×10^{-19} J d) 120 kJ e) 5.97×10^{-11} J

17. Calculate the energy per mole of photons of radio waves of frequency 92.1 MHz. Planck's constant is 6.63×10^{-34} J · s.
a) 1.39×10^{41} J b) 0.0368 J c) 7.20×10^{-42} J d) 6.11×10^{-26} J e) 1.83×10^{-17} J

18. Calculate the energy per mole of photons of radio waves of frequency 94.1 MHz. Planck's constant is 6.63×10^{-34} J · s.
a) 1.42×10^{41} J b) 0.0376 J c) 7.05×10^{-42} J d) 6.24×10^{-26} J e) 1.87×10^{-17} J

19. Calculate the energy per mole of photons of red light of frequency 4.30×10^{14} Hz. Planck's constant is 6.63×10^{-34} J · s.
a) 172 kJ b) 8.55×10^{-11} J c) 6.49×10^{47} J d) 2.85×10^{-19} J e) 1.54×10^{-48} J

20. Which of the following has the greatest frequency?
a) red light b) ultraviolet radiation c) blue light d) microwaves e) radiowaves

21. Which of the following has the lowest frequency?
a) blue light b) infrared radiation c) green light d) ultraviolet radiation e) x-rays

22. Which of the following has the longest wavelength?
a) ultraviolet radiation b) red light c) x-rays d) γ-rays e) violet light

23. Which of the following has the shortest wavelength?
a) infrared radiation b) green light c) ultraviolet rays d) radiowaves e) microwaves

24. The energy of a photon is equal to a constant times the frequency of the photon. The constant is called
a) the Balmer constant.
b) the Rydberg constant.
c) the speed of light.
d) the photoelectric constant.
e) Planck's constant.

25. The absorbance of a dye at 420 nm in a 1 cm cell is 0.827. If the concentration of the dye is 2.35×10^{-4} M, calculate the molar absorption coefficient of the dye. Describe the color and intensity of the solution of the dye.

26. A ruby laser produces radiation of 694 nm. If the output is 0.300 J/pulse, how many photons are produced per pulse?

27. A CO_2 laser has an output of 8.00×10^3 watts. How many photons are produced in 1.00 s?

28. Calculate the de Broglie wavelength of a 358 g soccer ball travelling at 45.0 miles per hour.

29. Which of the following is true?
a) An *s*-orbital has two lobes on opposite sides of the nucleus.
b) An electron in a *p*-orbital has zero probability of being found at the nucleus.
c) An electron in an *s*-orbital has zero probability of being found at the nucleus.
d) A *p*-orbital has a spherical boundary surface.
e) An *s*-orbital becomes more dense as the distance from the nucleus increases.

30. Calculate the frequency of a photon emitted by a hydrogen atom when an electron makes a transition from the fourth to the second principal quantum level.
[$\mathfrak{R} = 3.29 \times 10^{15}$ Hz; $h = 6.63 \times 10^{-34}$ J · s]
a) 2.99×10^3 Hz b) 1.83×10^8 Hz c) 6.17×10^{14} Hz d) 2.18×10^{-18} Hz e) 1.31×10^3 Hz

31. Calculate the frequency of a photon emitted by a hydrogen atom when an electron makes a transition from the state with $n = 6$ to $n = 2$.
[$\mathfrak{R} = 3.29 \times 10^{15}$ Hz; $h = 6.63 \times 10^{-34}$ J · s]
a) 2.99×10^3 Hz b) 1.31×10^3 Hz c) 2.18×10^{-18} Hz d) 1.83×10^8 Hz e) 7.31×10^{14} Hz

32. How many orbitals are there in the shell with $n = 4$?
a) 16 b) 9 c) 32 d) 18 e) 4

33. An electron in a hydrogen atom has the quantum numbers $n = 4$, $l = 2$, $m_\ell = 0$. In what type of orbital is the electron located?
a) 3*p* b) 4*d* c) 4*s* d) 4*d* e) 3*d*

34. An electron in a hydrogen atom has the quantum numbers $n = 3$, $l = 2$, $m_\ell = 0$. In what type of orbital is the electron located?
a) 3*p* b) 2*s* c) 3*s* d) 2*p* e) 3*d*

35. How many orbitals are there with $l = 3$?
a) 7 b) 5 c) 3 d) 2 e) 9

36. An electron in an atom has a magnetic quantum number of –3. What is the lowest possible value for the principal quantum number of this electron?
 a) 4 b) 3 c) 5 d) 1 e) –3

37. Which of the following is a possible set of quantum numbers for a 4*d*-electron?
 a) $n = 4, l = 1, m_\ell = -1, m_s = -\frac{1}{2}$
 b) $n = 4, l = 2, m_\ell = -2, m_s = -\frac{1}{2}$
 c) $n = 4, l = 3, m_\ell = 2, m_s = \frac{1}{2}$
 d) $n = 4, l = 1, m_\ell = 0, m_s = \frac{1}{2}$
 e) $n = 4, l = 4, m_\ell = -2, m_s = \frac{1}{2}$

38. Which of the following is a possible set of quantum numbers for a 4*s*-electron?
 a) $n = 4, l = 1, m_\ell = 0, m_s = \frac{1}{2}$
 b) $n = 4, l = 1, m_\ell = 1, m_s = \frac{1}{2}$
 c) $n = 4, l = 1, m_\ell = -1, m_s = -\frac{1}{2}$
 d) $n = 4, l = 0, m_\ell = 0, m_s = -\frac{1}{2}$
 e) $n = 4, l = 2, m_\ell = 2, m_s = \frac{1}{2}$

39. Which of the following is a possible set of quantum numbers for a 3*d*-electron?
 a) $n = 3, l = 2, m_\ell = 2, m_s = -\frac{1}{2}$
 b) $n = 3, l = 1, m_\ell = 1, m_s = \frac{1}{2}$
 c) $n = 3, l = 1, m_\ell = 0, m_s = \frac{1}{2}$
 d) $n = 3, l = 1, m_\ell = -1, m_s = -\frac{1}{2}$
 e) $n = 3, l = 0, m_\ell = 0, m_s = \frac{1}{2}$

40. Which of the following represents a possible set of quantum numbers for an electron in an atom?
 a) $n = 3, l = 3, m_\ell = -1, m_s = -\frac{1}{2}$
 b) $n = 3, l = 2, m_\ell = 0, m_s = 0$
 c) $n = 3, l = 0, m_\ell = 1, m_s = -\frac{1}{2}$
 d) $n = 3, l = 0, m_\ell = 1, m_s = \frac{1}{2}$
 e) $n = 3, l = 0, m_\ell = 0, m_s = -\frac{1}{2}$

41. Which subshell can hold the greatest number of electrons?
 a) 5*p* b) 4*p* c) 3*d* d) 3*p* e) 4*s*

42. For a 5*d* subshell, what is the most positive value of m_ℓ?
 a) +2 b) +1 c) +5 d) +4 e) 0

43. The atomic orbital for an electron in an atom with the quantum numbers $n = 4, l = 1, m_\ell = 0, m_s = -\frac{1}{2}$ is
 a) 4*p* b) 4*s* c) 4*f* d) 4*d* e) 3*d*

44. If the azimuthal quantum number is 2, the type and number of orbitals is
 a) 3*d*, 3 b) 3*p*, 3 c) 3*d*, 5 d) 3*s*, 1 e) 3*f*, 7

45. Which of the following is true?
 a) A 4*s*-electron in calcium is less strongly shielded than a 3*d*-electron.
 b) A 4*s*-electron in calcium is less strongly attracted by the nucleus than a 3*d*-electron.
 c) A *p*-electron in calcium penetrates more than an *s*-electron.
 d) The effective nuclear charge is the same for a 3*s*- and a 3*p*-electron in calcium.
 e) A *p*-electron in calcium can be found closer to the nucleus than an *s*-electron.

46. What is the ground-state electron configuration of a sulfur atom?
a) [Ne]$3s^23p^4$ b) [Ne]$3s^23p^5$ c) [Ne]$3s^23p^3$ d) [Ne]$3s^23p^2$ e) [Ne]$3s^23p^1$

47. What is the ground-state electron configuration of a silicon atom?
a) [Ne]$3s^23p^5$ b) [Ne]$3s^23p^4$ c) [Ne]$3s^23p^3$ d) [Ne]$3s^2$ e) [Ne]$3s^23p^2$

48. What is the ground-state electron configuration of a copper atom?
a) [Ar]$3d^{10}4s^1$ b) [Ar]$3d^84s^2$ c) [Ar]$3d^94s^1$ d) [Ar]$3d^94s^2$ e) [Ar]$3d^{10}4s^2$

49. What is the ground-state electron configuration of a titanium atom?
a) [Ar]$3d^24s^2$ b) [Ar]$3d^44s^2$ c) [Ar]$3d^4$ d) [Ar]$3d^34s^2$ e) [Ar]$3d^14s^2$

50. The highest-energy electron in an atom has quantum numbers $n = 3$, $l = 0$, $m_\ell = 0$, $m_s = ½$. This atom could be
a) magnesium b) potassium c) calcium d) bromine e) arsenic

51. Which atom contains an electron with quantum numbers $n = 3$, $l = 0$, $m_\ell = 0$, $m_s = ½$?
a) magnesium b) neon c) nitrogen d) lithium e) boron

52. How many unpaired electrons are there in the ground state of the ion Cr^{2+}?
a) 4 b) 1 c) 5 d) 3 e) 0

53. What is the electron configuration of the Fe^{3+} ion?
a) [Ar]$3d^34s^2$ b) [Ar]$3d^5$ c) [Ar]$3d^44s^1$ d) [Ar]$3d^4$ e) [Ar]$3d^54s^1$

54. What is the electron configuration of the P^{3-} ion?
a) [Ar] b) [Ar]$4s^1$ c) [Ar]$4s^2$ d) [Ne]$3s^23p^3$ e) [Ne]$3s^23p^5$

55. What is the electron configuration of the Cu^{3+} ion?
a) [Ar]$3d^74s^2$ b) [Ar]$3d^94s^1$ c) [Ar]$3d^9$ d) [Ar]$3d^8$ e) [Ar]$3d^74s^1$

56. A certain ion has the ground-state electron configuration [Ar]$3d^{10}$. The ion is likely
a) Cu^{2+} b) Ni^{2+} c) Cu^+ d) Ga^+ e) Cl^-

57. What main group element forms a positive ion having the ground-state electron configuration [Ar]?
a) calcium b) sulfur c) argon d) scandium e) barium

58. What is the valence-electron configuration of the atoms of the group that contains barium?
a) ns^2 b) np^2 c) ns^1 d) np^3 e) ns^2np^1

59. Which of the following atoms has the greatest number of valence-shell electrons?
a) selenium b) antimony c) tin d) gallium e) cesium

60. What is the valence-electron configuration of the alkaline earth elements?
a) np^2 b) ns^2 c) np^1 d) ns^1 e) ns^2np^1

61. Which of the following atoms has the largest radius?
a) aluminum b) carbon c) chlorine d) boron e) oxygen

62. Which of the following species has the smallest radius?
a) K^+ b) K c) Cl^- d) S^{2-} e) Rb

63. Which of the following species has the largest radius?
a) Cl^- b) K^+ c) Al d) Cl e) Na

64. Arrange K, Mg, and S in order of increasing radii.
a) S < Mg < K b) Mg < S < K c) Mg < K < S d) S < K < Mg e) K < Mg < S

65. Which of the following has the highest first ionization energy?
a) F b) N c) C d) Be e) Cl

66. Which of the following represents the first electron affinity of oxygen?
a) $O(g) + 2e^-(g) \rightarrow O^{2-}(g)$
b) $O(g) + e^-(g) \rightarrow O^-(g)$
c) $O(g) \rightarrow O^+(g) + e^-(g)$
d) $O(s) + e^-(g) \rightarrow O^-(g)$
e) $O(s) + 2e^- \rightarrow O^{2-}(g)$

67. For the elements Mg, Al, P, and S, the first ionization energy increases in the order
a) Mg < Al < P < S
b) Al < Mg < P < S
c) Al < Mg < S < P
d) Al < S < Mg < P
e) Mg < Al < S < P

68. Which of the following has the highest first ionization energy?
a) chlorine b) sulfur c) silicon d) iodine e) potassium

69. Which of the following has the lowest first ionization energy?
a) boron b) beryllium c) nitrogen d) carbon e) oxygen

70. Which of the following has the most metallic character?
a) Rb b) Na c) Sr d) Mg e) Be

71. Which of the following has the lowest second ionization energy?
a) Mg b) Ne c) Na d) Li e) Ar

72. If lithium reacts directly with nitrogen to form its nitride, which of the following is also likely to react directly with nitrogen to form a nitride?
a) magnesium b) beryllium c) calcium d) aluminum e) boron

73. Which of the following processes, with all species in the gas phase, would require the most energy?
a) $Rb \rightarrow Rb^+ + e^-$
b) $Ca \rightarrow Ca^+ + e^-$
c) $Ca^{2+} \rightarrow Ca^{3+} + e^-$
d) $Ca \rightarrow Ca^{2+} + 2e^-$
e) $Ca^+ \rightarrow Ca^{2+} + e^-$

74. Which of the following ions would least likely form?
a) Cr^{2+} b) Ca^{2+} c) Cu^{2+} d) K^{2+} e) Ti^{2+}

75. Magnesium has similar properties to
a) lithium. b) aluminum. c) sodium. d) boron. e) silicon.

76. Which of the following has the largest electron affinity?
a) Cl b) S c) S^- d) P e) Cl^-

77. Which of the following elements does not belong with the others?
a) Li b) Si c) As d) Ge e) Sb

78. Which of the following is an *s*-block element?
a) barium b) iron c) sulfur d) silicon e) bromine

Chapter 8: Inside Materials: Chemical Bonds
Form A

1. How many dots should one place around the symbol for gallium, Ga, when writing the Lewis symbol for the atom?

a) 3 b) 2 c) 13 d) 8 e) 1

2. How many dots should one place around the symbol for phosphorus, P, when writing the Lewis symbol for the atom?

a) 5 b) 15 c) 8 d) 4 e) 3

3. How many valence electrons are there in a sulfur atom?

a) 6 b) 16 c) 4 d) 8 e) 5

4. The lattice enthalpy of potassium iodide is the energy change for the reaction

a) $K(g) + I(g) \rightarrow KI(s)$
b) $KI(s) \rightarrow K^+(g) + I^-(g)$
c) $KI(s) \rightarrow K(g) + I(g)$
d) $KI(s) \rightarrow K(g) + \frac{1}{2}I_2(g)$
e) $K(s) + \frac{1}{2}I_2(s) \rightarrow KI(s)$

5. The lattice enthalpy of lithium bromide is the energy change for the reaction

a) $Li(g) + Br(g) \rightarrow LiBr(s)$
b) $LiBr(s) \rightarrow Li^+(g) + Br^-(g)$
c) $Li(s) + \frac{1}{2}Br_2(l) \rightarrow LiBr(s)$
d) $LiBr(s) \rightarrow Li(g) + Br(g)$
e) $LiBr(s) \rightarrow Li(g) + \frac{1}{2}Br_2(g)$

6. The lattice enthalpy of barium chloride is the energy change for the reaction

a) $BaCl_2(s) \rightarrow Ba^{2+}(g) + 2Cl^-(g)$
b) $Ba(s) + Cl_2(g) \rightarrow BaCl_2(s)$
c) $BaCl_2(s) \rightarrow Ba(g) + 2Cl(g)$
d) $BaCl_2(s) \rightarrow Ba(g) + Cl_2(g)$
e) $Ba(g) + 2Cl(g) \rightarrow BaCl_2(g)$

7. The lattice enthalpy of calcium bromide is the energy change for the reaction

a) $Ca(g) + 2Br(g) \rightarrow CaBr_2(g)$
b) $CaBr_2(s) \rightarrow Ca(g) + 2Br(g)$
c) $Ca(s) + Br_2(l) \rightarrow CaBr_2(s)$
d) $CaBr_2(s) \rightarrow Ca(g) + Br_2(g)$
e) $CaBr_2(s) \rightarrow Ca^{2+}(g) + 2Br^-(g)$

8. The lattice enthalpy of calcium oxide is the energy change for the reaction

a) $CaO(s) \rightarrow Ca^{2+}(g) + O^{2-}(g)$
b) $CaO(s) \rightarrow Ca(g) + \frac{1}{2}O_2(g)$
c) $CaO(s) \rightarrow Ca(g) + O(g)$
d) $Ca(s) + \frac{1}{2}O_2(g) \rightarrow CaO(s)$
e) $Ca(g) + O(g) \rightarrow CaO(g)$

9. The standard enthalpy of formation of KCl(s) is $-437\ kJ \cdot mol^{-1}$. In a Born-Haber cycle for the formation of KCl(s), which enthalpy change(s) are exothermic?
a) the lattice enthalpy and the electron affinity of chlorine
b) the electron affinity of chlorine and the reverse of the lattice enthalpy
c) the formation of Cl(g) from $Cl_2(g)$
d) the sum of the sublimation enthalpy of K(s) and the first ionization energy of K(g)
e) the lattice enthalpy

10. Which of the following has the largest lattice enthalpy?
a) $MgCl_2(s)$ b) KCl(s) c) LiCl(s) d) NaCl(s) e) AgCl(s)

11. Calculate the lattice enthalpy of silver chloride from the following data.
enthalpy of formation of Ag(g): $284\ kJ \cdot mol^{-1}$
first ionization energy of Ag(g): $731\ kJ \cdot mol^{-1}$
enthalpy of formation of Cl(g): $122\ kJ \cdot mol^{-1}$
electron affinity of Cl(g): $-349\ kJ \cdot mol^{-1}$
enthalpy of formation of AgCl(s): $-127\ kJ \cdot mol^{-1}$
a) $1037\ kJ \cdot mol^{-1}$ b) $661\ kJ \cdot mol^{-1}$ c) $915\ kJ \cdot mol^{-1}$ d) $1613\ kJ \cdot mol^{-1}$ e) $1359\ kJ \cdot mol^{-1}$

12. Calculate the lattice enthalpy of sodium chloride from the following data.
enthalpy of formation of Na(g): $107\ kJ \cdot mol^{-1}$
first ionization energy of Na(g): $494\ kJ \cdot mol^{-1}$
enthalpy of formation of Cl(g): $122\ kJ \cdot mol^{-1}$
electron affinity of Cl(g): $-349\ kJ \cdot mol^{-1}$
enthalpy of formation of NaCl(s): $-411\ kJ \cdot mol^{-1}$
a) $436\ kJ \cdot mol^{-1}$ b) $907\ kJ \cdot mol^{-1}$ c) $1483\ kJ \cdot mol^{-1}$ d) $785\ kJ \cdot mol^{-1}$ e) $661\ kJ \cdot mol^{-1}$

13. Calculate the lattice enthalpy of silver iodide from the following data.
enthalpy of formation of Ag(g): $284\ kJ \cdot mol^{-1}$
first ionization energy of Ag(g): $731\ kJ \cdot mol^{-1}$
enthalpy of formation of I(g): $107\ kJ \cdot mol^{-1}$
electron affinity of I(g): $-295\ kJ \cdot mol^{-1}$
enthalpy of formation of AgI(s): $-62\ kJ \cdot mol^{-1}$
a) $996\ kJ \cdot mol^{-1}$ b) $889\ kJ \cdot mol^{-1}$ c) $1479\ kJ \cdot mol^{-1}$ d) $1355\ kJ \cdot mol^{-1}$ e) $765\ kJ \cdot mol^{-1}$

14. Calculate the lattice enthalpy of potassium fluoride from the following data.
enthalpy of formation of K(g): $89\ kJ \cdot mol^{-1}$
first ionization energy of K(g): $418\ kJ \cdot mol^{-1}$
enthalpy of formation of F(g): $79\ kJ \cdot mol^{-1}$
electron affinity of F(g): $-328\ kJ \cdot mol^{-1}$
enthalpy of formation of KF(s): $-567\ kJ \cdot mol^{-1}$
a) $1481\ kJ \cdot mol^{-1}$ b) $825\ kJ \cdot mol^{-1}$ c) $904\ kJ \cdot mol^{-1}$ d) $347\ kJ \cdot mol^{-1}$ e) $497\ kJ \cdot mol^{-1}$

15. Calculate the lattice enthalpy of magnesium chloride from the following data.
    enthalpy of formation of Mg(g): 148 $kJ \cdot mol^{-1}$
    first ionization energy of Mg(g): 736 $kJ \cdot mol^{-1}$
    second ionization energy of Mg(g): 1450 $kJ \cdot mol^{-1}$
    enthalpy of formation of Cl(g): 122 $kJ \cdot mol^{-1}$
    electron affinity of Cl(g): –349 $kJ \cdot mol^{-1}$
    enthalpy of formation of $MgCl_2$(s): –641 $kJ \cdot mol^{-1}$

a) 2521 $kJ \cdot mol^{-1}$ b) 1239 $kJ \cdot mol^{-1}$ c) 2748 $kJ \cdot mol^{-1}$ d) 3917 $kJ \cdot mol^{-1}$ e) 2635 $kJ \cdot mol^{-1}$

16. Calculate the lattice enthalpy of calcium oxide from the following data.
    enthalpy of formation of Ca(g): 178 $kJ \cdot mol^{-1}$
    first ionization energy of Ca(g): 590 $kJ \cdot mol^{-1}$
    second ionization energy of Ca(g): 1150 $kJ \cdot mol^{-1}$
    enthalpy of formation of O(g): 249 $kJ \cdot mol^{-1}$
    first electron affinity of O(g): –141 $kJ \cdot mol^{-1}$
    second electron affinity of O(g): 844 $kJ \cdot mol^{-1}$
    enthalpy of formation of CaO(s): –635 $kJ \cdot mol^{-1}$

a) 1391 $kJ \cdot mol^{-1}$ b) 3505 $kJ \cdot mol^{-1}$ c) 2235 $kJ \cdot mol^{-1}$ d) 3754 $kJ \cdot mol^{-1}$ e) 1817 $kJ \cdot mol^{-1}$

17. Calculate the electron affinity of chlorine from the following data.
    enthalpy of formation of Mg(g): 148 $kJ \cdot mol^{-1}$
    first ionization energy of Mg(g): 736 $kJ \cdot mol^{-1}$
    second ionization energy of Mg(g): 1450 $kJ \cdot mol^{-1}$
    enthalpy of formation of Cl(g): 122 $kJ \cdot mol^{-1}$
    lattice enthalpy of $MgCl_2$(s): 2521 $kJ \cdot mol^{-1}$
    enthalpy of formation of $MgCl_2$(s): –641 $kJ \cdot mol^{-1}$

18. Given the following data, calculate the second electron affinity of oxygen.
    enthalpy of formation of Ca(g): 178 $kJ \cdot mol^{-1}$
    first ionization energy of Ca(g): 590 $kJ \cdot mol^{-1}$
    second ionization energy of Ca(g): 1150 $kJ \cdot mol^{-1}$
    enthalpy of formation of O(g): 249 $kJ \cdot mol^{-1}$
    first electron affinity of O(g): –141 $kJ \cdot mol^{-1}$
    lattice enthalpy of CaO(s): 3505 $kJ \cdot mol^{-1}$
    enthalpy of formation of CaO(s): –635 $kJ \cdot mol^{-1}$

19. Draw a Born-Haber cycle for the formation MgO(s) from Mg(s) and O_2(g). Using symbols for the enthalpies in the cycle, determine the equation for the lattice enthalpy.

20. How many lone pairs of electrons does the bromine atom possess in the Lewis structure of HBr?
a) 3 b) 2 c) 1 d) 6 e) 4

21. How many lone pairs of electrons does the oxygen atom possess in the Lewis structure of H_2O?
a) 2 b) 4 c) 3 d) 1 e) 0

22. How many valence electrons are there in HCN?
a) 8 b) 2 c) 10 d) 12 e) 14

23. How many valence electrons are there in CH_3COOH?
a) 24 b) 20 c) 14 d) 12 e) 26

24. How many valence electrons are there in the sulfite ion, SO_3^{2-}?
a) 26 b) 24 c) 22 d) 18 e) 20

25. How many valence electrons are there in the perchlorate ion?
a) 36 b) 34 c) 30 d) 40 e) 32

26. How many valence electrons are there in the cyanide ion?
a) 14 b) 10 c) 8 d) 12 e) 16

27. How many valence electrons are there in the ammonium ion?
a) 8 b) 10 c) 6 d) 12 e) 14

28. How many lone pairs of electrons are there in the Lewis structure of acetic acid, CH_3COOH?
a) 4 b) 5 c) 6 d) 7 e) 3

29. The Lewis structure of C_2^{2-}
a) has a double bond between the two carbon atoms.
b) does not obey the octet rule.
c) contains 4 lone pairs of electrons.
d) contains no lone pairs of electrons.
e) has a triple bond between the two carbon atoms.

30. How many lone pairs are there in the Lewis structure of urea, $(NH_2)_2CO$?
a) 4 b) 2 c) 5 d) 6 e) 0

31. Which of the following species can have resonance structures?
a) HCO_2^- b) CH_3NH_2 c) BF_4^- d) CH_3COOH e) H_3O^+

32. Which of the following species can have resonance structures?
a) NO_2^- b) H_3O^+ c) CH_3NH_2 d) CH_3COOH e) BF_4^-

33. Which of the following species can have resonance structures?
a) H_3O^+ b) CH_3NH_2 c) CH_3COOH d) O_3 e) BF_4^-

34. The Lewis structure of the nitrite ion shows around the central nitrogen atom
a) 3 single-bonded oxygens and 1 lone pair of electrons.
b) 2 single-bonded oxygens and 1 double-bonded oxygen.
c) 1 single-bonded oxygen, 1 double-bonded oxygen, and 1 lone pair of electrons.
d) 2 single-bonded oxygens and 2 lone pairs of electrons.
e) 2 double-bonded oxygens.

35. Resonance forms can be written for
a) NH_3 b) H_2O c) SO_2 d) H_2S e) OF_2

36. Resonance forms can be written for all of the following except
a) HCO_2^- b) SO_2 c) O_3 d) CO_3^{2-} e) OF_2

37. How many resonance forms can be written for the NCS^- ion?
a) 3 b) 0 c) 1 d) 2 e) 4

38. The formal charge on the chlorine atom in the Lewis structure of the perchlorate ion is
a) +3 b) +7 c) −1 d) 0 e) +1

39. How many lone pairs of electrons does the nitrogen atom possess in the Lewis structure of HCN?
a) 1 b) 4 c) 2 d) 0 e) 3

40. For the Lewis structure below, the formal charges on N, C, and S, respectively, are

$[\mathrm{N{=}C{=}S}]^-$

a) −1, 0, +1 b) +2, −1, −2 c) 0, 0, −1 d) −1, 0, 0 e) −1, −1, +1

41. For the Lewis structure below, the formal charges on C, S, and N, respectively, are

$[\mathrm{C{=}S{=}N}]^-$

a) −1, +1, −1 b) −2, +2, −1 c) −1, +2, −2 d) −2, +1, 0 e) 0, 0, −1

42. For the Lewis structure below, the formal charges on C, N, and O, respectively, are

$[\mathrm{C{=}N{=}O}]^-$

a) −1, 0, 0 b) −2, 0, +1 c) 0, +1. −2 d) −2, +1, 0 e) −1, −1, +1

43. For the Lewis structure below, considering the atoms from left to right, the formal charges on N, N, and O, respectively, are

$[\mathrm{N{=}N{=}O}]$

a) 0, +1, −1 b) −1, +1, 0 c) −1, 0, +1 d) −2, +1, +1 e) −2, 0, +2

44. For the Lewis structure below, considering the atoms from left to right, the formal charges on O, N, and O, respectively, are

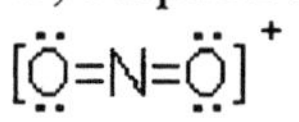

a) −1, +2, 0 b) +1, −1, +1 c) 0, +2, −1 d) 0, +1, 0 e) −1, +3, −1

45. For the Lewis structure below, the formal charges on Cl, N, and O, respectively, are

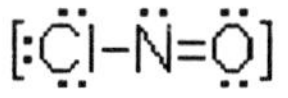

a) 0, 0, 0 b) 0, +1, −1 c) −1, 0, +1 d) −1, +1, 0 e) 0, −1, +1

46. For the Lewis structure below, the formal charges on the nitrogen atoms from left to right are

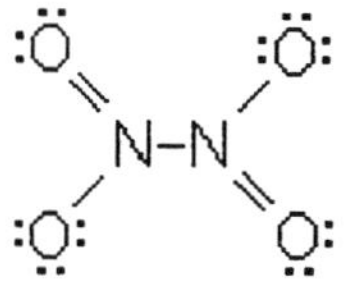

a) 0 and 0 b) +2 and 0 c) +1 and +1 d) −1 and −1 e) +1 and −1

47. A Lewis structure that obeys the octet rule can be drawn for
a) XeF_2 b) XeO_2 c) XeF_3^+ d) XeF_4 e) $XeOF_2$

48. The formula of dimethylsulfoxide is $(CH_3)_2SO$. Determine the arrangement of the atoms in dimethylsulfoxide and write the Lewis structure. Also, include all possible resonance forms and assign formal charges to all atoms.

49. All of the following have Lewis structures that obey the octet rule except
a) NO b) NO_2^+ c) N_2O_4 d) N_2O e) N_3^-

50. Which of the following is a radical?
a) O_2^{2-} b) SO_2^- c) Cl–O–O–Cl d) NO^+ e) $S_2O_4^{2-}$

51. The hydroxide ion and the hydroxyl radical
a) have 8 and 7 valence electrons, respectively.
b) have 7 and 8 valence electrons, respectively.
c) are two different names for the same species.
d) have a −1 and a +1 charge, respectively.
e) contain ^1H and ^2H, respectively.

52. All of the following have Lewis structures that obey the octet rule except
a) ClO_2 b) HOCl c) ClO_2^- d) BrCl e) ClO_4^-

53. Which of the following is a free radical?
a) NO b) NO_2^- c) CO d) CO_2 e) CN^-

54. Which of the following is a free radical?
a) ICl_2^+ b) OCl^- c) O_2^{2-} d) ClO_2 e) I_3^-

55. How many electron pairs (both bonding and lone pairs) are on the sulfur atom in SF_4?

a) 5 b) 4 c) 6 d) 7 e) 8

56. The formula of diazomethane is CH_2N_2. Determine the arrangement of the atoms in diazomethane and write the Lewis structure. Also, include all possible resonance forms and assign formal charges to all atoms.

57. The number of lone pairs around Xe in XeO_2 is

a) 2 b) 0 c) 4 d) 1 e) 3

58. The number of lone pairs around the central iodine atom in I_3^- is

a) 3 b) 0 c) 2 d) 5 e) 1

59. The number of lone pairs around the central sulfur atom in S_3^{2-} is

a) 2 b) 4 c) 0 d) 5 e) 3

60. Which of the following would have a Lewis structure most like that of ICl_2^+?

a) CO_2 b) I_3^- c) XeF_2 d) O_3 e) XeO_2

61. Which of the following would have a Lewis structure most like that of OF_2?

a) O_3 b) IF_2^- c) I_3^- d) CO_2 e) ClO_2^-

62. Which of the following would have a Lewis structure most like that of XeO_3?

a) CO_3^{2-} b) $XeOF_2$ c) SO_3^{2-} d) NO_3^- e) IF_4^+

63. Which of the following would have a Lewis structure most like that of SeF_3^+?

a) CO_3^{2-} b) IO_3^- c) IF_4^+ d) NO_3^- e) $XeOF_2$

64. Which of the following would have a Lewis structure most like that of $AlCl_4^-$?

a) ICl_4^- b) PO_4^{3-} c) IF_4^+ d) XeF_4 e) SF_4

65. Which of the following would have a Lewis structure most like that of SO_4^{2-}?

a) XeF_4 b) XeO_4 c) ICl_4^- d) IF_4^+ e) SF_4

66. All of the following have similar Lewis structures except

a) SO_4^{2-} b) $AlCl_4^-$ c) IF_4^+ d) XeO_4 e) CCl_4

67. All of the following have similar Lewis structures except

a) XeF_2 b) ClO_2^- c) S_3^{2-} d) XeO_2 e) SCl_2

68. How many lone pairs of electrons are there in the Lewis structure of XeF_4?
a) 14 b) 12 c) 2 d) 10 e) 4

69. In the Lewis structure of SeF_3^+, the central atom has
a) 2 lone pairs and 3 bonding pairs of electrons.
b) 1 lone pair of electrons.
c) 3 lone pairs of electrons.
d) 2 lone pairs of electrons.
e) 1 lone pair and 6 bonding pairs of electrons.

70. In the Lewis structure of $ClNO_2$, the central nitrogen atom has
a) 2 lone pairs and 3 bonding pairs of electrons.
b) 1 lone pair and 4 bonding pairs of electrons.
c) 5 bonding pairs of electrons.
d) 1 lone pair and 3 bonding pairs of electrons.
e) 4 bonding pairs of electrons.

71. How many lone pairs of electrons are in the Lewis structure of Al_2Cl_6?
a) 16 b) 18 c) 8 d) 12 e) 14

72. Which of the following are Lewis acids?
1: CN^-
2: Al^{3+}
3: Cl^-
4: PCl_3
5: BF_3
a) 2, 4, and 5 b) 2 and 5 c) 4 and 5 d) 1, 3, and 5 e) 1 and 3

73. Which of the following has bonds with the most ionic character?
a) $CaCl_2$ b) MgI_2 c) $CaBr_2$ d) $MgBr_2$ e) $MgCl_2$

74. Elements that have a high electronegativity have
a) a high ionization energy and a high electron affinity.
b) a low ionization energy and a high electron affinity.
c) a low ionization energy and a low electron affinity.
d) a high ionization energy and a low electron affinity.
e) an ionization energy that is approximately equal to its electron affinity.

75. Select the elements that have the lowest and the highest electronegativities, respectively.
S, Li, Cl, Na, Al
a) Li and Cl b) Na and Al c) Na and Cl d) Al and S e) Li and S

76. Which of the following would have both the largest ionization energy and largest electron affinity?
a) Cl b) P c) As d) Ca e) Na

77. For the following elements, predict the order of electronegativity, from lowest to highest.
    O, As, Rb, Se, Ca
    a) Rb < Ca < Se < As < O
    b) Rb < Ca < As < Se < O
    c) Ca < Rb < Se < As < O
    d) Rb < Ca < Se < O < As
    e) Ca < Rb < As < Se < O

78. Which of the following has bonds with the most covalent character?
    a) AgI b) AgCl c) AgF d) NaF e) AgBr

79. Which of the following pairs of elements would be most likely to form an ionic compound?
    a) As and Br b) K and Br c) P and Br d) C and S e) C and As

80. Which of the following compounds is ionic?
    a) BaO b) CO c) BF_3 d) SO_2 e) Cl_2O

Chapter 8: Inside Materials: Chemical Bonds
Form B

1. How many dots should one place around the symbol for aluminum, Al, when writing the Lewis symbol for the atom?
a) 3 b) 2 c) 13 d) 8 e) 1

2. How many dots should one place around the symbol for barium, Ba, when writing the Lewis symbol for the atom?
a) 2 b) 8 c) 3 d) 6 e) 4

3. How many valence electrons are there in a silicon atom?
a) 4 b) 14 c) 8 d) 2 e) 5

4. The lattice enthalpy of sodium bromide is the energy change for the reaction
a) $NaBr(s) \rightarrow Na(g) + Br(g)$
b) $Na(g) + Br(g) \rightarrow NaBr(s)$
c) $Na(s) + \frac{1}{2}Br_2(l) \rightarrow NaBr(s)$
d) $NaBr(s) \rightarrow Na(g) + \frac{1}{2}Br_2(g)$
e) $NaBr(s) \rightarrow Na^+(g) + Br^-(g)$

5. The lattice enthalpy of lithium iodide is the energy change for the reaction
a) $LiI(s) \rightarrow Li(g) + I(g)$
b) $Li(s) + \frac{1}{2}I_2(s) \rightarrow LiI(s)$
c) $LiI(s) \rightarrow Li^+(g) + I^-(g)$
d) $LiI(s) \rightarrow Li(g) + \frac{1}{2}I_2(g)$
e) $Li(g) + I(g) \rightarrow LiI(s)$

6. The lattice enthalpy of calcium chloride is the energy change for the reaction
a) $Ca(g) + 2Cl(g) \rightarrow CaCl_2(g)$
b) $CaCl_2(s) \rightarrow Ca^{2+}(g) + 2Cl^-(g)$
c) $CaCl_2(s) \rightarrow Ca(g) + Cl_2(g)$
d) $Ca(s) + Cl_2(g) \rightarrow CaCl_2(s)$
e) $CaCl_2(s) \rightarrow Ca(g) + 2Cl(g)$

7. The lattice enthalpy of magnesium bromide is the energy change for the reaction
a) $MgBr_2(s) \rightarrow Mg^{2+}(g) + 2Br^-(g)$
b) $Mg(s) + Br_2(l) \rightarrow MgBr_2(s)$
c) $Mg(g) + 2Br(g) \rightarrow MgBr_2(g)$
d) $MgBr_2(s) \rightarrow Mg(g) + 2Br(g)$
e) $MgBr_2(s) \rightarrow Mg(g) + Br_2(g)$

8. The lattice enthalpy of barium oxide is the energy change for the reaction
a) $BaO(s) \rightarrow Ba(g) + \frac{1}{2}O_2(g)$
b) $Ba(s) + \frac{1}{2}O_2(g) \rightarrow BaO(s)$
c) $BaO(s) \rightarrow Ba(g) + O(g)$
d) $BaO(s) \rightarrow Ba^{2+}(g) + O^{2-}(g)$
e) $Ba(g) + O(g) \rightarrow BaO(g)$

9. The standard enthalpy of formation of KCl(s) is $-437\ \text{kJ} \cdot \text{mol}^{-1}$. In a Born-Haber cycle for the formation of KCl(s), which enthalpy change is exothermic?

a) $K(g) \rightarrow K^+(g)$
b) $K^+(g) + Cl^-(g) \rightarrow KCl(s)$
c) $\frac{1}{2}Cl_2(g) \rightarrow Cl(g)$
d) $KCl(s) \rightarrow K^+(g) + Cl^-(g)$
e) $K(s) \rightarrow K(g)$

10. Which of the following has the largest lattice enthalpy?

a) MgF_2(s) b) KF(s) c) LiF(s) d) NaF(s) e) AgF(s)

11. Calculate the lattice enthalpy of lithium chloride from the following data.

enthalpy of formation of Li(g): $159\ \text{kJ} \cdot \text{mol}^{-1}$
first ionization energy of Li(g): $519\ \text{kJ} \cdot \text{mol}^{-1}$
enthalpy of formation of Cl(g): $122\ \text{kJ} \cdot \text{mol}^{-1}$
electron affinity of Cl(g): $-349\ \text{kJ} \cdot \text{mol}^{-1}$ 1
enthalpy of formation of LiCl(s): $-409\ \text{kJ} \cdot \text{mol}^{-1}$

a) $42.0\ \text{kJ} \cdot \text{mol}^{-1}$ b) $982\ \text{kJ} \cdot \text{mol}^{-1}$ c) $860\ \text{kJ} \cdot \text{mol}^{-1}$ d) $1558\ \text{kJ} \cdot \text{mol}^{-1}$ e) $740\ \text{kJ} \cdot \text{mol}^{-1}$

12. Calculate the lattice enthalpy of potassium chloride from the following data.

enthalpy of formation of K(g): $89\ \text{kJ} \cdot \text{mol}^{-1}$
first ionization energy of K(g): $418\ \text{kJ} \cdot \text{mol}^{-1}$
enthalpy of formation of Cl(g): $122\ \text{kJ} \cdot \text{mol}^{-1}$
electron affinity of Cl(g): $-349\ \text{kJ} \cdot \text{mol}^{-1}$
enthalpy of formation of KCl(s): $-437\ \text{kJ} \cdot \text{mol}^{-1}$

a) $1415\ \text{kJ} \cdot \text{mol}^{-1}$ b) $717\ \text{kJ} \cdot \text{mol}^{-1}$ c) $839\ \text{kJ} \cdot \text{mol}^{-1}$ d) $541\ \text{kJ} \cdot \text{mol}^{-1}$ e) $890\ \text{kJ} \cdot \text{mol}^{-1}$

13. Calculate the lattice enthalpy of potassium iodide from the following data.

enthalpy of formation of K(g): $89\ \text{kJ} \cdot \text{mol}^{-1}$
first ionization energy of K(g): $418\ \text{kJ} \cdot \text{mol}^{-1}$
enthalpy of formation of I(g): $107\ \text{kJ} \cdot \text{mol}^{-1}$
electron affinity of I(g): $-295\ \text{kJ} \cdot \text{mol}^{-1}$
enthalpy of formation of KI(s): $-328\ \text{kJ} \cdot \text{mol}^{-1}$

a) $1532\ \text{kJ} \cdot \text{mol}^{-1}$ b) $581\ \text{kJ} \cdot \text{mol}^{-1}$ c) $1237\ \text{kJ} \cdot \text{mol}^{-1}$ d) $754\ \text{kJ} \cdot \text{mol}^{-1}$ e) $647\ \text{kJ} \cdot \text{mol}^{-1}$

14. Calculate the lattice enthalpy of sodium fluoride from the following data.

enthalpy of formation of Na(g): $107\ \text{kJ} \cdot \text{mol}^{-1}$
first ionization energy of Na(g): $494\ \text{kJ} \cdot \text{mol}^{-1}$
enthalpy of formation of F(g): $79\ \text{kJ} \cdot \text{mol}^{-1}$
electron affinity of F(g): $-328\ \text{kJ} \cdot \text{mol}^{-1}$
enthalpy of formation of NaF(s): $-574\ \text{kJ} \cdot \text{mol}^{-1}$

a) $1582\ \text{kJ} \cdot \text{mol}^{-1}$ b) $106\ \text{kJ} \cdot \text{mol}^{-1}$ c) $926\ \text{kJ} \cdot \text{mol}^{-1}$ d) $1005\ \text{kJ} \cdot \text{mol}^{-1}$ e) $434\ \text{kJ} \cdot \text{mol}^{-1}$

15. Calculate the lattice enthalpy of calcium chloride from the following data.
enthalpy of formation of Ca(g): 178 kJ · mol^{-1}
first ionization energy of Ca(g): 590 kJ · mol^{-1}
second ionization energy of Ca(g): 1150 kJ · mol^{-1}
enthalpy of formation of Cl(g): 122 kJ · mol^{-1}
electron affinity of Cl(g): –349 kJ · mol^{-1}
enthalpy of formation of $CaCl_2$(s): –796 kJ · mol^{-1}
a) 668 kJ · mol^{-1} b) 3656 kJ · mol^{-1} c) 2064 kJ · mol^{-1} d) 2487 kJ · mol^{-1} e) 2260 kJ · mol^{-1}

16. Calculate the lattice enthalpy of barium oxide from the following data.
enthalpy of formation of Ba(g): 180 kJ · mol^{-1}
first ionization energy of Ba(g): 502 kJ · mol^{-1}
second ionization energy of Ba(g): 966 kJ · mol^{-1}
enthalpy of formation of O(g): 249 kJ · mol^{-1}
first electron affinity of O(g): –141 kJ · mol^{-1}
second electron affinity of O(g): 844 kJ · mol^{-1}
enthalpy of formation of BaO(s): –554 kJ · mol^{-1}
a) 2310 kJ · mol^{-1} b) 2046 kJ · mol^{-1} c) 3154 kJ · mol^{-1} d) 1466 kJ · mol^{-1} e) 3403 kJ · mol^{-1}

17. Calculate the heat of formation of $CaCl_2$(s) from the following data.
enthalpy of formation of Ca(g): 178 kJ · mol^{-1}
first ionization energy of Ca(g): 590 kJ · mol^{-1}
second ionization energy of Ca(g): 1150 kJ · mol^{-1}
enthalpy of formation of Cl(g): 122 kJ · mol^{-1}
electron affinity of Cl(g): –349 kJ · mol^{-1}
lattice enthalpy of $CaCl_2$(s): 2260 kJ · mol^{-1}

18. Calculate the electron affinity of iodine from the following data.
enthalpy of formation of K(g): 89 kJ · mol^{-1}
first ionization energy of K(g): 418 kJ · mol^{-1}
enthalpy of formation of I(g): 107 kJ · mol^{-1}
lattice enthalpy of KI(s): 647 kJ · mol^{-1}
enthalpy of formation of KI(s): –328 kJ · mol^{-1}

19. Explain each of the following.
(a) The lattice enthalpy of CaO(s) is 3505 kJ · mol^{-1} but the lattice enthalpies of $CaCl_2$(s) –and KCl(s) are only 2260 and 717 kJ · mol^{-1}, respectively.
(b) The second electron affinity of oxygen is positive whereas the first is negative.

20. How many lone pairs of electrons does the iodide atom possess in the Lewis structure of HI?
a) 3 b) 2 c) 1 d) 6 e) 4

21. How many lone pairs of electrons does the oxygen atom possess in the Lewis structure of H_2S?
a) 3 b) 4 c) 2 d) 1 e) 0

22. How many valence electrons are there in HNO_2?
a) 18 b) 10 c) 16 d) 20 e) 8

23. How many valence electrons are there in HNO_3?
a) 12 b) 18 c) 20 d) 24 e) 14

24. How many valence electrons are there in the carbonate ion, CO_3^{2-}?
a) 24 b) 26 c) 22 d) 18 e) 20

25. How many valence electrons are there in the acetate ion?
a) 24 b) 26 c) 28 d) 20 e) 32

26. How many valence electrons are there in the phosphate ion?
a) 42 b) 32 c) 34 d) 36 e) 30

27. How many valence electrons are there in the nitrate ion?
a) 24 b) 32 c) 26 d) 22 e) 28

28. How many lone pairs of electrons are there in the Lewis structure of formic acid, HCOOH?
a) 4 b) 5 c) 6 d) 7 e) 3

29. The Lewis structure of CN^-
a) has a double bond between the two bonded atoms.
b) contains 4 lone pairs of electrons.
c) has a triple bond between the two bonded atoms.
d) does not obey the octet rule.
e) contains no lone pairs of electrons.

30. How many lone pairs are there in the Lewis structure of dinitrogen tetroxide, $O_2N–NO_2$?
a) 10 b) 12 c) 8 d) 4 e) 14

31. Which of the following species can have resonance structures?
a) CH_3NH_2 b) SO_3^{2-} c) BF_4^- d) CH_3COOH e) H_3O^+

32. Which of the following species can have resonance structures?
a) H_3O^+ b) CH_3COOH c) BF_4^- d) $ClNO_2$ e) CH_3NH_2

33. Which of the following species can have resonance structures?
a) BF_4^- b) CO_3^{2-} c) H_3O^+ d) CH_3COOH e) CH_3NH_2

34. The Lewis structure of the nitrate ion shows around the central nitrogen atom
   a) 3 single-bonded oxygens and 1 lone pair of electrons.
   b) 2 single-bonded oxygens and 1 double-bonded oxygen.
   c) 3 single-bonded oxygens and 2 lone pairs of electrons.
   d) 2 double-bonded oxygens and 1 single-bonded oxygen.
   e) 2 single-bonded oxygens and 2 lone pairs of electrons.

35. Resonance forms can be written for
   a) N_2F_4 b) OF_2 c) C_2H_2 d) SO_3 e) H_2O

36. Resonance forms can be written for all of the following except
   a) NO_3^- b) PCl_3 c) NO_2^- d) CO_3^{2-} e) SO_3

37. How many resonance forms can be written for the NCO^- ion?
   a) 3 b) 0 c) 1 d) 2 e) 4

38. The formal charge on the xenon atom in the Lewis structure of XeF_2 is
   a) 0 b) −2 c) +2 d) −1 e) +1

39. How many lone pairs of electrons does the carbon atom possess in the Lewis structure of CO_2?
   a) 0 b) 4 c) 2 d) 1 e) 3

40. For the Lewis structure below, the formal charges on C, N, and S, respectively, are
   $[\ddot{\underset{..}{C}}=N=\ddot{\underset{..}{S}}]^-$
   a) 0, +1, −2 b) −1, 0, 0 c) −2, +1, 0 d) −1, +1, −1 e) −2, +2, −1

41. For the Lewis structure below, the formal charges on N, C, and O, respectively, are
   $[\ddot{\underset{..}{N}}=C=\ddot{\underset{..}{O}}]^-$
   a) −1, +1, −1 b) +1, 0, −2 c) −2, 0, +1 d) −1, 0, 0 e) +1, −1, −1

42. For the Lewis structure below, the formal charges on C, O, and N, respectively, are
   $[\ddot{\underset{..}{C}}=O=\ddot{\underset{..}{N}}]^-$
   a) 0, +1, −2 b) −1, +1, −1 c) −1, 0, 0 d) −2, +1, 0 e) −2, +2, −1

43. For the Lewis structure below, considering the atoms from left to right, the formal charges on N, N, and O, respectively, are
   $[:N\equiv N-\ddot{\underset{..}{O}}:]$
   a) 0, 0, 0 b) −1, 0, 1 c) 0, +2, −2 d) −1, +1, 0 e) 0, +1, −1

44. For the Lewis structure below, considering the atoms from left to right, the formal charges on N, N, and N, respectively, are

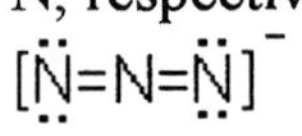

a) −1, +1, −1 b) 0, 0, −1 c) +1, −3, +1 d) −2, +3, −2 e) 0, −1, 0

45. For the Lewis structure below, the formal charges on the oxygen atoms from left to right, respectively, are

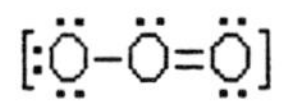

a) −1, 0, +1 b) 0, +1, −1 c) −1, +1, 0 d) 0, −1. +1 e) +1, −1, 0

46. For the Lewis structure below, the formal charges on the oxygen atoms at the top left and top right of the molecule, respectively, are

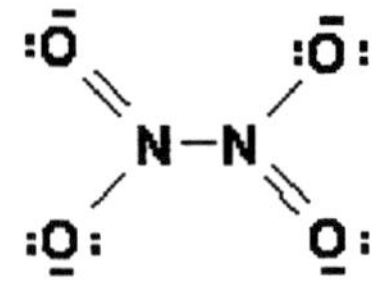

a) +1 and −1 b) 0 and −1 c) −2 and −2 d) −1 and +1 e) 0 and +1

47. A Lewis structure that obeys the octet rule can be drawn for

a) XeF_2 b) XeF_3^+ c) XeO_3 d) $XeOF_2$ e) XeF_4

48. Write all resonance structures for SO_3. Include those structures which use an expanded octet on the sulfur atom. Assign formal charges to all atoms.

49. Which of the following is a radical?

a) XeO_4 b) CH_3O c) Cl–O–O–Cl d) NO^+ e) $S_2O_4^{2-}$

50. Which of the following is a radical?

a) O_2^- b) ClO^- c) Cl–O–O–Cl d) NO^+ e) $S_2O_4^{2-}$

51. All of the following have Lewis structures that obey the octet rule except

a) NO_2^- b) N_2O c) NO_2 d) N_3^- e) N_2O_4

52. All of the following have Lewis structures that obey the octet rule except

a) ClO b) ClO_2^- c) ClO_4^- d) BrCl e) CN^-

53. Which of the following is a free radical?

a) NO_2 b) CO c) OF_2 d) H_2S e) NO^+

54. Which of the following is a free radical?

a) OCl b) ClO_2^- c) I_3^- d) ICl_2^+ e) NO_2^+

55. How many electron pairs (both bonding and lone pairs) are on the central iodine atom in I_3^-?
a) 5 b) 2 c) 3 d) 4 e) 6

56. The formula of cyanogen is C_2N_2. Determine the arrangement of the atoms in cyanogen and write the Lewis structure. Also, include all possible resonance forms and assign formal charges to all atoms.

57. The number of lone pairs around Xe in XeO_3 is
a) 1 b) 0 c) 4 d) 2 e) 3

58. The number of lone pairs around the central iodine atom in I_3^+ is
a) 2 b) 0 c) 3 d) 5 e) 1

59. The number of lone pairs around the central iodine atom in ICl_2^+ is
a) 2 b) 4 c) 0 d) 3 e) 1

60. Which of the following would have a Lewis structure most like that of ClO_2^-?
a) IF_2^- b) OF_2 c) O_3 d) XeF_2 e) CO_2

61. Which of the following would have a Lewis structure most like that of XeO_2?
a) XeF_2 b) CO_2 c) ICl_2^+ d) I_3^- e) O_3

62. Which of the following would have a Lewis structure most like that of SO_3^{2-}?
a) NO_3^- b) IF_4^+ c) $XeOF_2$ d) CO_3^{2-} e) XeO_3

63. Which of the following would have a Lewis structure most like that of IO_3^-?
a) NO_3^- b) $XeOF_2$ c) IF_4^+ d) CO_3^{2-} e) SeF_3^+

64. Which of the following would have a Lewis structure most like that of PO_4^{3-}?
a) $AlCl_4^-$ b) SF_4 c) IF_4^+ d) ICl_4^- e) XeF_4

65. Which of the following would have a Lewis structure most like that of XeO_4?
a) XeF_4 b) ICl_4^- c) SO_4^{2-} d) SF_4 e) IF_4^+

66. All of the following have similar Lewis structures except
a) $AlCl_4^-$ b) XeO_4 c) CCl_4 d) SO_4^{2-} e) SF_4

67. All of the following have similar Lewis structures except
a) I_3^- b) ClO_2^- c) SCl_2 d) S_3^{2-} e) XeO_2

68. How many lone pairs of electrons are there in the Lewis structure of XeF_2?
a) 9 b) 6 c) 3 d) 7 e) 8

69. In the Lewis structure of ozone, the central oxygen atom has
a) 1 lone pair and 4 bonding pairs of electrons.
b) 1 lone pair and 3 bonding pairs of electrons.
c) 2 lone pairs of electrons.
d) 1 lone pair and 2 bonding pairs of electrons.
e) 2 lone pairs and 2 bonding pairs of electrons.

70. In the Lewis structure of the nitrate ion, the central atom has
a) 4 bonding pairs of electrons.
b) 1 lone pair and 3 bonding pairs of electrons.
c) 2 lone pairs and 3 bonding pairs of electrons.
d) 1 lone pair and 4 bonding pairs of electrons.
e) 5 bonding pairs of electrons.

71. How many lone pairs of electrons are in the Lewis structure of NH_3BF_3?
a) 9 b) 12 c) 3 d) 7 e) 18

72. Which of the following are Lewis bases?
1: CN^-
2: Al^{3+}
3: Cl^-
4: PCl_3
5: BF_3
a) 1, 3, and 5 b) 1, 3, and 4 c) 1 and 3 d) 2 and 5 e) 3 and 4

73. Which of the following has bonds with the most ionic character?
a) KBr b) NaI c) NaBr d) KI e) RbI

74. Elements that have a low electronegativity have
a) a low ionization energy and a low electron affinity.
b) a low ionization energy and a high electron affinity.
c) a high ionization energy and a high electron affinity.
d) a high ionization energy and a low electron affinity.
e) an ionization energy that is approximately equal to its electron affinity.

75. Select the elements that have the lowest and the highest electronegativities, respectively.
Na, Ga, K, Br, Se
a) Ga and Se b) K and Ga c) K and Br d) Na and Br e) Na and Se

76. Which of the following would have both the largest ionization energy and largest electron affinity?
a) S b) P c) Si d) Ca e) K

77. For the following elements, predict the order of electronegativity, from lowest to highest.
K, As, Cl, Ca, P
a) Ca < K < P < As < Cl
b) Ca < K < As < P < Cl
c) K < Ca < P < As < Cl
d) K < Ca < As < P < Cl
e) K < Ca < Cl < P < As

78. Which of the following has bonds with the most covalent character?
a) $SrCl_2$ b) $CaCl_2$ c) $BeCl_2$ d) NaCl e) KCl

79. Which of the following pairs of elements would be most likely to form an ionic compound?
a) Mg and Cl b) Ca and Br c) Al and Cl d) Sr and Cl e) Be and Cl

80. Which of the following compounds is ionic?
a) N_2O b) SO_2 c) Na_2CO_3 d) CO_2 e) SO_3

Chapter 9: Molecules: Shape, Size, and Bond Strength Form A

1. What is the shape of SF_2?
a) angular b) tetrahedral c) linear d) trigonal planar e) T-shaped

2. What is the shape of $AlCl_4^-$?
a) tetrahedral b) seesaw c) square planar d) T-shaped e) square pyramidal

3. What is the shape of BrO_4^-?
a) tetrahedral b) seesaw c) square planar d) T-shaped e) square pyramidal

4. What is the shape of $H_2PO_2^-$?
a) tetrahedral b) seesaw c) square planar d) T-shaped e) square pyramidal

5. What is the shape of SOF_4?
a) trigonal bipyramidal b) square planar c) square pyramidal d) seesaw e) octahedral

6. What is the shape of PCl_5?
a) square pyramidal b) trigonal bipyramidal c) square planar d) seesaw e) octahedral

7. What is the shape of $ClOF_4^+$?
a) square planar b) trigonal bipyramidal c) square pyramidal d) seesaw e) octahedral

8. What is the shape of AlF_6^{3-}?
a) octahedral
b) pentagonal bipyramidal
c) square pyramidal
d) square planar
e) trigonal bipyramidal

9. What is the shape of SF_6?
a) octahedral
b) pentagonal bipyramidal
c) square pyramidal
d) square planar
e) trigonal bipyramidal

10. The O–Cl–O bond angle in ClO_4^- is
a) <109°. b) equal to 109°. c) equal to 90°, 120°, or 180°. d) equal to 90°. e) <120° but >109°.

11. The O–C–O bond angle in CO_3^{2-} is
a) equal to 90° or 120°.
b) equal to 120°.
c) <109°.
d) equal to 90° or 180°.
e) much less than 120°.

12. What is the shape of OF_2?
a) angular b) tetrahedral c) linear d) trigonal planar e) T-shaped

13. If a molecule has a VSEPR formula AX_3E, then the shape of the molecule is
a) trigonal planar. b) trigonal pyramidal. c) tetrahedral. d) seesaw. e) square planar.

14. The shape of a molecule with no multiple bonds and a central atom with five bonding pairs of electrons is
a) square planar. b) trigonal bipyramidal. c) octahedral. d) seesaw. e) square pyramidal.

15. The shape of a molecule with no multiple bonds and a central atom with 2 bonding pairs and 1 lone pair of electrons is
a) angular. b) tetrahedral. c) trigonal planar. d) linear. e) trigonal pyramidal.

16. What is the shape of CO_2?
a) linear b) angular c) trigonal planar d) T-shaped e) trigonal bipyramidal

17. What is the shape of IF_2^-?
a) linear b) angular c) trigonal planar d) T-shaped e) trigonal bipyramidal

18. What is the shape of ICl_2^-?
a) linear b) angular c) trigonal planar d) T-shaped e) trigonal bipyramidal

19. What is the shape of PCl_3?
a) trigonal pyramidal b) trigonal planar c) T-shaped d) tetrahedral e) seesaw

20. What is the shape of SeF_3^+?
a) trigonal pyramidal b) trigonal planar c) T-shaped d) tetrahedral e) seesaw

21. What is the shape of SO_3^{2-}?
a) trigonal planar b) trigonal pyramidal c) T-shaped d) tetrahedral e) seesaw

22. What is the shape of $XeOF_2$?
a) T-shaped b) trigonal pyramidal c) seesaw d) trigonal planar e) trigonal bipyramidal

23. What is the shape of ICl_4^+?
a) seesaw b) square planar c) tetrahedral d) square pyramidal e) trigonal bipyramidal

24. What is the shape of BrF_5?
a) trigonal bipyramidal
b) square pyramidal
c) octahedral
d) pentagonal bipyramidal
e) square planar

25. If a molecule has a VSEPR formula AX_4E, then the shape of the molecule is
a) seesaw. b) square planar. c) tetrahedral. d) trigonal planar. e) trigonal pyramidal.

26. What is the shape of S_3^{2-}?
a) angular b) tetrahedral c) linear d) trigonal planar e) T-shaped

27. What is the shape of SO_3?
a) trigonal pyramidal b) trigonal planar c) T-shaped d) tetrahedral e) seesaw

28. The F–S–F bond angle in SF_2 is
a) equal to 180°. b) <120° but >109°. c) <109°. d) equal to 109°. e) equal to 120°.

29. The F–Se–F bond angle in SeF_3^+ is
a) <120° but >109°. b) equal to 109°. c) <109°. d) equal to 90° or 180°. e) equal to 120°.

30. What is the shape of CO_3^{2-}?
a) trigonal pyramidal b) trigonal planar c) T-shaped d) tetrahedral e) seesaw

31. If a molecule has a VSEPR formula AX_5E, then the shape of the molecule is
a) square pyramidal. b) square planar. c) octahedral. d) trigonal bipyramidal. e) seesaw.

32. What is the shape of NO_2Cl?
a) trigonal pyramidal b) trigonal planar c) T-shaped d) tetrahedral e) seesaw

33. The F–I–F bond angle in IF_2^- is
a) equal to 109°. b) equal to 120°. c) <109°. d) <120°. e) equal to 180°.

34. The Cl–P–Cl bond angle in PCl_5 is
a) equal to 90° or 109°.
b) equal to 90° only.
c) equal to 90°, 120°, or 180°.
d) equal to 120° or 72°.
e) <109°.

35. The F–S–F bond angle in SF_6 is
a) equal to 90° only.
b) equal to 90° or 120°.
c) equal to 90° or 180°.
d) equal to 72° or 180°.
e) equal to 90°, 180°, or 72°.

36. The F–Cl–F bond angle in ClF_3 is
a) equal to 120°. b) equal to 109°. c) equal to 90° and 180°. d) <109°. e) equal to 90° only.

37. Estimate the C–O–O bond angle in peroxyacetylnitrate.

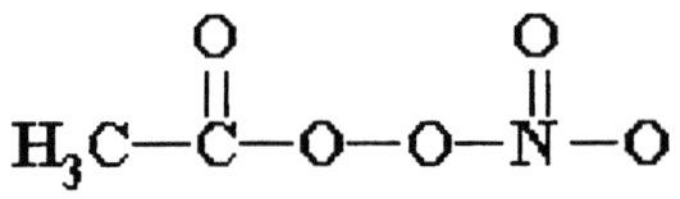

a) $>120^\circ$ b) $<120^\circ$ but $>109^\circ$ c) $<109^\circ$ d) 90^0 e) 180°

38. Which of the following molecules is polar?
a) COS b) S_2 c) CO_2 d) CS_2 e) O_2

39. Which of the following molecules is polar?
a) CH_4 b) XeF_4 c) SF_6 d) IF_5 e) PCl_5

40. Which of the following molecules is polar?
a) XeF_4 b) BF_3 c) CH_4 d) CCl_4 e) CH_2F_2

41. All of the following molecules are polar except
a) XeF_4 b) XeO_2F_2 c) BrF_5 d) SF_4 e) $XeOF_4$

42. All of the following molecules are polar except
a) OF_2 b) SF_4 c) SF_2 d) XeF_2 e) ClNO

43. Which of the following molecules is polar?
a) SO_3 b) XeF_4 c) ClF_3 d) BF_3 e) AsF_5

44. Which of the following is least stable at 25°C and 1 atm?
a) SiH_4 b) PbH_4 c) GeH_4 d) CH_4 e) SnH_4

45. Estimate the standard enthalpy of formation of hydrazine, $N_2H_4(g)$, from the following data.

$\Delta H_B(H_2) = 436$ kJ/mol

$\Delta H_B(N_2) = 944$ kJ/mol

$\Delta H_B(N\text{–}N) = 163$ kJ/mol

$\Delta H_B(N\text{–}H) = 388$ kJ/mol

a) –802 kJ/mol b) –1715 kJ/mol c) +264 kJ/mol d) +101 kJ/mol e) –335 kJ/mol

46. Estimate the standard enthalpy of formation of NO(g) from the following data.

$\Delta H_B(O_2) = 496$ kJ/mol

$\Delta H_B(N_2) = 944$ kJ/mol

$\Delta H_B(N{=}O) = 630$ kJ/mol

a) +180 kJ/mol b) –540 kJ/mol c) –270 kJ/mol d) +90 kJ/mol e) +810 kJ/mol

47. Which of the following has the largest bond enthalpy?
a) H_2 b) N_2 c) O_2 d) HI e) HBr

48. Consider the following:
(1) H_2CO; (2) CH_3OH; (3) HCO_2^-
List the carbon-oxygen bonds in order of increasing length.
a) (1) < (3) < (2) b) (1) < (2) < (3) c) (2) < (3) < (1) d) (3) < (1) < (2) e) (2) < (1) < (3)

49. Consider the following:
(1) H_2CO; (2) CH_3OH; (3) HCO_2^-
List the carbon–oxygen bonds in order of decreasing average bond enthalpy.
a) (1) > (3) > (2) b) (3) > (1) > (2) c) (2) > (3) > (1) d) (2) > (1) > (3) e) (1) > (2) > (3)

50. Estimate the reaction enthalpy for
$CH_4(g) + 4Cl_2(g) \rightarrow CCl_4(g) + 4HCl(g)$
from the data below.
$\Delta H_B(HCl) = 431$ kJ/mol
$\Delta H_B(C\text{-}Cl) = 326$ kJ/mol
$\Delta H_B(C\text{-}H) = 412$ kJ/mol
$\Delta H_B(Cl_2) = 242$ kJ/mol
a) –94 kJ/mol b) –1148 kJ/mol c) –412 kJ/mol d) +1148 kJ/mol e) –890 kJ/mol

51. If the average O–H bond enthalpy is 463 kJ/mol, estimate the enthalpy change when water vapor dissociates to gaseous atoms.
a) 926 kJ/mol b) 1389 kJ/mol c) 463 kJ/mol d) –926 kJ/mol e) –463 kJ/mol

52. Estimate the average C ≡N bond enthalpy from the following data.
standard enthalpy of formation of HCN(g): 135 kJ/mol
standard enthalpy of formation of C(g): 717 kJ/mol
$\Delta H_B(H_2) = 436$ kJ/mol
$\Delta H_B(N_2) = 944$ kJ/mol
$\Delta H_B(C\text{–}H) = 412$ kJ/mol

53. What type of orbitals best describe the bond between the two oxygen atoms in peroxyacetylnitrate?

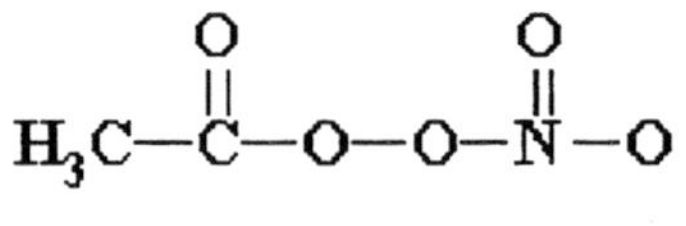

a) sp^2–sp^2 b) sp^3–sp^3 c) $2p_x$–$2p_x$ d) sp^2–sp^3 e) sp–sp

54. What type of orbitals best describe the bond between the two oxygen atoms in peroxymonosulfate?
    H–O–O–SO_3^-
 a) sp–sp^3 b) sp^3–sp^3 c) sp–sp^2 d) $2p_x$–$2p_x$ e) sp^2–sp^2

55. What type of orbitals best describe the bond between carbon and oxygen atom 1 in peroxyacetic acid?

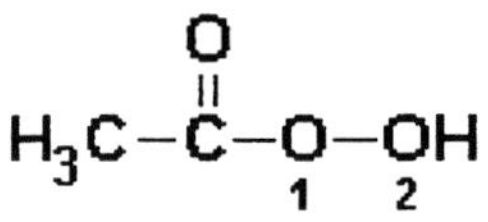

 a) sp^2–sp^3 b) sp–sp c) sp^2–sp^2 d) sp^3–sp^3 e) sp^2–sp

56. One of the dominant resonance forms of diazomethane is

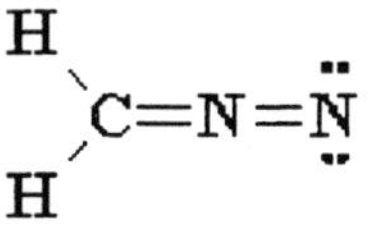

(a) Describe the hybrid orbitals used to form the C–N bond and the N–N bond.
(b) What are the values of the H–C–N and C–N–N bond angles?

57. Consider the structure of the peroxydisulfate ion below.

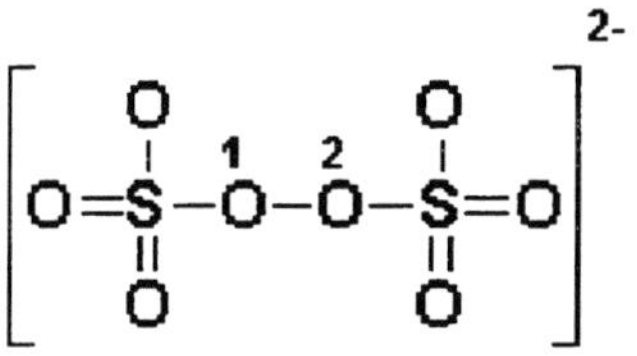

(a) Complete the Lewis structure and indicate the formal charges on all atoms.
(b) What are the O–S–O and S–O(1)–O(2) bond angles?

58. What hybrid orbitals are used by the nitrogens in N_2F_2?
 a) sp^2 b) $d\,sp^3$ c) sp^3 d) $d\,sp^2$ e) sp

59. What set of hybrid orbitals is used by Xe in XeF_4?
 a) $d\,sp^2$ b) $d\,sp^3$ c) $d^2\,sp$ d) d^2sp^3 e) sp^3

60. What set of hybrid orbitals is used by Xe in $XeOF_4$?
 a) $d\,sp^3$ b) $d^2\,sp^3$ c) $d\,sp^2$ d) $d^2\,sp$ e) sp^3

61. What set of hybrid orbitals is used by Si in SiF_6^{2-}?
 a) sp^3 b) $d\,sp^2$ c) $d\,sp^3$ d) $d^2\,sp^3$ e) $d^2\,sp$

62. What set of hybrid orbitals is used by As in AsF_5?
 a) sp^3 b) $d\,sp^3$ c) $d\,sp^2$ d) $d^2\,sp$ e) $d^2\,sp^3$

63. What set of hybrid orbitals is used by Xe in $XeOF_2$?
a) sp^2 b) $d^2 sp^3$ c) $d sp^3$ d) $d^2 sp$ e) sp^3

64. What set of hybrid orbitals is used by S in SF_4?
a) sp^3 b) $d sp^3$ c) $d^2 sp^3$ d) sp^2 e) $d^2 sp$

65. What set of hybrid orbitals is used by I in I_3^-?
a) $d sp^3$ b) sp^2 c) sp d) sp^3 e) $d^2 sp^3$

66. What set of hybrid orbitals is used by Si in SiH_4?
a) sp^3 b) $d sp^3$ c) $d^2 sp^3$ d) $d^2 sp$ e) $d sp^2$

67. What set of hybrid orbitals is used by Xe in XeO_3?
a) $d sp^3$ b) sp^3 c) $d sp^2$ d) d^2sp^3 e) sp^2

68. What set of hybrid orbitals is used by I in I_3^+?
a) sp^2 b) d^2sp^3 c) sp d) sp^3 e) $d sp^3$

69. What set of hybrid orbitals is used by C in $COCl_2$?
a) d^2sp^3 b) $d sp^3$ c) sp d) sp^2 e) sp^3

70. What set of hybrid orbitals is used by N in NO_2^-?
a) sp^2 b) sp c) $d sp^3$ d) d^2sp^3 e) sp^3

71. What set of hybrid orbitals is used by C in CH_2^{2-}?
a) sp^3 b) d^2sp^3 c) $d sp^3$ d) sp^2 e) sp

72. What set of hybrid orbitals is used by the central N in N_3^-?
a) sp^2 b) $d sp^3$ c) sp d) d^2sp^3 e) sp^3

73. What type of orbitals are used to form the O–O bond in hydrogen peroxide?
a) sp^3–sp^3 b) sp–sp c) sp–sp^3 d) sp–sp^2 e) sp^2–sp^2

74. The hybridization of X in XY_4 is $d^2 sp^3$. The shape of the molecule is
a) square planar b) trigonal bipyramidal c) tetrahedral d) trigonal pyramidal e) seesaw

75. How many sigma and how many pi bonds are present in H_3C–CH=CH–CH_2–OH?
a) 14 sigma and 0 pi
b) 4 sigma and 1 pi
c) 12 sigma and 1 pi
d) 13 sigma and 1 pi
e) 12 sigma and 2 pi

76. How many sigma and how many pi bonds are present in CH_2=C=CH_2?
a) 6 sigma and 1 pi
b) 8 sigma and 0 pi
c) 4 sigma and 4 pi
d) 6 sigma and 2 pi
e) 4 sigma and 2 pi

77. How many sigma and how many pi bonds are present in H_3C–SCN?
a) 8 sigma and 2 pi
b) 6 sigma and 0 pi
c) 4 sigma and 2 pi
d) 8 sigma and 0 pi
e) 6 sigma and 2 pi

78. How many sigma and how many pi bonds are present in diazomethane, H_2CNN?
a) 4 sigma and 2 pi
b) 4 sigma and 0 pi
c) 6 sigma and 0 pi
d) 3 sigma and 2 pi
e) 4 sigma and 1 pi

79. The structure of the amino acid histidine is shown below.

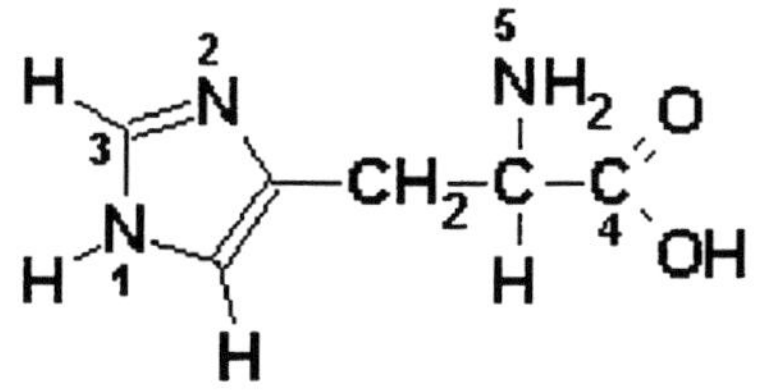

(a) What hybrid orbitals are used on ring nitogen 1 and 2?
(b) Describe the hybrid orbitals used to form the bond between ring nitrogen 1 and ring carbon 3.
(c) What hybrid orbitals are used on the amino nitrogen 5?
(d) What hybrid orbitals are used on carbon 4?

80. Describe the bonding in acetonitrile, CH_3CN.

Chapter 9: Molecules: Shape, Size, and Bond Strength Form B

1. What is the shape of OF_2?
a) angular b) tetrahedral c) linear d) trigonal planar e) T-shaped

2. What is the shape of BF_4^-?
a) tetrahedral b) seesaw c) square planar d) T-shaped e) square pyramidal

3. What is the shape of ClO_4^-?
a) tetrahedral b) seesaw c) square planar d) T-shaped e) square pyramidal

4. What is the shape of $POCl_3$?
a) tetrahedral b) seesaw c) square planar d) T-shaped e) square pyramidal

5. What is the shape of XeO_3F_2?
a) square pyramidal b) square planar c) trigonal bipyramidal d) seesaw e) octahedral

6. What is the shape of IO_2F_3?
a) square pyramidal b) square planar c) trigonal bipyramidal d) seesaw e) octahedral

7. What is the shape of AsF_5?
a) trigonal bipyramidal b) square pyramidal c) square planar d) seesaw e) octahedral

8. What is the shape of AsF_6?
a) octahedral
b) pentagonal bipyramidal
c) square pyramidal
d) square planar
e) trigonal bipyramidal

9. What is the shape of XeO_2F_4?
a) octahedral
b) pentagonal bipyramidal
c) square pyramidal
d) square planar
e) trigonal bipyramidal

10. The H–N–H bond angle in NH_4^+ is
a) equal to 109°. b) equal to 90°. c) $<109^\circ$. d) $<120^\circ$ but $>109^\circ$. e) equal to 90°, 120°, or 180°.

11. The O–N–O bond angle in NO_3^- is
a) equal to 90° or 180°.
b) $<109^\circ$.
c) equal to 90° or 120°.
d) equal to 120°.
e) much less than 120°.

12. What is the shape of I_3^+?
a) angular b) tetrahedral c) linear d) trigonal planar e) T-shaped

13. If a molecule has a VSEPR formula AX_4E_2, then the shape of the molecule is
a) square planar. b) seesaw. c) tetrahedral. d) octahedral. e) square pyramidal.

14. The shape of a molecule with no multiple bonds and a central atom with 2 bonding pairs and 2 lone pairs of electrons is
a) angular. b) tetrahedral. c) linear. d) trigonal planar. e) T-shaped.

15. The shape of a molecule with no multiple bonds and a central atom with 4 bonding pairs of electrons and 1 lone pair of electrons is
a) seesaw. b) square planar. c) tetrahedral. d) trigonal bipyramidal. e) square pyramidal.

16. What is the shape of CS_2?
a) trigonal planar b) angular c) linear d) T-shaped e) trigonal bipyramidal

17. What is the shape of I_3^-?
a) linear b) angular c) trigonal planar d) T-shaped e) trigonal bipyramidal

18. What is the shape of XeF_2?
a) linear b) angular c) trigonal planar d) T–shaped e) trigonal bipyramidal

19. What is the shape of PO_3^{3-}?
a) trigonal pyramidal b) trigonal planar c) T-shaped d) tetrahedral e) seesaw

20. What is the shape of XeO_3?
a) trigonal planar b) trigonal pyramidal c) T-shaped d) tetrahedral e) seesaw

21. What is the shape of $SOCl_2$?
a) trigonal planar b) trigonal pyramidal c) T-shaped d) tetrahedral e) seesaw

22. What is the shape of ClF_3?
a) T-shaped b) trigonal pyramidal c) seesaw d) trigonal planar e) trigonal bipyramidal

23. What is the shape of XeO_2F_2?
a) seesaw b) square planar c) tetrahedral d) square pyramidal e) trigonal bipyramidal

24. What is the shape of IF_5?
a) square pyramidal
b) trigonal bipyramidal
c) octahedral
d) pentagonal bipyramidal
e) square planar

25. If a molecule has a VSEPR formula AX_2E_3, then the shape of the molecule is
a) linear. b) angular. c) square planar. d) trigonal planar. e) trigonal bipyramidal.

26. What is the shape of SO_2?
a) angular b) tetrahedral c) linear d) trigonal planar e) T-shaped

27. What is the shape of NO_3^-?
a) trigonal pyramidal b) trigonal planar c) T-shaped d) tetrahedral e) seesaw

28. The Cl–I–Cl bond angle in ICl_2^+ is
a) <109°. b) <120° but >109°. c) equal to 180°. d) equal to 109°. e) equal to 120°.

29. The O–S–O bond angle in SO_3^{2-} is
a) equal to 90° or 180°. b) equal to 109°. c) <109°. d) equal to 120°. e) <120° but >109°.

30. What is the shape of $COCl_2$?
a) trigonal planar b) trigonal pyramidal c) T-shaped d) tetrahedral e) seesaw

31. If a molecule has a VSEPR formula AX_2E_2, then the shape of the molecule is
a) angular. b) tetrahedral. c) linear. d) T-shaped. e) trigonal planar.

32. What is the shape of H_2CO?
a) trigonal planar b) trigonal pyramidal c) T-shaped d) tetrahedral e) seesaw

33. The F–Xe–F bond angle in XeF_2 is
a) <109°. b) equal to 180°. c) equal to 120°. d) equal to 109°. e) <120°.

34. The F–As–F bond angle in AsF_5 is
a) equal to 90°, 120°, or 180°.
b) <109°.
c) equal to 120° or 72°.
d) equal to 90° or 109°.
e) equal to 90° only.

35. The F–P–F bond angle in PF_6^- is
a) equal to 90°, 180°, or 72°.
b) equal to 90° or 120°.
c) equal to 90° only.
d) equal to 72° or 180°.
e) equal to 90° or 180°.

36. The F–I–F bond angle in IF_5 is
a) <90° but >120°.
b) equal to 90°, 120°, or 180°.
c) equal to 90° or 180°.
d) equal to 90° or 120°.
e) equal to 72°.

37. Estimate the O–C–O bond angle in peroxyacetylnitrate.

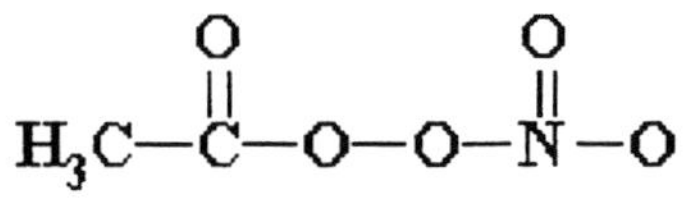

a) 120° b) $<120^{\circ}$ but $>109^{\circ}$ c) 90° d) 109° e) $<109^{\circ}$ but $>90^{\circ}$

38. Which of the following molecules is polar?
a) SF_4 b) CCl_4 c) PCl_5 d) BF_3 e) SF_6

39. Which of the following molecules is polar?
a) CO_2 b) SO_2 c) XeF_2 d) O_3 e) CS_2

40. Which of the following molecules is polar?
a) $COCl_2$ b) BF_3 c) CS_2 d) PCl_5 e) SF_6

41. All of the following molecules are polar except
a) BrF_5 b) PCl_4F c) SF_4 d) SO_3 e) XeO_2F_2

42. All of the following molecules are polar except
a) SF_2 b) SF_4 c) OF_2 d) CS_2 e) ClNO

43. Which of the following molecules is polar?
a) SiH_4 b) $TeCl_4$ c) XeF_4 d) AsF_5 e) TeF_6

44. Which of the following has the largest bond enthalpy?
a) C–H b) Pb–H c) Sn–H d) Ge–H e) Si–H

45. Estimate the standard enthalpy of formation of methane from the following data.
$\Delta H_B(H_2) = 436$ kJ/mol
$\Delta H_B(C–H) = 412$ kJ/mol
standard enthalpy of formation od C(g): 717 kJ/mol
a) –59 kJ/mol b) –495 kJ/mol c) +813 kJ/mol d) –774 kJ/mol e) +741 kJ/mol

46. Estimate the standard enthalpy of formation of carbon monoxide from the following data.
$\Delta H_B(O_2) = 496$ kJ/mol
$\Delta H_B(CO) = 1074$ kJ/mol
standard enthalpy of formation of C(g): 717 kJ/mol
a) +965 kJ/mol b) –1074 kJ/mol c) –109 kJ/mol d) +139 kJ/mol e) –218 kJ/mol

47. Which of the following has the largest bond enthalpy?
a) CO b) O_2 c) H_2 d) HI e) HBr

48. Consider the following:
$(1)\ C_2H_4$; $(2)\ C_2H_2$; $(3)\ C_6H_6$
List the carbon-carbon bonds in order of increasing length.
a) (2) < (3) < (1) b) (3) < (2) < (1) c) (3) < (1) < (2) d) (1) < (2) < (3) e) (2) < (1) < (3)

49. Consider the following:
$(1)\ C_2H_4$; $(2)\ C_2H_2$; $(3)\ C_6H_6$
List the carbon-carbon bonds in order of decreasing average bond enthalpy.
a) (2) > (3) > (1) b) (1) > (2) > (3) c) (3) > (1) > (2) d) (3) > (2) > (1) e) (2) > (1) > (3)

50. Estimate the reaction enthalpy for the combustion of methane from the data below.
$\Delta H_B(\text{O-H}) = 463$ kJ/mol
$\Delta H_B(\text{C=O}) = 743$ kJ/mol
$\Delta H_B(\text{C-H}) = 412$ kJ/mol
$\Delta H_B(O_2) = 496$ kJ/mol
a) –698 kJ/mol b) –298 kJ/mol c) –3338 kJ/mol d) –1194 kJ/mol e) –228 kJ/mol

51. If the average N–H bond enthalpy is 388 kJ/mol, estimate the enthalpy change when ammonia dissociates to gaseous atoms.
a) –1164 kJ/mol b) 776 kJ/mol c) –388 kJ/mol d) 388 kJ/mol e) 1164 kJ/mol

52. Estimate the average C ≡O bond enthalpy from the following data.
standard enthalpy of formation of CO(g): –111 kJ/mol
standard enthalpy of formation of C(g): 717 kJ/mol
$\Delta H_B(O_2) = 496$ kJ/mol

53. What type of orbitals best describe the bond between the two carbon atoms in peroxyacetylnitrate?

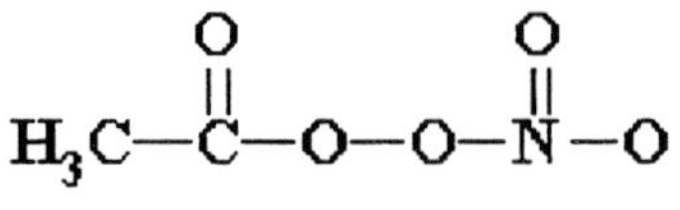

a) $2p_x$–$2p_x$ b) sp^3–sp^2 c) sp–sp d) sp^3–sp^3 e) sp^2–sp^2

54. What type of orbitals best describe the bond between the sulfur atom and the peroxy oxygen atom (underlined) in peroxymonosulfate?
H–O–<u>O</u>–SO_3^-
a) sp^2–sp^2 b) $2p_x$–$2p_x$ c) sp^3–sp^3 d) sp–sp^3 e) sp–sp^2

55. What type of orbitals best describe the bond between oxygen atom 1 and 2 in peroxyacetic acid?

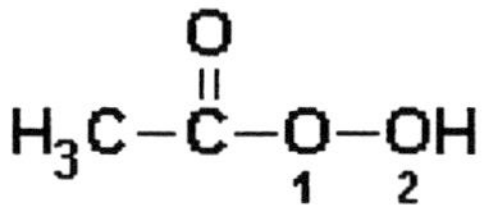

a) sp^2–sp^2 b) sp^3–sp^3 c) sp^2–sp d) sp–sp e) sp^2–sp^3

56. One of the dominant resonance forms of diazomethane is

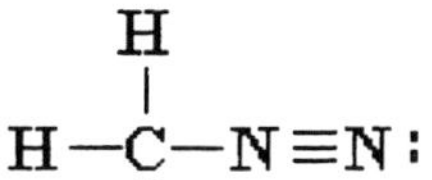

(a) Describe the hybrid orbitals used to form the C–N bond and the N–N bond.
(b) What are the values of the H–C–N and C–N–N bond angles?

57. Consider the structure of the peroxydisulfate ion below.

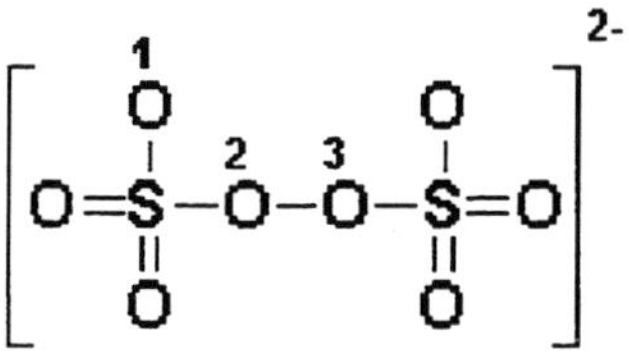

(a) Describe the hybrid orbitals used to form the sulfur-oxygen(1) bond.
(b) Describe the hybrid orbitals used to form the O(2)–O(3) bond.
(c) What is the O(1)–S–O(2) bond angle?

58. What hybrid orbitals are used by the fluorines in N_2F_2?
a) sp^2 b) sp^3 c) d^2sp^3 d) dsp^3 e) sp

59. What set of hybrid orbitals is used by I in ICl_4^-?
a) dsp^2 b) dsp^3 c) d^2sp^3 d) sp^3 e) d^2sp

60. What set of hybrid orbitals is used by I in IF_5?
a) sp^3 b) d^2sp c) dsp^2 d) d^2sp^3 e) dsp^3

61. What set of hybrid orbitals is used by I in IOF_5?
a) dsp^3 b) d^2sp c) dsp^2 d) sp^3 e) d^2sp^3

62. What set of hybrid orbitals is used by Xe in XeO_3F_2?
a) dsp^2 b) dsp^3 c) d^2sp^3 d) d^2sp e) sp^3

63. What set of hybrid orbitals is used by Cl in ClF_3?
a) d^2sp^3 b) sp^2 c) sp^3 d) d^2sp e) dsp^3

64. What set of hybrid orbitals is used by I in IF_4^+?
a) d^2sp b) dsp^3 c) sp^2 d) sp^3 e) d^2sp^3

65. What set of hybrid orbitals is used by I in IF_2^-?
a) sp^2 b) d^2sp^3 c) sp d) dsp^3 e) sp^3

66. What set of hybrid orbitals is used by S in SO_2Cl_2?
a) d^2sp^3 b) sp^3 c) dsp^2 d) dsp^3 e) d^2sp

67. What set of hybrid orbitals is used by Br in BrO_3^-?
a) dsp^2 b) sp^3 c) d^2sp^3 d) sp^2 e) dsp^3

68. What set of hybrid orbitals is used by O in HOCl?
a) sp b) d^2sp^3 c) sp^2 d) sp^3 e) dsp^3

69. What set of hybrid orbitals is used by C in CO_3^{2-}?
a) sp^2 b) d^2sp^3 c) dsp^3 d) sp e) sp^3

70. What set of hybrid orbitals is used by the central O in O_3?
a) dsp^3 b) sp c) sp^2 d) sp^3 e) d^2sp^3

71. What set of hybrid orbitals is used by C in ClCN?
a) dsp^3 b) sp c) sp^3 d) sp^2 e) d^2sp^3

72. What set of hybrid orbitals is used by the central N in N_2O?
a) sp^3 b) sp c) d^2sp^3 d) sp^2 e) dsp^3

73. What type of orbitals are used to form the N–N bond in hydrazine, N_2H_4?
a) sp^3–sp^3 b) sp–sp c) sp–sp^3 d) sp–sp^2 e) sp^2–sp^2

74. The hybridization of X in XY_4 is dsp^3. The shape of the molecule is
a) seesaw b) trigonal bipyramidal c) tetrahedral d) trigonal pyramidal e) square planar

75. How many sigma and how many pi bonds are present in CH_2=CH–CH=CH–CHO?
a) 9 sigma and 3 pi
b) 8 sigma and 6 pi
c) 10 sigma and 3 pi
d) 8 sigma and 4 pi
e) 11 sigma and 3 pi

76. How many sigma and how many pi bonds are present in H_2N–CH_2–COOH?
a) 7 sigma and 2 pi
b) 8 sigma and 1 pi
c) 8 sigma and 2 pi
d) 10 sigma and 0 pi
e) 9 sigma and 1 pi

77. How many sigma and how many pi bonds are present in formamide, H_2NCHO?
a) 5 sigma and 2 pi
b) 5 sigma and 0 pi
c) 4 sigma and 2 pi
d) 4 sigma and 1 pi
e) 5 sigma and 1 pi

78. How many sigma and how many pi bonds are present in acrylonitrile, $H_2C{=}CH{-}CN$?
a) 6 sigma and 2 pi
b) 9 sigma and 0 pi
c) 6 sigma and 1 pi
d) 6 sigma and 3 pi
e) 5 sigma and 3 pi

79. The structure of imidazole is shown below.

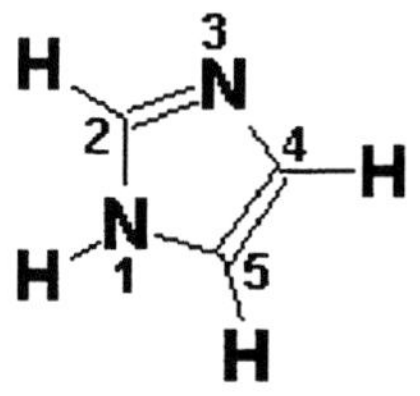

(a) Describe the hybrid orbitals used to form the bond between nitrogen 1 and carbon 5.
(b) Describe the hybrid orbitals used to form the bond between nitrogen 3 and carbon 4.
(c) What is the N(3)–C(4)–C(5) bond angle?

80. Sketch all possible shapes for compounds whose central atom uses $d\,sp^3$ hybrid orbitals.

Chapter 10: Liquid and Solid Materials
Form A

1. Which of the following is most volatile?
a) CBr_3H b) CCl_4 c) CBr_2H_2 d) CCl_3H e) CBr_4

2. Which of the following has the highest boiling point?
a) Br_2 b) F_2 c) Cl_2 d) I_2 e) Ar

3. Which of the following has the smallest standard molar enthalpy of vaporization?
a) CH_4 b) H_2O c) NH_3 d) HF e) SnH_4

4. Which of the following has the highest boiling point?
a) *n*-propane, $CH_3CH_2CH_3$
b) *n*-pentane, $CH_3(CH_2)_3CH_3$
c) *n*-heptane, $CH_3(CH_2)_5CH_3$
d) *n*-hexane, $CH_3(CH_2)_4CH_3$
e) *n*-butane, $CH_3(CH_2)_2CH_3$

5. Consider the following molecules, all with molecular formula C_6H_{14}:

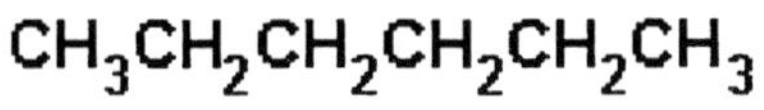

1

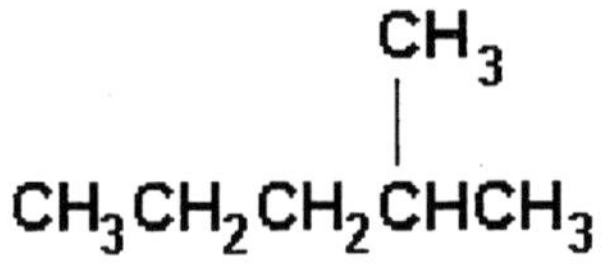

2

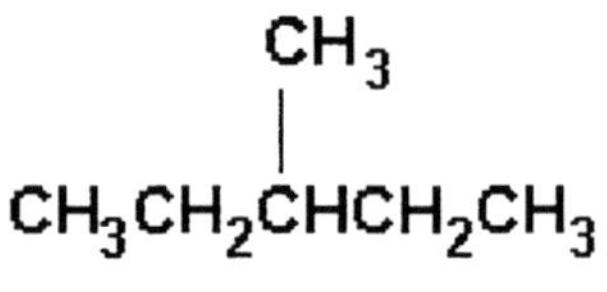

3

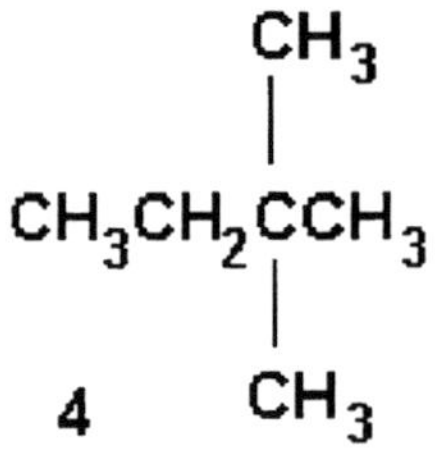

4

Which has the highest boiling point?
a) 1
b) 2
c) 3
d) 4
e) All have the same boiling point because they have the same molecular formula.

6. Which of the following has the highest boiling point?
a) Br_2 b) N_2 c) GeH_4 d) I_2 e) SiH_4

7. Which of the following is correct?
a) The energy of dipole-dipole interactions is about 20 $kJ \cdot mol^{-1}$.
b) The energy of ion-dipole interactions is about 2 $kJ \cdot mol^{-1}$.
c) The energy of London forces is about 2 $kJ \cdot mol^{-1}$.
d) The energy of a hydrogen bond is about 2 $kJ \cdot mol^{-1}$.
e) The energy of ion-ion interactions is about 2 $kJ \cdot mol^{-1}$.

8. Which of the following has the higest boiling point?
a) SiH_4 b) SbH_3 c) AsH_3 d) PH_3 e) CH_4

9. Which of the following has the largest standard molar enthalpy of vaporization?
a) HF b) Ne c) Ar d) He e) CH_4

10. Which of the following has the largest standard molar enthalpy of vaporization?
a) CF_4 b) F_2 c) CCl_4 d) CH_4 e) CBr_4

11. Which of the following has the highest boiling point?
a) H_2S b) H_2O c) HF d) NH_3 e) CH_4

12. The boiling point of O_2 is higher than N_2 due to
a) London forces.
b) hydrogen bonding.
c) dipole-dipole forces.
d) ion-ion forces.
e) ion-dipole forces.

13. The boiling point of HF is higher than HCl due to
a) hydrogen bonding.
b) dipole-dipole forces.
c) ion-dipole forces.
d) ion-ion forces.
e) London forces.

14. Arrange the following substances in order of decreasing boiling point.
H_2O, PH_3, H_2S, CH_4
a) $H_2S > H_2O > CH_4 > PH_3$
b) $CH_4 > PH_3 > H_2S > H_2O$
c) $H_2O > H_2S > CH_4 > PH_3$
d) $H_2S > H_2O > PH_3 > CH_4$
e) $H_2O > H_2S > PH_3 > CH_4$

15. For which of the following substances would hydrogen bonding be most important?
a) NH_3 b) HI c) CH_4 d) H_2 e) GeH_4

16. Which of the following species is likely to forn hydrogen bonds in the pure state?
a) CH_3OCH_3 b) $CH_3C(O)CH_3$ c) $(CH_3)_3N$ d) CH_4 e) $(CH_3)_2NH$

17. Which of the following species is likely to forn hydrogen bonds in the pure state?
a) $(CH_3)_3N$ b) CH_3CH_2OH c) $CH_3C(O)CH_3$ d) CH_4 e) CH_3OCH_3

18. Which of the following species is likely to forn hydrogen bonds in the pure state?
a) HF b) CH_3OCH_3 c) $(CH_3)_3N$ d) $CH_3C(O)CH_3$ e) CH_4

19. List the following in order of increasing boiling point:
argon, hydrogen iodide, hydrogen chloride
a) HCl < HI < Ar b) Ar < HCl < HI c) HCl < Ar < HI d) Ar < HI < HCl e) HI < Ar < HCl

20. Which of the following is correct?
a) The energy of ion-ion interactions is about 2 $kJ \cdot mol^{-1}$.
b) The energy of dipole-dipole interactions is about 2 $kJ \cdot mol^{-1}$.
c) The energy of London forces is about 20 $kJ \cdot mol^{-1}$.
d) The energy of ion-dipole interactions is about 2 $kJ \cdot mol^{-1}$.
e) The energy of a hydrogen bond is about 2 $kJ \cdot mol^{-1}$.

21. Which of the following has the greatest viscosity?
a) water b) benzene, C_6H_6 c) propane, C_3H_8 d) HF e) NH_3

22. The meniscus of mercury curves downward in a glass tube forming a "$\cap$" shape. This means
a) the cohesive forces between mercury atoms are not as strong as the adhesive forces between mercury and glass.
b) the cohesive forces between mercury atoms are stronger than the adhesive forces between mercury and glass.
c) mercury has a low surface tension.
d) mercury has a large capillary action.
e) mercury tends to cover the greatest possible area of the glass.

23. Which of the following has the greatest surface tension?
a) mercury b) water c) benzene d) propane, C_3H_8 e) NH_3

24. Gold has a face-centered cubic unit cell. If the density of gold is 19.4 $g \cdot cm^{-3}$, calculate the atomic radius of gold.

25. Rhodium has a density of 12.4 $g \cdot cm^{-3}$ and an atomic radius of 134 pm. What is the structure of the solid?

26. If the radius of K^+ is 133 pm and that of Br^- is 195 pm, calculate the density of potassium bromide.

27. The density of CsI is 4.51 $g \cdot cm^{-3}$ and the radii of Cs^+ and I^- are 169 and 216 pm, respectively. Calculate Avogadro's number.

28. If gold has a face-centered cubic unit cell, how many atoms of gold does the unit cell contain?
a) 4 b) 7 c) 2 d) 5 e) 14

29. If silver has a face-centered cubic unit cell, how many atoms of silver does the unit cell contain?
a) 4 b) 7 c) 2 d) 5 e) 14

30. In a body-centered cubic lattice, an atom lying at the corner of a unit cell is shared equally by how many unit cells?
a) 8 b) 1 c) 6 d) 4 e) 16

31. A hexagonally close-packed structure has an
a) ABABAB... pattern of layers and a coordination number of 12.
b) ABCDABCD... pattern of layers and a coordination number of 4.
c) ABCABC... pattern of layers and a coordination number of 12.
d) ABBA... pattern of layers and a coordination number of 8.
e) AABB... pattern of layers and a coordination number of 8.

32. A structure with an ABABAB... pattern of layers is called
a) body-centered cubic.
b) primitive cubic.
c) hexagonally close-packed.
d) face-centered cubic.
e) cubic close-packed.

33. The coordination numbers of metal ions in hexagonal close-packed and cubic close-packed structures, respectively, are
a) 6 and 6. b) 12 and 6. c) 12 and 12. d) 12 and 8. e) 8 and 8.

34. Aluminum has a face-centered cubic unit cell. If the atomic radius of aluminum is 125 pm, what is the edge length of the unit cell?
a) 500 pm b) 354 pm c) 177 pm d) 250 pm e) 217 pm

35. Silver has a face-centered cubic unit cell. If the atomic radius of silver is 145 pm, what is the edge length of the unit cell?
a) 205 pm b) 410 pm c) 251 pm d) 290 pm e) 580 pm

36. Aluminum has a face-centered cubic unit cell. If the edge length of the unit cell is 354 pm, what is the atomic radius of aluminum?
a) 88.5 pm b) 125 pm c) 177 pm d) 204pm e) 250 pm

37. Silver has a face-centered cubic unit cell. If the edge length of the unit cell is 410 pm, what is the atomic radius of silver?
a) 290 pm b) 145 pm c) 103 pm d) 205 pm e) 237 pm

38. Iron has a body-centered cubic unit cell. If the atomic radius of iron is 126 pm, what is the edge length of the unit cell?
a) 356 pm b) 504 pm c) 291 pm d) 178 pm e) 252 pm

39. Potassium has a body-centered cubic unit cell. If the atomic radius of potassium is 220 pm, what is the edge length of the unit cell?
 a) 880 pm b) 440 pm c) 311 pm d) 508 pm e) 622 pm

40. Iron has a body-centered cubic unit cell. If the edge length of the unit cell is 291 pm, what is the atomic radius of iron?
 a) 72.8 pm b) 146 pm c) 206 pm d) 103 pm e) 126 pm

41. Potassium has a body-centered cubic unit cell. If the edge length of the unit cell is 508 pm, what is the atomic radius of potassium?
 a) 127 pm b) 359 pm c) 220 pm d) 254 pm e) 180 pm

42. Gold has a face-centered cubic unit cell. If the edge length of the unit cell is 407 pm, what is the density of gold?
 a) $6.86\ g \cdot cm^{-3}$ b) $9.70\ g \cdot cm^{-3}$ c) $4.85\ g \cdot cm^{-3}$ d) $19.4\ g \cdot cm^{-3}$ e) $13.7\ g \cdot cm^{-3}$

43. Silver has a face-centered cubic unit cell. If the edge length of the unit cell is 410 pm, what is the density of silver?
 a) $5.20\ g \cdot cm^{-3}$ b) $7.35\ g \cdot cm^{-3}$ c) $2.60\ g \cdot cm^{-3}$ d) $3.68\ g \cdot cm^{-3}$ e) $10.4\ g \cdot cm^{-3}$

44. Iron has a body-centered cubic unit cell. If the edge length of the unit cell is 291 pm, what is the density of iron?
 a) $5.32\ g \cdot cm^{-3}$ b) $2.66\ g \cdot cm^{-3}$ c) $3.76\ g \cdot cm^{-3}$ d) $15.1\ g \cdot cm^{-3}$ e) $7.53\ g \cdot cm^{-3}$

45. Potassium has a body-centered cubic unit cell. If the edge length of the unit cell is 508 pm, what is the density of potassium?
 a) $0.495\ g \cdot cm^{-3}$ b) $1.98\ g \cdot cm^{-3}$ c) $0.700\ g \cdot cm^{-3}$ d) $0.990\ g \cdot cm^{-3}$ e) $0.350\ g \cdot cm^{-3}$

46. The cubic unit cell of an ionic compound has A cations at the corners and at the body center, and B anions at the face centers. What is the empirical formula of the compound?
 a) A_2B_3 b) AB c) AB_2 d) A_9B_6 e) A_5B_6

47. One form of the mineral perovskite has a unit cell that has calcium cations at the corners of a cube, a titanium cation in the cube center, and oxygen anions in the cube faces. What is the empirical formula of perovskite?
 a) Ca_2TiO_3 b) $CaTiO_3$ c) $CaTiO_6$ d) Ca_8TiO_6 e) Ca_4TiO_3

48. The following unit cell is the

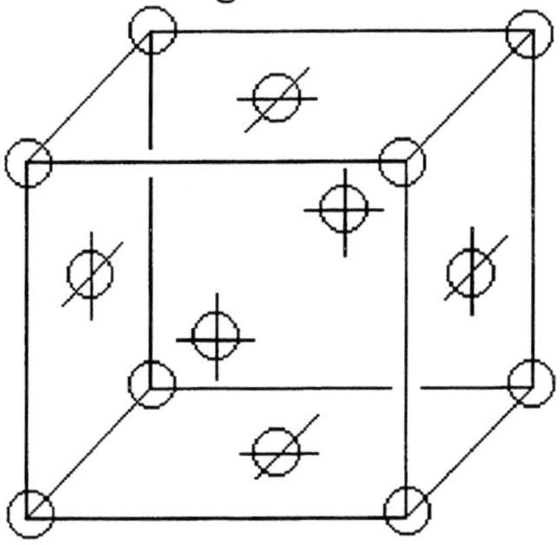

a) hexagonally close-packed unit cell.
b) coordination number 8 cubic unit cell.
c) primitive cubic unit cell.
d) cubic close-packed unit cell.
e) body-centered cubic unit cell.

49. Which of the following apply to the unit cell shown below?

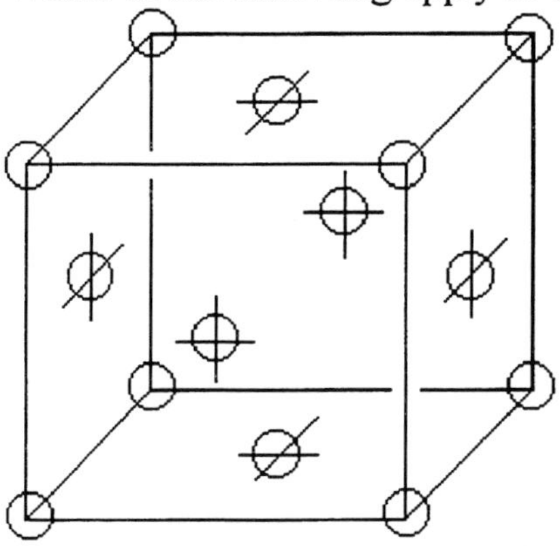

a) The unit cell is body-centered cubic.
b) The coordination number of each atom is 8.
c) The coordination number of each atom is 4.
d) The coordination number of each atom is 14.
e) The coordination number of each atom is 12.

50. Which of the following apply to the unit cell shown below?

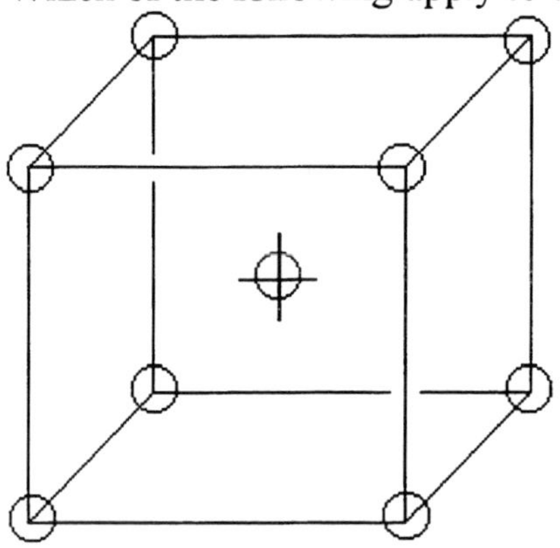

a) The unit cell is cubic close-packed and has a coordination number of 8.
b) The unit cell is face-centered cubic and has a coordination number of 12.
c) The unit cell is cubic close-packed and has a coordination number of 12.
d) The unit cell is body-centered cubic and has a coordination number of 8.
e) The unit cell is face-centered cubic and has a coordination number of 8.

51. The coordination number of cesium chloride is
a) (8,8) b) (8,4) c) (1,1) d) (6,6) e) (4,4)

52. All of the following have the rock-salt structure except
a) CsI b) MgO c) AgCl d) KBr e) RbI

53. If the radii of Mg^{2+} and O^{2-} are 65 and 140 pm, respectively, what is the coordination number of Mg^{2+} in MgO?
a) 6 b) 4 c) 8 d) 12 e) 1

54. If the radii of Cs^{+} and I^{-} are 169 and 216 pm, respectively, what is the coordination number of Cs^{+} in CsI?
a) 8 b) 4 c) 6 d) 12 e) 1

55. If the radii of cation and anion are 199 and 210 pm, respectively, in an ionic compound, predict the coordination number of the cation.
a) 8 b) 4 c) 6 d) 12 e) 1

56. Which of the following is an amorphous solid?
a) rubber b) ice c) graphite d) glucose e) black phosphorus

57. Which of the following is a network solid?
a) diamond b) vanadium c) $CaTiO_3$ d) benzene, C_6H_6 e) mercury

58. Which of the following has the highest vapor pressure at room temperature?
 a) water, H_2O
 b) methanol, CH_3OH
 c) acetic acid, CH_3COOH
 d) ethanol, CH_3CH_2OH
 e) dimethylether, CH_3OCH_3

59. The vapor pressure of water above 50 mL of water in a 60-mL closed container is 23.8 Torr at 25°C. What is the vapor pressure of water if the volume of the container is changed to 100 mL?
 a) 11.9 Torr b) 15.9 Torr c) 39.7 Torr d) about 26 Torr e) 23.8 Torr

60. The vapor pressure of benzene above 50 mL of benzene in a 60-mL container is 94.6 Torr at 25°C. What is the vapor pressure of benzene if the volume of the container is changed to 100 mL?
 a) 94.6 Torr b) about 98 Torr c) 56.8 Torr d) 47.3 Torr e) 158 Torr

61. The normal boiling point of a liquid
 a) is the temperature at which its vapor pressure is 1 atm.
 b) is the temperature at which its vapor pressure is 2 atm.
 c) is the temperature at which there is a single uniform phase.
 d) is the temperature at which the gas cannot be condensed.
 e) is the temperature calculated from the ideal gas equation for a pressure 2 atm.

62. In which of the following would the boiling point of water be highest?
 a) in New York city where the pressure is about 760 Torr
 b) in New Mexico where the pressure is about 710 Torr
 c) in a pressure cooker where the pressure is 1400 Torr
 d) at the peak of Mt. Everest
 e) in the "mile high" city of Denver

63. If the density of the solid is greater than the density of the liquid form of a substance, the slope of the solid/liquid phase boundary
 a) is negative.
 b) cannot be determined without the critical point.
 c) is zero.
 d) is positive.
 e) cannot be determined without the triple point.

64. The phase diagram for a pure compound is given below. The triple point occurs at

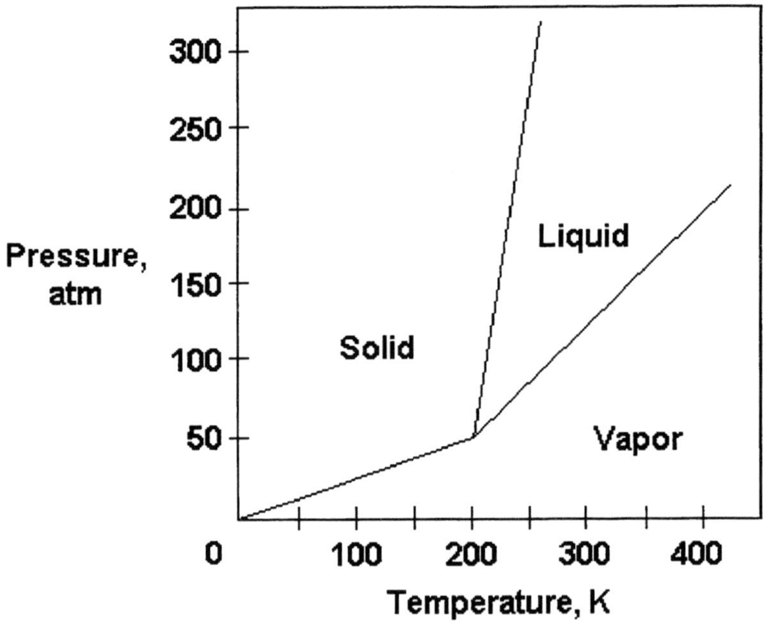

a) greater than 50 atm and greater than 200 K.
b) 200 atm and 400 K.
c) 0 atm and 200 K.
d) 320 atm and 250 K.
e) 50 atm and 200 K.

65. The phase diagram for a pure substance is given below. The solid sublimes

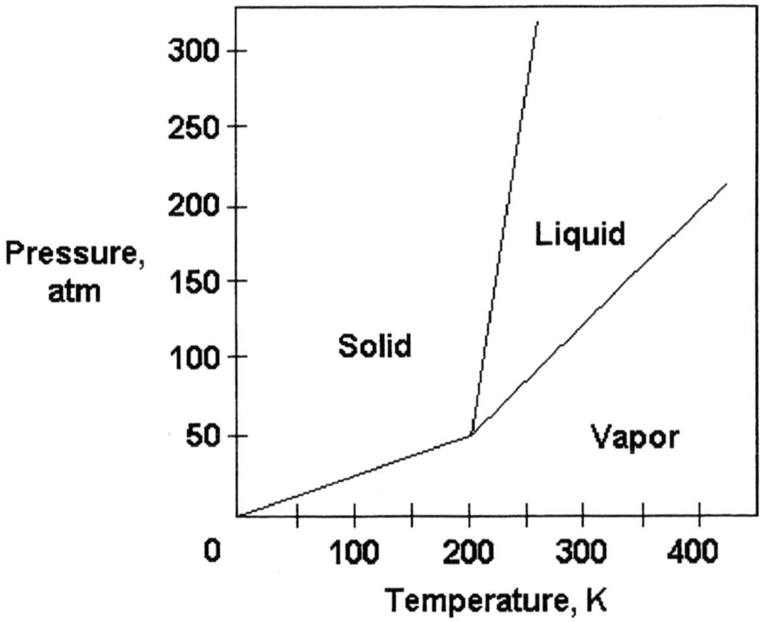

a) at 300 K and 75 atm.
b) at 200 K and 100 atm.
c) at 300 K and 100 atm.
d) if warmed at any pressure below 50 atm.
e) at 400 K and 200 atm.

66. The phase diagram for sulfur is given below.

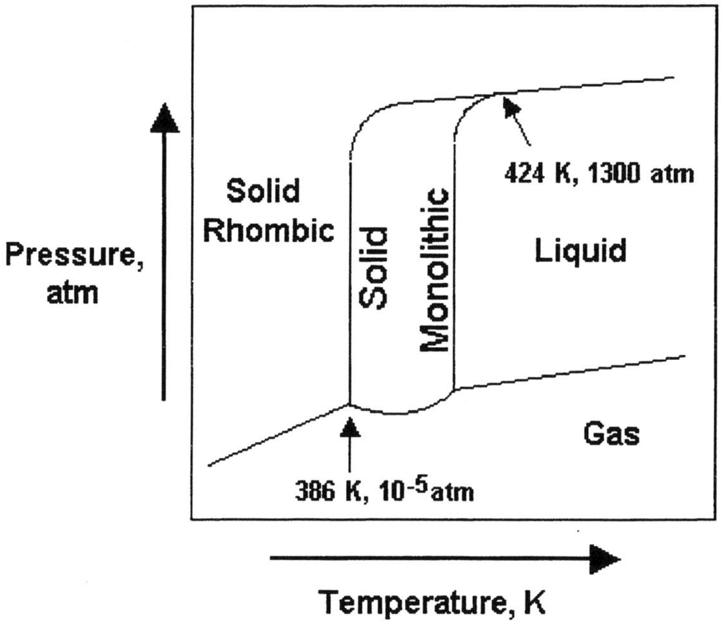

What conditions are required to convert rhombic sulfur to monoclinic sulfur?

a) heat rhombic sulfur at 1400 atm
b) reduce the pressure on rhombic sulfur at 160°C
c) heat rhombic sulfur at pressures between 1 and 1000 atm
d) heat rhombic sulfur at 1×10^{-6} atm
e) reduce the pressure on rhombic sulfur at 90°C

67. The phase diagram for sulfur is given below.

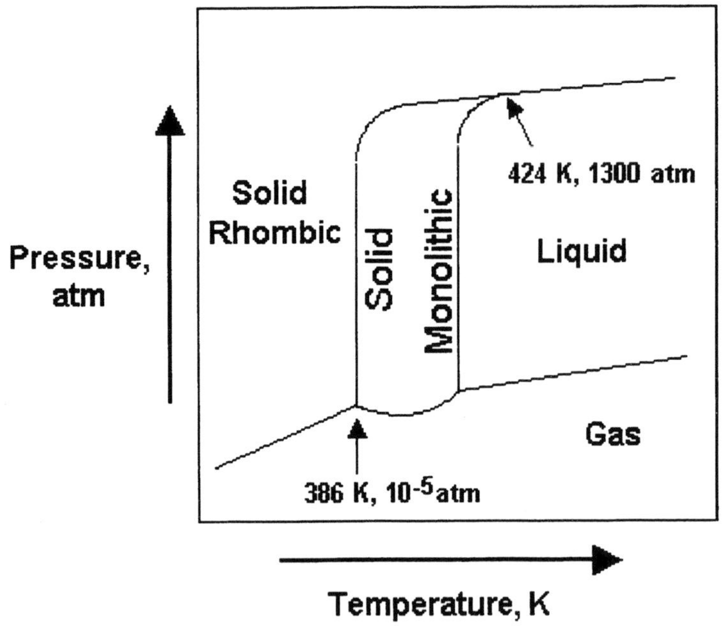

At 430 K, what pressure should be applied to completely melt rhombic sulfur?

a) 1300 atm b) 10^{-5} atm c) 1400 atm d) rhombic sulfur sublimes instead of melting e) 1000 atm

68. Using the phase diagram for water, describe how to freeze-dry foods.

69. The triple point of carbon dioxide is –56°C and 5.1 atm. The critical temperature is 31°C. Liquid carbon dioxide can exist at

a) 10 atm and –25°C.
b) 5.1 atm and –25°C.
c) 10 atm and –56°C.
d) 10 atm and –60°C.
e) 10 atm and 33°C.

70. Which of the following gases cannot be liquefied at 25°C?

a) CH_4, critical temperature 190 K
b) C_6H_6, critical temperature 562 K
c) NH_3, critical temperature 405 K
d) CO_2, critical temperature 304 K
e) Cl_2, critical temperature 419 K

71. The critical temperature of HCl is 52°C. At temperatures above 52°C,

a) HCl can be liquefied if the pressure applied is greater than the critical pressure.
b) HCl exists exclusively as a liquid.
c) HCl decomposes into its constituent elements.
d) HCl cannot be liquefied no matter what the pressure.
e) HCl decomposes into the atoms H and Cl.

72. The phase diagram for a pure substance is given below. What is the critical temperature?

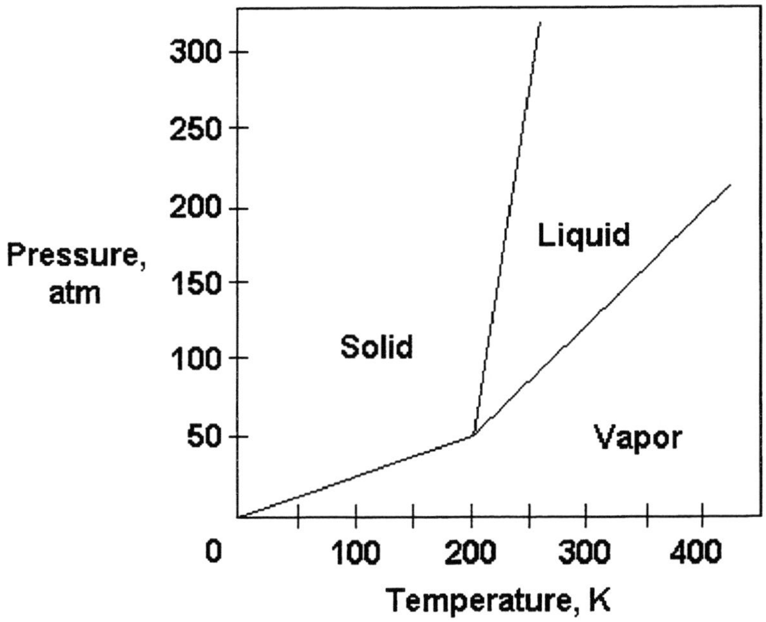

a) 0 K b) 200 K c) 400 K d) 250 K e) 300 K

73. The phase diagram for a pure substance is shown below.

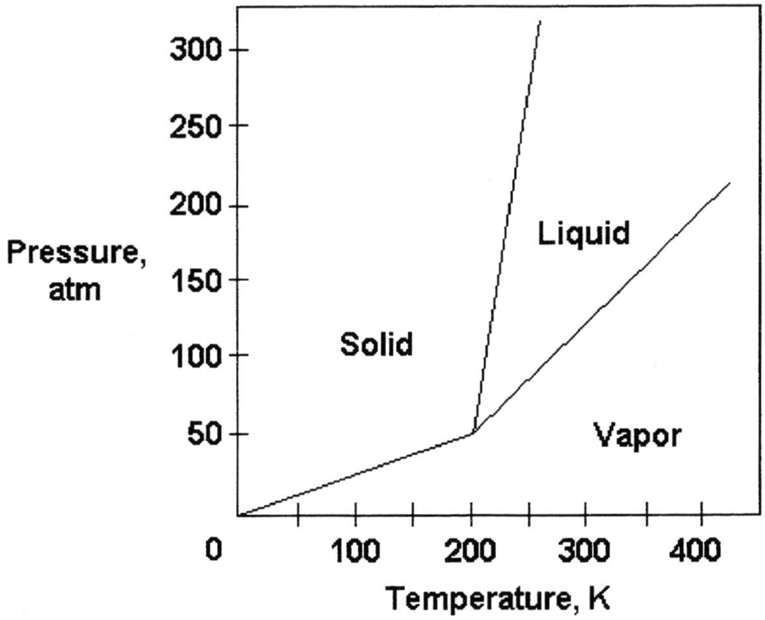

What is the highest temperature at which the substance can exist as a liquid?

a) any temperature above 200 K b) 400 K c) 200 K d) 350 K e) 250 K

74. The phase diagram for a pure substance is given below.

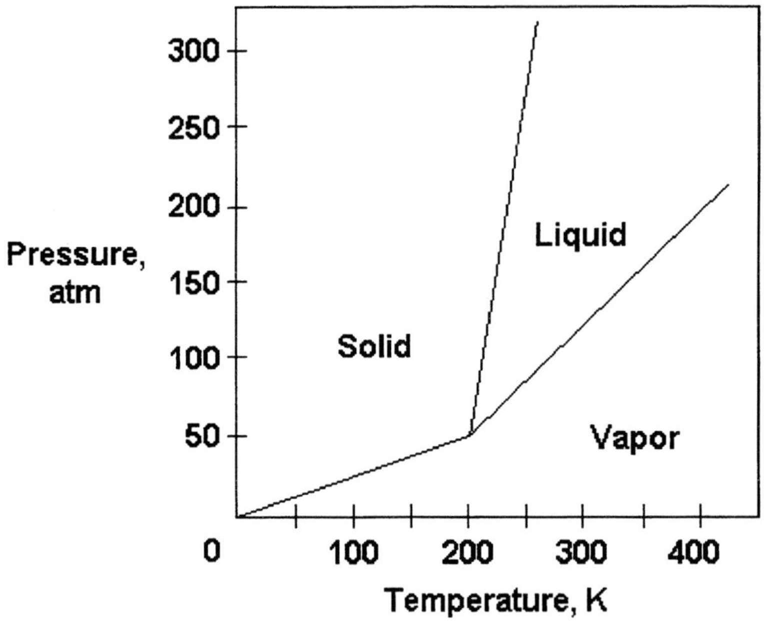

What pressure must be applied to liquefy a sample at 400 K, the critical temperature?

a) 150 atm b) 175 atm c) 250 atm d) 50 atm e) The sample cannot be liquefied at 400 K.

75. The critical temperature and pressure of carbon dioxide are 31°C and 73 atm. What phase of CO_2 exists at 32°C and 73 atm?

a) one phase called supercritical CO_2
b) vapor
c) vapor in equilibrium with liquid
d) liquid
e) supercritical CO_2 in equilibrium with liquid

76. The triple point of water is at 0.01°C and 4.6 Torr. Which of the following conditions would be most suitable for freeze-drying of food? The initial pressure is atmospheric.

a) The critical temperature and pressure should be used.
b) –10°C and 50 Torr
c) –10°C and 1×10^{-2} Torr
d) 0.01°C and 4.6 Torr
e) 0.01°C and 50 Torr

77. The vapor pressure of water at 25°C is 24 Torr. The vapor pressure at 50°C is about

a) 24 Torr b) 760 Torr c) 25 Torr d) 23 Torr e) 100 Torr

78. Which is a possible method to produce liquid carbon dioxide from the gas? The critical temperature and pressure are 31°C and 73 atm, and the triple point is –57°C, 5.1 atm.

a) Compress the gas at room temperature.
b) Cool the gas to –57°C at atmospheric pressure.
c) Apply 73 atm pressure at 32°C.
d) Compress the gas at 100°C.
e) Compress the gas at 35°C.

79. Which of the following has a critical temperature of about 5K?
a) He b) Ne c) Ar d) Kr e) Xe

80. Which of the following has the highest critical temperature?
a) CH_4 b) H_2O c) NH_3 d) CO_2 e) O_2

Chapter 10: Liquid and Solid Materials
Form B

1. Which of the following is most volatile?
a) CBr_2H_2 b) CCl_3H c) CBr_4 d) CCl_4 e) CF_4

2. Which of the following has the highest boiling point?
a) Xe b) Kr c) Ar d) Ne e) He

3. Which of the following has the smallest standard molar enthalpy of vaporization?
a) PH_3 b) SiH_4 c) HCl d) H_2S e) SnH_4

4. Which of the following has the lowest boiling point?
a) *n*-propane, $CH_3CH_2CH_3$
b) *n*-pentane, $CH_3(CH_2)_3CH_3$
c) *n*-heptane, $CH_3(CH_2)_5CH_3$
d) *n*-hexane, $CH_3(CH_2)_4CH_3$
e) *n*-butane, $CH_3(CH_2)_2CH_3$

5. Consider the following molecules, all with molecular formula C_6H_{14}:

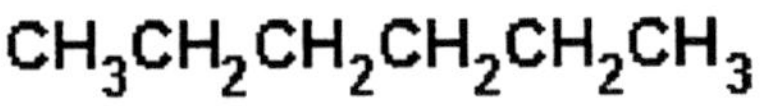
1

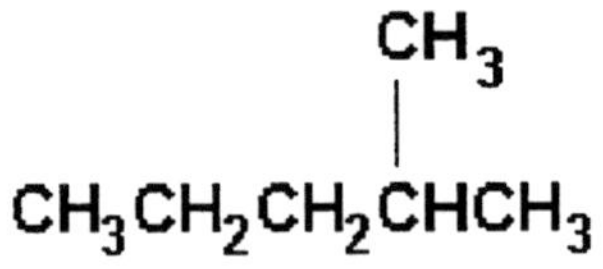
2

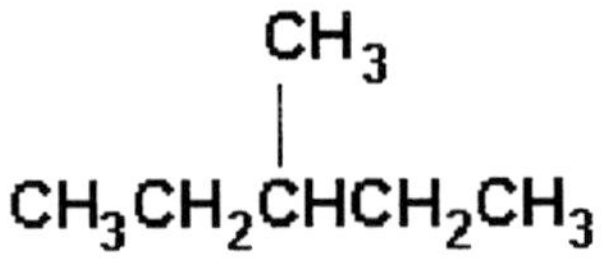
3

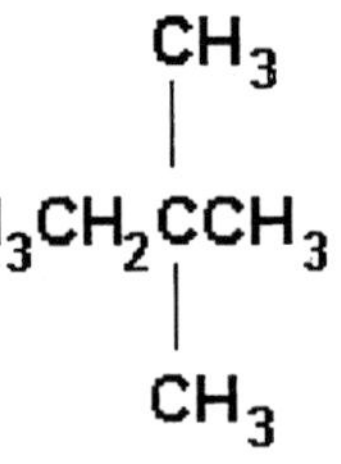
4

Which has the lowest boiling point?
a) 4
b) All have the same boiling point because they have the same molecular formula.
c) 2
d) 3
e) 1

6. Which of the following has the lowest boiling point?
a) I_2 b) Br_2 c) SiH_4 d) GeH_4 e) N_2

7. Which of the following is correct?
a) The energy of ion-ion interactions is about 250 $kJ \cdot mol^{-1}$.
b) The energy of London forces is about 20 $kJ \cdot mol^{-1}$.
c) The energy of dipole-dipole interactions is about 20 $kJ \cdot mol^{-1}$.
d) The energy of a hydrogen bond is about 2 $kJ \cdot mol^{-1}$.
e) The energy of ion-dipole interactions is about 2 $kJ \cdot mol^{-1}$.

8. Which of the following has the highest boiling point?
a) SiH_4 b) H_2S c) PH_3 d) H_2Se e) H_2Te

9. Which of the following has the largest standard molar enthalpy of vaporization?
a) Xe b) He c) Ne d) Kr e) Ar

10. Which of the following has the largest standard molar enthalpy of vaporization?
a) H_2S b) H_2O c) HF d) NH_3 e) CH_4

11. Which of the following has the highest boiling point?
a) HF b) HCl c) PH_3 d) SiH_4 e) GeH_4

12. The boiling point of Br_2 is higher than Cl_2 due to
a) dipole-dipole forces.
b) London forces.
c) hydrogen bonding.
d) ion-dipole forces.
e) ion-ion forces.

13. The boiling point of H_2O is higher than H_2S due to
a) ion-dipole forces.
b) London forces.
c) hydrogen bonding.
d) dipole-dipole forces.
e) ion-ion forces.

14. Arrange the following substances in order of decreasing boiling point.
SiH_4, HCl, CH_4, HI
a) HCl > HI > SiH_4 > CH_4
b) CH_4 > SiH_4 > HCl > HI
c) HI > HCl > SiH_4 > CH_4
d) HI > HCl > CH_4 > SiH_4
e) HCl > HI > CH_4 > SiH_4

15. For which of the following substances would hydrogen bonding be most important?
a) H_2Se b) H_2S c) H_2Te d) H_2O e) HBr

16. Which of the following species is likely to forn hydrogen bonds in the pure state?
a) $CH_3C(O)CH_3$ b) CH_3OCH_3 c) $(CH_3)_3N$ d) CH_3CH_2COOH e) CH_4

17. Which of the following species is likely to forn hydrogen bonds in the pure state?
a) CH_4 b) CH_3OCH_3 c) H_2O d) $(CH_3)_3N$ e) $CH_3C(O)CH_3$

18. Which of the following species is likely to forn hydrogen bonds in the pure state?
a) CH_3OCH_3 b) $CH_3C(O)CH_3$ c) $CH_2(OH)CH(OH)CH_2(OH)$ d) CH_4 e) $(CH_3)_3N$

19. List the following in order of increasing boiling point:
hydrogen chloride, hydrogen fluoride, argon
a) HF < Ar < HCl b) Ar < HCl < HF c) Ar < HF < HCl d) HCl < Ar < HF e) HF < HCl < Ar

20. Which of the following is correct?
a) The energy of London forces is about 20 $kJ \cdot mol^{-1}$.
b) The energy of a hydrogen bond is about 20 $kJ \cdot mol^{-1}$.
c) The energy of dipole-dipole interactions is about 20 $kJ \cdot mol^{-1}$.
d) The energy of a ion-dipole interactions is about 2 $kJ \cdot mol^{-1}$.
e) The energy of ion-ion interactions is about 2 $kJ \cdot mol^{-1}$.

21. Which of the following has the greatest viscosity?
a) HF b) benzene, C_6H_6 c) propane, C_3H_8 d) HCl e) NH_3

22. The meniscus of water curves upward in a glass tube forming a " ∪" shape. This means
a) mercury has a low surface tension.
b) mercury tends to cover the greatest possible area of the glass.
c) the adhesive forces between water molecules and glass are stronger than the cohesive forces between water molecules.
d) the cohesive forces between water molecules are stronger than the adhesive forces between water and glass.
e) mercury has a large capillary action.

23. Which of the following has the greatest surface tension?
a) water b) HBr c) benzene d) propane, C_3H_8 e) NH_3

24. Copper has a face-centered cubic unit cell. If the density of copper is 8.90 $g \cdot cm^{-3}$, calculate the atomic radius of copper.

25. Vanadium has a density of 6.08 $g \cdot cm^{-3}$ and an atomic radius of 131 pm. What is the structure of the solid?

26. If the radius of Cs^+ is 169 pm and that of Cl^- is 181 pm, calculate the density of cesium chloride.

27. The density of AgCl is 5.56 $g \cdot cm^{-3}$ and the radii of Ag^+ and Cl^- are 126 pm and 181 pm, respectively. Determine the number of AgCl formula units in a unit cell. What is the structure of AgCl?

28. If iron has a body-centered cubic unit cell, how many atoms of iron does the unit cell contain?
a) 2 b) 9 c) 5 d) 1 e) 3

29. If molybdenum has a body-centered cubic unit cell, how many atoms of molybdenum does the unit cell contain?
a) 2 b) 9 c) 5 d) 1 e) 3

30. In a face-centered cubic lattice, an atom lying in the face of a unit cell is shared equally by how many unit cells?
a) 2 b) 1 c) 6 d) 4 e) 8

31. A cubic close-packed structure has an
a) ABCABC... pattern of layers and a coordination number of 12.
b) ABABAB... pattern of layers and a coordination number of 12.
c) ABCDABCD... pattern of layers and a coordination number of 4.
d) ABBA... pattern of layers and a coordination number of 8.
e) AABB... pattern of layers and a coordination number of 8.

32. A structure with an ABCABC... pattern of layers is called
a) cubic close-packed.
b) hexagonally close-packed.
c) primitive cubic.
d) face-centered cubic.
e) body-centered cubic.

33. The coordination numbers of metal ions in face-centered cubic and body-centered cubic structures, respectively, are
a) 6 and 6. b) 8 and 8. c) 12 and 8. d) 12 and 12. e) 12 and 6.

34. Copper has a face-centered cubic unit cell. If the atomic radius of copper is 128 pm, what is the edge length of the unit cell?
a) 512 pm b) 362 pm c) 181 pm d) 256 pm e) 222 pm

35. Gold has a face-centered cubic unit cell. If the atomic radius of gold is 144 pm, what is the edge length of the unit cell?
a) 204 pm b) 576 pm c) 407 pm d) 249 pm e) 288 pm

36. Copper has a face-centered cubic unit cell. If the edge length of the unit cell is 362 pm, what is the atomic radius of copper?
a) 90.5 pm b) 181 pm c) 128 pm d) 209pm e) 256 pm

37. Gold has a face-centered cubic unit cell. If the edge length of the unit cell is 407 pm, what is the atomic radius of gold?
a) 235 pm b) 204 pm c) 288 pm d) 102 pm e) 144 pm

38. Sodium has a body-centered cubic unit cell. If the atomic radius of sodium is 180 pm, what is the edge length of the unit cell?
a) 255 pm b) 360 pm c) 720 pm d) 509 pm e) 416 pm

39. Vanadium has a body-centered cubic unit cell. If the atomic radius of vanadium is 131 pm, what is the edge length of the unit cell?
a) 186 pm b) 162 pm c) 324 pm d) 371 pm e) 303 pm

40. Sodium has a body-centered cubic unit cell. If the edge length of the unit cell is 416 pm, what is the atomic radius of sodium?
a) 147 pm b) 180 pm c) 208 pm d) 104 pm e) 294 pm

41. Vanadium has a body-centered cubic unit cell. If the edge length of the unit cell is 303 pm, what is the atomic radius of vanadium?
a) 107 pm b) 75.8 pm c) 131 pm d) 214 pm e) 152 pm

42. Copper has a face-centered cubic unit cell. If the edge length of the unit cell is 362 pm, what is the density of copper?
a) $8.90\ g \cdot cm^{-3}$ b) $2.23\ g \cdot cm^{-3}$ c) $6.29\ g \cdot cm^{-3}$ d) $4.45\ g \cdot cm^{-3}$ e) $3.15\ g \cdot cm^{-3}$

43. Aluminum has a face-centered cubic unit cell. If the edge length of the unit cell is 354 pm, what is the density of aluminum?
a) $1.01\ g \cdot cm^{-3}$ b) $2.86\ g \cdot cm^{-3}$ c) $4.04\ g \cdot cm^{-3}$ d) $2.02\ g \cdot cm^{-3}$ e) $1.43\ g \cdot cm^{-3}$

44. Sodium has a body-centered cubic unit cell. If the edge length of the unit cell is 416 pm, what is the density of sodium?
a) $0.530\ g \cdot cm^{-3}$ b) $1.06\ g \cdot cm^{-3}$ c) $0.375\ g \cdot cm^{-3}$ d) $2.12\ g \cdot cm^{-3}$ e) $0.750\ g \cdot cm^{-3}$

45. Vanadium has a body-centered cubic unit cell. If the edge length of the unit cell is 303 pm, what is the density of vanadium?
a) $12.2\ g \cdot cm^{-3}$ b) $2.15\ g \cdot cm^{-3}$ c) $6.08\ g \cdot cm^{-3}$ d) $4.30\ g \cdot cm^{-3}$ e) $3.04\ g \cdot cm^{-3}$

46. The cubic unit cell of an ionic compound has A cations at the corners and the face centers, and B anions in the center of the edges. What is the empirical formula of the compound?
a) AB_3 b) A_7B_6 c) A_5B_6 d) A_4B e) A_4B_3

47. One form of the mineral perovskite has a unit cell that has barium cations at the corners of a cube, a titanium cation in the cube center, and oxygen anions in the center of the edges. What is the empirical formula of perovskite?
a) Ba_4TiO_3 b) $BaTiO_6$ c) Ba_8TiO_6 d) Ba_2TiO_3 e) $BaTiO_3$

48. The following unit cell is the

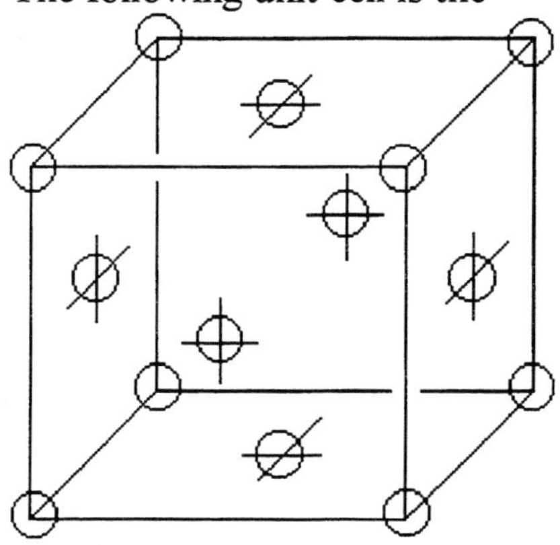

a) body-centered cubic unit cell.
b) hexagonally close-packed unit cell.
c) face-centered cubic unit cell.
d) primitive cubic unit cell.
e) coordination number 8 cubic unit cell.

49. The following unit cell is the

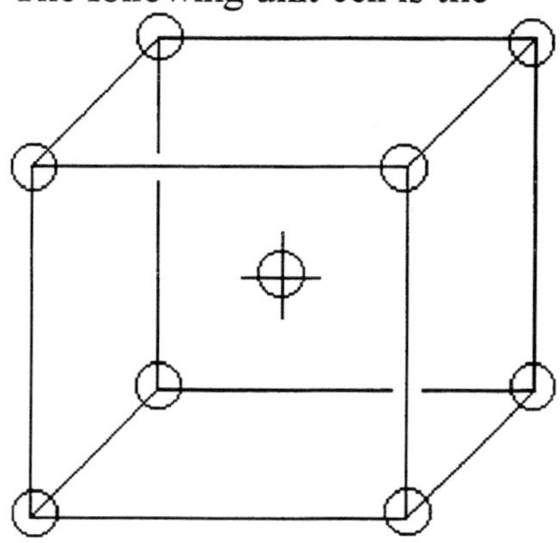

a) body-centered cubic unit cell.
b) face-centered cubic unit cell.
c) cubic close-packed unit cell.
d) hexagonally close-packed unit cell.
e) coordination number 4 unit cell.

50. How many atoms are in the unit cell below?

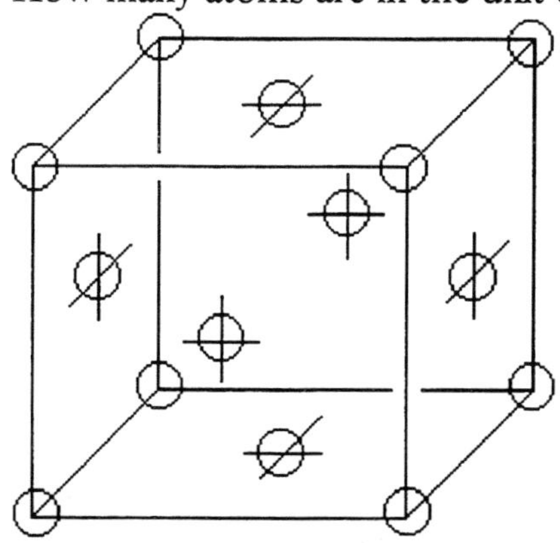

a) 14 b) 6 c) 4 d) 5 e) 7

51. The coordination number of sodium chloride is
a) (6,6) b) (4,4) c) (8,8) d) (8,4) e) (1,1)

52. All of the following have the rock-salt structure except
a) CsCl b) AgCl c) CaO d) KBr e) RbI

53. If the radii of Mg^{2+} and O^{2-} are 65 and 140 pm, respectively, what is the coordination number of O^{2-} in MgO?
a) 6 b) 4 c) 8 d) 12 e) 1

54. If the radii of Cs^+ and I^- are 169 and 216 pm, respectively, what is the coordination number of I^- in CsI?
a) 8 b) 4 c) 6 d) 12 e) 1

55. If the radii of cation and anion are 60 and 115 pm, respectively, in an ionic compound, predict the coordination number of the cation.
a) 6 b) 4 c) 8 d) 12 e) 1

56. Which of the following is a molecular solid?
a) ice b) graphite c) diamond d) TiO_2 e) NaCl

57. All of the following are network solids except
a) S_8 b) BN c) graphite d) SiO_2 e) black phosphorus

58. Which of the following has the highest vapor pressure at room temperature?
a) methanol, CH_3OH b) water, H_2O c) ethanol, CH_3CH_2OH d) benzene, C_6H_6 e) mercury, Hg

59. The vapor pressure of ethanol above 50 mL of ethanol in a 60-mL container is 58.9 Torr at 25°C. What is the vapor pressure of ethanol if the volume of the container is changed to 100 mL?
a) 35.3 Torr b) 58.9 Torr c) 29.5 Torr d) about 65 Torr e) 98.2 Torr

60. The vapor pressure of methanol above 50 mL of methanol in a 60-mL container is 122.7 Torr at 25°C. What is the vapor pressure of methanol if the volume of the container is changed to 100 mL?
a) 122.7 Torr b) 73.6 Torr c) 61.4 Torr d) about 125 Torr e) 204.5 Torr

61. In a closed container, a liquid and its vapor reach the dynamic equilibrium,
rate of evaporation = rate of condensation
when the pressure of the vapor has risen to a definite value. This value depends on
a) the liquid and the temperature.
b) the normal boiling point of the liquid and the amount of liquid.
c) the liquid and is independent of the temperature.
d) the temperature and is independent of the liquid.
e) the amount of liquid.

62. In which of the following would the boiling point of water be lowest?
a) in New York city where the pressure is about 760 Torr
b) in a pressure cooker where the pressure is 1400 Torr
c) in the "mile high" city of Denver
d) in New Mexico where the pressure is about 710 Torr
e) at the peak of Mt. Everest

63. If the density of the liquid is greater than the density of the solid form of a substance, this indicates that
a) the slope of the solid/liquid phase boundary is zero.
b) the solid will not melt under pressure.
c) the solid will melt under pressure.
d) the slope of the solid/liquid phase boundary is positive.
e) the solid will never melt.

64. The phase diagram for a pure substance is given below.

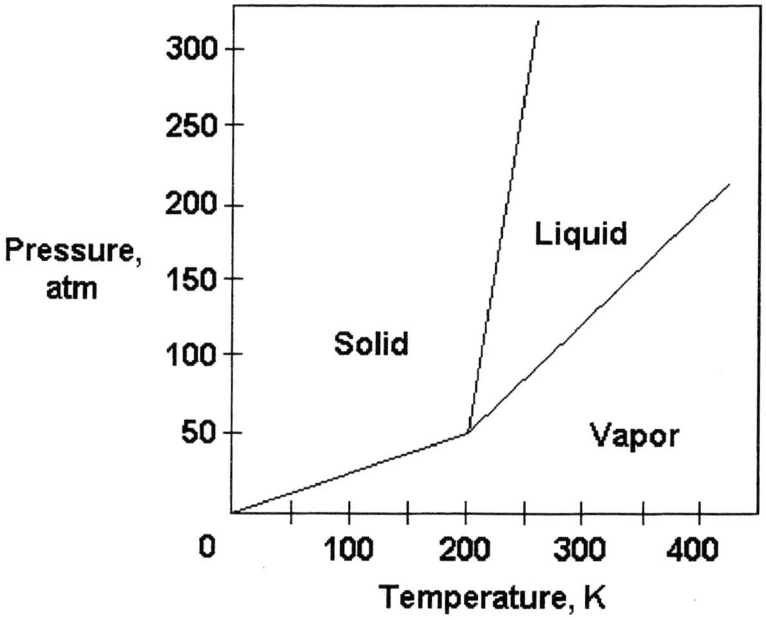

What is the lowest temperature at which liquid can exist?
a) 150 K b) 0 K c) 250 K d) 400 K e) 200 K

65. The phase diagram for a pure substance is given below.

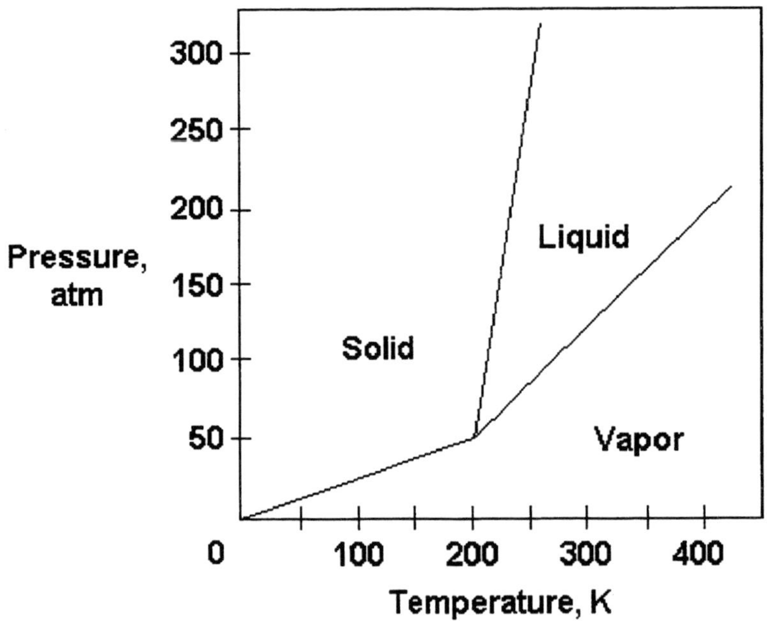

The substance is stored in a container at 150 atm at 25°C. Describe what happens if the container is opened at 25°C.

a) The liquid in the container vaporizes.
b) The solid in the container sublimes.
c) The liquid in the container freezes.
d) The solid in the container melts.
e) The vapor in the container escapes.

66. The phase diagram for sulfur is given below.

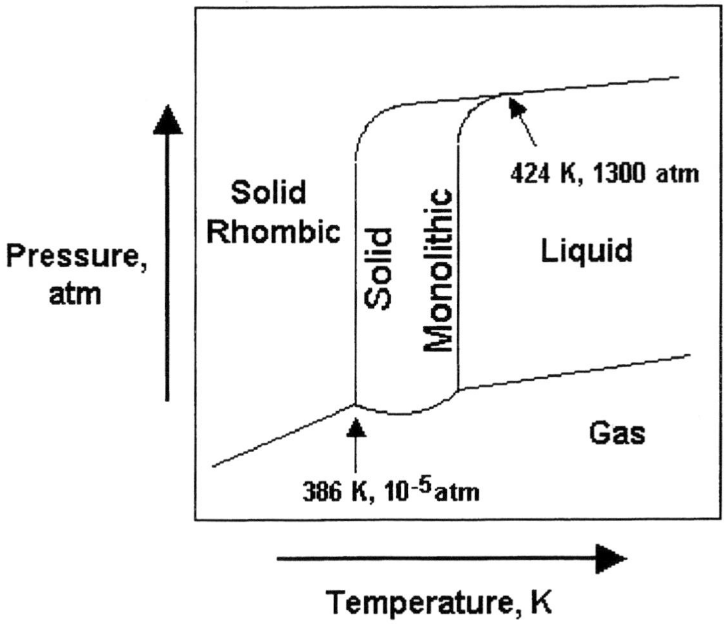

Which of the following is true?

a) Rhombic sulfur cannot be directly converted to liquid.
b) Sulfur has 2 triple points.
c) Sulfur has 3 triple points.
d) Monoclinic sulfur does not sublime.
e) Sulfur has 0 triple points.

67. The phase diagram for sulfur is given below.

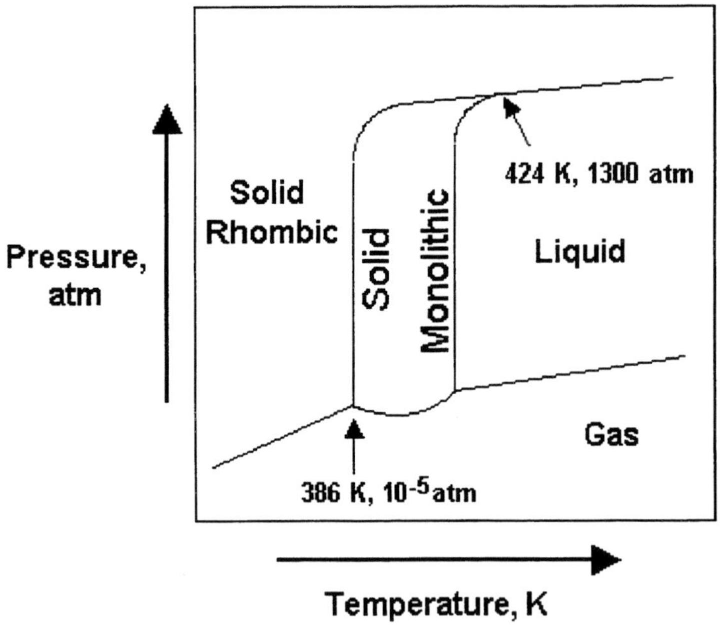

At 368 K and 10^{-5} atm,

a) only rhombic sulfur is present.
b) rhombic sulfur, monoclinic sulfur, and sulfur gas exist in equilibrium.
c) rhombic sulfur, monoclinic sulfur, sulfur liquid, and sulfur gas exist in equilibrium.
d) only rhombic sulfur and sulfur gas exist in equilibrium.
e) only monoclinic sulfur is present.

68. Explain why the slope of the solid/liquid phase boundary line is negative for the phase diagram of water but positive for the carbon dioxide phase diagram. Also, explain why it is easier to cut a block of ice by pulling a wire through it than to cut a block of carbon dioxide in the same manner.

69. The triple point of carbon dioxide is –56°C and 5.1 atm. The critical temperature is 31°C. Gaseous carbon dioxide can exist at

a) 5.1 atm and –60°C.
b) 6 atm and –56°C.
c) 6 atm and 25°C.
d) 10 atm and –50°C.
e) 25 atm and 30°C.

70. Which of the following gases cannot be liquefied at 25°C?

a) C_6H_6, critical temperature 562 K
b) NH_3, critical temperature 405 K
c) CO_2, critical temperature 304 K
d) Cl_2, critical temperature 419 K
e) O_2, critical temperature 155 K

71. The critical temperature of Xe is 17°C. At 17°C,
  a) Xe exists exclusively as a liquid.
  b) Xe cannot be liquefied no matter what the pressure.
  c) Xe can be liquefied if the critical pressure of 58 atm is applied.
  d) liquid and gaseous Xe exist no matter what the pressure.
  e) Xe can be liquefied at any pressure below the critical pressure.

72. The phase diagram for a pure substance is given below. What is the critical pressure?

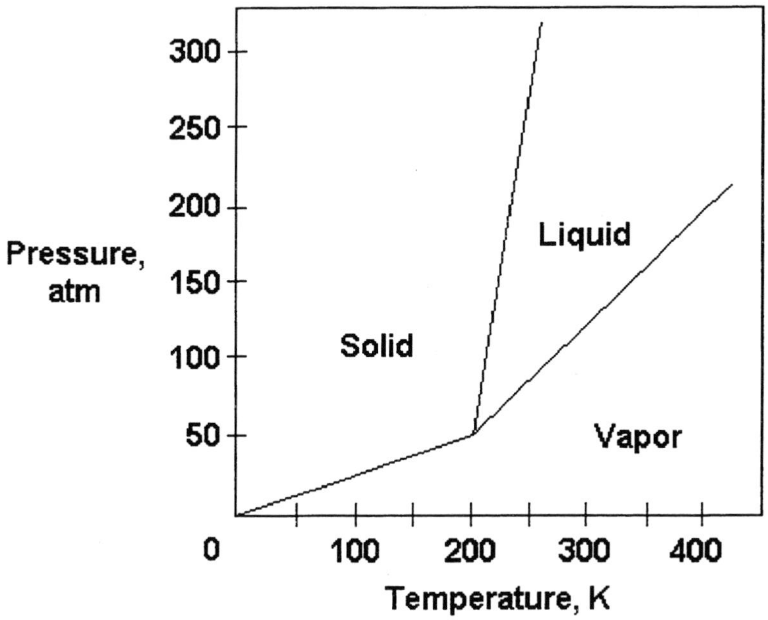

a) 150 atm b) any pressure above 325 atm c) 50 atm d) 325 atm e) 200 atm

73. The phase diagram for a pure substance is given below.

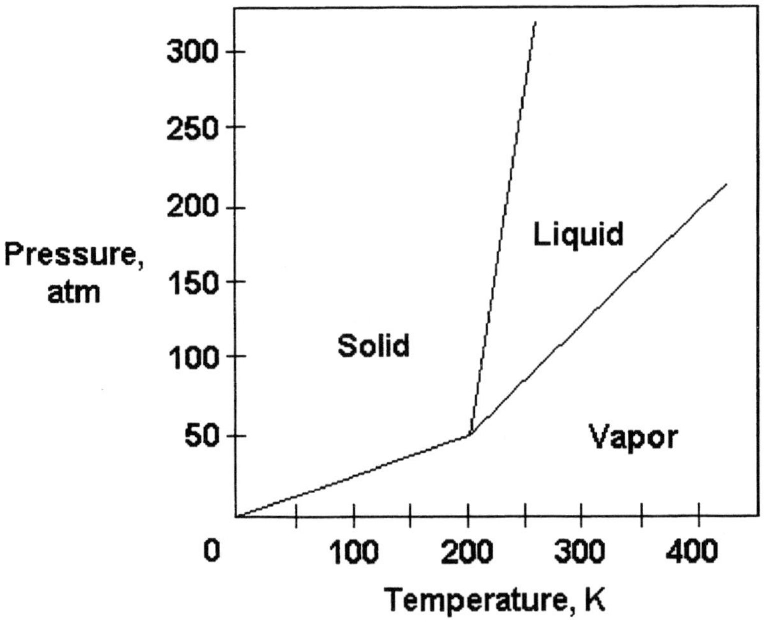

At 1 atm and 150 K, the substance exists as

a) both vapor and liquid in equilibrium.
b) liquid.
c) solid.
d) both vapor and solid in equilibrium.
e) vapor.

74. The phase diagram for a pure substance is given below.

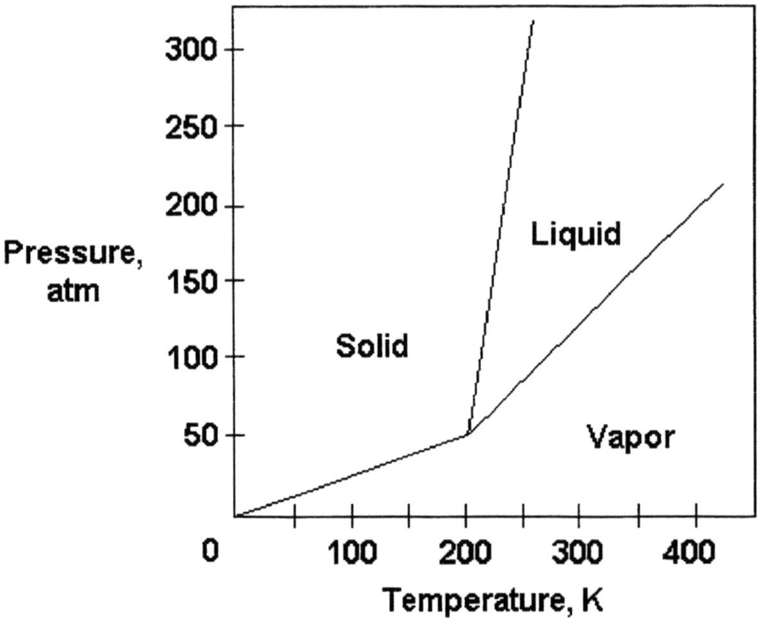

What pressure must be applied to liquefy a sample at 425 K?

a) 350 atm b) 150 atm c) 250 atm d) The sample cannot be liquefied at 425 K. e) 50 atm

75. The critical temperature and pressure of carbon dioxide are 31°C and 73 atm. Which of the following is true?
  a) At 31°C, CO_2 cannot be liquefied no matter how great the pressure.
  b) At 31°C, CO_2 can be solidified at 150 atm.
  c) At 31°C, CO_2 can be liquefied at pressures below 73 atm.
  d) At 31°C, CO_2 can be liquefied at pressures greater than 73 atm.
  e) At 31°C, CO_2 exists as an equilibrium mixture of solid, liquid and vapor.

76. The triple point of water is at 0.01°C and 4.6 Torr. What change in conditions represents the disappearance of frost without any water being formed?
  a) –5°C and 6 Torr to 5°C and 6 Torr
  b) –5°C and 3 Torr to 5°C and 3 Torr
  c) –5°C and 10 Torr to 25°C and 10 Torr
  d) 0°C and 4.6 Torr to 5°C and 4.6 Torr
  e) 0°C and 10 Torr to 25°C and 10 Torr

77. The vapor pressure of chloroform, $CHCl_3$, is 170 Torr at 20°C. The vapor pressure at 30°C is about
  a) 165 Torr b) 175 Torr c) 170 Torr d) 300 Torr e) 760 Torr

78. Which is a possible method to liquefy helium? The critical temperature and pressure are 5.2 K and 2.3 atm, and the triple point is 2.2 K, 0.07 atm.
  a) Cool the gas to 5.2 K at 2 atm.
  b) Compress the gas at room temperature.
  c) Cool the gas to 2.2 K at 0.06 atm.
  d) Compress the gas at 6 K.
  e) Compress the gas at 4 K.

79. Which of the following has a critical temperature about 30°C?
  a) N_2 b) CO_2 c) CH_4 d) O_2 e) Kr

80. Which of the following has the highest critical temperature?
  a) O_2 b) H_2 c) N_2 d) Ne e) He

Chapter 11: Carbon-Based Materials
Form A

1. Which of the following is an unsaturated hydrocarbon?
a) C_5H_{12} b) C_4H_{10} c) C_3H_4 d) C_4H_8 e) C_3H_8

2. All of the following are saturated hydrocarbons except
a) C_5H_{12} b) C_3H_8 c) C_3H_4 d) C_4H_8 e) C_4H_{10}

3. The tenth member of the alkanes has
a) 22 hydrogen atoms
b) 12 hydrogen atoms
c) 10 hydrogen atoms
d) 20 hydrogen atoms
e) 18 hydrogen atoms

4. The seventh member of the alkanes has the molecular formula
a) C_7H_{10} b) C_7H_{12} c) C_7H_{14} d) C_7H_8 e) C_7H_{16}

5. Which of the following could be a cycloalkane?
a) C_6H_{14} b) C_6H_{12} c) C_3H_8 d) C_5H_{12} e) C_4H_{10}

6. All of the following could be cycloalkanes except
a) C_6H_{12} b) C_3H_8 c) C_3H_6 d) C_4H_8 e) C_5H_{10}

7. Which of the following has the highest boiling point?
a) $CH_3(CH_2)_4CH_3$ b) $CH_3(CH_2)_5CH_3$ c) $CH_3(CH_2)_3CH_3$ d) C_3H_8 e) $CH_3(CH_2)_7CH_3$

8. Butane reacts with
a) boiling aqueous sodium hydroxide.
b) boiling nitric acid.
c) concentrated sulfuric acid.
d) chlorine when heated.
e) the strong oxidizing agent $KMnO_4$.

9. Which of the following pairs react?
a) $CH_4(g)$ and $Cl_2(g)$
b) $CH_4(g)$ and $NaOH(aq)$
c) $CH_4(g)$ and $H_2SO_4(l)$
d) $CH_4(g)$ and $KMnO_4(aq)$
e) $CH_4(g)$ and $HNO_3(l)$

10. Consider a molecule of molecular formula $C_{10}H_{16}$. What total number of rings plus π bonds are present in the molecule?
a) 3 b) 2 c) 1 d) 0 e) 4

11. For $(CH_3)_2CHCH(CH_3)CH_2CH(CH_2CH_3)CH_2CH_2CH_3$, the longest unbranched chain of carbon atoms is
a) 8 b) 7 c) 6 d) 9 e) 12

12. Which of the following is most volatile?
a) $CH_3CH_2CH_2CH_3$
b) $(CH_3)_3COH$
c) $CH_3(CH_2)_3OH$
d) $(CH_3)_2C(OH)CH_3$
e) $CH_3CH(OH)CH_2CH_3$

13. Which of the following could be an alkyne?
a) C_4H_6 b) C_2H_4 c) C_3H_6 d) C_4H_8 e) C_2H_6

14. The most characteristic reaction of alkenes is
a) addition b) elimination c) substitution d) dehydrogenation e) dimerization

15. There are three dichlorobenzenes(o, m, and p). Suppose that we have samples of each, labeled A, B, and C, but don't know which is which. Each reacts with $HNO_3 + H_2SO_4$ in the same way that benzene does to give one or more mononitrodichlorobenzenes, $C_6H_3Cl_2NO_2$. Compound A gives only one mononitrodichlorobenzene, compound B gives two, and compound C gives three. Write the structures of A, B, and C and their mononitration products.

16. The compound $(CH_3)_2CHCH(CH_3)CH_2CH(CH_2CH_3)CH_2CH_2CH_3$ is named as
a) a decane. b) a heptane. c) a hexane. d) a nonane. e) an octane.

17. The compound $(CH_3)_3CCH_2CH(CH_3)_2$ is
a) named as a pentane but is an isomer of octane.
b) named as a pentane but is an isomer of heptane.
c) named as a butane but is an isomer of octane.
d) named as a pentane but is an isomer of hexane.
e) named as a hexane but is an isomer of octane.

18. Name the compound $(CH_3)_2CHCH(CH_3)CH_2CH(CH_2CH_3)CH_2CH_2CH_3$.
a) 2-methyl-3-methyl-5-ethyloctane
b) 5-ethyl-2,3-dimethyloctane
c) 6,7-dimethyl-4-ethyloctane
d) 2,3-dimethyl-5-propylheptane
e) 4-ethyl-6,7-dimethyloctane

19. Name the compound $(CH_3)_2CHCH(CH_3)C(CH_3)_3$.
a) 2,2,3,4-tetramethylpentane
b) 2,3,4,4-tetramethylpentane
c) nonane
d) 1,1,2,3,3-pentamethylbutane
e) 1,1,1,2,3-pentamethylbutane

20. Name the compound $C(CH_3)_4$.
a) 2,2-dimethylpropane b) isopropylmethane c) 2-methylbutane d) isobutylmethane e) pentane

21. Name the following compound.

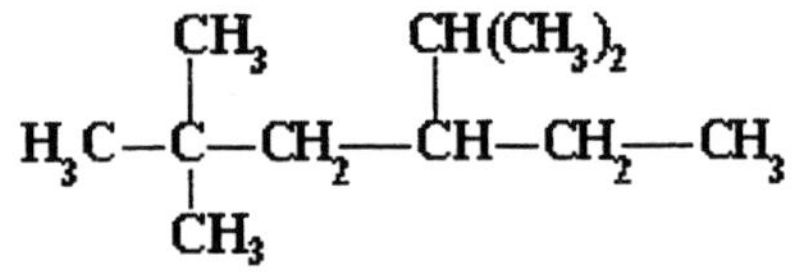

a) 4-ethyl-2,2,5-trimethylhexane
b) 3-ethyl-2,5,5-trimethylhexane
c) undecane
d) 1,1,1,4-tetramethyl-3-ethylpentane
e) 1-tert-butyl-4-ethyl-5-methylbutane

22. Name the following compound.

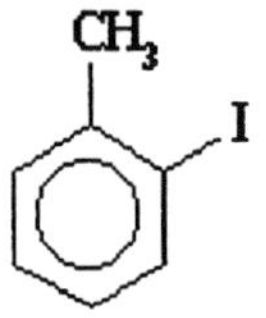

a) 1-iodo-2-methylbenzene
b) 1-methyl-2-iodobenzene
c) m-iodobenzene
d) m-toluene
e) 2-iodo-3-methylbenzene

23. Name the compound $H_2C{=}C(Cl)CH_2CH_2CH_3$.

a) 2-chloro-1-pentene
b) 2-chloro-2-pentene
c) 2-chloropentene
d) 4-chloro-4-pentene
e) chloropentene

24. Name the following compound.

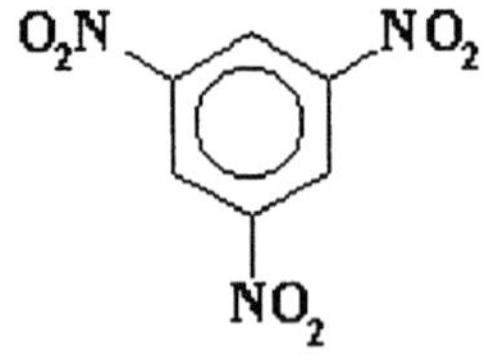

a) 1,3,5-trinitrobenzene
b) 3,5-dinitrobenzene
c) di-m-nitrobenzene
d) 2,4,6-trinitrobenzene
e) trinitrobenzene

25. The ester $CH_3(CH_2)_2COO(CH_2)_4CH_3$ is responsible for the odor of apricots. It can be prepared from

a) $CH_3(CH_2)_2COOH$ and $CH_3(CH_2)_3CHO$
b) $CH_3(CH_2)_2CH_2OH$ and $CH_3(CH_2)_3COOH$
c) $CH_3(CH_2)_2COOH$ and $CH_3(CH_2)_3CH_2OH$
d) $CH_3(CH_2)_2CHO$ and $CH_3(CH_2)_3CH_2OH$
e) $CH_3(CH_2)_2CHO$ and $CH_3(CH_2)_3COOH$

26. The formation of an ester is

a) an oxygenation reaction.
b) an irreversible reaction.
c) a condensation reaction.
d) a substitution reaction.
e) an addition reaction.

27. The product of the reaction of ethanol and acetic acid is
a) $CH_3COOCH_2CH_3$
b) $CH_3CH_2COCH_3$
c) CH_3CHO and CH_3CH_3
d) CH_3CH_2CHO
e) CH_3CH_2COOH

28. Which of the following is a tertiary alcohol?
a) $C_6H_5CH_2OH$ b) $C(CH_3)_3OH$ c) $CH(CH_3)_2OH$ d) $CH_3(CH_2)_3OH$ e) CH_3CH_2OH

29. The reaction of 1-pentanol with acetic acid gives
a) $CH_3(CH_2)_3CHO$
b) $CH_3CO(CH_2)_4CH_3$
c) $CH_3COO(CH_2)_4CH_3$
d) $CH_3(CH_2)_3COOCH_3$
e) $CH_3(CH_2)_3COOH$

30. Name the compound $CH_3COO(CH_2)_7CH_3$.
a) octyl acetate b) methyl octanoate c) heptyl acetate d) 2-decanone e) decanoic acid

31. Which of the following can produce esters?
a) an aldehyde plus $KMnO_4(aq)$ in acidic solution
b) an acid plus an aldehyde
c) a primary alcohol plus $K_2Cr_2O_7(aq)$, H_2SO_4
d) an alcohol plus an aldehyde
e) an acid plus an alcohol

32. The compound $CH_3CH_2OC_2CH_3$ is
a) a ketone. b) an aldehyde. c) an ester. d) an ether. e) an alcohol.

33. Oxidation of secondary alcohols produces
a) ketones. b) ethers. c) aldehydes. d) acids. e) Secondary alcohols cannot be oxidized.

34. Oxidation of 2-pentanol gives
a) 2-pentanone. b) 2-pentanal. c) 3-pentanone. d) pentanal. e) No reaction occurs.

35. To produce butanone, which of the following should be reacted with $Na_2Cr_2O_7(aq)$, $H_2SO_4(aq)$?
a) $CH_3CH_2CH_2CH_2OH$
b) $(CH_3)_2CHCHO$
c) $CH_3CH_2CH(COOH)CH_3$
d) $CH_3CH_2CH(OH)CH_3$
e) $CH_3CH_2CH(CH_2OH)CH_3$

36. Treatment of ethanol with $Na_2Cr_2O_7(aq)$, $H_2SO_4(aq)$ yields compound A which reacts further to give B. Identify A and B, respectively.
a) CH_3CH_3 and CH_2CH_2
b) CH_3COCH_3 and CH_3COOH
c) CH_3OCH_3 and CH_3CH_3
d) CH_3CHO and CH_3COOH
e) CH_3CHO and CH_3OCH_3

37. All of the following produce a silver mirror with Tollen's reagent except
a) $CH_3CH_2COCH_3$ b) CH_3CH_2CHO c) HCHO d) $CH_3CH_2CH(CHO)CH_3$ e) $CH_3(CH_2)_3CHO$

38. When aldehydes react with Tollen's reagent,
a) the aldehyde reduces silver ions.
b) silver ions are produced.
c) a red precipitate forms.
d) an alcohol is produced.
e) a ketone is produced.

39. Write structures for the principal product or products expected in each of the following reactions.
(a) $CH_2CH_2 + H_2$ (Ni, 500°C) →
(b)

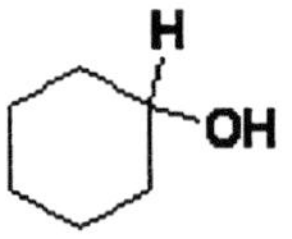

$+ Na_2Cr_2O_7(aq), H_2SO_4(aq) \rightarrow$

40. Write an equation for the synthesis of
(a) 2-methylbutyl butyrate
(b) 2-butanone
(c) N-(4-hydroxyphenyl)acetamide, Tylenol.

41. The hydroxyl group occurs in
a) phenols and ketones.
b) alcohols only.
c) ketones and carboxylic acids.
d) aldehydes and alcohols.
e) alcohols, phenols, and carboxylic acids.

42. The final product of the reaction of $CH_2(CH_2OH)_2$ with excess $KMnO_4(aq)$ in acidic solution is
a) CH_3CH_2COOH b) $CH_2(CHO)_2$ c) $CH_2(CHO)CH_2CH_2CHO$ d) $CH_2(COOH)_2$ e) CH_3CHO

43. Predict the product of the reaction of acetic acid with methylamine.
a) CH_3CONH_2 b) no reaction occurs c) $CH_3CONHCH_3$ d) $CH_3CONH_2(CH_3)^+$ e) CH_3COCH_3

44. Predict the product of the reaction of acetic acid with trimethylamine.
a) CH_3CONH_2 b) no reaction occurs c) $CH_3CONHCH_3$ d) $CH_3CON(CH_3)_2$ e) $CH_3CON(CH_3)_3^+$

45. The systematic name of $CH_3CH_2CH(CHO)CH_3$ is
a) 2-methylbutanal. b) 3-methylbutanal. c) 2-pentanone. d) 2-pentanal. e) 2-ethylpropanal.

46. Polyester is formed by the condensation polymerization of terephthalic acid, shown below, and ethylene glycol. The systematic name for terephthalic acid is

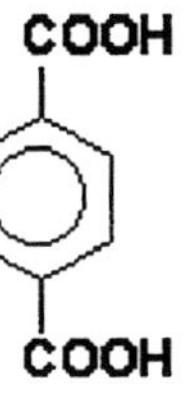

a) 1,4-dicarboxybenzene.
b) p-carboxybenzoic acid.
c) 3,6-benzenedicarboxylic acid.
d) 3-carboxybenzoic acid.
e) 1,4-phthalic acid.

47. (a) Draw structures for the five isomers of formula C_6H_{14}.
(b) Give the systematic name of each compound.
(c) Label all of the hydrogen atoms in each structure as primary, secondary, or tertiary.

48. How many structural isomers are possible for hexane?
a) 5 b) 4 c) 3 d) 2 e) 6

49. All of the following are structural isomers of C_6H_{14} except
a) $(CH_3)_2CHCH_2CH_3$
b) $CH_3(CH_2)_4CH_3$
c) $CH_3CH_2C(CH_3)_3$
d) $CH_3(CH_2)_2CH(CH_3)_2$
e) $(CH_3)_2CHCH(CH_3)_2$

50. The compound 1-chloro-1-pentene
a) has the formula $C_5H_{11}Cl$.
b) is an alkyne.
c) has 3 structural isomers.
d) cannot exist as cis and trans isomers.
e) can exist as cis and trans isomers.

51. The product of the reaction of cis-2-butene with hydrogen gas is
a) butane b) 2-butyne c) trans-2-butane d) cis-butane e) trans-butane

52. The product of the reaction of 2-butene with Cl_2 is
a) 2,3-dichlorobutane b) 2,2-dichlorobutane c) 3,3-dichlorobutane d) 2-chlorobutane e) butane

53. The compound 4-methyl-2-hexene
a) is incorrectly named
b) can exist as cis and trans isomers
c) cannot exist as cis and trans isomers.
d) has 3 structural isomers.
e) has the formula C_7H_{16}.

54. Name the following compound.

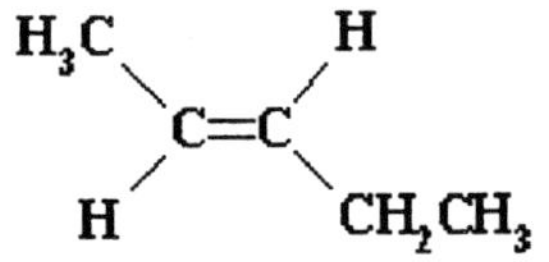

a) trans-2-pentene
b) ethylmethylethene
c) trans-1-ethyl-1-propene
d) cis-2-pentene
e) trans-1-methyl-1-butene

55. Which of the following alkenes shows cis-trans isomerism?
a) $CH(Cl)C(CH_2CH_3)_2$
b) $CH(Cl)CHCH(CH_2CH_3)_2$
c) $(CH_3)_2CCH(CH_2CH_3)$
d) $CCl_2CH(CH_2CH_3)$
e) $CH_2CH(CH_2CH_3)$

56. Which of the following is optically active?
a) $C(CH_3)_4$
b) $CH_3CHCH(Cl)$
c) NH_2CH_2COOH
d) $NH_2CH(CH_3)COOH$
e) $(CH_3)_2C(NH_2)COOH$

57. Which of the following monomers is used to produce Teflon?
a) CH_2CH_2 b) $CH(CN)CH_2$ c) $CH(CH_3)CH_2$ d) CF_2CF_2 e) $CHClCH_2$

58. Which of the following monomers is used to produce polystyrene?
a) $CHClCH_2$ b) CH_2CH_2 c) $CH(CH_3)CH_2$ d) $CH(C_6H_5)CH_2$ e) $CH(CN)CH_2$

59. The polymer that is formed from acrylonitrile is
a) Orlon. b) Lucite. c) Teflon. d) PVC. e) polystyrene.

60. Write the structure for a segment of the condensation polymer that would arise from the following monomers.
(a) $HO–CH(CH_3)CH_2COOH$
(b) $HOCH_2CH_2OH + HOOCCH_2CH_2COOH$
(c) $HOOCCH_2(CH_2)_3CH_2NH_2$

61. The following polymer is called

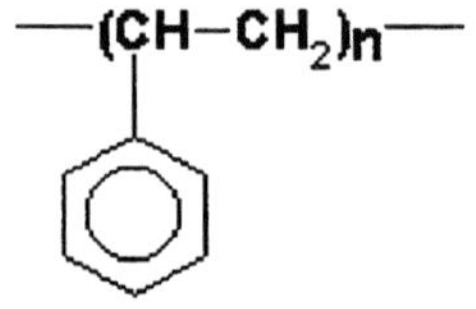

a) polystyrene. b) PVC. c) polyethylene. d) Teflon. e) polypropylene.

62. Condensation polymerization involves the use of
 a) a monomer with a triple bond.
 b) two monomers with reactive groups at each end of the molecules.
 c) two monomers, each with a nitrile group.
 d) two monomers, each with one reactive group.
 e) a monomer with a double bond.

63. Which of the following is serine?
 a) $(CH_3)_2CHCH(NH_2)COOH$
 b) $CH_2(NH_2)COOH$
 c) $CH_3CH(NH_2)COOH$
 d) $HOCH_2CH(NH_2)COOH$
 e) $HSCH_2CH(NH_2)COOH$

64. What is the formula of the side chain X in methionine?
 a) $-CH_2(CH_2)_3NH_2$ b) $-CH_3$ c) $-CH_2COOH$ d) $-CH_2CH_2SCH_3$ e) $-CH_2SH$

65. Which of the following is a peptide bond?
 a) –C=N– b) –C-NH– c) –C(O)O– d) –CONH– e) –C(O)–

66. Which of the following refers to the primary structure of a protein?
 a) The number of aromatic side chains in the peptide chain.
 b) The number of peptide bonds in the peptide chain.
 c) The size of the peptide chain.
 d) The number of amino acid residues in the peptide chain.
 e) The sequence of amino acid residues in the peptide chain.

67. How many different dipeptides can be formed from 2 different amino acids?
 a) 2 b) 1 c) 3 d) 4 e) 0

68. The primary structure of aspartame is shown below.

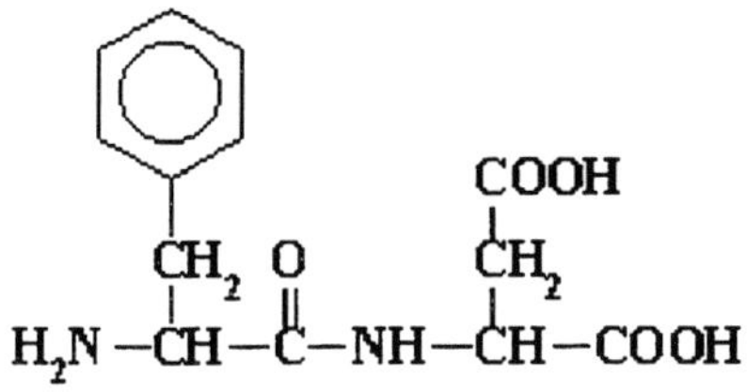

This dipeptide is
 a) Phe-Asp b) Tyr-Glu c) Trp-Glu d) Tyr-Asn e) Phe-Glu

69. The formula of the dipeptide formed from two molecules of glycine is
 a) $NH_2CH(CH_3)CH=N-CH(CH_3)COOH$
 b) $NH_2CH_2CH=N-CH_2COOH$
 c) $NH_2CH_2C(O)NHCH_2COOH$
 d) $NH_2CH(CH_3)C(O)NHCH(CH_3)COOH$
 e) $NH_2CH_2C(O)CH_2COOH$

70. Which of the following amino acids has a side group that is capable of H-bonding?
a) methionine b) alanine c) glycine d) leucine e) valine

71. Which of the following amino acids has a nonpolar side group?
a) phenylalanine b) tyrosine c) serine d) lysine e) glutamine

72. The primary structure of a peptide is shown below.

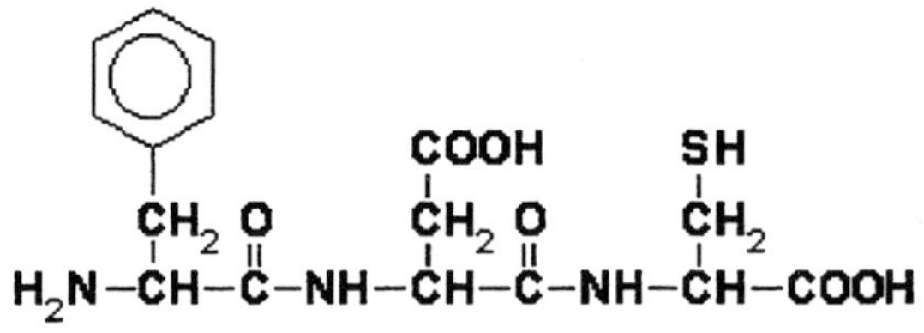

This peptide is
a) Phe-Asp-Cys b) Tyr-Glu-Met c) Tyr-Asp-Cys d) Phe-Asp-Met e) His-Glu-Cys

73. A peptide contains five peptide bonds. How many amino acids make up the peptide?
a) 6 b) 5 c) 4 d) 7 e) 10

74. Both DNA and RNA have the same bases except
a) RNA has uracil instead of adenine.
b) RNA has guanine instead of thymine.
c) RNA has cytosine instead of thymine.
d) RNA has uracil instead of thymine.
e) RNA has adenine instead of thymine.

75. Name the following base.

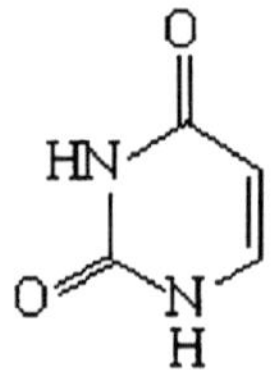

a) uracil b) thymine c) cytosine d) adenine e) guanine

76. The α helix results from the pairing of specific bases by hydrogen bonding. Which pairs form hydrogen bonds?
a) AT and GC b) GT c) AC and GC d) AC e) AT, GC, AC, and GT

77. If the base sequence along a portion of one strand of a double helix is CTACACG, the corresponding sequence on the other strand is
a) GATGTGC b) CTACACG c) CTUCUCG d) CUACACG e) TCGTGTC

78. How many hydrogen bonds are formed between A and T?
a) 2 b) 1 c) 0 d) 3 e) 4

79. How many hydrogen bonds are formed between G and A?
a) 0 b) 2 c) 1 d) 3 e) 4

80. Nucleotides are formed from a monosaccharide, a base, and a phosphate group. Which of the following is found in DNA?
a) 2-deoxyribose b) sucrose c) lactose d) amylose e) fructose

Chapter 11: Carbon-Based Materials
Form B

1. Which of the following is an unsaturated hydrocarbon?
a) C_3H_8 b) C_3H_4 c) C_4H_6 d) C_5H_{12} e) C_4H_{10}

2. All of the following are saturated hydrocarbons except
a) C_5H_{12} b) C_4H_{10} c) C_4H_6 d) C_3H_8 e) C_3H_4

3. The sixth member of the alkanes has
a) 6 hydrogen atoms
b) 16 hydrogen atoms
c) 10 hydrogen atoms
d) 14 hydrogen atoms
e) 12 hydrogen atoms

4. The ninth member of the alkanes has the molecular formula
a) C_9H_{18} b) C_9H_{14} c) C_9H_{16} d) C_9H_{20} e) C_9H_{12}

5. Which of the following could be a cycloalkane?
a) C_6H_{14} b) C_3H_8 c) C_4H_{10} d) C_5H_{10} e) C_5H_{12}

6. All of the following could be cycloalkanes except
a) C_5H_{10} b) C_3H_6 c) C_6H_{12} d) C_4H_8 e) C_6H_{14}

7. Which of the following has the highest boiling point?
a) $CH_3CH(CH_3)CH(CH_3)CH_2CH_3$
b) $CH_3(CH_2)_4CH_3$
c) $CH_3(CH_2)_5CH_3$
d) $CH_3CH(CH_3)(CH_2)_3CH_3$
e) C_3H_8

8. Alkanes react with
a) boiling nitric acid.
b) concentrated sulfuric acid.
c) boiling aqueous sodium hydroxide to give alcohols.
d) the strong oxidizing agent $KMnO_4$.
e) oxygen to give carbon dioxide and water.

9. Which of the following pairs react?
a) $CH_4(g)$ and $NaOH(aq)$
b) $CH_4(g)$ and $HNO_3(l)$
c) $CH_4(g)$ and $H_2SO_4(l)$
d) $CH_4(g)$ and $O_2(g)$
e) $CH_4(g)$ and $KMnO_4(aq)$

10. Consider a molecule of molecular formula $C_{10}H_{18}$. What total number of rings plus π bonds are present in the molecule?
a) 2 b) 3 c) 1 d) 0 e) 4

11. For $CH_3CH(CH_2CH_3)CH(CH_2CH_3)CH(CH_3)CH(CH_3)_2$, the longest unbranched chain of carbon atoms is

a) 7 b) 6 c) 8 d) 5 e) 9

12. Which of the following is most volatile?

a) $(CH_3)_3COH$
b) $CH_3CH_2OCH_2CH_3$
c) $CH_3CH(OH)CH_2CH_3$
d) $(CH_3)_2C(OH)CH_3$
e) $CH_3(CH_2)_3OH$

13. Which of the following could be an alkyne?

a) C_5H_{12} b) C_6H_{12} c) C_4H_{10} d) C_6H_6 e) C_6H_{14}

14. Arenes predominantly undergo

a) substitution b) addition c) elimination d) hydrogenation e) dehydrogenation

15. In the reaction

$$CH_3CH(CH_3)CH_2CH_2CH_3 \xrightarrow[\text{light}]{Br_2}$$

(a) How many different monobromination products result?
(b) What is the most likely product?

16. The compound $(CH_3)_3CCH_2CH(CH_3)_2$ is named as

a) a hexane. b) a heptane. c) a pentane. d) an octane. e) a butane.

17. The compound $(CH_3)_2CHCH(CH_3)C(CH_3)_3$ is

a) named as a pentane but is an isomer of nonane.
b) named as a heptane but is an isomer of nonane.
c) named as a pentane but is an isomer of octane.
d) named as a hexane but is an isomer of nonane.
e) named as a hexane but is an isomer of octane.

18. Name the compound $CH_3CH(CH_2CH_3)CH(CH_2CH_3)CH(CH_3)CH(CH_3)_2$.

a) 2,3-diethyl-4,5-dimethylhexane
b) 2,3-dimethyl-4,5-diethylhexane
c) 2,3-diethy-4-isopropylpentane
d) dodecane
e) 2-ethyl-3-ethyl-4-methyl-5-methylhexane

19. Name the compound $(CH_3)_3CCH_2CH(CH_3)_2$.

a) 2,2,4-trimethylpentane
b) octane
c) 2,4,4-trimethylpentane
d) 1,1,1,3-tetramethylbutane
e) 2-isopropyl-4-methylpentane

20. Name the compound $CH_3C(CH_3)_2C(CH_3)(CH_2CH_3)CH(CH_3)_2$.

a) 3-ethyl-2,2,3,4-tetramethylpentane
b) undecane
c) 3-ethyl-2,3,4,4-tetramethylpentane
d) 3-isopropyl-3,4,4-trimethylpenatane
e) 3-ethyl-1,1,1,2,3-pentamethylbutane

21. Name the following compound.

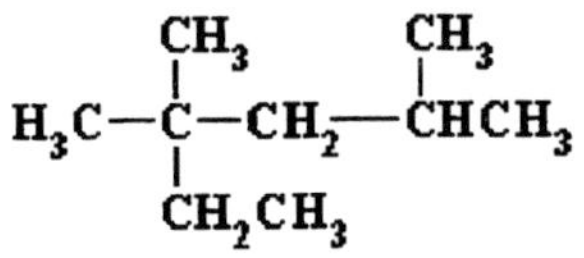

a) 2-ethyl-2,4-dimethylpentane
b) 2,4,4-trimethylhexane
c) 2-ethyl-2-methyl-4-methylpentane
d) 2,4-dimethyl-4-ethylpentane
e) 2,4-dimethyl-2-ethylpentane

22. Name the following compound.

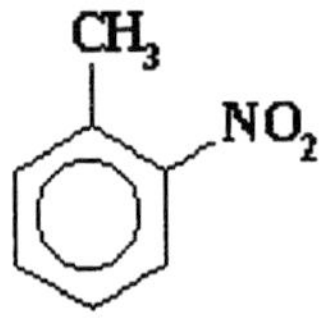

a) 1-methyl-2-nitrobenzene
b) 1-nitro-2-methylbenzene
c) nitrotoluene
d) 2-methyl-3-nitrobenzene
e) nitrobenzene

23. Name the compound $(CH_3CH_2)_2CHCH{=}CHCH_3$.
a) 4-ethyl-2-hexene b) 3-ethyl-4-hexene c) 4,4-diethyl-2-butene d) ethylhexene e) diethylbutene

24. Name the following compound.

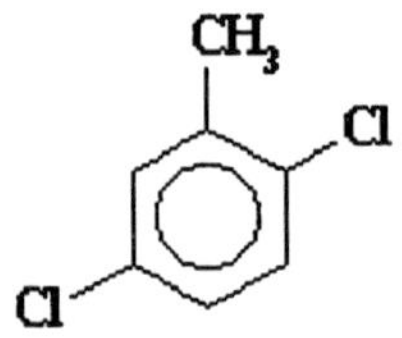

a) 2,5-dichloro-1-methylbenzene
b) dichlorotoluene
c) 1,4-dichlorotoluene
d) p-chloro-1-methylbenzene
e) 1,4-dichloro-6-methylbenzene

25. The ester $CH_3COO(CH_2)_4CH_3$ is responsible for the odor of bananas. It can be prepared from
a) CH_3COOH and $CH_3(CH_2)_3CHO$
b) CH_3CHO and $CH_3(CH_2)_3COOH$
c) CH_3COOH and $CH_3(CH_2)_3CH_2OH$
d) CH_3CHO and $CH_3(CH_2)_3CH_2OH$
e) CH_3CH_2OH and $CH_3(CH_2)_3COOH$

26. When an ester is formed via a condensation reaction with elimination of water, the oxygen atom in the water molecule comes from
a) the carbonyl group of the acid.
b) the aldehyde.
c) the alcohol.
d) the solvent.
e) the hydroxyl group of the acid.

27. Which of the following is a secondary alcohol?
a) $C(CH_3)_3OH$ b) $CH(CH_3)_2OH$ c) $CH_3(CH_2)_3OH$ d) CH_3CH_2OH e) $C_6H_5CH_2OH$

28. The product of the reaction of ethanol and acetic acid is
a) ethyl acetate b) propanoic acid c) 2-butanone d) propanal e) acetaldehyde and ethane

29. The reaction of 1-octanol with acetic acid gives
a) $CH_3(CH_2)_6COOH$
b) $CH_3CO(CH_2)_7CH_3$
c) $CH_3COO(CH_2)_7CH_3$
d) $CH_3(CH_2)_6COOCH_3$
e) $CH_3(CH_2)_6CHO$

30. Name the compound $CH_3COO(CH_2)_4CH_3$.
a) pentyl acetate b) methyl pentanoate c) butyl acetate d) 2-hexanone e) hexanoic acid

31. The reaction of an alcohol and a carboxylic acid gives
a) a dicarboxylic acid b) an aldehyde c) an ether d) an ester e) a ketone

32. The compound CH_3COCH_3 is
a) an aaldehyde. b) a ketone. c) an ester. d) an ether. e) an alcohol.

33. Mild oxidation of primary alcohols produces
a) aldehydes. b) ketones. c) acids. d) ethers. e) esters.

34. Oxidation of 2-propanol gives
a) acetone. b) acetic acid. c) dimethyl ether. d) acetaldehyde. e) propanoic acid.

35. To produce acetone, which of the following should be reacted with $Na_2Cr_2O_7(aq)$, $H_2SO_4(aq)$?
a) CH_3CH_2CHO
b) $(CH_3)_2CHCHO$
c) $CH_3CH(COOH)CH_3$
d) $CH_3CH_2CH_2OH$
e) $CH_3CH(OH)CH_3$

36. Treatment of propanol with $Na_2Cr_2O_7(aq)$, $H_2SO_4(aq)$ yields compound A which reacts further to give B. Identify A and B, respectively.
a) CH_3CH_2CHO and $CH_3CH_2OCH_3$
b) $CH_3CH_2OCH_3$ and $CH_3CH_2CH_3$
c) CH_3CH_2CHO and CH_3CH_2COOH
d) CH_3COCH_3 and CH_3CH_2COOH
e) $CH_3CH_2CH_3$ and CH_3CHCH_2

37. Which of the following produces a silver mirror with Tollen's reagent?
a) $CH_3C(O)OCH_3$ b) CH_3COOH c) CH_3COCH_3 d) CH_3OCH_3 e) CH_3CHO

38. When aldehydes react with Tollen's reagent,
a) an alcohol is produced.
b) a ketone is produced.
c) a red precipitate forms.
d) silver ions are produced.
e) a carboxylic acid is produced.

39. Write structures for the principal product or products expected in each of the following reactions.
(a) $(CH_3)_3COH + Na_2Cr_2O_7(aq), H_2SO_4(aq) \rightarrow$
(b) $CH_3CH_2COOH + (CH_3)_2CHCH_2CH_2OH\ (H^+) \rightarrow$
(c) $CH_3CH_2COOH + (CH_3CH_2)_2NH \rightarrow$

40. Write an equation for the synthesis of
(a) 4-methylpentanal
(b) 1,4-benzenedicarboxylic acid

41. The carbonyl group occurs in all of the following except
a) phenols. b) amides. c) carboxylic acids. d) aldehydes. e) ketones.

42. The product of the reaction of CH_3CH_2CHO with excess $KMnO_4(aq)$ in acidic solution is
a) CH_3CH_2COOH b) CH_3COOH c) $CH_3CH_2CH_3$ d) CH_3CH_2OH e) $CH_3CH_2COCH_2CH_3$

43. Predict the product of the reaction of acetic acid with dimethylamine.
a) CH_3CONH_2
b) $CH_3CON(CH_3)_2$
c) $CH_3CONH(CH_3)_2{}^+$
d) no reaction occurs
e) $CH_3CONHCH_3$

44. Predict the product of the reaction of acetic acid with triphenylamine.
a) $CH_3CON(C_6H_5)_2$
b) no reaction occurs
c) CH_3CONH_2
d) $CH_3CON(C_6H_5)_3{}^+$
e) $CH_3CONH(C_6H_5)$

45. The systematic name of $CH_3CH_2COCH_3$ is
a) 2-butanone. b) 3-butanone. c) methypropylether. d) 2-butanal. e) 3-butanal.

46. Rubber can be viewed as being formed from isoprene monomers, $CH_2C(CH_3)CHCH_2$. The systematic name for isoprene is
a) 2-methyl-1,3-butene.
b) isopropylpropene.
c) methylene-1-butene.
d) methylbutene.
e) 1-ethylpropene.

47. Write structures for all compounds of molecular formula C_8H_{18} that do not contain any secondary hydrogen atoms.

48. How many structural isomers are possible for heptane?
a) 9 b) 7 c) 5 d) 6 e) 4

49. All of the following are structural isomers of C_4H_9Cl except
a) $CH_3CH(Cl)CH_2CH_3$
b) $CH_2(Cl)CH_2CH_2CH_3$
c) $CH_2(Cl)CH_2CH_3$
d) $CH_3C(Cl)(CH_3)CH_3$
e) $CH_3CH(CH_3)CH_2(Cl)_2$

50. The compound 2-chloro-1-pentene
a) cannot exist as cis and trans isomers.
b) has the formula $C_5H_{11}Cl$.
c) has 3 structural isomers.
d) is an alkyne.
e) can exist as cic and trans isomers.

51. The product of the reaction of 2-butene with HBr is
a) 2-bromobutane b) butane c) 1-bromobutane d) 2-bromo-1-butene e) 2-bromo-2-butene

52. The product of the reaction of 2-butene with Br_2 is
a) 2,3-dibromobutane b) 2,2-dibromobutane c) 3,3-dibromobutane d) 2-bromobutane e) butane

53. The compound 3-methyl-3-hexene
a) can exist as cis or trans isomers.
b) cannot exist as cis and trans isomers.
c) has 3 structural isomers.
d) is an alkyne.
e) has the formula C_7H_{16}.

54. Name the following compound.

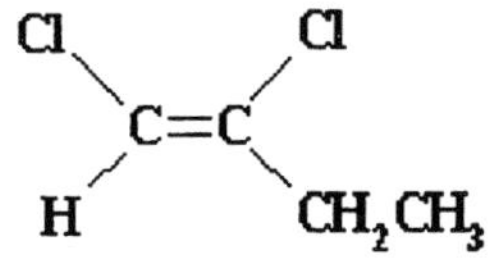

a) cis-1,2-dichloro-1-butene
b) cis-2-ethyl-1,2-dichloroethene
c) cis-1,2-dichloro-2-ethylethene
d) trans-1,2-dichloro-1-butene
e) dichlorobutene

55. Which of the following alkenes shows cis-trans isomerism?
a) $CCl_2CH(CH_2CH_3)$
b) $(CH_3)_2CCH(CH_2CH_3)$
c) $CH(Cl)C(CH_2CH_3)_2$
d) $CH(Cl)CCl(CH_2CH_3)$
e) $CH_2CH(CH_2CH_3)$

56. Which of the following is optically active?
a) $C(CH_3)_4$
b) $CH_3CHCH(Cl)$
c) $(CH_3)_2C(NH_2)COOH$
d) $CH_3CH_2CH(NH_2)COOH$
e) NH_2CH_2COOH

57. Which of the following monomers is used to produce Orlon?
a) $CH(CH_3)CH_2$ b) CH_2CH_2 c) $CH(CN)CH_2$ d) $CHClCH_2$ e) CF_2CF_2

58. Which of the following monomers is used to produce polypropylene?
a) CH_2CH_2 b) $CH(CH_3)CH_2$ c) $CHClCH_2$ d) $CH(CN)CH_2$ e) $CH(C_6H_5)CH_2$

59. The polymer that is formed from tetrafluoroethene is
a) Teflon. b) Orlon. c) Lucite. d) PVC. e) polystyrene.

60. Identify the monomer or monomers used in the preparation of each of the following polymers and indicate whether each is an addition or condensation polymer.

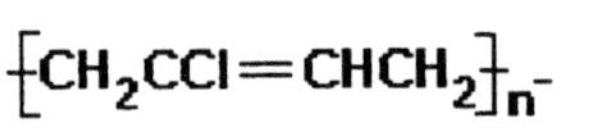

(A)

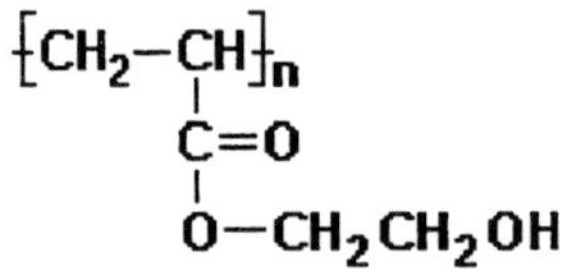

(B)

61. The following polymer is called

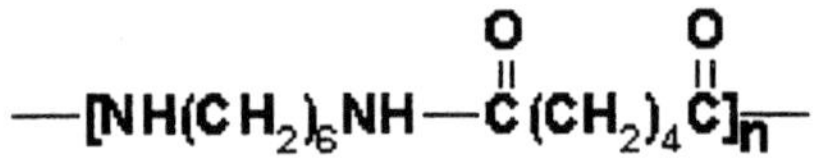

a) nylon-66 b) polyester c) kevlar d) Dacron e) Teflon

62. All of the following polymers are produced by addition polymerization except
a) Dacron. b) PVC. c) Teflon. d) polyethylene. e) polypropylene.

63. Which of the following is cysteine?
a) $CH_3CH(NH_2)COOH$
b) $HSCH_2CH(NH_2)COOH$
c) $HOCH_2CH(NH_2)COOH$
d) $(CH_3)_2CHCH(NH_2)COOH$
e) $CH_2(NH_2)COOH$

64. What is the formula of the side chain X in lysine?
a) $-CH_2COOH$ b) $-CH_3$ c) $-CH_2CH_2SCH_3$ d) $-CH_2SH$ e) $-CH_2(CH_2)_3NH_2$

65. Which of the following is an ester linkage?
a) –C(O)O– b) –CONH– c) –C(O)– d) –C=N– e) –C-NH–

66. Which of the following refers to the secondary structure of a protein?
a) α helix b) base pairs c) denaturation d) disulfide linkages e) amino acid sequence

67. How many different tripeptides can be formed from 3 different amino acids?
a) 6 b) 12 c) 3 d) 2 e) 4

68. The primary structure of a dipeptide is shown below.

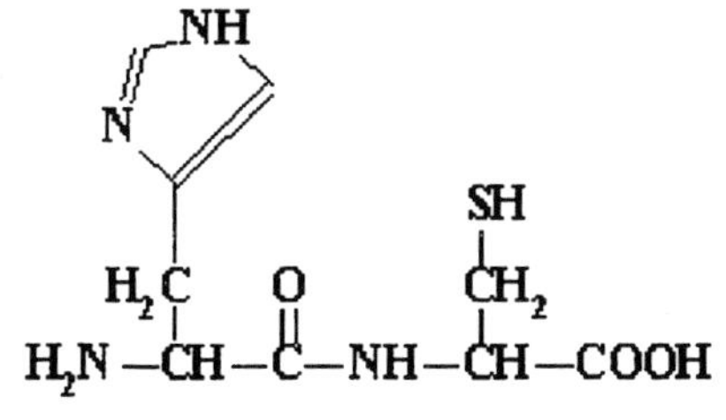

This dipeptide is
a) His-Cys b) Trp-Cys c) His-Ser d) Tyr-Ser e) Phe-Cys

69. The formula of the dipeptide formed from two molecules of alanine is
a) $NH_2CH_2C(O)CH_2COOH$
b) $NH_2CH_2CH{=}N\text{-}CH_2COOH$
c) $NH_2CH(CH_3)CH{=}N\text{-}CH(CH_3)COOH$
d) $NH_2CH(CH_3)C(O)NHCH(CH_3)COOH$
e) $NH_2CH_2C(O)NHCH_2COOH$

70. Which of the following amino acids has a side group that is capable of H-bonding?
a) asparagine b) alanine c) glycine d) leucine e) valine

71. Which of the following amino acids has a nonpolar side group?
a) leucine b) tyrosine c) serine d) lysine e) glutamine

72. The primary structure of a peptide is shown below.

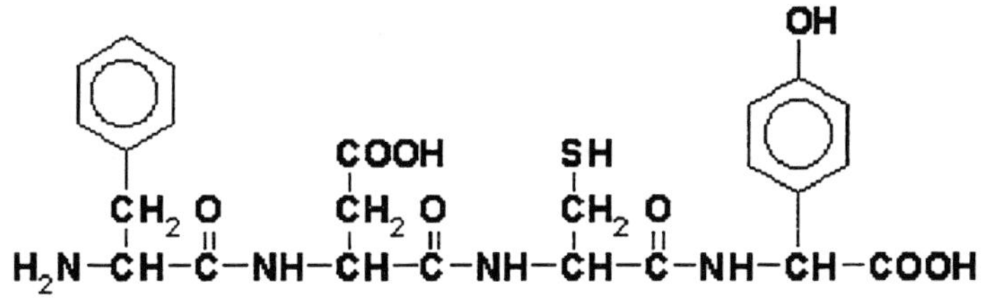

This peptide is
a) Phe-Asp-Cys-Tyr
b) Tyr-Glu-Cys-Phe
c) Phe-Glu-Met-Tyr
d) Tyr-Asp-Met-Phe
e) Tyr-Asn-Cys-Phe

73. A peptide contains six peptide bonds. How many amino acids make up the peptide?
a) 7 b) 6 c) 12 d) 5 e) 10

74. Both DNA and RNA have the same bases except
a) DNA has uracil instead of adenine.
b) DNA has guanine instead of thymine.
c) DNA has thymine instead of uracil.
d) DNA has cytosine instead of thymine.
e) DNA has adenine instead of thymine.

75. Name the following base.

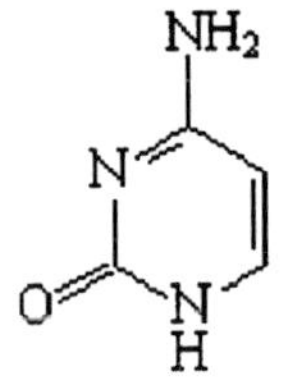

a) cytosine b) uracil c) thymine d) adenine e) guanine

76. If the base sequence along a portion of one strand of a double helix is GATG, the corresponding sequence on the other strand is
a) CTAC b) CTUC c) TCGT d) GATG e) CUAC

77. If the base sequence along a portion of one strand of a double helix is GATGTGC, the corresponding sequence on the other strand is
a) CTACACG b) CTACACG c) CTUCUCG d) CUACACG e) TCGTGTC

78. How many hydrogen bonds are formed between G and C?
a) 3 b) 2 c) 1 d) 0 e) 4

79. How many hydrogen bonds are formed between C and T?
a) 0 b) 2 c) 1 d) 3 e) 4

80. Nucleotides are formed from a monosaccharide, a base, and a phosphate group. Which of the following is found in RNA?
a) ribose b) 2-deoxyribose c) lactose d) amylose e) fructose

Chapter 12: The Properties of Solutions
Form A

1. The deposit of "scale" inside hot water pipes occurs because
a) calcium chloride is very insoluble
b) calcium bicarbonate precipitates readily.
c) carbon dioxide is very soluble in hot water.
d) carbonates are less soluble than bicarbonates.
e) carbonic acid is driven from the hot water.

2. Which of the following would likely be most soluble?
a) $CaHPO_4$ b) $Ca_5(PO_4)_3OH$ c) $Ca(NO_3)_2$ d) $Ca_3(PO_4)_2$ e) $CaCO_3$

3. Which of the following would you expect to be least soluble?
a) $Ca(ClO_4)_2$ b) $Ca_3(PO_4)_2$ c) $CaHPO_4$ d) $Ca(H_2PO_4)_2$ e) $CaCO_3$

4. Which of the following would likely dissolve in water?
a) Cl_2CCCl_2 b) C_6H_6, benzene c) $Ca(HCO_3)_2$ d) CCl_4 e) C_6H_5Cl

5. Calculate the solubility of carbon dioxide in $g \cdot L^{-1}$ at 20°C in a closed container of water if the partial pressure of $CO_2(g)$ is 3.6 atm. Henry's constant is 0.023 $mol \cdot L^{-1} \cdot atm^{-1}$ for $CO_2(g)$ at 20°C.
a) 3.7 $g \cdot L^{-1}$ b) 160 $g \cdot L^{-1}$ c) 530 $g \cdot L^{-1}$ d) 0.28 $g \cdot L^{-1}$ e) 1.0 $g \cdot L^{-1}$

6. Calculate Henry's constant for oxygen if the solubility of oxygen in a lake is 8.3×10^{-3} $g \cdot L^{-1}$ at a partial pressure of oxygen of 152 Torr.
a) 0.20 $mol \cdot L^{-1} \cdot atm^{-1}$
b) 0.042 $mol \cdot L^{-1} \cdot atm^{-1}$
c) 0.00065 $mol \cdot L^{-1} \cdot atm^{-1}$
d) 0.0013 $mol \cdot L^{-1} \cdot atm^{-1}$
e) 0.0026 $mol \cdot L^{-1} \cdot atm^{-1}$

7. Henry's constants($mol \cdot L^{-1} \cdot atm^{-1}$) for $CO_2(g)$, He(g), Ar(g), $N_2(g)$, and $O_2(g)$ are 0.023, 0.00037, 0.0015, 0.0007, and 0.0013, respectively, in water at 20°C. Which gas would be least soluble at a pressure of 25 atm?
a) He(g) b) $O_2(g)$ c) Ar(g) d) $N_2(g)$ e) $CO_2(g)$

8. Calculate the solubility of carbon dioxide at 20°C in a closed container of water if the partial pressure of $CO_2(g)$ is 3.6 atm. Henry's constant is 0.023 $mol \cdot L^{-1} \cdot atm^{-1}$ for $CO_2(g)$ at 20°C.
a) 12 M b) 6.4×10^{-3} M c) 3.7 M d) 8.3×10^{-2} M e) 2.3×10^{-2} M

9. The equilibrium

$$Ca(HCO_3)_2(aq) \leftrightarrow CaCO_3(s) + CO_2(g) + H_2O(l)$$

exists during the formation of stalactites and stalagmites, which are mainly $CaCO_3(s)$, in underground caves. If a cave was opened to the atmosphere, predict the effect of the increased partial pressure of carbon dioxide.

a) No change would occur.
b) The stalactites and stalagmites would become smaller due to dissolution of $CaCO_3(s)$.
c) The $Ca(HCO_3)_2(aq)$ concentration would decrease.
d) More $CO_2(g)$ and $H_2O(l)$ would be produced.
e) More $CaCO_3(s)$ would be produced and the stalactites and stalagmites would become larger.

10. Calculate the concentration of CO_2 in champagne at 20°C if the partial pressure of carbon dioxide is 5.0 atm and the Henry's law constant for CO_2 is 0.023 mol · L^{-1} · atm^{-1}.

a) 0.051 M b) 0.12 M c) 0.0.0046 M d) 0.29 M e) 5.0 M

11. The lattice enthalpy of AgF is 971 kJ · mol^{-1} and the hydration enthalpy is –993 kJ · mol^{-1}. Estimate the enthalpy of solution of AgF.

a) +22 kJ · mol^{-1} b) –971 kJ · mol^{-1} c) –1964 kJ · mol^{-1} d) –22 kJ · mol^{-1} e) –993 kJ · mol^{-1}

12. The lattice enthalpy of $CaCl_2$ is 2260 kJ · mol^{-1} and the hydration enthalpy is –2340 kJ · mol^{-1}. Estimate the enthalpy of solution of $CaCl_2$.

a) –4600 kJ · mol^{-1} b) +2340 kJ · mol^{-1} c) +4600 kJ · mol^{-1} d) +80 kJ · mol^{-1} e) –80 kJ · mol^{-1}

13. The lattice enthalpy of KBr is 689 kJ · mol^{-1} and the hydration enthalpy is –669 kJ · mol^{-1}. Estimate the enthalpy of solution of KBr.

a) –1360 kJ · mol^{-1} b) +669 kJ · mol^{-1} c) +20 kJ · mol^{-1} d) +1360 kJ · mol^{-1} e) –20 kJ · mol^{-1}

14. The lattice enthalpy of AgF is 971 kJ · mol^{-1} and the enthalpy of solution is –22 kJ · mol^{-1}. Estimate the enthalpy of hydration of AgF.

a) –949 kJ · mol^{-1} b) +993 kJ · mol^{-1} c) +22 kJ · mol^{-1} d) –993 kJ · mol^{-1} e) +949 kJ · mol^{-1}

15. The lattice enthalpy of $CaCl_2$ is 2260 kJ · mol^{-1} and the enthalpy of solution is about –80 kJ · mol^{-1}. Estimate the enthalpy of hydration of $CaCl_2$.

a) +80 kJ · mol^{-1} b) –2180 kJ · mol^{-1} c) –2340 kJ · mol^{-1} d) +2180 kJ · mol^{-1} e) +2340 kJ · mol^{-1}

16. For AgCl, the lattice enthalpy is larger than the absolute value of the enthalpy of hydration. This means that for AgCl

a) the enthalpy of solution is endothermic.
b) the enthalpy of hydration is endothermic.
c) the solubility increases when the temperature decreases.
d) the lattice enthalpy is exothermic.
e) the enthalpy of solution is exothermic.

17. For AgI, the lattice enthalpy is larger than the absolute value of the enthalpy of hydration. This means that for AgI
a) ΔH_{hyd} is positive.
b) ΔH_{sol} is negative.
c) ΔH_L is negative.
d) ΔH_{sol} is positive.
e) the solubility increases when the temperature decreases.

18. For $CaCl_2$, the absolute value of the enthalpy of hydration is larger than the lattice enthalpy. This means that for $CaCl_2$
a) the lattice enthalpy is negative.
b) the enthalpy of hydration is positive.
c) the enthalpy of solution is endothermic.
d) the enthalpy of solution is exothermic.
e) the solubility increases when the temperature increases.

19. For MgF_2, the absolute value of the enthalpy of hydration is larger than the lattice enthalpy. This means that for MgF_2
a) ΔH_{sol} is negative.
b) ΔH_L is negative.
c) the solubility increases when the temperature increases.
d) ΔH_{hyd} is positive.
e) ΔH_{sol} is positive.

20. Which of the following is likely to have the largest exothermic hydration enthalpy?
a) Ca^{2+} b) Li^+ c) Al^{3+} d) Na^+ e) Mg^{2+}

21. Which of the following is likely to have the largest exothermic hydration enthalpy?
a) Li^+ b) H^+ c) Na^+ d) K^+ e) Ag^+

22. Which of the following is likely to have the largest exothermic hydration enthalpy?
a) F^- b) Cl^- c) I^- d) NO_3^- e) Br^-

23. All of the following are colligative properties except
a) solubility.
b) vapor pressure lowering.
c) boiling point elevation.
d) freezing point depression.
e) osmotic pressure.

24. When the mole fraction of solute is 1.0, there is
a) 1.0 mole of solute and 99 moles of solvent.
b) only solute present.
c) only solvent present.
d) a 1:1 ratio of solute to solvent.
e) 1.0 g of solute per 100 g of solution.

25. Calculate the mole fraction of methanol, CH_3OH, in a mixture of equal masses of methanol and water. The molar mass of methanol is 32.0 $g \cdot mol^{-1}$ and that of water is 18.0 $g \cdot mol^{-1}$.
a) 0.360 b) 0.500 c) 0.640 d) 0.0313 e) 1.00

26. Calculate the mole fraction of water in a mixture of equal masses of acetone and water. The molar mass of acetone is 58.0 $g \cdot mol^{-1}$ and that of water is 18.0 $g \cdot mol^{-1}$.
a) 0.763 b) 0.237 c) 0.310 d) 0.500 e) 0.690

27. Calculate the mole fraction of ethanol in a mixture of equal masses of ethanol and water. The molar mass of ethanol is 46.0 $g \cdot mol^{-1}$ and that of water is 18.0 $g \cdot mol^{-1}$.
a) 0.281 b) 0.719 c) 0.500 d) 0.391 e) 0.609

28. If 75.0 g glucose (molar mass, 180.2 $g \cdot mol^{-1}$) are dissolved in 625 g water (molar mass, 18.0 $g \cdot mol^{-1}$), what is the mole fraction of water in the solution?
a) 0.989 b) 0.416 c) 1.00 d) 0.0118 e) 0.584

29. Calculate the mole fraction of sucrose (molar mass, 342.3 $g \cdot mol^{-1}$) in a solution prepared by dissolving 50.0 g of sucrose in 200.0 g of water (molar mass, 18.0 $g \cdot mol^{-1}$).
a) 0.0130 b) 0.987 c) 0.146 d) 0.854 e) 1.00

30. How many moles of glucose should be added to 450 g of water to prepare a 0.300 *m* solution?
a) 0.135 b) 0.300 c) 1.50 d) 0.667 e) 2.92

31. How many moles of glucose should be added to 450 g of water to prepare a 0.600 *m* solution?
a) 0.270 b) 0.600 c) 3.00 d) 1.33 e) 5.84

32. How many moles of sodium sulfate should be added to 1750 g of water in order to prepare 0.0500 *m* $Na_2SO_4(aq)$?
a) 0.286 mol b) 0.0500 mol c) 3.50 mol d) 0.0875 mol e) 7.04 mol

33. What is the molality of a solution prepared by dissolving 15.1 g KNO_3 (molar mass, 101 $g \cdot mol^{-1}$) in 250 g water?
a) 0.0604 *m* b) 0.598 *m* c) 0.0375 *m* d) 0.150 *m* e) 1.67 *m*

34. The mole fraction of glucose in an aqueous solution is 0.200. Calculate the molality of glucose in the solution. The molar mass of glucose is 342 $g \cdot mol^{-1}$ and that of water is 18.0 $g \cdot mol^{-1}$.
a) 13.9 *m* b) 55.6 *m* c) 2.92 *m* d) 2.42 *m* e) 0.200 *m*

35. The mole fraction of glucose in an aqueous solution is 0.100. Calculate the molality of glucose in the solution. The molar mass of glucose is 342 $g \cdot mol^{-1}$ and that of water is 18.0 $g \cdot mol^{-1}$.
a) 0.100 *m* b) 1.98 *m* c) 6.17 *m* d) 55.6 *m* e) 2.92 *m*

36. Calculate the mole fraction of glucose in 6.00 m glucose(aq). The molar mass of glucose is 342 $g \cdot mol^{-1}$ and that of water 18.0 $g \cdot mol^{-1}$.
 a) 0.0975 b) 0.903 c) 0.0556 d) 0.00600 e) 0.108

37. Calculate the mole fraction of glucose in 12.0 m glucose(aq). The molar mass of glucose is 342 $g \cdot mol^{-1}$ and that of water 18.0 $g \cdot mol^{-1}$.
 a) 0.178 b) 0.822 c) 0.216 d) 0.0120 e) 0.0556

38. Commercial nitric acid (molar mass, 63.0 $g \cdot mL^{-1}$) has a density of 1.42 g/mL. If the molarity of HNO_3(aq) is 16.0 M, calculate the molality of HNO_3(aq).
 a) 38 *m* b) 39 *m* c) 89 *m* d) 23 *m* e) 16 *m*

39. Concentrated aqueous ammonia is 14.8 M and has a density of 0.900 $g \cdot mL^{-1}$. Calculate the molality of NH_3(aq) in this solution. The molar mass of ammonia is 17.0 $g \cdot mol^{-1}$.
 a) 22.8 *m* b) 13.3 *m* c) 14.8 *m* d) 16.4 *m* e) 58.7 *m*

40. Calculate the molality of 9.98 M NaOH(aq). The density of the solution is 1.33 g mL^{-1}.

41. Calculate the molality of ethyl alcohol in a bottle of wine that is 19.0% C_2H_5OH by mass. The molar mass of ethyl alcohol is 46.1 $g \cdot mol^{-1}$.
 a) 5.09 *m* b) 3.73 *m* c) 1.54 *m* d) 0.412 *m* e) 1.76 *m*

42. In order to calculate the molality of 1.00 M glucose(aq), the
 a) molar mass of water is required.
 b) volume of solution must be known.
 c) mole fraction of water must be given.
 d) mole fraction of glucose must be given.
 e) density of the solution must be known.

43. Calculate the vapor pressure at 80°C of a solution prepared by dissolving 0.030 moles of glucose in 100 g of water. The vapor pressure of water at 80°C is 355 Torr and the molar mass of water is 18.0 $g \cdot mol^{-1}$.
 a) 355 Torr b) 760 Torr c) 353 Torr d) 756 Torr e) 357 Torr

44. Calculate the vapor pressure at 80°C of a solution prepared by dissolving 3.00 moles of glucose in 200 g of water. The vapor pressure of water at 80°C is 355 Torr and the molar mass of water is 18.0 $g \cdot mol^{-1}$.
 a) 760 Torr b) 279 Torr c) 431 Torr d) 75.5 Torr e) 355 Torr

45. Calculate the vapor pressure at 25°C of a mixture of benzene and toluene in which the mole fraction of benzene is 0.650. The vapor pressure at 25°C of benzene is 94.6 Torr and that of toluene is 29.1 Torr.
 a) 84.4 Torr b) 124 Torr c) 71.7 Torr d) 61.5 TOrr e) 51.3 Torr

46. The vapor pressures of pure carbon disulfide and carbon tetrachloride are 360 and 99.8 Torr, respectively, at 296 K. What is the vapor pressure of a solution containing 50.0 g of each compound?
 a) 460 Torr b) 274 Torr c) 260 Torr d) 33.0 Torr e) 241 Torr

47. A solution made up of 50.0 g each of CS_2 (molar mass, 76.13 g · mol^{-1}) and CCl_4 (molar mass, 153.8 g · mol^{-1}) has a vapor pressure of 274 Torr at 296 K. If the vapor pressure of pure carbon disulfide is 360 Torr at 296 K, what is the vapor pressure of pure carbon tetrachloride?
a) 241 Torr b) 119 Torr c) 86 Torr d) 100 Torr e) 33 Torr

48. Calculate the vapor pressure lowering of an aqueous solution containing 50.0 g sucrose (molar mass, 342.3 g · mol^{-1}) dissolved in 200 g water (molar mass, 18.0 g · mol^{-1}) at 100°C.
a) 750 Torr b) 10 Torr c) 190 Torr d) 40 Torr e) 110 Torr

49. Which of the following 1.0 M solutions contains the most particles?
a) magnesium sulfate b) sodium sulfate c) glucose d) ethylene glycol e) potassium chloride

50. All of the following have a van't Hoff *i* factor of 3 except
a) calcium nitrate.
b) magnesium chloride.
c) ethylene glycol.
d) barium hydroxide.
e) sodium sulfate.

51. A 1.0-molal solution of methanol (CH_3OH) in water, and a 1.0-molal solution of calcium chloride in water have total solute molalities of
a) 2.0 and 3.0, respectively.
b) 1.0 and 2.0, respectively.
c) 2.0 and 2.0.
d) 1.0 and 3.0, respectively.
e) 1.0 and 1.0.

52. Which of the following has the lowest freezing point and the highest boiling point?
a) 1.0 *m* sodium chloride
b) 2.0 *m* potassium chloride
c) 1.0 *m* acetic acid
d) 1.5 *m* aluminum nitrate
e) 1.5 *m* calcium chloride

53. Calculate the freezing point of a 3.5 m aqueous solution of ethylene glycol. The k_f for water is 1.86 K · kg · mol^{-1}.
a) −0.53°C b) −6.5°C c) −1.9°C d) 0°C e) −13°C

54. Calculate the freezing point of a 14.0 m aqueous solution of ethylene glycol. The k_f for water is 1.86 K · kg · mol^{-1}.
a) −52°C b) −26°C c) −7.5°C d) −0.11°C e) 0°C

55. The freezing point of seawater is about −1.85°C. If seawater is an aqueous solution of sodium chloride, calculate the molality of seawater. The k_f for water is 1.86 K · kg · mol^{-1}.
a) 1.99 *m* b) 3.70 *m* c) 0.995 *m* d) 0.497 *m* e) 3.44 *m*

56. A 1.0 *m* aqueous solution of an unknown solute has a freezing point of –9.30°C. Which of the following is the solute? The k_f for water is 1.86 K · kg · mol^{-1}.
a) calcium chloride
b) iron(III) chloride
c) sodium sulfate
d) aluminum sulfate
e) sodium chloride

57. Which of the following aqueous solutions would have the lowest freezing point?
a) 1.0 *m* glucose
b) 2.0 *m* silver nitrate
c) 2.0 *m* sodium chloride
d) 1.5 *m* calcium chloride
e) 2.5 *m* sucrose

58. The addition of 125 mg of caffeine to 100 g of cyclohexane lowered the freezing point by 0.13 K. Calculate the molar mass of caffeine. The k_f for cyclohexane is 20.1 K · kg · mol^{-1}.
a) 19.3 g/mol b) 193 g/mol c) 47.8 g/mol d) 481 g/mol e) 96.5 g/mol

59. The addition of 1.96 g of vitamin B_{12}, isolated as cyanocobalamin, to 100 g camphor lowers the freezing point by 0.57°C. What is the molar mass of vitamin B_{12}? The k_f for camphor is 39.7 K · kg · mol^{-1}.

60. Calculate the freezing point of a solution of 1.50 g caffeine, molar mass 194 g · mol^{-1}, in 25.0 g benzene. The k_f for benzene is 5.10 K · kg · mol^{-1} and the freezing point of benzene is 5.50°C.
a) 3.92°C b) 0°C c) 7.08°C d) 5.50°C e) 1.58°C

61. A 1.0 m aqueous solution of glucose boils at 100.51°C. What is the theoretical boiling point elevation of a 1.0 *m* aqueous solution of aluminum sulfate?
a) 1.02°C b) 0.26°C c) 2.55°C d) 1.53°C e) 0.51°C

62. The solubility of calcium chloride in water at 0°C is 5.4 m.Calculate the freezing point depression of this solution. The k_f for water is 1.86 K · kg · mol^{-1}.
a) 10°C b) 20°C c) 30°C d) 27°C e) 5.0°C

63. The solubility of sodium chloride in water at 100°C is about 6.7 *m*. Calculate the boiling point elevation of this solution. The k_b for water is 0.51 K · kg · mol^{-1}.
a) 14°C
b) 6.8 °C
c) 3.4°C
d) The boiling point elevation is zero since the oppositely charged ions cancel each other.
e) 13.6°C

64. Calculate the boiling point elevation of an aqueous solution which is 2.20 *m* calcium perchlorate. The k_b for water is 0.51 K · kg · mol^{-1}.
a) 0.56°C b) 0.51°C c) 1.12°C d) 2.24°C e) 3.37°C

65. The freezing point depression of a 0.10 *m* NaCl(aq) solution is 0.37 K. What is the experimental van't Hoff *i* factor for NaCl(aq)? The k_f for water is 1.86 K · kg mol^{-1}.
a) 2.0 b) 0.10 c) 1.86 d) 1.0 e) 0.20

66. How many grams of calcium chloride (molar mass, 111 g mol^{-1}) must be added to 100 g water to give a total solute molality (effective molality of particles) of 1.80 *m*?
a) 6.66 g b) 3.33 g c) 2.22 g d) 66.6 g e) 18.5 g

67. When 8.80 grams of sulfur are dissolved in 60.0 grams of carbon disulfide, the boiling point of the solution is 47.64°C. The normal boiling point of CS_2 is 46.30°C and k_b for CS_2 is 2.34 K · kg mol^{-1}. Determine the molecular formula of sulfur.

68. The freezing point depression of a 0.10 m HF(aq) solution is 0.20°C. Estimate the experimental van't Hoff *i* factor for HF(aq) in this solution. Explain.

69. Calculate the osmotic pressure of 0.010 M aluminum sulfate at 25°C. The gas constant is 0.0821 L · atm · K^{-1} · mol^{-1}. (Hint: The effective molarity of all particles must be used.)
a) 0.72 atm b) 0.96 atm c) 1.20 atm d) 0.48 atm e) 0.24 atm

70. Blood, sweat, and tears are about 0.15 M in sodium chloride. Estimate the osmotic pressure of these solutions at 37°C. The gas constant is 0.0821 L · atm · K^{-1} · mol^{-1}.(Hint: The effective molarity of all particles must be used.)
a) 11 atm b) 7.6 atm c) 3.8 atm d) 1.8 atm e) 0.91 atm

71. What is the minimum pressure that must be applied at 25°C to obtain pure water by reverse osmosis from seawater which is 2.0 M in sodium chloride? (Hint: The effective molarity of all particles must be used.)
a) 49 atm b) 1 atm c) 147 atm d) 24 atm e) 98 atm

72. From which of the following solutions would it be most difficult to obtain pure water by reverse osmosis? (Hint: The effective molarity of all particles must be used.)
a) 3.0 M ethylene glycol b) 4.0 M CH_3OH c) 2.0 M NaCl d) 2.0 M $Cu(ClO_4)_2$ e) 1.0 M $FeCl_3$

73. From which of the following solutions would it be easiest to obtain water by reverse osmosis? (Hint: The effective molarity of all particles must be used.)
a) 2.0 M KCl, 1.0 M $MgCl_2$
b) 2.0 M NaCl, 1.0 M KCl
c) All the solutions given would require the same pressure.
d) 1.0 M $CaCl_2$, 1.0 M $Al_2(SO_4)_3$
e) 2.0 M $CaCl_2$, 1.0 M glucose

74. From which of the following solutions would it be easiest to obtain water by reverse osmosis? (Hint: The effective molarity of all particles must be used.)
 a) All the solutions given would require the same pressure.
 b) 1.0 M KCl, 1.0 M NaCl
 c) 0.5 M $CaCl_2$, 0.5 M $Al_2(SO_4)_3$
 d) 1.0 M $CaCl_2$, 1.0 M glucose
 e) 1.0 M NaCl, 0.5 M $FeCl_3$

75. The osmotic pressure of 0.25 g of the protein cytochrome c dissolved in enough water to make 50.0 mL of solution is 1.52 kPa at 25°C. Estimate the molar mass of cytochrome c. The gas law constant is 0.0821 $L \cdot atm \cdot K^{-1} \cdot mol^{-1}$.
 a) 4,000 $g \cdot mol^{-1}$ b) 800,000 $g \cdot mol^{-1}$ c) 79 $g \cdot mol^{-1}$ d) 79,000 $g \cdot mol^{-1}$ e) 8,000 $g \cdot mol^{-1}$

76. The osmotic pressure of 1.00 g of a polymer dissolved in benzene to give 200 mL of solution is 1.50 kPa at 25°C.Estimate the average molar mass of the polymer. The gas law constant is 0.0821 $L \cdot atm \cdot K^{-1} \cdot mol^{-1}$.
 a) 693 $g \cdot mol^{-1}$ b) 62,000 $g \cdot mol^{-1}$ c) 1650 $g \cdot mol^{-1}$ d) 8260 $g \cdot mol^{-1}$ e) 41,300 $g \cdot mol^{-1}$

77. A 1.00-g sample of the protein horseradish peroxidase was dissolved in water to make 0.100 L of solution. The osmotic pressure of this solution was 4.57 Torr at 20°C. Estimate the molar mass of horseradish peroxidase. The gas law constant is 0.0821 $L \cdot atm \cdot K^{-1} \cdot mol^{-1}$.
 a) 40,000 g/mol b) 80,000 g/mol c) 50,000 g/mol d) 4,000 g/mol e) 20,000 g/mol

78. An animal cell assumes its normal volume when it is placed in a solution with a total solute molarity (molarity of nonelectrolytes plus ions) of 0.3 M. If the cell is placed in a solution with a total solute molarity of 0.6 M,
 a) the escaping tendency of water in the solution increases.
 b) water enters the cell causing expansion.
 c) water leaves the cell causing contraction.
 d) no movement of water takes place.
 e) the escaping tendency of water in the cell decreases.

79. All of the following are independent of temperature except
 a) molarity.
 b) molality.
 c) mole fraction.
 d) mass percent.
 e) grams of solute per total mass of solute plus solvent.

80. The osmotic pressure of 2.0 g of a polymer dissolved in enough benzene to give 200 mL of solution is 1.50 kPa at 25°C. Calculate the average molar mass of the polymer. The gas law constant is 0.0821 $L \cdot atm \cdot K^{-1} \cdot mol^{-1}$.

Chapter 12: The Properties of Solutions
Form B

1. As water runs through the ground, which one of the following is the predominant reaction?

a) $CaCO_3(s) + H_2CO_3(aq) \rightarrow Ca(HCO_3)_2(aq)$

b) $CaCO_3(s) + 2H^+(aq) \rightarrow Ca^{2+}(aq) + CO_2(g) + H_2O(l)$

c) $CO_2(aq) \rightarrow CO_2(g)$

d) $2HCO_3^-(aq) \rightarrow CO_3^{2-}(aq) + CO_2(g) + H_2O(l)$

e) $Ca(HCO_3)_2(aq) \rightarrow CaCO_3(s) + H_2CO_3(aq)$

2. Which of the following would likely be most soluble?

a) $CaHPO_4$ b) $Ca_3(PO_4)_2$ c) $CaCO_3$ d) $Ca_5(PO_4)_3OH$ e) $Ca(ClO_4)_2$

3. Which of the following would you expect to be least soluble?

a) $Ca_5(PO_3)_3OH$ b) $Ca(ClO_4)_2$ c) $CaHPO_4$ d) $Ca(H_2PO_4)_2$ e) $CaCO_3$

4. Which of the following would likely dissolve in benzene?

a) $C_6H_{12}O_6$ b) $Ca(ClO_4)_2$ c) CCl_4 d) Na_2CO_3 e) NaCl

5. Calculate the solubility of oxygen in ppm at 20°C in a lake where the partial pressure of oxygen is 152 Torr. Henry's constant is 0.0013 $mol \cdot L^{-1} \cdot atm^{-1}$ for oxygen at 20°C.

a) 42 ppm b) 4.2 ppm c) 0.28 ppm d) 8.3 ppm e) 210 ppm

6. Calculate Henry's constant for carbon dioxide if the solubility of $CO_2(g)$ in water in a closed container is 3.7 $g \cdot L^{-1}$ where the partial pressure of $CO_2(g)$ is 3.6 atm.

a) 0.023 $mol \cdot L^{-1} \cdot atm^{-1}$

b) 0.28 $mol \cdot L^{-1} \cdot atm^{-1}$

c) 8.3×10^{-2} $mol \cdot L^{-1} \cdot atm^{-1}$

d) 1.0 $mol \cdot L^{-1} \cdot atm^{-1}$

e) 5.2×10^{-4} $mol \cdot L^{-1} \cdot atm^{-1}$

7. Henry's constants($mol \cdot L^{-1} \cdot atm^{-1}$) for $CO_2(g)$, He(g), Ar(g), $N_2(g)$, and $O_2(g)$ are 0.023, 0.00037, 0.0015, 0.0007, and 0.0013, respectively, in water at 20°C. Which gas would be most soluble at a pressure of 25 atm?

a) $O_2(g)$ b) $CO_2(g)$ c) Ar(g) d) $N_2(g)$ e) He(g)

8. Calculate the solubility of oxygen at 20°C in a lake where the partial pressure of oxygen is 152 Torr. Henry's constant is 0.0013 $mol \cdot L^{-1} \cdot atm^{-1}$ for oxygen at 20°C.

a) 1.3×10^{-3} M b) 0.20 M c) 8.6×10^{-6} M d) 2.6×10^{-4} M e) 6.5×10^{-3} M

9. The equilibrium

$$Ca(HCO_3)_2(aq) \leftrightarrow CaCO_3(s) + CO_2(g) + H_2O(l)$$

exists during the formation of stalactites and stalagmites, which are mainly $CaCO_3(s)$, in underground caves. If the atmosphere in the cave suddenly changed and the partial pressure of $CO_2(g)$ reduced, predict the result.

a) The stalactites and stalagmites would become smaller because $H_2O(l)$ would be used up.
b) The stalactites and stalagmites would become smaller due to dissolution of $CaCO_3(s)$.
c) The stalactites and stalagmites would become larger due to formation of more $CaCO_3(s)$.
d) The $Ca(HCO_3)_2(aq)$ concentration would increase.
e) No change would occur.

10. The "bends," experienced by divers who ascend too rapidly, can be explained by

a) Raoult's law.
b) Henry's law.
c) vapor pressure changes.
d) colligative molarity.
e) osmotic pressure.

11. The lattice enthalpy of AgBr is 903 kJ/mol and the hydration enthalpy is $-819\ kJ \cdot mol^{-1}$. Estimate the enthalpy of solution of AgBr.

a) $+1722\ kJ \cdot mol^{-1}$ b) $+819\ kJ \cdot mol^{-1}$ c) $+84\ kJ \cdot mol^{-1}$ d) $-84\ kJ \cdot mol^{-1}$ e) $-1722\ kJ \cdot mol^{-1}$

12. The lattice enthalpy of AgI is $887\ kJ \cdot mol^{-1}$ and the hydration enthalpy is $-775\ kJ \cdot mol^{-1}$. Estimate the enthalpy of solution of AgI.

a) $+775\ kJ \cdot mol^{-1}$ b) $+1660\ kJ \cdot mol^{-1}$ c) $+112\ kJ \cdot mol^{-1}$ d) $-112\ kJ \cdot mol^{-1}$ e) $-1660\ kJ \cdot mol^{-1}$

13. The lattice enthalpy of MgF_2 is $2960\ kJ \cdot mol^{-1}$ and the hydration enthalpy is $-2980\ kJ \cdot mol^{-1}$. Estimate the enthalpy of solution of MgF_2.

a) $-20\ kJ \cdot mol^{-1}$ b) $+20\ kJ \cdot mol^{-1}$ c) $+2980\ kJ \cdot mol^{-1}$ d) $-5940\ kJ \cdot mol^{-1}$ e) $+5940\ kJ \cdot mol^{-1}$

14. The lattice enthalpy of AgBr is $903\ kJ \cdot mol^{-1}$ and the enthalpy of solution is $+84\ kJ \cdot mol^{-1}$. Estimate the enthalpy of hydration of AgBr.

a) $+987\ kJ \cdot mol^{-1}$ b) $-84\ kJ \cdot mol^{-1}$ c) $+819\ kJ \cdot mol^{-1}$ d) $-819\ kJ \cdot mol^{-1}$ e) $-987\ kJ \cdot mol^{-1}$

15. The lattice enthalpy of MgF_2 is $2960\ kJ \cdot mol^{-1}$ and the enthalpy of solution is about $-20\ kJ \cdot mol^{-1}$. Estimate the enthalpy of hydration of MgF_2.

a) $+2980\ kJ \cdot mol^{-1}$ b) $-2940\ kJ \cdot mol^{-1}$ c) $-2980\ kJ \cdot mol^{-1}$ d) $+20\ kJ \cdot mol^{-1}$ e) $+2940\ kJ \cdot mol^{-1}$

16. For LiF, the lattice enthalpy is larger than the absolute value of the enthalpy of hydration. This means that for LiF

a) the solubility increases when the temperature decreases.
b) the enthalpy of hydration is endothermic.
c) the lattice enthalpy is exothermic.
d) the enthalpy of solution is endothermic.
e) the enthalpy of solution is exothermic.

17. For KBr, the lattice enthalpy is larger than the absolute value of the enthalpy of hydration. This means that for KBr
   a) ΔH_L is negative.
   b) ΔH_{hyd} is positive.
   c) ΔH_{sol} is positive.
   d) the solubility increases when the temperature decreases.
   e) ΔH_{sol} is negative.

18. For MgF_2, the absolute value of the enthalpy of hydration is larger than the lattice enthalpy. This means that for MgF_2
   a) the solubility increases when the temperature increases.
   b) the enthalpy of solution is endothermic.
   c) the enthalpy of hydration is positive.
   d) the enthalpy of solution is exothermic.
   e) the lattice enthalpy is negative.

19. For $CaCl_2$, the absolute value of the enthalpy of hydration is larger than the lattice enthalpy. This means that for $CaCl_2$
   a) ΔH_{hyd} is positive.
   b) ΔH_{sol} is positive.
   c) ΔH_{sol} is negative.
   d) the solubility increases when the temperature increases.
   e) ΔH_L is negative.

20. Which of the following is likely to have the largest exothermic hydration enthalpy?
   a) Li^+ b) Sr^{2+} c) Ca^{2+} d) Na^+ e) Mg^{2+}

21. Which of the following is likely to have the largest exothermic hydration enthalpy?
   a) Li^+ b) Cs^+ c) Na^+ d) K^+ e) Rb^+

22. Which of the following is likely to have the smallest exothermic hydration enthalpy?
   a) Cl^- b) Br^- c) I^- d) NO_3^- e) F^-

23. Colligative properties are those that
   a) result from a change in vapor pressure of the solute when it is dissolved in the solvent.
   b) change the concentration of a solute in a solution.
   c) are only observed for pure solvents.
   d) are those that do not depend on the temperature and pressure but do depend on the type of solute which is dissolved in the solvent.
   e) depend on the number of solute particles in a solution and not on the type of particle.

24. When the mole fraction of a solvent is 1.0, there is
a) only solute present.
b) only pure solvent present.
c) a 1:1 ratio of solute to solvent.
d) 1.0 mole of solvent and 99 moles of solute.
e) 1.0 g of solute per 100 g of solution.

25. Calculate the mole fraction of water in a mixture of equal masses of methanol and water. The molar mass of methanol is 32.0 $g \cdot mol^{-1}$ and that of water is 18.0 $g \cdot mol^{-1}$.
a) 0.640 b) 0.500 c) 0.360 d) 0.0556 e) 1.00

26. Calculate the mole fraction of acetone in a mixture of equal masses of acetone and water. The molar mass of acetone is 58.0 $g \cdot mol^{-1}$ and that of water is 18.0 $g \cdot mol^{-1}$.
a) 0.237 b) 0.763 c) 0.310 d) 0.500 e) 0.690

27. Calculate the mole fraction of water in a mixture of equal masses of ethanol and water. The molar mass of ethanol is 46.0 $g \cdot mol^{-1}$ and that of water is 18.0 $g \cdot mol^{-1}$.
a) 0.719 b) 0.281 c) 0.500 d) 0.391 e) 0.609

28. If 75.0 g glucose (molar mass, 180.2 $g \cdot mol^{-1}$) are dissolved in 625 g water (molar mass, 18.0 $g \cdot mol^{-1}$), what is the mole fraction of glucose in the solution?
a) 0.0118 b) 0.416 c) 1.00 d) 0.989 e) 0.584

29. Calculate the mole fraction of water (molar mass, 18.0 $g \cdot mol^{-1}$) in a solution prepared by dissolving 50.0 g of sucrose (molar mass, 342.3 $g \cdot mol^{-1}$) in 200.0 g of water.
a) 0.987 b) 0.0130 c) 0.146 d) 0.854 e) 1.00

30. How many moles of glucose should be added to 650 g of water to prepare a 0.300 *m* solution?
a) 0.195 b) 0.300 c) 2.17 d) 0.462 e) 0.105

31. How many moles of glucose should be added to 650 g of water to prepare a 0.600 *m* solution?
a) 0.390 b) 0.600 c) 4.34 d) 0.924 e) 0.210

32. How many moles of sodium sulfate should be added to 1750 g of water in order to prepare 0.100 *m* $Na_2SO_4(aq)$?
a) 0.100 mol b) 14.1 mol c) 0.175 mol d) 0.572 mol e) 7.00 mol

33. What is the molality of a solution prepared by dissolving 15.1 g KNO_3 (molar mass, 101 $g \cdot mol^{-1}$) in 500 g water?
a) 0.300 *m* b) 0.121 *m* c) 1.20 *m* d) 3.34 *m* e) 0.0750 *m*

34. The mole fraction of glucose in an aqueous solution is 0.400. Calculate the molality of glucose in the solution. The molar mass of glucose is 342 $g \cdot mol^{-1}$ and that of water is 18.0 $g \cdot mol^{-1}$.
a) 37.0 *m* b) 2.92 *m* c) 0.400 *m* d) 2.71 *m* e) 55.6 *m*

35. The mole fraction of glucose in an aqueous solution is 0.0500. Calculate the molality of glucose in the solution. The molar mass of glucose is 342 g · mol^{-1} and that of water is 18.0 g · mol^{-1}.
a) 1.46 *m* b) 27.8 *m* c) 0.0500 *m* d) 55.6 *m* e) 2.92 *m*

36. Calculate the mole fraction of water in 6.00 m glucose(aq). The molar mass of glucose is 342 g · mol^{-1} and that of water 18.0 g · mol^{-1}.
a) 0.903 b) 0.0975 c) 0.0556 d) 0.00600 e) 0.108

37. Calculate the mole fraction of water in 12.0 m glucose(aq). The molar mass of glucose is 342 g · mol^{-1} and that of water 18.0 g · mol^{-1}.
a) 0.822 b) 0.178 c) 0.216 d) 0.0120 e) 0.0556

38. Commercial perchloric acid (molar mass, 100.5 g · mol^{-1}) has a density of 1.67 g · mL^{-1}. If the molarity of $HClO_4$(aq) is 11.7 M, calculate the molality of $HClO_4$(aq).
a) 11.7 *m* b) 19.5 *m* c) 23.7 *m* d) 9.95 *m* e) 7.01 *m*

39. Concentrated sulfuric acid is 18.1 M and has a density of 1.84 g · mL^{-1}.Calculate the molality of sulfuric acid in this solution. The molar mass of sulfuric acid is 98.1 g · mol^{-1}.
a) 283 *m* b) 33.3 *m* c) 9.8 *m* d) 10.2 *m* e) 18.1 *m*

40. Calculate the molality of 30.0 M N_2H_4(aq). The density of the solution is 1.011 g mL^{-1}.

41. Calculate the molality of methanol in a bottle of "methyl hydrate" used to help start automobiles on damp days, if the solution is 90.0% CH_3OH by mass. The molar mass of methanol is 32.0 g · mol^{-1}.
a) 0.320 *m* b) 28.8 *m* c) 2.81 *m* d) 0.900 *m* e) 281 *m*

42. The molality of a solution of benzene in toluene is
a) mol toluene per kg benzene.
b) mol benzene per L toluene.
c) kg benzene per kg toluene.
d) mol benzene per kg toluene.
e) mol benzene per 100 g toluene.

43. Calculate the vapor pressure at 80°C of a solution prepared by dissolving 3.00 moles of glucose in 100 g of water. The vapor pressure of water at 80°C is 355 Torr and the molar mass of water is 18.0 g · mol^{-1}.
a) 480 Torr b) 760 Torr c) 355 Torr d) 125 Torr e) 230 Torr

44. Calculate the vapor pressure at 80°C of a solution prepared by dissolving 0.0300 moles of glucose in 200 g of water. The vapor pressure of water at 80°C is 355 Torr and the molar mass of water is 18.0 g · mol^{-1}.
a) 710 Torr b) 760 Torr c) 356 Torr d) 354 Torr e) 355 Torr

45. Calculate the vapor pressure at 25°C of a mixture of benzene and toluene in which the mole fraction of benzene is 0.800. The vapor pressure at 25°C of benzene is 94.6 Torr and that of toluene is 29.1 Torr.
a) 81.5 Torr b) 69.9 Torr c) 105 Torr d) 75.7 TOrr e) 124 Torr

46. The vapor pressures of acetone and cyclohexane are 23.5 and10.2 kPa, respectively, at a certain temperature. What is the vapor pressure of a solution containing 25.0 g of each compound?
a) 33.7 kPa b) 13.9 kPa c) 4.2 kPa d) 9.7 kPa e) 18.1 kPa

47. A solution made up of 50.0 g each of CS_2 (molar mass, 76.13 $g \cdot mol^{-1}$) and CCl_4 (molar mass, 153.8 $g \cdot mol^{-1}$) has a vapor pressure of 274 Torr at 296 K. If the vapor pressure of pure carbon tetrachloride is 99.8 Torr at 296 K, what is the vapor pressure of pure carbon disulfide?
a) 33 Torr b) 241 Torr c) 360 Torr d) 410 Torr e) 728 Torr

48. Calculate the vapor pressure lowering of an aqueous solution containing 10.0 g naphthalene (molar mass, 128 $g \cdot mol^{-1}$) dissolved in 100 g benzene (molar mass, 78.1 $g \cdot mol^{-1}$) at 25°C. At 25°C, the vapor pressure of pure benzene is 97.0 Torr.
a) 0.76 Torr b) 91.3 Torr c) 5.57 Torr d) 9.70 Torr e) 75.8 Torr

49. Which of the following 1.0 M solutions contains the most particles?
a) magnesium sulfate b) aluminum sulfate c) glucose d) ethylene glycol e) potassium chloride

50. Which of the following pairs have a van't Hoff *i* factor of 2?
a) magnesium sulfate and ethylene glycol
b) sodium sulfate and potassium chloride
c) perchloric acid and barium hydroxide
d) glucose and sodium chloride
e) sodium chloride and magnesium sulfate

51. A 1.0-molal solution of ethanol (CH_3CH_2OH) in water, and a 1.0-molal solution of aluminum sulfate in water have total solute molalities of
a) 1.0 and 3.0, respectively.
b) 1.0 and 5.0, respectively.
c) 1.0 and 4.0, respectively.
d) 2.0 and 4.0, respectively.
e) 1.0 and 2.0, respectively.

52. Which of the following has the lowest freezing point and the highest boiling point?
a) 1.5 *m* magnesium phosphate
b) 1.0 *m* sodium chloride
c) 1.5 *m* aluminum nitrate
d) 1.5 *m* calcium chloride
e) 2.0 *m* potassium chloride

53. Calculate the freezing point of a 7.0 m aqueous solution of ethylene glycol. The k_f for water is 1.86 $K \cdot kg \cdot mol^{-1}$.
a) –13°C b) –3.8°C c) –26°C d) 0°C e) –0.27°C

54. Calculate the freezing point of a 28.0 m aqueous solution of ethylene glycol. The k_f for water is 1.86 $K \cdot kg \cdot mol^{-1}$.
a) –15°C b) –104°C c) –52°C d) –0.066°C e) 0OC

55. The freezing point of an aqueous ethylene glycol solution is –52°C. Calculate the molality of ethylene glycol in this solution. The k_f for water is 1.86 $K \cdot kg \cdot mol^{-1}$.
a) 97 *m* b) 56 *m* c) 7.0 *m* d) 14 *m* e) 28 *m*

56. A 1.0 m aqueous solution of an unknown solute has a freezing point of –7.44°C. Which of the following is the solute? The k_f for water is 1.86 K · kg · mol^{-1}.
a) sodium sulfate
b) calcium chloride
c) aluminum sulfate
d) iron(III) chloride
e) sodium chloride

57. Which of the following aqueous solutions would have the lowest freezing point?
a) 2.0 m silver nitrate
b) 2.5 m sucrose
c) 2.0 m sodium chloride
d) 1.0 m glucose
e) 1.5 m barium nitrate

58. A solution of 1.50 g caffeine in 25.0 g benzene freezes at 3.89°C. Calculate the molar mass of caffeine given that the freezing point of benzene is 5.50°C and the k_f for benzene is 5.10 K · kg · mol^{-1}.
a) 191 g · mol^{-1} b) 316 g · mol^{-1} c) 78.9 g · mol^{-1} d) 789 g · mol^{-1} e) 526 g · mol^{-1}

59. The addition of 3.62 g of a-tocopherol, vitamin E, to 100 g of camphor lowers the freezing point by 3.33 K. What is the molar mass of vitamin E? The k_f for camphor is 39.7 K · kg mol^{-1}.

60. Calculate the freezing point of a solution of 125 mg of caffeine, molar mass 194 g · mol^{-1} l, in 100 g cyclohexane. The k_f for cyclohexane is 20.1 K · kg · mol^{-1} and the freezing point of cyclohexane is 6.50°C.
a) 4.07°C b) 2.43°C c) 6.37°C d) 0°C e) 0.13°C

61. A 1.0 m aqueous solution of glucose boils at 100.51°C. What is the theoretical boiling point elevation of a 1.0 m aqueous solution of calcium chloride?
a) 2.55°C b) 1.53°C c) 0.26°C d) 0.51°C e) 1.02°C

62. Calculate the freezing point depression of 1.60 m $Mg(OH)_2$(aq). The k_f for water is 1.86 K · kg · mol^{-1}.
a) 5.96°C b) 8.94°C c) 1.49°C d) 1.86°C e) 2.98°C

63. Calculate the boiling point elevation of 5.5 m NaCl(aq). The k_b for water is 0.51 K · kg · mol^{-1}.
a) 5.6°C b) 2.8°C c) 0.51°C d) 8.4°C e) 1.4°C

64. Calculate the boiling point elevation of an aqueous solution which is 3.20 m calcium nitrate. The k_b for water is 0.51 K · kg · mol^{-1}.
a) 3.26°C b) 0.82°C c) 1.63°C d) 4.90°C e) 0.51°C

65. The freezing point depression of a 0.10 m HF(aq) solution is 0.20 K. What is the experimental van't Hoff i factor for HF(aq)? The k_f for water is 1.86 K · kg mol^{-1}.
a) 1.1 b) 0.10 c) 1.86 d) 2.0 e) 0.20

66. How many grams of calcium chloride (molar mass, 111 g mol^{-1}) must be added to 100 g water to give a total solute molality (effective molality of particles) of 2.40 *m*?
a) 2.22 g b) 13.9 g c) 8.88 g d) 88.8 g e) 4.44 g

67. A solution of 0.640 g nicotine in 25.0 g benzene freezes at 4.69°C. Calculate the molar mass of nicotine, given the freeezing point of benzene, 5.50°C, and k_f for benzene of 5.12 K · kg mol^{-1}.

68. The freezing point depression of a 0.010 m HF(aq) solution is only 0.02°C. Estimate the experimental van't Hoff *i* factor for HF(aq) in this solution. Explain. Compare the results with those of question 153. Explain.

69. Calculate the osmotic pressure of 0.010 M iron(III) chloride at 25°C. The gas constant is 0.0821 L · atm · K^{-1} · mol^{-1}.(Hint: The effective molarity of all particles must be used.)
a) 0.72 atm b) 1.22 atm c) 0.24 atm d) 0.96 atm e) 0.48 atm

70. Calculate the osmotic pressure of 0.010 M calcium chloride at 25°C. The gas constant is 0.0821 L · atm · K^{-1} · mol^{-1}.(Hint: The effective molarity of all particles must be used.)
a) 0.96 atm b) 0.12 atm c) 0.72 atm d) 0.24 atm e) 0.48 atm

71. What is the minimum pressure that must be applied at 25°C to obtain pure water by reverse osmosis from seawater which is 1.0 M in sodium chloride? (Hint: The effective molarity of all particles must be used.)
a) 98 atm b) 1 atm c) 49 atm d) 24 atm e) 147 atm

72. From which of the following solutions would it be most difficult to obtain pure water by reverse osmosis? (Hint: The effective molarity of all particles must be used.)
a) 1.5 M $MgSO_4$ b) 3.0 M acetic acid c) 1.0 M $Fe(NO_3)_3$ d) 3.5 M sucrose e) 1.0 M $CaCl_2$

73. From which of the following solutions would it be easiest to obtain water by reverse osmosis? (Hint: The effective molarity of all particles must be used.)
a) All the solutions given would require the same pressure.
b) 2.0 M $CaCl_2$, 1.0 M glucose
c) 4.0 M glucose, 1.0 M KCl
d) 1.0 M $CaCl_2$, 1.0 M $Al_2(SO_4)_3$
e) 2.0 M KCl, 1.0 M $MgCl_2$

74. From which of the following solutions would it be easiest to obtain water by reverse osmosis? (Hint: The effective molarity of all particles must be used.)
a) 1.0 M $CaCl_2$, 1.0 M glucose
b) 0.5 M $CaCl_2$, 0.5 M $Al_2(SO_4)_3$
c) 1.0 M glucose, 0.5 M $FeCl_3$
d) 1.0 M KCl, 1.0 M NaCl
e) All the solutions given would require the same pressure.

75. The osmotic pressure of 125 mg of a newly-discovered protein dissolve in enough water to give 25.0 mL of solution is 2.25 kPa at 25°C. Estimate the molar mass of this protein. The gas law constant is 0.0821 L · atm · K^{-1} · mol^{-1}.
a) 138 g/mol b) 558,000 g · mol^{-1} c) 66,000 g · mol^{-1} d) 7340 g · mol^{-1} e) 5510 g · mol^{-1}

76. The osmotic pressure of 1.85 g of a polymer dissolved in benzene to give 100 mL of solution is 1.25 kPa at 25°C.Estimate the average molar mass of the polymer. The gas law constant is 0.0821 L · atm · K^{-1} · mol^{-1}.
a) 19,800 g · mol^{-1} b) 275,000 g · mol^{-1} c) 3080 g · mol^{-1} d) 36,700 g · mol^{-1} e) 3670 g · mol^{-1}

77. A 0.10-g sample of the protein myoglobin was dissolved in water to make 0.100 L of solution. The osmotic pressure of this solution was 1.08 Torr at 20°C. Estimate the molar mass of myoglobin. The gas law constant is 0.0821 L · atm · K^{-1} · mol^{-1}.
A) 35,000 g/mol
B) 9,000 g/mol
C) 23,000 g/mol
D) 64,000 g/mol
E) 17,000 g/mol
a) E

78. An animal cell assumes its normal volume when it is placed in a solution with a total solute molarity (molarity of nonelectrolytes plus ions) of 0.3 M. If the cell is placed in a solution with a total solute molarity of 0.15 M,
a) water leaves the cell causing expansion.
b) the escaping tendency of water in the cell increases.
c) no movement of water takes place.
d) the escaping tendency of water in the solution decreases.
e) water enters the cell causing contraction.

79. All of the following are independent of temperature except
a) osmotic pressure.
b) molality.
c) mole fraction.
d) mass percent.
e) grams of solute per total mass of solute plus solvent.

80. The osmotic pressure of 250 mg of the protein carbonic anhydrase dissolved in enough water to give 25.0 mL of solution is 6.20 Torr at 25°C. Calculate the molar mass of carbonic anhydrase. The gas law constant is 0.0821 L · atm · K^{-1} · mol^{-1}.

Chapter 13: Chemical Equilibrium
Form A

1. Consider the following reaction:

   $NH_4(NH_2CO_2)(s) \leftrightarrow 2NH_3(g) + CO_2(g)$

   When 0.200 moles of $NH_4(NH_2CO_2)(s)$ are placed in a 1.50-L flask and allowed to reach equilibrium at 400 K, 33.0% of the solid remains. Calculate the values of K_c and K_p for the reaction.

2. Consider the following reaction:

   $4PCl_3(g) \leftrightarrow P_4(g) + 6Cl_2(g)$

   If the initial concentration of $PCl_3(g)$ is 1.0 M, and "x" is the equilibrium concentration of $P_4(g)$, what is the correct equilibrium relation?

a) $K_c = x^7/(1.0 - x)^4$
b) $K_c = 6x^7/(1.0 - x)^4$
c) $K_c = 6x^7/(1.0 - 4x)^4$
d) $K_c = (x)(6x)^6/(1.0 - 4x)^4$
e) $K_c = 6x^7$

3. Consider the following reaction:

   $NH_4(NH_2CO_2)(s) \leftrightarrow 2NH_3(g) + CO_2(g)$

   If the initial concentration of $NH_3(g)$ is 1.0 M, and "x" is the equilibrium concentration of $NH_3(g)$, what is the correct equilibrium relation? (The solid is in excess.)

a) $K_c = (1.0 - x)^2(x/2)$ b) $K_c = x^3$ c) $K_c = x^3/2$ d) $K_c = (1.0 - x)^2(x)$ e) $K_c = 4x^3$

4. Consider the following reaction:

   $P_4(g) + 5O_2(g) \leftrightarrow P_4O_{10}(s)$

   If the initial concentration of $O_2(g)$ is 1.0 M, and "x" is the equilibrium concentration of $O_2(g)$, what is the correct equilibrium relation?

a) $K_c = 5x^6$ b) $K_c = 5/x^6$ c) $K_c = x^6$ d) $K_c = 1/x^6$ e) $K_c = 1/x^5$

5. Consider the following reaction:

   $H_2O(g) + C(s) \leftrightarrow H_2(g) + CO(g)$

   If the value of K_p for this reaction is 3.72 at 1000 K, and the equilibrium partial pressures of $H_2(g)$ and CO(g) are 1.50 atm, calculate the equilibrium partial pressure of $H_2O(g)$.

a) 2.48 atm b) 1.50 atm c) 0.605 atm d) 0.403 atm e) 1.65 atm

6. Consider the following reaction:

   $H_2(g) + I_2(g) \leftrightarrow 2HI(g)$

   If the value of K_c for this reaction is 25 at 1100 K, and initially only 4.00 M HI(g) is present, what is the equilibrium concentration of $I_2(g)$?

a) 2.00 M b) 0.667 M c) 0.148 M d) 0.571 M e) 0.363 M

7. Consider the following reaction:

   $Ni(CO)_4(g) \leftrightarrow Ni(s) + 4CO(g)$ $K_c = 0.0309$

   A mixture of $Ni(CO)_4(g)$ and $CO(g)$, each with a concentration of 0.800 M, and an excess of $Ni(s)$ were confined in a 1.0-L container at 300 K. Which of the following is true?

a) K_c changes to 0.51
b) some $Ni(CO)_4(g)$ decomposes to $CO(g)$ and $Ni(s)$
c) no change occurs since we are at equilibrium
d) $Ni(s)$ is formed
e) $Ni(CO)_4(g)$ forms until equilibrium is reached

8. Consider the reaction

   $3Fe(s) + 4H_2O(g) \leftrightarrow 4H_2(g) + Fe_3O_4(s)$

   If the total pressure is increased suddenly by reducing the volume,

a) no change occurs.
b) more $Fe(s)$ is produced.
c) the equilibrium constant increases.
d) more $H_2O(g)$ is produced.
e) more $H_2(g)$ is produced.

9. For the reaction

   $C(s) + H_2O(g) \leftrightarrow CO(g) + H_2(g)$

   if the total pressure is decreased by increasing the volume,

a) more $H_2O(g)$ is produced.
b) no change occurs.
c) the equilibrium constant decreases.
d) the equilibrium constant increases.
e) more products are produced.

10. Consider the following reaction:

    $3H_2(g) + N_2(g) \leftrightarrow 2NH_3(g)$

    All of the following will lead to production of more $NH_3(g)$ except

a) removal of $NH_3(g)$.
b) a decrease in volume of the container.
c) an increase in pressure by addition of $N_2(g)$.
d) an increase in pressure by addition of argon.
e) addition of $H_2(g)$.

11. Consider the following reaction:

    $4NH_3(g) + 7O_2(g) \leftrightarrow 2N_2O_4(g) + 6H_2O(g)$

    If the initial concentrations of $N_2O_4(g)$ and $H_2O(g)$ are 3.60 M, at equilibrium the concentration of $H_2O(g)$ is 0.60 M. Calculate the equilibrium concentration of $NH_3(g)$.

a) 3.5 M b) 1.5 M c) 2.0 M d) 0.50 M e) 2.6 M

12. Consider the following reaction:

$2NOCl(g) \leftrightarrow 2NO(g) + Cl_2(g)$

If the initial concentration of NOCl(g) is 2.5 M, at equilibrium $[Cl_2] = 0.60$ M. Calculate the equilibrium concentration of NOCl(g).

a) 1.2 M b) 1.3 M c) 1.9 M d) 2.2 M e) 0.70 M

13. What is the equilibrium constant expression for the reaction below?

$2CaSO_4(s) \leftrightarrow 2CaO(s) + 2SO_2(g) + O_2(g)$

a) $K_c = [CaO]^2[SO_2]^2[O_2]/[CaSO_4]^2$
b) $K_c = 1/[SO_2]^2[O_2]$
c) $K_c = [CaO]/[CaSO_4]$
d) $K_c = [SO_2]^2[O_2]$
e) $K_c = [SO_2][O_2]$

14. Given: $N_2(g) + 3H_2(g) \leftrightarrow 2NH_3(g)$

At equilibrium at a certain temperature, the concentration of $NH_3(g)$, $H_2(g)$ and $N_2(g)$ are 0.980 M, 1.53 M and 0.510 M, respectively. Calculate the value of K_c for this reaction.

a) 0.526 b) 0.837 c) 1.26 d) 3.02 e) 1.96

15. When $NH_4HS(s)$ is placed in a 1.0-L container at 24°C, the following equlibrium is reached:

$NH_4HS(s) \leftrightarrow NH_3(g) + H_2S(g)$ $K_c = 1.6 \times 10^{-4}$

Calculate the equilibrium concentration of $NH_3(g)$.

a) 1.6×10^{-4} b) 3.2×10^{-4} c) 8.0×10^{-5} d) 1.3×10^{-2} e) 2.6×10^{-8}

16. Consider the following reaction:

$NH_4HS(s) \leftrightarrow NH_3(g) + H_2S(g)$

If the value of K_p for this reaction is 0.11 at 300 K, calculate the equilibrium partial pressure of $NH_3(g)$ starting from pure $NH_4HS(s)$.

a) 0.33 atm b) 0.22 atm c) 0.012 atm d) 0.055 atm e) 0.11 atm

17. What is the equilibrium constant expression for the reaction below?

$2TiCl_3(s) + 2HCl(g) \leftrightarrow 2TiCl_4(g) + H_2(g)$

a) $K_c = [HCl]^2[TiCl_3]^2/[H_2][TiCl_4]^2$
b) $K_c = 1/[TiCl_3]^2$
c) $K_c = [H_2][TiCl_4]^2/[HCl]^2[TiCl_3]^2$
d) $K_c = [HCl]^2/[H_2][TiCl_4]^2$
e) $K_c = [H_2][TiCl_4]^2/[HCl]^2$

18. Consider the following reaction:

$3N_2H_4(g) + 4ClF_3(g) \leftrightarrow 12HF(g) + 3N_2(g) + 2Cl_2(g)$

A mixture, initially consisting of 0.880 M $N_2H_4(g)$ and 0.880 M $ClF_3(g)$, reacts at a certain temperature. At equilibrium, the concentration of $N_2(g)$ is 0.525 M. Calculate the concentration of HF(g) at equilibrium.

a) 0.525 M b) 0.700 M c) 2.10 M d) 0.705 M e) 0.175 M

19. Consider the following reaction:

$2SO_2(g) + O_2(g) \leftrightarrow 2SO_3(g)$

If the value of K_c for this reaction is 13 at 900 K, calculate the value of K_p at 900 K.

a) 0.18 b) 960 c) 13 d) 0.0024 e) 0.077

20. What is the relationship between K_p and K_c for the reaction below?

$2HgO(s) \leftrightarrow 2Hg(l) + O_2(g)$

a) $K_c = RTK_p$ b) $K_p = K_c$ c) $K_p = (RT)^2K_c$ d) $K_p = RTK_c$ e) $K_c = (RT)^2K_p$

21. At 600°C, the equilibrium constant for the reaction

$2HgO(s) \leftrightarrow 2Hg(l) + O_2(g)$

is 2.8. Calculate the equilibrium constant for the reaction

$1/2O_2(g) + Hg(l) \leftrightarrow HgO(s)$

a) 0.60 b) 0.36 c) 1.7 d) −1.7 e) 1.1

22. Given:

$4NH_3(g) + 5O_2(g) \leftrightarrow 4NO(g) + 6H_2O(g)$

Calculate the equilibrium constant for the following reaction.

$2NH_3(g) + 5/2O_2(g) \leftrightarrow 2NO(g) + 3H_2O(g)$

a) K b) $K^{1/2}$ c) $2K$ d) K^{-1} e) $0.5K$

23. Consider the following reaction:

$4NH_3(g) + 7O_2(g) \leftrightarrow 2N_2O_4(g) + 6H_2O(g)$

If the initial concentrations of $NH_3(g)$ and $O_2(g)$ are 3.60 M, at equilibrium the concentration of $N_2O_4(g)$ is 0.60 M. Calculate the equilibrium concentration of $NH_3(g)$.

a) 1.8 M b) 3.3 M c) 1.5 M d) 2.4 M e) 3.0 M

24. Consider the following reaction:

$4NH_3(g) + 7O_2(g) \leftrightarrow 2N_2O_4(g) + 6H_2O(g)$

If the initial concentrations of $NH_3(g)$ and $O_2(g)$ are 3.60 M, at equilibrium the concentration of $N_2O_4(g)$ is 0.60 M. Calculate the equilibrium concentration of $O_2(g)$.

a) 1.5 M b) 1.8 M c) 3.3 M d) 3.0 M e) 2.1 M

25. Consider the following reaction:

$2SO_3(g) \leftrightarrow 2SO_2(g) + O_2(g)$

For an initial concentration of $SO_3(g)$ equal to 0.0264 M, at equilibrium it is found that the concentration of $O_2(g)$ equals 0.00280 M. Calculate the concentrations of $SO_3(g)$ and $SO_2(g)$ at equilibrium.

26. Consider the following reaction:

$$4NH_3(g) + 6NO(g) \leftrightarrow 6H_2O(g) + 5N_2(g)$$

If the initial concentrations of $NH_3(g)$ and NO(g) are 6.0 M, at equilibrium $[N_2] = 2.50$ M. Calculate the equilibrium concentration of NO(g).

a) 2.5 M b) 2.0 M c) 4.0 M d) 3.5 M e) 3.0 M

27. Given: $2SO_2(g) + O_2(g) \leftrightarrow 2SO_3(g)$

At equilibrium at a certain temperature, the concentrations of $SO_3(g)$, $SO_2(g)$ and $O_2(g)$ are 0.12 M, 0.86 M and 0.33 M, respectively. Calculate the value of K_c for this reaction.

a) 0.059 b) 0.42 c) 1.31 d) 0.87 e) 0.014

28. Consider the following reaction:

$$COCl_2(g) \leftrightarrow CO(g) + Cl_2(g)$$

For an initial concentration of 0.38 M $COCl_2(g)$, the equilibrium concentration of $Cl_2(g)$ is 0.080 M at 1100 K. Calculate the value of K_c.

a) 0.021 b) 0.017 c) 0.029 d) 0.21 e) 59.4

29. Consider the following reaction:

$$4NH_3(g) + 6NO(g) \leftrightarrow 6H_2O(g) + 5N_2(g)$$

If the initial concentrations of $NH_3(g)$ and NO(g) are 4.0 M, at equilibrium [NO] = 1.0 M. Calculate the equilibrium concentration of $N_2(g)$.

a) 3.0 M b) 2.5 M c) 2.0 M d) 0.50 M e) 5.0 M

30. Consider the following reaction:

$$4NH_3(g) + 6NO(g) \leftrightarrow 6H_2O(g) + 5N_2(g)$$

If the initial concentrations of $NH_3(g)$ and NO(g) are 4.0 M, at equilibrium [NO] = 1.0 M. Calculate the equilibrium concentration of $NH_3(g)$.

a) 3.0 M b) 5.0 M c) 2.0 M d) 2.5 M e) 0.50 M

31. Consider the following reaction:

$$2NO(g) + Cl_2(g) \leftrightarrow 2NOCl(g)$$

At a certain temperature, if the initial concentrations of NO(g) and $Cl_2(g)$ are 1.0 M, at equilibrium [NOCl] = 0.20 M. Calculate the value of K_c at this temperature.

a) 0.069 b) 0.040 c) 0.28 d) 0.14 e) 14

32. Consider the following reaction:

$$2NOCl(g) \leftrightarrow 2NO(g) + Cl_2(g)$$

If the initial concentration of NOCl(g) is 4.00 M, the equilibrium concentration of NO(g) is 1.32 M at a certain temperature. Calculate the value of K_c at this temperature.

a) 0.16 b) 0.020 c) 2.85 d) 0.33 e) 0.072

33. Given: $C(s) + CO_2(g) \leftrightarrow 2CO(g)$
At equilibrium at a certain temperature, the partial pressures of CO(g) and CO_2(g) are 1.22 atm and 0.780 atm, respectively. Calculate the value of K_p for this reaction.
a) 1.91 b) 3.13 c) 1.56 d) 0.640 e) 2.00

34. What is the equilibrium constant expression for the reaction below?
$CaO(s) + CO_2(g) \leftrightarrow CaCO_3(s)$
a) $K_c = [CO_2]$
b) $K_c = [CaCO_3]/[CaO]$
c) $K_c = [CaCO_3]/[CaO][CO_2]$
d) $K_c = 1/[CO_2]$
e) $K_c = [CaO][CO_2]/[CaCO_3]$

35. Consider the following reaction:
$2SO_2(g) + O_2(g) \leftrightarrow 2SO_3(g)$
If, initially, the SO_2 concentration is 3.2 mM and at equilibrium $[SO_3] = 3.0$ mM, calculate the equilibrium concentration of SO_2.
a) 3.0 mM b) 1.7 mM c) 0.20 mM d) 1.5 mM e) 1.8 mM

36. Consider the following reaction:
$2SO_2(g) + O_2(g) \leftrightarrow 2SO_3(g)$
If, initially, the SO_3 concentration is 4.8 mM and at equilibrium $[O_2] = 0.80$ mM, calculate the equilibrium concentration of SO_2.
a) 1.6 mM b) 0.80 mM c) 4.0 mM d) 2.4 mM e) 3.2 mM

37. Consider the following reaction:
$3N_2H_4(g) + 4ClF_3(g) \leftrightarrow 12HF(g) + 3N_2(g) + 2Cl_2(g)$
A mixture, initially consisting of 0.880 M N_2H_4(g) and 0.880 M ClF_3(g), reacts at a certain temperature. At equilibrium, the concentration of N_2(g) is 0.525 M. Calculate the concentration of ClF_3(g) at equilibrium.
a) 0.180 M b) 0.355 M c) 0.175 M d) 0.700 M e) 0.705 M

38. Given: $2CO_2(g) \leftrightarrow 2CO(g) + O_2(g)$
When 1.00 mole of CO_2(g) is placed in a 1.00-L container and allowed to come to equilibrium, the concentration of CO(g) is 0.50 M. What is the value of K_c for this reaction?
a) 0.25 b) 0.13 c) 0.50 d) 1.0 e) 0.063

39. The equilibrium constant for the reaction below
$2SO_2(g) + O_2(g) \leftrightarrow 2SO_3(g)$
is 11.7 at a certain temperature. What is the equilibrium constant for the reverse reaction?
a) 0.0855 b) 0.292 c) 3.42 d) 5.85 e) 0.00731

40. The equilibrium constant for the reaction

$MgCl_2(s) + 1/2O_2(g) \leftrightarrow MgO(s) + Cl_2(g)$

is 2.98 at a certain temperature. What is the equilibrium constant for the reaction below?

$2Cl_2(g) + 2MgO(s) \leftrightarrow 2MgCl_2(s) + O_2(g)$

a) 0.113 b) 1.73 c) 0.336 d) 0.579 e) 8.88

41. A mixture consisting of 0.250 M $N_2(g)$ and 0.500 M $H_2(g)$ reaches equilibrium according to the equation below:

$N_2(g)\ 3H_2(g) \leftrightarrow 2NH_3(g)$

At equilibrium, the concentration of ammonia is 0.150 M. Calculate the concentration of $H_2(g)$ at equilibrium.

a) 0.275 M b) 0.250 M c) 0.0500 M d) 0.225 M e) 0.425 M

42. Calculate the value of K_p at 700 K for the reaction

$H_2(g) + I_2(g) \leftrightarrow 2HI(g)$

given that $K_c = 54$ at the same temperature.

a) 54 b) 3100 c) 9.3 d) 1300 e) 2.2

43. What is the equilibrium constant expression for the reaction below?

$2NaCl(s) + H_2SO_4(l) \leftrightarrow Na_2SO_4(s) + 2HCl(g)$

a) $K_p = (P_{HCl})^2/[H_2SO_4]$ b) $K_p = P^{1/2}{}_{HCl}$ c) $K_c = [HCl]^2/[H_2SO_4]$ d) $K_p = 1/(P_{HCl})^2$ e) $K_p = P^2{}_{HCl}$

44. The equilibrium constant, K_p, for the reaction

$2HgO(s) \leftrightarrow 2Hg(l) + O_2(g)$

is 1.2×10^{-30}. Calculate K_p for the reaction

$1/2O_2(g) + Hg(l) \leftrightarrow HgO(s)$.

a) 9.1×10^{14} b) 8.3×10^{29} c) 1.1×10^{-15} d) 4.2×10^{29} e) -1.1×10^{-15}

45. What is the relationship between K_p and K_c for the reaction below?

$NH_4(NH_2CO_2)(s) \leftrightarrow 2NH_3(g) + CO_2(g)$

a) $K_c = (RT)^3K_p$ b) $K_c = (RT)^2K_p$ c) $K_p = (RT)^3K_c$ d) $K_p = RTK_c$ e) $K_p = (RT)^2K_c$

46. What is the relationship between K_p and K_c for the reaction below?

$N_2(g) + 3H_2(g) \leftrightarrow 2NH_3(g)$

a) $K_p = (RT)^6K_c$ b) $K_p = (RT)^{-2}K_c$ c) $K_p = (RT)^2K_c$ d) $K_c = (RT)^{-2}K_p$ e) $K_c = (RT)^2K_p$

47. At 25°C, $K_p = 6.9 \times 10^5$ for the reaction

$N_2(g) + 3H_2(g) \leftrightarrow 2NH_3(g)$

Calculate K_c at 25°C for this reaction.

a) 6.8×10^5 b) 2.8×10^4 c) 1.1×10^3 d) 4.1×10^8 e) 1.7×10^7

48. For the following reaction

$$H_2(g) + I_2(g) \leftrightarrow 2HI(g)$$

K_c = 54 at 700 K. At equilibrium the concentration of H_2 is 0.021 M and of I_2 is 0.013 M. Calculate the equilibrium concentration of HI.

a) 0.80 M b) 0.017 M c) 0.015 M d) 0.12 M e) 0.0020 M

49. For the following reaction

$$2BrCl(g) \leftrightarrow Br_2(g) + Cl_2(g)$$

K_c = 32 at 500 K. At equilibrium, the concentration of Br_2 is 0.22 M and of Cl_2 is 0.015 M. Calculate the equilibrium concentration of BrCl.

a) 0.010 M b) 0.00010 M c) 0.0033 M d) 0.0064 M e) 0.21 M

50. For the following reaction

$$2SO_2(g) + O_2(g) \leftrightarrow 2SO_3(g)$$

K_c = 11.7 at 1100 K. At equilibrium, the concentration of SO_2 is 2.3 M and of O_2 is 0.88 M. Calculate the equilibrium concentration of SO_3.

a) 4.9 M b) 7.4 M c) 1.4 M d) 3.4 M e) 4.0 M

51. The equilibrium constant for the reaction

$$2BrCl(g) \leftrightarrow Br_2(g) + Cl_2(g)$$

is 32 at 500 K. A mixture of BrCl, Br_2 and Cl_2, each at 0.050 M, was introduced into a container at 500 K. Which of the following is true?

a) Br_2(g) is used up.
b) The reaction proceeds until all BrCl(g) is used up.
c) The reaction proceeds to the right, forming more products.
d) The reaction proceeds to the left, forming more BrCl(g).
e) The system is at equilibrium and therefore no net change occurs.

52. The equilibrium constant for the reaction

$$2SO_2(g) + O_2(g) \leftrightarrow 2SO_3(g)$$

is 11.7 at 1100 K. A mixture of SO_2, O_2 and SO_3, each with a concentration of 0.22 M, was introduced into a container at 1100 K. Which of the following is true?

a) $[SO_3]$ = 0.11 M at equilibrium.
b) $[SO_3]$ = 0.47 M at equilibrium.
c) SO_2(g) and O_2(g) will be formed until equilibrium is reached.
d) SO_3(g) will be formed until equilibrium is reached.
e) $[SO_3] = [SO_2] = [O_2]$ at equilibrium.

53. The equilibrium constant for the reaction

$$2NOCl(g) \leftrightarrow 2NO(g) + Cl_2(g)$$

is 0.51 at a certain temperature. A mixture of NOCl, NO and Cl_2 with concentrations 1.3, 1.2 and 0.60 M, respectively, was introduced into a container at this temperature. Which of the following is true?

a) The concentrations do not change.
b) NOCl(g) is produced until equilibrium is reached.
c) $[Cl_2] = 0.30$ M at equilibrium.
d) $Cl_2(g)$ is produced until equilibrium is reached.
e) $[NOCl] = [NO] = [Cl_2]$ at equilibrium.

54. The equilibrium constant for the reaction

$$H_2(g) + I_2(g) \leftrightarrow 2HI(g)$$

is 54 at 700 K. A mixture of H_2, I_2 and HI, each at 0.020 M, was introduced into a container at 700 K. Which of the following is true?

a) At equilibrium, [HI] = 0.010 M.
b) The reaction proceeds to the right producing more HI(g).
c) The reaction proceeds to the left producing more $H_2(g)$ and $I_2(g)$.
d) At equilibrium, $[H_2] = [I_2] = [HI]$.
e) No net change occurs because the system is at equilibrium.

55. Which one of the following statements about chemical equilibrium is incorrect?

a) At equilibrium, the forward reaction rate equals the reverse reaction rate.
b) At equilibrium, the reactant and the product concentrations show no further change with time.
c) A true chemical equilibrium can only be attained starting from the reactant side of the reaction.
d) At equilibrium, the reactant and product concentrations are constant.
e) The same equilibrium state is attained starting either from the reactant or product side of the equation.

56. Consider the following reaction at 298 K:

$$PCl_5(g) \leftrightarrow PCl_3(g) + Cl_2(g) \qquad K_p = 1.2$$

If the concentrations of $PCl_3(g)$ and $Cl_2(g)$ are doubled from their equilibrium values, then the value of K_p

a) equals 0.60 b) equals 1.2 c) equals 2.4 d) equals 0.24 e) equals 4.8

57. For the following reaction

$$NH_3(g) + H_2S(g) \leftrightarrow NH_4HS(s)$$

$K_c = 9.7$ at 900 K. If the initial concentrations of $NH_3(g)$ and $H_2S(g)$ are 1.0 M, what is the equilibrium concentration of $NH_3(g)$?

a) 0.46 M b) 0.32 M c) 0.10 M d) 0.90 M e) 0.68 M

58. For the following reaction

$$NH_3(g) + H_2S(g) \leftrightarrow NH_4HS(s)$$

$K_c = 9.7$ at 900 K. If 2.0 moles of $NH_3(g)$ and 2.0 moles of $H_2S(g)$ are placed in a 1.0-L container at 900 K, how many moles of $NH_4HS(s)$ exist at equilibrium?

a) Since solids do not appear in K_c, this quantity cannot be calculated.
b) 4.0 mol
c) 0.32 mol
d) 1.7 mol
e) 2.0 mol

59. For the reaction

$$NH_4HS(s) \leftrightarrow NH_3(g) + H_2S(g)$$

$K_p = 560$ at 900 K. Starting with a large excess of pure $NH_4HS(s)$ in a closed container, what is the equilibrium partial pressure of $H_2S(g)$?

a) 560 atm b) 280 atm c) 24 atm d) 12 atm e) 1.8×10^{-3} atm

60. For the reaction

$$NH_4(NH_2CO_2)(s) \leftrightarrow 2NH_3(g) + CO_2(g)$$

$K_p = 2.3 \times 10^{-4}$ at 298 K. Starting with a large excess of pure $NH_4(NH_2CO_2)(s)$ in a closed container, what is the equilibrium partial pressure of $NH_3(g)$?

a) 0.039 atm b) 0.077 atm c) 0.015 atm d) 0.0076 atm e) 0.011 atm

61. For the reaction below

$$2CaSO_4(s) \leftrightarrow 2CaO(s) + 2SO_2(g) + O_2(g)$$

$K_p = 0.032$ at 700 K. What is the equilibrium partial pressure of $SO_2(g)$ starting from pure $CaSO_4(s)$?

a) 0.22 atm b) 0.20 atm c) 0.60 atm d) 0.011 atm e) 0.40 atm

62. Consider the following reaction:

$$NH_4(NH_2CO_2)(s) \leftrightarrow 2NH_3(g) + CO_2(g)$$

If the total gas pressure at equilibrium at 400 K is 0.843 atm, calculate the value of K_p.

a) 0.0888 b) 0.843 c) 0.599 d) 0.0222 e) 0.00701

63. Consider the following reaction:

$$2NO_2(g) \leftrightarrow 2NO(g) + O_2(g)$$

Starting with pure $NO_2(g)$ at a pressure of 0.500 atm, at equilibrium the total pressure was 0.732 atm. Calculate the equilibrium partial pressure of NO(g).

a) 0.0360 atm b) 0.232 atm c) 0.464 atm d) 0.400 atm e) 0.493 atm

64. When excess $NH_4HS(s)$ and 0.800 mol of $NH_3(g)$ are placed in a 1.0-L container at 24°C, the following equilibrium is reached:

$NH_4HS(s) \leftrightarrow NH_3(g) + H_2S(g)$ $\qquad K_c = 1.6 \times 10^{-4}$

What is the approximate equilibrium concentration of $NH_3(g)$?

a) 0.40 M b) 2.0×10^{-4} M c) 1.6×10^{-4} M d) 1.3×10^{-2} M e) 0.80 M

65. Consider the following reaction:

$SO_2Cl_2(g) \leftrightarrow SO_2(g) + Cl_2(g)$

At equilibrium at 200°C in a 1.00-L container, $[SO_2Cl_2] = 0.80$ M, $[SO_2] = 0.20$ M, and $[Cl_2] = 0.20$ M. If 0.20 moles of $SO_2(g)$ is added, calculate the new equilibrium concentrations.

66. Consider the following reaction:

$H_2(g) + I_2(g) \leftrightarrow 2HI(g)$

At equilibrium at 710 K, $[H_2] = [I_2] = 0.010$ M and $[HI] = 0.074$ M. If the HI(g) concentration is increased to 0.100 M, calculate the concentrations of all gases after equilibrium is re-established.

67. Consider the following reaction at a certain temperature:

$PCl_5(g) \leftrightarrow PCl_3(g) + Cl_2(g)$ $\qquad K_c = 0.0736$

At equilibrium, $[PCl_5] = 0.110$ M and $[PCl_3] = [Cl_2] = 0.0900$ M. If suddenly 0.100 M $PCl_5(g)$, $PCl_3(g)$ and $Cl_2(g)$ were added, which of the following is true?

a) More products will be formed.
b) $[Cl_2] = 0.190$ M at equilibrium.
c) Since the equilibrium constant does not change, nothing happens.
d) More $PCl_5(g)$ will be formed.
e) $K_c = 0.17$

68. Consider the following reaction:

$CO(g) + 3H_2(g) \leftrightarrow CH_4(g) + H_2O(g)$

The result of removing some $CH_4(g)$ and $H_2O(g)$ from the system is that

a) K_c decreases.
b) more $CH_4(g)$ and $H_2O(g)$ are produced.
c) no change occurs.
d) more CO(g) is produced.
e) $H_2O(g)$ is consumed.

69. All of the following reactions are affected by an increase in pressure except

a) $H_2(g) + I_2(s) \leftrightarrow 2HI(g)$
b) $C(s) + H_2O(g) \leftrightarrow CO(g) + H_2(g)$
c) $3Fe(s) + 4H_2O(g) \leftrightarrow Fe_3O_4(s) + 4H_2(g)$
d) $NH_4HS(s) \leftrightarrow NH_3(g) + H_2S(g)$
e) $NH_3(g) + HCl(g) \leftrightarrow NH_4Cl(s)$

70. All of the following reactions are affected by an increase in pressure except
a) $H_2(g) + I_2(s) \leftrightarrow 2HI(g)$
b) $N_2(g) + O_2(g) \leftrightarrow 2NO(g)$
c) $C(s) + H_2O(g) \leftrightarrow CO(g) + H_2(g)$
d) $NH_4HS(s) \leftrightarrow NH_3(g) + H_2S(g)$
e) $NH_3(g) + HCl(g) \leftrightarrow NH_4Cl(s)$

71. Consider the following reaction:
$C(s) + H_2O(g) \leftrightarrow CO(g) + H_2(g)$
What is the result of a decrease in container volume on the reaction?
a) no change occurs
b) more $H_2O(g)$ is formed
c) more $CO(g)$ is formed
d) K_c decreases
e) more $C(s)$ reacts

72. Consider the following reaction:
$Cl_2(g) \leftrightarrow 2Cl(g)$
What conditions favor the production of chlorine atoms, $Cl(g)$?
a) high temperature and low pressure
b) high temperature and high pressure
c) low temperature and low pressure
d) low temperature and high pressure
e) low temperature and a small container volume

73. For the reaction
$CO(g) + 3H_2(g) \leftrightarrow CH_4(g) + H_2O(g)$
$\Delta H° = -206$ kJ. What conditions favor maximum conversion of reactants to products?
a) low pressure and low temperature
b) removal of $H_2(g)$ and low temperature
c) high pressure and high temperature
d) high pressure and low temperature
e) low pressure and high temperature

74. For the reaction
$NH_3(g) + HCl(g) \leftrightarrow NH_4Cl(s)$
$\Delta H° = -177$ kJ. What conditions would best prevent this reaction?
a) low temperature and high pressure
b) addition of $HCl(g)$
c) high temperature and low pressure
d) low temperature and low pressure
e) high temperature and high pressure

75. For the reaction
$2SO_3(g) \leftrightarrow 2SO_2(g) + O_2(g)$
$\Delta H° = +198$ kJ. What conditions will favor $SO_2(g)$ production?
a) low temperature and high pressure
b) low temperature and small container volume
c) low temperature and low pressure
d) high temperature and high pressure
e) high temperature and low pressure

76. The reaction

$$2SO_3(g) \leftrightarrow 2SO_2(g) + O_2(g)$$

is endothermic. This means that

a) K_c will not change if the temperature is increased.
b) the forward reaction is much more sensitive to temperature than the reverse reaction.
c) K_c will decrease as the temperature is increased.
d) the activation energy of the reverse reaction is greater than that for the forward reaction.
e) K_c will be negative for the reverse reaction.

77. The reaction

$$2NO_2(g) \leftrightarrow N_2O_4(g)$$

is exothermic. This means that

a) K_c will not change if the temperature is increased.
b) K_c will be negative.
c) the forward reaction is more sensitive to temperature than the reverse reaction.
d) K_c will decrease as the temperature is increased.
e) K_c will increase as the temperature is increased.

78. Consider the reaction below:

$$Ni(s) + 4CO(g) \leftrightarrow Ni(CO)_4(g)$$

At 30°C and P_{CO} = 1 atm, Ni reacts with CO(g) to form $Ni(CO)_4$(g). At 200°C, $Ni(CO)_4$(g) decomposes to Ni(s) and CO(g). This means

a) K_p at 30°C is greater than K_p at 200°C.
b) a decrease in pressure favors the forward reaction.
c) the reaction is endothermic.
d) the activation energy for the forward reaction is greater than for the reverse reaction.
e) adding an inert gas like argon favors the forward reaction.

79. Consider the following reaction:

$$CO(g) + 3H_2(g) \leftrightarrow CH_4(g) + H_2O(g)$$

If the enthalpy change for the reaction is –206 kJ, predict the result of decreasing the temperature.

a) H_2O(g) is consumed
b) more CH_4(g) is produced
c) more CO(g) is produced
d) no change occurs
e) K_c decreases

80. Consider the following reaction:

$$CO_2(g) + 2NH_3(g) \leftrightarrow CO(NH_2)_2(s) + H_2O(g)$$

If K_c at a certain temperature is 2.9, and at a higher temperature K_c is 0.18,
(a) is the reaction endothermic or exothermic?
(b) what change in the concentration of NH_3(g) occurs as a result of an increase in temperature?

Chapter 13: Chemical Equilibrium
Form B

1. Consider the following reaction:

   $NH_4(NH_2CO_2)(s) \leftrightarrow 2NH_3(g) + CO_2(g)$

   When 0.400 moles of $NH_4(NH_2CO_2)(s)$ are placed in a 2.00-L flask and allowed to reach equilibrium at a certain temperature, 12.0% of the solid remains. Calculate the value of K_c for the reaction.

2. Consider the following reaction:

   $Ni(CO)_4(g) \leftrightarrow Ni(s) + 4CO(g)$

   If the initial concentration of $Ni(CO)_4(g)$ is 1.0 M, and "x" is the equilibrium concentration of CO(g), what is the correct equilibrium relation?

a) $K_c = x^4/(1.0 - x/4)$
b) $K_c = x^4/(1.0 - 4x)$
c) $K_c = 4x/(1.0 - 4x)$
d) $K_c = x^5/(1.0 - x/4)$
e) $K_c = x/(1.0 - x/4)$

3. Consider the following reaction:

   $P_4(g) + 5O_2(g) \leftrightarrow P_4O_{10}(s)$

   If the initial concentration of $O_2(g)$ is 1.0 M, and "x" is the equilibrium concentration of $P_4(g)$, what is the correct equilibrium relation?

a) $K_c = 1/(x)(1.0 - x)^5$
b) $K_c = 1/(x)(1.0 - 5x)^5$
c) $K_c = (x)(1.0 - x)^5$
d) $K_c = 1/(x)(1.0 - 5x)$
e) $K_c = (x)(1.0 - 5x)^5$

4. Consider the following reaction:

   $2NO(g) \leftrightarrow N_2(g) + O_2(g)$

   If the initial concentration of NO(g) is 2.0 M, and "x" is the equilibrium concentration of $N_2(g)$, what is the correct equilibrium relation?

a) $K_c = x^2/(2.0)^2$ b) $K_c = x^2$ c) $K_c = 2x/4.0$ d) $K_c = x^2/(2.0 - x)^2$ e) $K_c = x^2/(2.0 - 2x)^2$

5. Consider the following reaction:

   $H_2O(g) + C(s) \leftrightarrow H_2(g) + CO(g)$

   If the value of K_p for this reaction is 3.72 at 1000 K, and the equilibrium partial pressures of $H_2(g)$ and CO(g) are 3.00 atm, calculate the equilibrium partial pressure of $H_2O(g)$.

a) 2.42 atm b) 0.413 atm c) 4.66 atm d) 0.214 atm e) 3.00 atm

6. Consider the following reaction:

   $H_2(g) + I_2(g) \leftrightarrow 2HI(g)$

   If the value of K_c for this reaction is 25 at 1100 K, and initially only 3.00 M HI(g) is present, what is the equilibrium concentration of $I_2(g)$?

a) 1.00 M b) 2.00 M c) 1.73 M d) 0.214 M e) 0.429 M

7. Consider the following reaction:

   $Ni(CO)_4(g) \leftrightarrow Ni(s) + 4CO(g)$ $\quad K_c = 0.0309$

   A mixture of $Ni(CO)_4(g)$ and CO(g), each with a concentration of 0.200 M, and an excess of Ni(s) were confined in a 1.0-L container at 300 K. Which of the following is true?

a) Ni(s) is consumed
b) no change occurs since we are at equilibrium
c) $Ni(CO)_4(g)$ forms until equilibrium is reached
d) K_c changes to 0.0080
e) some $Ni(CO)_4(g)$ decomposes to CO(g) and Ni(s)

8. Consider the reaction

   $3Fe(s) + 4H_2O(g) \leftrightarrow 4H_2(g) + Fe_3O_4(s)$

   If the total pressure is decreased suddenly by increasing the volume,

a) the equilibrium constant increases.
b) no change occurs.
c) more Fe(s) is produced.
d) more $H_2(g)$ is produced.
e) more $H_2O(g)$ is produced.

9. For the reaction

   $C(s) + H_2O(g) \leftrightarrow CO(g) + H_2(g)$

   if the total pressure is increased by decreasing the volume,

a) the equilibrium constant decreases.
b) more products are produced.
c) more $H_2O(g)$ is produced.
d) the equilibrium constant increases.
e) no change occurs.

10. Consider the following reaction:

    $CO(g) + 3H_2(g) \leftrightarrow CH_4(g) + H_2O(g)$

    Predict the result of a volume decrease on the reaction.

a) more CO(g) is produced
b) K_c decreases
c) K_c increases
d) more $CH_4(g)$ is produced
e) no change occurs

11. Consider the following reaction:

    $4NH_3(g) + 7O_2(g) \leftrightarrow 2N_2O_4(g) + 6H_2O(g)$

    If the initial concentrations of $N_2O_4(g)$ and $H_2O(g)$ are 3.60 M, at equilibrium the concentration of $H_2O(g)$ is 0.60 M. Calculate the equilibrium concentration of $N_2O_4(g)$.

a) 2.6 M b) 3.5 M c) 1.5 M d) 2.0 M e) 0.50 M

12. Consider the following reaction:

    $2NOCl(g) \leftrightarrow 2NO(g) + Cl_2(g)$

    If the initial concentration of NOCl(g) is 2.5 M, at equilibrium $[Cl_2] = 0.60$ M. Calculate the equilibrium concentration of NOCl(g).

a) 1.9 M b) 1.3 M c) 1.2 M d) 0.70 M e) 2.2 M

13. What is the equilibrium constant expression for the reaction below?

$3Fe(s) + 4H_2O(g) \leftrightarrow Fe_3O_4(s) + 4H_2(g)$

a) $K_c = [Fe_3O_4]/[Fe]^3$
b) $K_c = [H_2]/[H_2O]$
c) $K_c = [H_2]^4/[H_2O]^4$
d) $K_c = [Fe_3O_4][Fe]$
e) $K_c = [Fe_3O_4][H_2]^4/[Fe]^3[H_2O]^4$

14. Given: $N_2(g) + 3H_2(g) \leftrightarrow 2NH_3(g)$

At equilibrium at a certain temperature, the concentration of $NH_3(g)$, $H_2(g)$ and $N_2(g)$ are 0.940 M, 1.60 M and 0.520 M, respectively. Calculate the value of K_c for this reaction.

a) 0.415 b) 1.13 c) 1.06 d) 0.664 e) 1.27

15. When $NH_4(NH_2CO_2)(s)$ is placed in a 1.0-L container at 400K, the following equilibrium is reached:

$NH_4(NH_2CO_2)(s) \leftrightarrow 2NH_3(g) + CO_2(g)$ $K_p = 0.089$

Calculate the equilibrium concentration of $NH_3(g)$.

a) 0.56 b) 0.28 c) 0.89 d) 0.45 e) 0.71

16. Consider the following reaction:

$CaCO_3(s) \leftrightarrow CaO(s) + CO_2(g)$

$K_p = 15.9$ at 1300 K for this reaction. What is the equilibrium partial pressure of $CO_2(g)$, starting from pure $CaCO_3(s)$?

a) 15.9 atm b) 0.0629 atm c) 3.99 atm d) 1.99 atm e) 253 atm

17. What is the equilibrium constant expression for the reaction below?

$2MgCl_2(s) + O_2(g) \leftrightarrow 2MgO(s) + 2Cl_2(g)$

a) $K_c = [Cl_2]/[O_2]$
b) $K_c = [MgO]^2/[MgCl_2]^2$
c) $K_c = [MgO][Cl_2]/[MgCl_2][O_2]$
d) $K_c = [Cl_2]^2/[O_2]$
e) $K_c = [Cl_2]^2[MgO]^2/[O_2][MgCl_2]^2$

18. Consider the following reaction:

$3N_2H_4(g) + 4ClF_3(g) \leftrightarrow 12HF(g) + 3N_2(g) + 2Cl_2(g)$

A mixture, initially consisting of 0.880 M $N_2H_4(g)$ and 0.880 M $ClF_3(g)$, reacts at a certain temperature. At equilibrium, the concentration of $N_2(g)$ is 0.525 M. Calculate the concentration of $Cl_2(g)$ at equilibrium.

a) 0.350 M b) 0.525 M c) 0.705 M d) 0.700 M e) 0.175 M

19. Consider the following reaction:

$N_2(g) + 3H_2(g) \leftrightarrow 2NH_3(g)$

If the value of K_c for this reaction is 170 at 500 K, calculate the value of K_p at the same temperature.

a) 0.10 b) 4.1 c) 7000 d) 170 e) 0.22

20. At 600°C, $K_c = 2.8$ for the reaction

$2HgO(s) \leftrightarrow 2Hg(l) + O_2(g)$

Calculate K_p at 600°C for this reaction.

a) 200 b) 138 c) 2.8 d) 6800 e) 1.4×10^4

21. Given:

$4NH_3(g) + 5O_2(g) \leftrightarrow 4NO(g) + 6H_2O(g)$

Calculate the equilibrium constant for the following reaction.

$2NO(g) + 3H_2O(g) \leftrightarrow 2NH_3(g) + 5/2O_2(g)$

a) K^{-1} b) $K^{-1/2}$ c) $-0.5K$ d) $-K$ e) $-2K$

22. Given:

$P_4(s) + 6Cl_2(g) \leftrightarrow 4PCl_3(l)$ K

Calculate the equilibrium constant for the following reaction.

$2PCl_3(l) \leftrightarrow 3Cl_2(g) + 1/2P_4(s)$

a) $1/K^{1/2}$ b) $-K^{1/2}$ c) $1/K$ d) $K^{1/2}$ e) $1/K^2$

23. Consider the following reaction:

$4NH_3(g) + 7O_2(g) \leftrightarrow 2N_2O_4(g) + 6H_2O(g)$

If the initial concentrations of $NH_3(g)$ and $O_2(g)$ are 3.60 M, at equilibrium the concentration of $N_2O_4(g)$ is 0.60 M. Calculate the equilibrium concentration of $H_2O(g)$.

a) 3.0 M b) 1.5 M c) 1.8 M d) 2.4 M e) 3.3 M

24. Consider the following reaction:

$4NH_3(g) + 7O_2(g) \leftrightarrow 2N_2O_4(g) + 6H_2O(g)$

If the initial concentrations of $N_2O_4(g)$ and $H_2O(g)$ are 3.60 M, at equilibrium the concentration of $H_2O(g)$ is 0.60 M. Calculate the equilibrium concentration of $O_2(g)$.

a) 3.5 M b) 0.50 M c) 2.0 M d) 1.5 M e) 2.6 M

25. Consider the following reaction:

$2SO_3(g) \leftrightarrow 2SO_2(g) + O_2(g)$

For an initial concentration of $SO_3(g)$ equal to 0.344 M, at equilibrium it is found that the concentration of $O_2(g)$ equals 0.0432 M. Calculate the concentrations of $SO_3(g)$ and $SO_2(g)$ at equilibrium.

26. Consider the following reaction:

$4NH_3(g) + 6NO(g) \leftrightarrow 6H_2O(g) + 5N_2(g)$

If the initial concentrations of $NH_3(g)$ and $NO(g)$ are 6.0 M, at equilibrium $[N_2] = 2.50$ M. Calculate the equilibrium concentration of $H_2O(g)$.

a) 3.0 M b) 6.0 M c) 2.5 M d) 2.0 M e) 3.5 M

27. Given: $2SO_2(g) + O_2(g) \leftrightarrow 2SO_3(g)$

At equilibrium at a certain temperature, the concentrations of $SO_3(g)$, $SO_2(g)$ and $O_2(g)$ are 0.24 M, 0.82 M and 0.33 M, respectively. Calculate the value of K_c for this reaction.

a) 0.26 b) 0.89 c) 0.21 d) 1.04 e) 0.79

28. Consider the following reaction:

$Ni(CO)_4(g) \leftrightarrow Ni(s) + 4CO(g)$

For an initial concentration of $Ni(CO)_4(g)$ equal to 0.800 M, equilibrium is established after 12.0% of the $Ni(CO)_4(g)$ has formed products at 300 K. Calculate the value of K_c at 300 K.

a) 0.0309 b) 0.0119 c) 0.00297 d) 0.000121 e) 0.0524

29. Consider the following reaction:

$4NH_3(g) + 6NO(g) \leftrightarrow 6H_2O(g) + 5N_2(g)$

If the initial concentrations of $NH_3(g)$ and $NO(g)$ are 4.0 M, at equilibrium [NO] = 1.0 M. Calculate the equilibrium concentration of $H_2O(g)$.

a) 3.0 M b) 0.50 M c) 2.0 M d) 5.0 M e) 2.5 M

30. Consider the following reaction:

$4NH_3(g) + 6NO(g) \leftrightarrow 6H_2O(g) + 5N_2(g)$

If the initial concentrations of $NH_3(g)$ and $NO(g)$ are 6.0 M, at equilibrium $[N_2] = 2.50$ M. Calculate the equilibrium concentration of $NH_3(g)$.

a) 2.0 M b) 3.0 M c) 4.0 M d) 2.5 M e) 3.5 M

31. Consider the following reaction:

$PCl_5(g) \leftrightarrow PCl_3(g) + Cl_2(g)$

At a certain temperature, if the initial concentration of $PCl_5(g)$ is 2.0 M, at equilibrium the concentration of $Cl_2(g)$ is 0.30 M. Calculate the value of K_c at this temperature.

a) 0.053 b) 0.090 c) 0.045 d) 0.064 e) 19

32. Consider the following reaction:

$NH_4(NH_2CO_2)(s) \leftrightarrow 2NH_3(g) + CO_2(g)$

When 2 moles of ammonium carbamate are placed in an evacuated 250-mL flask at 298 K, at equilibrium 3.16×10^{-3} moles of $NH_3(g)$ are present. Calculate the value of K_c.

a) 1.58×10^{-8} b) 1.26×10^{-7} c) 7.89×10^{-9} d) 9.99×10^{-6} e) 3.16×10^{-8}

33. Given: $C(s) + CO_2(g) \leftrightarrow 2CO(g)$

At equilibrium at a certain temperature, the partial pressures of $CO(g)$ and $CO_2(g)$ are 1.44 atm and 0.820 atm, respectively. Calculate the value of K_p for this reaction.

a) 2.53 b) 1.76 c) 3.08 d) 3.51 e) 10.1

34. What is the equilibrium constant expression for the reaction below?

$2HgO(s) \leftrightarrow 2Hg(l) + O_2(g)$

a) $K_c = 1/[O_2]$ b) $K_c = [O_2][Hg]^2$ c) $K_c = [O_2][Hg]^2/[HgO]^2$ d) $K_c = [Hg]^2/[HgO]^2$ e) $K_c = [O_2]$

35. Consider the following reaction:

$2SO_2(g) + O_2(g) \leftrightarrow 2SO_3(g)$

If, initially, the SO_2 concentration is 3.2 mM and at equilibrium $[SO_3]$ = 3.0 mM, calculate the equilibrium concentration of O_2.

a) 3.0 mM b) 0.20 mM c) 1.7 mM d) 1.5 mM e) 1.8 mM

36. Consider the following reaction:

$2SO_2(g) + O_2(g) \leftrightarrow 2SO_3(g)$

If, initially, the SO_3 concentration is 4.8 mM and at equilibrium $[O_2]$ = 0.80 mM, calculate the equilibrium concentration of SO_3.

a) 1.6 mM b) 0.80 mM c) 4.0 mM d) 2.4 mM e) 3.2 mM

37. Consider the following reaction:

$3N_2H_4(g) + 4ClF_3(g) \leftrightarrow 12HF(g) + 3N_2(g) + 2Cl_2(g)$

A mixture, initially consisting of 0.880 M N_2H_4(g) and 0.880 M ClF_3(g), reacts at a certain temperature. At equilibrium, the concentration of N_2(g) is 0.525 M. Calculate the concentration of N_2H_4(g) at equilibrium.

a) 0.705 M b) 0.700 M c) 0.175 M d) 0.180 M e) 0.355 M

38. Given: $2CO_2(g) \leftrightarrow 2CO(g) + O_2(g)$

When 1.00 mole of CO_2(g) is placed in a 1.00-L container and allowed to come to equilibrium, the concentration of CO(g) is 0.600 M. What is the value of K_c for this reaction?

a) 0.675 b) 0.450 c) 0.600 d) 1.20 e) 0.270

39. The equilibrium constant for the reaction below

$PCl_5(g) \leftrightarrow PCl_3(g) + Cl_2(g)$

is 0.050 at a certain temperature. What is the equilibrium constant for the reverse reaction?

a) 20 b) 4.5 c) 0.22 d) 0.50 e) 0.0025

40. The equilibrium constant for the reaction

$2NOCl(g) \leftrightarrow 2NO(g) + Cl_2(g)$

is 0.51 at a certain temperature. What is the equilibrium constant for the reaction below?

$NO(g) + 1/2Cl_2(g) \leftrightarrow NOCl(g)$

a) 1.4 b) 2.0 c) 3.8 d) 0.71 e) 0.26

41. A mixture consisting of 0.250 M $N_2(g)$ and 0.500 M $H_2(g)$ reaches equilibrium according to the equation below:

$N_2(g)\ 3H_2(g) \leftrightarrow 2NH_3(g)$

At equilibrium, the concentration of ammonia is 0.150 M. Calculate the concentration of $N_2(g)$ at equilibrium.

a) 0.0500 M b) 0.150 M c) 0.175 M d) 0.0750 M e) 0.100 M

42. What is the relation between K_p and K_c for the reaction below?

$H_2(g) + I_2(g) \leftrightarrow 2HI(g)$

a) $K_p = (RT)^2K_c$ b) $K_c = RTK_p$ c) $K_p = RTK_c$ d) $K_c = (RT)^2K_p$ e) $K_p = K_c$

43. What is the equilibrium constant expression for the reaction below?

$H_2O_2(l) \leftrightarrow H_2O(l) + 1/2O_2(g)$

a) $K_c = [O_2]^{1/2}$
b) $K_c = [O_2]^{1/2}[H_2O]/[H_2O_2]$
c) $K_c = [O_2][H_2O][H_2O_2]$
d) $K_c = 1/[O_2]^{1/2}$
e) $K_c = [O_2]$

44. Given:

$SO_2(g) \leftrightarrow O_2(g) + S(s)$ $\quad K_c = 2.5 \times 10^{-53}$

$SO_3(g) \leftrightarrow 1/2O_2(g) + SO_2(g)$ $\quad K_c = 4.0 \times 10^{-13}$

Calculate K_c for the reaction

$2S(s) + 3O_2(g) \leftrightarrow 2SO_3(g)$

a) 1.6×10^{80} b) 1.6×10^{40} c) 1.0×10^{130} d) 1.0×10^{65} e) 1.6×10^{103}

45. At 25°C, $K_c = 1.58 \times 10^{-8}$ for the reaction

$NH_4(NH_2CO_2)(s) \leftrightarrow 2NH_3(g) + CO_2(g)$

Calculate K_p at 25°C for this reaction.

a) 3.87×10^{-7} b) 2.31×10^{-4} c) 5.69×10^{-3} d) 1.36×10^{-7} e) 9.45×10^{-5}

46. At 25°C, $K_c = 4.1 \times 10^{8}$ for the reaction

$N_2(g) + 3H_2(g) \leftrightarrow 2NH_3(g)$

Calculate K_p at 25°C for this reaction.

a) 4.1×10^{8} b) 2.5×10^{11} c) 1.7×10^{9} d) 9.7×10^{7} e) 6.9×10^{5}

47. At 700 K, $K_p = 54$ for the reaction

$H_2(g) + I_2(g) \leftrightarrow 2HI(g)$

Calculate K_c at 700 K for this reaction.

a) 54 b) 0.94 c) 3.2×10^{-4} d) 7.7×10^{-4} e) 0.45

48. For the following reaction

$$H_2(g) + I_2(g) \leftrightarrow 2HI(g)$$

$K_c = 54$ at 700 K. At equilibrium, the concentration of HI is 0.12 M and of I_2 is 0.013 M. Calculate the equilibrium concentration of H_2.

a) 0.021 M b) 0.017 M c) 0.013 M d) 0.0020 M e) 0.060 M

49. For the following reaction

$$2BrCl(g) \leftrightarrow Br_2(g) + Cl_2(g)$$

$K_c = 32$ at 500 K. At equilibrium, the concentration of Br_2 is 0.22 M and of BrCl is 0.015 M. Calculate the equilibrium concentration of Cl_2.

a) 0.15 M b) 2.2 M c) 0.033 M d) 0.46 M e) 0.21 M

50. For the following reaction

$$2SO_2(g) + O_2(g) \leftrightarrow 2SO_3(g)$$

$K_c = 11.7$ at 1100 K. At equilibrium, the concentration of SO_2 is 2.3 M and of SO_3 is 0.80 M. Calculate the equilibrium concentration of O_2.

a) 1.5 M b) 1.2 M c) 0.030 M d) 0.010 M e) 0.024 M

51. The equilibrium constant for the reaction

$$2BrCl(g) \leftrightarrow Br_2(g) + Cl_2(g)$$

is 32 at 500 K. A mixture of BrCl, Br_2 and Cl_2, was introduced into a container at 500 K. If the initial concentrations of BrCl, Br_2 and Cl_2 in the container are 0.10, 2.2 and 2.2 M, respectively, which of the following is true?

a) The system is at equilibrium and therefore no net change occurs.
b) The reaction proceeds to the left, forming more BrCl(g).
c) The reaction proceeds to the right, forming more products.
d) The concentration of $Br_2(g)$ remains constant at 2.2 M.
e) The reaction proceeds until all BrCl(g) is used up.

52. The equilibrium constant for the reaction

$$2SO_2(g) + O_2(g) \leftrightarrow 2SO_3(g)$$

is 11.7 at 1100 K. A mixture of SO_2, O_2 and SO_3, each with a concentration of 0.015 M, was introduced into a container at 1100 K. Which of the following is true?

a) $[SO_3] = 0.045$ M at equilibrium.
b) $[SO_3] = 0.015$ M at equilibrium.
c) $SO_3(g)$ will be formed until equilibrium is reached.
d) $[SO_3] = [SO_2] = [O_2]$ at equilibrium.
e) $SO_2(g)$ and $O_2(g)$ will be formed until equilibrium is reached.

53. The equilibrium constant for the reaction

$$2NOCl(g) \leftrightarrow 2NO(g) + Cl_2(g)$$

is 0.51 at a certain temperature. A mixture of NOCl, NO and Cl_2 with concentrations 1.3, 1.2 and 0.60 M, respectively, was introduced into a container at this temperature. Which of the following is true?

a) $Cl_2(g)$ is produced until equilibrium is reached.
b) We are at equilibrium and thus not net change takes place.
c) $[NOCl] = [NO] = [Cl_2]$ at equilibrium.
d) $[Cl_2] = 0.30$ M at equilibrium.
e) NOCl(g) is produced until equilibrium is reached.

54. The equilibrium constant for the reaction

$$H_2(g) + I_2(g) \leftrightarrow 2HI(g)$$

is 54 at 700 K. A mixture of H_2, I_2 and HI was introduced into a container at 700 K. If the initial concentrations in the container are $[H_2] = [I_2] = 0.30$ M, and $[HI] = 2.2$ M, which of the following is true?

a) The reaction proceeds to the left producing more $H_2(g)$ and $I_2(g)$.
b) At equilibrium, $[H_2] = [I_2] = [HI]$.
c) The reaction proceeds to the right producing more HI(g).
d) No net change occurs because the system is at equilibrium.
e) At equilibrium, $[HI] = 1.9$ M.

55. The concentration of a pure solid or liquid is left out of the equilibrium constant expression because

a) solids and liquids react slowly with gases.
b) their concentrations cannot be determined.
c) their concentrations in the gas phase are constant and independent of the amount of solid or liquid present.
d) solids and liquids drive the reaction.
e) solids and liquids do not react.

56. Given:

$$2H_2O(g) + 2Cl_2(g) \leftrightarrow 4HCl(g) + O_2(g) \qquad K_p = 8.0$$

If the initial partial pressure of each reactant is 0.10 atm and of each product 0.25 atm, which of the following is true?

a) a change in pressure does not affect the reaction
b) the reaction proceeds to the right
c) the reaction is at equilibrium
d) the reaction proceeds to the left
e) addition of argon causes the reaction to produce more $Cl_2(g)$

57. For the following reaction

$$NH_3(g) + H_2S(g) \leftrightarrow NH_4HS(s)$$

$K_c = 9.7$ at 900 K. If the initial concentrations of $NH_3(g)$ and $H_2S(g)$ are 2.0 M, what is the equilibrium concentration of $NH_3(g)$?

a) 0.20 M b) 0.10 M c) 1.9 M d) 1.7 M e) 0.32 M

58. For the following reaction

$$NH_3(g) + H_2S(g) \leftrightarrow NH_4HS(s)$$

$K_c = 9.7$ at 900 K. If 2.0 moles of $NH_3(g)$ and 1.0 moles of $H_2S(g)$ are placed in a 1.0-L container at 900 K, how many moles of $NH_4HS(s)$ exist at equilibrium?

a) 0.32 mol
b) 1.0 mol
c) 0.68 mol
d) 1.7 mol
e) Since solids do not appear in K_c, this quantity cannot be calculated.

59. For the reaction

$$NH_4(NH_2CO_2)(s) \leftrightarrow 2NH_3(g) + CO_2(g)$$

$K_p = 2.3 \times 10^{-4}$ at 298 K. Starting with a large excess of pure $NH_4(NH_2CO_2)(s)$ in a closed container, what is the equilibrium partial pressure of $CO_2(g)$?

a) 0.011 atm b) 0.061 atm c) 0.0076 atm d) 0.039 atm e) 0.015 atm

60. For the following reaction

$$CaO(s) + CO_2(g) \leftrightarrow CaCO_3(s)$$

$K_p = 7.7$ at 1000 K. In a closed container with excess $CaCO_3(s)$ present, what is the equilibrium partial pressure of $CO_2(g)$?

a) 2.6 atm b) 0.36 atm c) 0.13 atm d) 7.7 atm e) 2.8 atm

61. For the reaction below

$$2CaSO_4(s) \leftrightarrow 2CaO(s) + 2SO_2(g) + O_2(g)$$

$K_p = 0.032$ at 700 K. What is the equilibrium partial pressure of $O_2(g)$ starting from pure $CaSO_4(s)$?

a) 0.40 atm b) 0.22 atm c) 0.011 atm d) 0.20 atm e) 0.60 atm

62. Consider the following reaction:

$$P_4(g) + 5O_2(g) \leftrightarrow P_4O_{10}(s)$$

If the total gas pressure at equilibrium at 400 K is 1.86 atm, calculate the value of K_p.

a) 0.36 b) 0.65 c) 0.31 d) 0.73 e) 0.54

63. Consider the following reaction:

$2NO_2(g) \leftrightarrow 2NO(g) + O_2(g)$

Starting with pure $NO_2(g)$ at a pressure of 0.500 atm, at equilibrium the total pressure was 0.732 atm. Calculate the equilibrium partial pressure of $NO_2(g)$.

a) 0.0360 atm b) 0.232 atm c) 0.493 atm d) 0.200 atm e) 0.464 atm

64. When $NH_4HS(s)$ and 0.800 mol of $NH_3(g)$ are placed in a 1.0-L container at 24°C, the following equilibrium is reached:

$NH_4HS(s) \leftrightarrow NH_3(g) + H_2S(g)$ $\qquad K_c = 1.6 \times 10^{-4}$

What is the approximate equilibrium concentration of $H_2S(g)$?

a) 2.0×10^{-4} M b) 0.40 M c) 1.3×10^{-2} M d) 1.6×10^{-4} M e) 0.80 M

65. Consider the following reaction:

$SO_2Cl_2(g) \leftrightarrow SO_2(g) + Cl_2(g)$

At equilibrium at 200°C in a 1.00-L container, $[SO_2Cl_2] = 0.80$ M, $[SO_2] = 0.20$ M, and $[Cl_2] = 0.20$ M. If 0.30 moles of $Cl_2(g)$ is added, calculate the new equilibrium concentrations.

66. Consider the following reaction:

$H_2(g) + I_2(g) \leftrightarrow 2HI(g)$

At equilibrium at a certain temperature, $[H_2] = [I_2] = 0.125$ M and $[HI] = 0.210$ M. If the HI(g) concentration is increased to 0.400 M, calculate the concentrations of all gases after equilibrium is re-established.

67. Consider the following reaction at a certain temperature:

$PCl_5(g) \leftrightarrow PCl_3(g) + Cl_2(g)$ $\qquad K_c = 0.0736$

At equilibrium, $[PCl_5] = 0.110$ M and $[PCl_3] = [Cl_2] = 0.0900$ M. If suddenly 0.200 M $PCl_5(g)$, 0.0500 M $PCl_3(g)$ and 0.0400 M $Cl_2(g)$ were added, which of the following is true?

a) $K_c = 0.061$
b) More $PCl_5(g)$ will be formed.
c) $[Cl_2] = 0.130$ M at equilibrium.
d) More products will be formed.
e) Since the equilibrium constant does not change, nothing happens.

68. Consider the following reaction:

$CO(g) + 3H_2(g) \leftrightarrow CH_4(g) + H_2O(g)$

The result of adding some $CH_4(g)$ and $H_2O(g)$ to the system is that

a) no change occurs.
b) more $CH_4(g)$ and $H_2O(g)$ are produced.
c) more $CO(g)$ and $H_2(g)$ are consumed.
d) K_c decreases.
e) more $CO(g)$ and $H_2(g)$ are produced.

69. All of the following reactions are affected by an increase in pressure except
a) $H_2(g) + I_2(s) \leftrightarrow 2HI(g)$
b) $2BrCl(g) \leftrightarrow Br_2(g) + Cl_2(g)$
c) $NH_4HS(s) \leftrightarrow NH_3(g) + H_2S(g)$
d) $NH_3(g) + HCl(g) \leftrightarrow NH_4Cl(s)$
e) $C(s) + H_2O(g) \leftrightarrow CO(g) + H_2(g)$

70. The position of equilibrium would not be appreciably affected by changes in container volume for which of the following reactions?
a) $H_2(g) + I_2(g) \leftrightarrow 2HI(g)$
b) $NH_3(g) + HCl(g) \leftrightarrow NH_4Cl(s)$
c) $NH_4HS(s) \leftrightarrow NH_3(g) + H_2S(g)$
d) $4NH_3(g) + 5O_2(g) \leftrightarrow 4NO(g) + 6H_2O(g)$
e) $C(s) + H_2O(g) \leftrightarrow CO(g) + H_2(g)$

71. Consider the following reaction:

$$C(s) + H_2O(g) \leftrightarrow CO(g) + H_2(g)$$

What is the result of an increase in container volume on the reaction?
a) no change occurs
b) more $CO(g)$ and $H_2(g)$ are formed
c) more $H_2O(g)$ is formed
d) more $CO(g)$ reacts
e) K_c decreases

72. Consider the following reaction:

$$Br_2(g) \leftrightarrow 2Br(g)$$

What conditions favor the production of bromine atoms, $Br(g)$?
a) high temperature and low pressure
b) high temperature and high pressure
c) low temperature and low pressure
d) low temperature and high pressure
e) low temperature and a small container volume

73. For the reaction

$$NH_3(g) + HCl(g) \leftrightarrow NH_4Cl(s)$$

$\Delta H^o = -177$ kJ. What conditions favor production of ammonium chloride?
a) high temperature and low pressure
b) low temperature and low pressure
c) addition of ammonium chloride
d) low temperature and high pressure
e) high temperature and high pressure

74. Given:

$$NO(g) + CO(g) \leftrightarrow 1/2N_2(g) + CO_2(g) \qquad \Delta H^o = -374 \text{ kJ}$$

What conditions favor maximum conversion of reactants to products?
a) low temperature and low pressure
b) high temperature and high pressure
c) high temperature and large container volume
d) low temperature and high pressure
e) high temperature and low pressure

75. For the reaction

$N_2O_4(g) \leftrightarrow 2NO_2(g)$

$\Delta H° = +58.2$ kJ. What will cause an increase in $NO_2(g)$ concentration?

a) a decrease in temperature
b) an increase in pressure
c) a decrease in container volume
d) an increase in temperature
e) removal of some $N_2O_4(g)$

76. The reaction

$2SO_3(g) \leftrightarrow 2SO_2(g) + O_2(g)$

is endothermic. Predict what will happen if the temperature is increased.

a) the pressure decreases
b) K_c remains the same
c) K_c decreases
d) more $SO_3(g)$ is produced
e) K_c increases

77. The reaction

$2NO_2(g) \leftrightarrow N_2O_4(g)$

is exothermic. This means that

a) K_c will not change if the temperature is increased.
b) the forward reaction is more sensitive to temperature than the reverse reaction.
c) K_c at 25°C is larger than K_c at 95°C.
d) K_c will increase as the temperature is increased.
e) K_c will be negative.

78. Consider the reaction below:

$Ni(s) + 4CO(g) \leftrightarrow Ni(CO)_4(g)$

At 30°C and $P_{CO} = 1$ atm, Ni reacts with CO(g) to form $Ni(CO)_4(g)$. At 200°C, $Ni(CO)_4(g)$ decomposes to Ni(s) and CO(g). This means

a) a decrease in pressure favors the forward reaction.
b) K_p at 30°C is less than K_p at 200°C.
c) adding an inert gas like argon favors the forward reaction.
d) the reaction is exothermic.
e) the reaction is endothermic.

79. Consider the following reaction:

$N_2(g) + 3H_2(g) \leftrightarrow 2NH_3(g)$

If the enthalpy change for the reaction is –92 kJ, predict the result of decreasing the temperature.

a) more $NH_3(g)$ is produced
b) K_p decreases
c) more $N_2(g)$ is produced
d) $NH_3(g)$ is consumed
e) no change occurs

80. Answer the following questions:

(a) At 500 *K*, $K_c = 7.3 \times 10^{-13}$ for the reaction

$F_2(g) \leftrightarrow 2F(g)$ $\Delta H° = +158\ \text{kJ} \cdot \text{mol}^{-1}$

Predict the value of K_c at 1000 K.

(b) At 100 K, $K_c = 8.3 \times 10^{6}$ for the reaction

$2Cl(g) \leftrightarrow Cl_2(g)$ $\Delta H° = -243\ \text{kJ} \cdot \text{mol}^{-1}$

Predict the value of K_c at 1200 K.

Chapter 14: Protons in Transition: Acids and Bases Form A

1. Consider the following reaction:

   $HSO_4^-(aq) + H_2O(l) \leftrightarrow SO_4^{2-}(aq) + H_3O^+(aq)$

   Which of the following statements is true?

a) The reaction of HSO_4^- with H_2O produces OH^- rather than H_3O^+.
b) HSO_4^- is the acid and H_2O is the base.
c) H_2O is the acid and SO_4^{2-} is its conjugate base.
d) HSO_4^- is not an acid and thus the equation is incorrect.
e) HSO_4^- is the base and H_3O^+ is its conjugate acid.

2. When CaO(s) is dissolved in water which of the following is true?

a) The solution contains $OH^-(aq)$ and $Ca^{2+}(aq)$.
b) The solution contains $O^{2-}(aq)$ and $Ca^{2+}(aq)$.
c) The solution contains $O^{2-}(aq)$, $OH^-(aq)$, and $Ca^{2+}(aq)$.
d) The solution contains CaO(aq).
e) CaO(s) does not dissolve in water.

3. The conjugate acid of HPO_4^{2-} is

a) H_3O^+ b) PO_4^{3-} c) HPO_4^{2-} d) H_3PO_4 e) $H_2PO_4^-$

4. The conjugate acid of HSO_4^- is

a) H_3O^+ b) SO_4^{2-} c) HSO_4^- d) H_2O e) H_2SO_4

5. The conjugate acid of HCO_3^- is

a) H_3O^+ b) CO_3^{2-} c) HCO_3^- d) H_2O e) H_2CO_3

6. The conjugate base of ammonia is

a) OH^- b) NH_2OH c) NH_3 d) NH_2^- e) NH_4^+

7. The conjugate acid of hydrazine is

a) H_2O b) $N_2H_3^-$ c) $N_2H_5^+$ d) H_3O^+ e) N_2H_4

8. The conjugate acid of OH^- is

a) H^+ b) OH^- c) H_2O d) H_3O^+ e) O^{2-}

9. Calculate the value of K_b for $H_2PO_4^-$ given $pK_{a1} = 2.12$, $pK_{a2} = 7.21$, and $pK_{a3} = 12.67$ for phosphoric acid. Compare this value to the value of K_{a2} and predict whether $H_2PO_4^-$ is an acid or a base in aqueous solution.

10. At 37°C, K_w for water is 5.00×10^{-14}. What is the pH of a neutral aqueous solution at this temperature?
a) 6.65 b) 6.00 c) 7.35 d) 7.00 e) 7.50

11. Calculate the concentration of H_3O^+ for an aqueous solution with a pH of 8.20.
a) 1.6×10^{-6} M b) 0.91 M c) 1.4×10^{-3} M d) 6.3×10^{-9} M e) 1.0×10^{-14} M

12. The pH of a 0.0050 M aqueous solution of calcium hydroxide is
a) 12.00 b) 11.70 c) 11.40 d) 12.70 e) 2.00

13. The pH of a 0.050 M aqueous solution of calcium hydroxide is
a) 13.00 b) 12.70 c) 12.40 d) 12.00 e) 1.00

14. For a solution labeled "0.20 M barium hydroxide," which of the following is correct?
a) $[OH^-] = 0.40$ M, $[Ba^{2+}] = 0.20$ M
b) $[OH^-] = 0.20$ M, $[Ba^{2+}] = 0.10$ M
c) $[OH^-] = 0.20$ M, $[Ba^{2+}] = 0.40$ M
d) $[OH^-] = 0.40$ M, $[Ba^{2+}] = 0.40$ M
e) $[OH^-] = 0.20$ M, $[Ba^{2+}] = 0.20$ M

15. When 500 mL of 0.120 M KOH(aq) is mixed with 500 mL of 0.0480 M $Ba(OH)_2$(aq), the pH of the resulting solution is
a) 13.03 b) 14.00 c) 13.23 d) 12.92 e) 13.33

16. In a solution labeled "0.10 M HCl," which of the following is correct?
a) $[H_3O^+] = 0.10$ M, $[Cl^-] = 0.10$ M
b) $[H_3O^+] = 0.10$ M, $[OH^-] = 1.0 \times 10^{-7}$ M
c) $[H_3O^+] = 0.090$ M, $[Cl^-] = 0.010$ M
d) $[HCl] = 0.050$ M, $[H_3O^+] = 0.050$ M, $[Cl^-] = 0.050$ M
e) $[HCl] = 0.10$ M

17. Calculate the concentration of OH^- for an aqueous solution with a pH of 9.45.
a) 1.8×10^{-10} M b) 1.0×10^{-14} M c) 2.8×10^{-5} M d) 0.35 M e) 3.5×10^{-10} M

18. Which of the following is the strongest acid?
a) $HClO_2$ ($pK_a = 2.00$)
b) HF ($pK_a = 3.45$)
c) HNO_2 ($pK_a = 3.37$)
d) HCN ($pK_a = 9.31$)
e) CH_3COOH ($pK_a = 4.75$)

19. Which of the following produces the strongest conjugate base?
a) HClO ($pK_a = 7.53$)
b) HF ($pK_a = 3.45$)
c) HCOOH ($pK_a = 3.75$)
d) CH_3COOH ($pK_a = 4.75$)
e) HCN ($pK_a = 9.31$)

20. Given:

    acetic acid, $pK_a = 4.75$

    HSO_4^-, $pK_a = 1.92$

    HF, $pK_a = 3.45$

    The order of these acids from strongest to weakest is

a) $HSO_4^- >$ HF > acetic acid
b) HF > $HSO_4^- >$ acetic acid
c) $HSO_4^- >$ acetic acid > HF
d) acetic acid > HF > HSO_4^-
e) HF > acetic acid > HSO_4^-

21. If a small amount of HCl(aq) is added to 0.10 M NH_3(aq),

a) no change occurs.
b) the equilibrium concentration of NH_4^+(aq) is increased.
c) the equilibrium concentration of the ammonium ion is decreased.
d) the equilibrium concentration of ammonia increases.
e) K_a becomes larger.

22. If a small amount of CsOH(aq) is added to 0.10 M $(CH_3)_3N$(aq),

a) the equilibrium concentration of $(CH_3)_3NH^+$(aq) is decreased.
b) no change occurs.
c) K_b becomes smaller.
d) the equilibrium concentration of $(CH_3)_3N$(aq) is decreased.
e) K_b becomes larger.

23. If a small amount of CsOH(aq) is added to 0.10 M HCN(aq),

a) the equilibrium concentration of CN^-(aq) is increased.
b) no change occurs.
c) K_a becomes smaller.
d) the equilibrium concentration of CN^-(aq) is decreased.
e) the equilibrium concentration of HCN(aq) increases.

24. The pK_b for CH_3NH_2 is 3.44. This expression refers to which of the following reactions?

a) $CH_3NH_3^+(aq) + H_2O(l) \leftrightarrow CH_3NH_2(aq) + H_3O^+(aq)$
b) $CH_3NH_3^+(aq) + OH^-(aq) \leftrightarrow CH_3NH_2(aq) + H_2O(aq)$
c) $CH_3NH_2(aq) + H_3O^+(aq) \leftrightarrow CH_3NH_3^+(aq) + H_2O(l)$
d) $H_3O^+(aq) + OH^-(aq) \leftrightarrow 2H_2O(l)$
e) $CH_3NH_2(aq) + H_2O(l) \leftrightarrow CH_3NH_3^+(aq) + OH^-(aq)$

25. The pK_a for HF is 3.45. This expression refers to which of the following reactions?
a) $HF(aq) + H_2O(l) \leftrightarrow F^-(aq) + H_3O^+(aq)$
b) $HF(aq) + OH^-(aq) \leftrightarrow F^-(aq) + H_2O(l)$
c) $H_3O^+(aq) + OH^-(aq) \leftrightarrow 2H_2O(l)$
d) $F^-(aq) + H_2O(l) \leftrightarrow HF(aq) + OH^-(aq)$
e) $F^-(aq) + H_3O^+(aq) \leftrightarrow HF(aq) + H_2O(l)$

26. Consider the following reaction:

$$HSO_4^-(aq) + H_2O(l) \leftrightarrow SO_4^{2-}(aq) + H_3O^+(aq)$$

Which of the following statements is true?
a) The reaction of HSO_4^- with H_2O produces OH^- rather than H_3O^+.
b) H_2O is the acid and SO_4^{2-} is its conjugate base.
c) HSO_4^- is not an acid and thus the equation is incorrect.
d) HSO_4^- is the acid and SO_4^{2-} is its conjugate base.
e) HSO_4^- is the base and H_3O^+ is its conjugate acid.

27. Which of the following is a Bronsted base?
a) $CH_3NH_3^+$ b) H_2CO_3 c) SO_4^{2-} d) H_2S e) NH_4^+

28. All of the following are strong bases except
a) NH_2OH b) RbOH c) LiOH d) KOH e) $Ba(OH)_2$

29. Which of the following 0.10 M aqueous solutions has the highest pH?
a) NH_3 b) $(C_2H_5)_3N$ c) $B(OH)_3$ d) HIO e) C_6H_5COOH

30. All of the following are strong acids except
a) HF b) $HClO_4$ c) HI d) HBr e) HNO_3

31. Given: HClO, pK_a = 7.53
$CH_2ClCOOH$, pK_a = 2.85
H_2CO_3, pK_{a1} = 6.37 and pK_{a2} = 10.25
List the conjugate bases, ClO^-, CH_2ClCOO^-, and CO_3^{2-}, in order of decreasing basicity.
a) $CO_3^{2-} > CH_2ClCOO^- > ClO^-$
b) $CH_2ClCOO^- > ClO^- > CO_3^{2-}$
c) $ClO^- > CO_3^{2-} > CH_2ClCOO^-$
d) $CH_2ClCOO^- > CO_3^{2-} > ClO^-$
e) $CO_3^{2-} > ClO^- > CH_2ClCOO^-$

32. Which of the following is the strongest acid?
a) HNO_2 b) H_2SO_3 c) HNO_3 d) HCN e) H_3PO_4

33. Which of the following is the strongest base?
a) ammonia ($pK_b = 4.75$)
b) methylamine ($pK_b = 3.44$)
c) urea ($pK_b = 13.90$)
d) pyridine ($pK_b = 8.75$)
e) morphine ($pK_b = 5.79$)

34. Which of the following produces the strongest conjugate acid?
a) aniline ($pK_b = 9.37$)
b) pyridine ($pK_b = 8.75$)
c) nicotine ($pK_b = 5.98$)
d) ammonia ($pK_b = 4.74$)
e) methylamine ($pK_b = 3.44$)

35. If the value of K_a for HF is 3.5×10^{-4}, calculate the equilibrium constant for

$$F^-(aq) + H_2O(l) \leftrightarrow HF(aq) + OH^-(aq)$$

a) 2.9×10^{3} b) -3.5×10^{-4} c) 3.5×10^{-18} d) 2.9×10^{-11} e) 3.5×10^{-4}

36. If the value of K_b for NH_3 is 1.8×10^{-5}, calculate the equilibrium constant for

$$NH_4^+(aq) + H_2O(l) \leftrightarrow NH_3(aq) + H_3O^+(aq)$$

a) 1.8×10^{-12} b) 5.6×10^{4} c) 5.6×10^{-10} d) -1.8×10^{-5} e) 1.8×10^{-5}

37. Which of the following is the strongest acid in aqueous solution?
a) All the acids listed have the same strength in water. b) HCl c) $HClO_4$ d) HNO_3 e) H_2SO_4

38. Which of the following is the strongest acid?
a) H_2O b) CH_4 c) H_2S d) H_2Se e) H_2Te

39. Which of the following is the strongest acid?
a) CCl_3COOH b) $CH_2BrCOOH$ c) $CHCl_2COOH$ d) CH_3COOH e) $CH_2ClCOOH$

40. Which of the following has the highest pK_a?
a) HClO b) $HClO_2$ c) $HClO_3$ d) $HClO_4$ e) HCl

41. Which of the following aqueous solutions gives a pH greater than 7?
a) 10^{-8} M HCl b) 10^{-8} M CH_3COOH c) 10^{-8} M HNO_3 d) 10^{-8} M HCOOH e) 10^{-8} M NH_3

42. Which of the following acids gives the anion that is the strongest base in aqueous solution?
a) HBrO ($pK_a = 8.69$)
b) HClO ($pK_a = 7.53$)
c) $CH_2ClCOOH$ ($pK_a = 2.85$)
d) C_6H_5COOH ($pK_a = 4.19$)
e) HNO_2 ($pK_a = 3.37$)

43. Which of the following 0.10 M aqueous solutions has the lowest pH?
a) HI b) HF c) NH_3 d) $CO(NH_2)_2$ e) H_3PO_4

44. If the pK_a of acetic acid is 4.75, the pK_a of CH_3CH_2OH is
a) also 4.75. b) about 4. c) about 16. d) about 7. e) much less than 4.75.

45. Estimate the pH of 10^{-7} M $HClO_4$(aq).
a) 6.9 b) 8.0 c) 7.0 d) 1.0 e) 5.0

46. The pH of 1.0 M HCOOH(aq) is 1.87. What is the percent ionization of HCOOH?
a) 1.3% b) 1.8% c) 1.9% d) 94% e) 0%

47. The pH of 0.10 M CH_3COOH(aq) is 2.87. What is the percent ionization of CH_3COOH?
a) 5.0% b) 1.3% c) 10% d) 13% e) 0.13%

48. The pH of 1.0 M HCOOH(aq) is 1.87. What is the value of K_a for HCOOH?
a) 1.8×10^{-4} b) 1.8×10^{-5} c) 0.013 d) 8.6×10^{-7} e) 9.3×10^{-4}

49. The pH of 1.0 M aqueous benzoic acid is 2.09. What is the value of K_a for benzoic acid?
a) 6.6×10^{-5} b) 1.3×10^{-4} c) 1.6×10^{-3} d) 8.0×10^{-3} e) 6.6×10^{-6}

50. What is the pH of 1.0 M HCNO(aq) ($K_a = 2.2 \times 10^{-4}$)?
a) 7.32 b) 7.00 c) 1.83 d) 4.70 e) 3.66

51. What is the pH of 0.010 M HCN(aq) ($K_a = 4.9 \times 10^{-10}$)?
a) 5.65 b) 4.65 c) 9.31 d) 2.00 e) 6.69

52. What is the pH of 0.24 M HClO(aq) (pK_a = 7.52)?
a) 4.07 b) 3.45 c) 3.76 d) 8.14 e) 0.62

53. What is the pH of 0.30 M C_6H_5COOH(aq) (pK_a = 4.19)?
a) 2.35 b) 4.19 c) 3.66 d) 0.52 e) 9.42

54. A weak acid with $K_a = 2.2 \times 10^{-4}$ is 1.5% ionized in a 1.0 M solution. In a 0.010 M solution, the percent ionization would be
a) less than 1.5%. b) greater than 1.5%. c) 1.5%. d) zero. e) 1.0%.

55. For a 0.10 M solution of a weak acid, HA, with pK_a = 10, which of the following is true?
a) $[HA] = K_a$ b) $[HA] \cong 0.10$ M c) $[HA] \cong 0$ d) $[HA] = [A^-]$ e) $[HA] = [H_3O^+]$

56. If a small amount of HCl(aq) is added to 0.10 M CH_3COOH(aq),
a) K_a becomes larger.
b) the equilibrium concentration of CH_3COO^-(aq) is decreased.
c) the equilibrium concentration of acetic acid decreases.
d) the equilibrium concentration of the acetate ion is increased.
e) no change occurs.

57. If an aqueous solution of HF is diluted with water,
a) more HF is produced by reaction of F^-(aq) with H_2O(l).
b) no change occurs.
c) the percent ionization of HF(aq) increases.
d) the percent ionization of HF(aq) decreases.
e) the value of K_a changes.

58. Consider the follwing acids:
$HClO_2$, pK_a = 2.00
HCN, pK_a = 9.31
HF, pK_a = 3.45
HIO, pK_a = 10.64
HClO, pK_a = 7.53
Which acid will have the greatest percentage of solute ionized in a 0.10 M aqueous solution?
a) $HClO_2$ b) All solutes are ionized to the same degree. c) HIO d) HCN e) HF

59. When 0.648 g of an organic base of molar mass 162 g · mol^{-1} is dissolved in 50.0 mL of water, the pH is found to be 10.45. Calculate the pK_b of the base and the pK_a of its conjugate acid.

60. The pH of 0.0945 M NH_3(aq) is 11.12. What is the percent of NH_3 protonated?
a) 1.4% b) 28% c) 0.0018% d) 0.0047% e) 2.8%

61. The pH of 0.15 M NH_2OH(aq) is 9.60. What is the percent of NH_2OH protonated?
a) 0.027% b) 2.7% c) 0.27% d) 1.6% e) 5.4%

62. The pH of 0.10 M pyridine(aq) is 9.13. What is the value of K_b for pyridine?
a) 2.7×10^{-4} b) 7.4×10^{-10} c) 2.7×10^{-5} d) 1.8×10^{-10} e) 1.8×10^{-9}

63. The pH of 0.50 M aniline(aq) is 9.17. What is the value of K_b for aniline?
a) 1.4×10^{-9} b) 1.2×10^{-4} c) 4.4×10^{-10} d) 2.1×10^{-5} e) 6.8×10^{-10}

64. What is the pH of 0.010 M aniline(aq) (pK_b = 9.37)?
a) 8.32 b) 5.68 c) 9.37 d) 12.00 e) 10.30

65. What is the pH of 0.22 M aniline(aq) ($K_b = 4.3 \times 10^{-10}$)?
a) 8.99 b) 5.01 c) 13.34 d) 9.32 e) 9.65

66. What is the pH of 0.025 M NH_2OH(aq) ($K_b = 1.1 \times 10^{-8}$)?
a) 9.22 b) 10.82 c) 7.96 d) 9.56 e) 4.78

67. What is the pH of 0.35 M CH_3NH_2(aq) ($pK_b = 3.44$)?
a) 10.56 b) 11.01 c) 12.51 d) 7.80 e) 12.05

68. What is the pH of 0.010 M NH_2NH_2(aq) ($pK_b = 5.77$)?
a) 10.11 b) 11.11 c) 12.00 d) 8.23 e) 12.12

69. The equation for which one could write a K_{a2} expression for sulfurous acid is
a) $HSO_3^-(aq) + H_2O(l) \leftrightarrow SO_3^{2-}(aq) + H_3O^+(aq)$
b) $HSO_3^-(aq) + H_2O(l) \leftrightarrow H_2SO_3(aq) + OH^-(aq)$
c) $H_2SO_3(aq) + H_2O(l) \leftrightarrow HSO_3^-(aq) + H_3O^+(aq)$
d) $SO_3^{2-}(aq) + H_2O(l) \leftrightarrow HSO_3^-(aq) + OH^-(aq)$
e) $H_2SO_3(aq) + 2H_2O(l) \leftrightarrow SO_3^{2-}(aq) + 2H_3O^+(aq)$

70. For a solution labeled "0.10 M H_2SO_4(aq),"
a) $[HSO_4^-]$ is greater than 0.10 M.
b) $[SO_4^{2-}] = 0.10$ M
c) the pH is greater than 1.0.
d) the pH is less than 1.0.
e) the pH equals 1.0.

71. For a solution labeled "0.10 M H_3PO_4(aq),"
a) $[H_2PO_4^-]$ is greater than 0.10 M.
b) $[H^+] = 0.10$ M.
c) $[H^+]$ is less than 0.10 M.
d) $[PO_4^{3-}] = 0.10$ M.
e) $[H^+] = 0.30$ M.

72. Calculate $[H^+]$ for a solution labeled "0.0500 M H_2SO_3(aq)" ($pK_{a1} = 1.81$, $pK_{a2} = 6.91$).
a) 0.021 M b) 0.025 M c) 0.029 M d) 0.050 M e) 0.015 M

73. Calculate the equilibrium concentration of sulfurous acid in a solution labeled "0.100 M H_2SO_3(aq)" ($pK_{a1} = 1.81$, $pK_{a2} = 6.91$).
a) 0.068 M b) 0.100 M c) 0.032 M d) 0.015 M e) 0.050 M

74. Calculate $[H^+]$ in a solution labeled "0.100 M $(COOH)_2$(aq), oxalic acid" ($pK_{a1} = 1.23$, $pK_{a2} = 4.19$).
a) 0.047 M b) 0.050 M c) 0.200 M d) 0.053 M e) 0.100 M

75. Calculate the pH of 0.050 M H_2S(aq) ($pK_{a1} = 6.88$, $pK_{a2} = 14.15$).
a) 4.09 b) 6.88 c) 7.12 d) 2.79 e) 3.44

76. Which of the following 0.10 M aqueous solutions gives the lowest pH?
 a) CH_3COOH (pK_a = 4.75)
 b) H_3PO_4 (pK_{a1} = 2.12)
 c) HIO_3 (pK_a = 0.77)
 d) HF (pK_a = 3.45)
 e) Since all are acids, the pH is the same for all solutions.

77. If pK_{a1} and pK_{a2} for H_2CO_3 are 6.37 and 10.25, respectively, calculate the equilibrium constant for the reaction below:

$$H_2CO_3(aq) + 2H_2O(l) \leftrightarrow 2H_3O^+(aq) + CO_3^{2-}(aq)$$

 a) 5.6×10^{-11} b) 4.1×10^{-11} c) 4.3×10^{-7} d) 2.3×10^{-8} e) 2.4×10^{-17}

78. The predominant acid responsible for the acidity of "acid rain" is H_2SO_3 (pK_{a1} = 1.81 and pK_{a2} = 6.91). If a sample of water from a lake has a pH of 3.20, calculate the stoichiometric concentration of H_2SO_3 in the lake.

79. Calculate the pH of 0.0010 M H_2SO_4(aq) (K_a for HSO_4^- is 0.012).

80. Calculate [H^+], [OH^-], [HSO_3^-], [H_2SO_3], and the pH of 0.0500 M H_2SO_3(aq) (pK_{a1} = 1.81 and pK_{a2} =6.91).

Chapter 14: Protons in Transition: Acids and Bases Form B

1. Consider the following reaction:

   $HCO_3^-(aq) + H_2O(l) \leftrightarrow H_2CO_3(aq) + OH^-(aq)$

   Which of the following statements is true?

a) The reaction of HCO_3^- with H_2O produces H_3O^+ rather than OH^-.
b) HCO_3^- is not an acid and thus the equation is incorrect.
c) HCO_3^- is the base and H_2O is the acid.
d) H_2O is the base and H_2CO_3 is its conjugate acid.
e) HCO_3^- is the acid and OH^- is its conjugate base.

2. When BaO(s) is dissolved in water which of the following is true?

a) The solution contains $OH^-(aq)$ and $Ba^{2+}(aq)$.
b) BaO(s) does not dissolve in water.
c) The solution contains BaO(aq).
d) The solution contains $O^{2-}(aq)$, $OH^-(aq)$, and $Ba^{2+}(aq)$.
e) The solution contains $O^{2-}(aq)$ and $Ba^{2+}(aq)$.

3. The conjugate base of $H_2PO_4^-$ is

a) HPO_4^{2-} b) OH^- c) $H_2PO_4^-$ d) H_3PO_4 e) PO_4^{3-}

4. The conjugate base of HSO_4^- is

a) OH^- b) SO_4^{2-} c) H_2SO_4 d) HSO_4^- e) H_2O

5. The conjugate base of HCO_3^- is

a) CO_3^{2-} b) H_2CO_3 c) HCO_3^- d) H_2O e) H_3O^+

6. The conjugate acid of ammonia is

a) NH_3 b) NH_4OH c) NH_2^- d) NH_4^+ e) H_3O^+

7. The conjugate base of hydrazine is

a) $N_2H_3^-$ b) N_2H_4 c) $N_2H_5^+$ d) OH^- e) H_2O

8. The conjugate base of OH^- is

a) OH^- b) H_2O c) H^+ d) H_3O^+ e) O^{2-}

9. Calculate the value of K_b for HCO_3^- given $pK_{a1} = 6.37$ and $pK_{a2} = 10.25$ for carbonic acid. Compare this value to the value of K_{a2} and predict whether HCO_3^- is an acid or a base in aqueous solution.

10. At 0°C, K_w for water is 1.52×10^{-15}. What is the pH of a neutral aqueous solution at this temperature?
a) 7.41 b) 6.59 c) 7.56 d) 7.00 e) 7.26

11. Calculate the concentration of H_3O^+ for an aqueous solution with a pH of 6.20.
a) 4.0×10^{-4} M b) 1.6×10^{-8} M c) 1.0×10^{-14} M d) 0.0.79 M e) 6.3×10^{-7} M

12. The pH of a 0.010 M aqueous solution of calcium hydroxide is
a) 12.30 b) 12.00 c) 11.70 d) 12.60 e) 1.70

13. The pH of a 0.0025 M aqueous solution of calcium hydroxide is
a) 11.70 b) 12.40 c) 11.10 d) 12.00 e) 2.30

14. For a solution labeled "0.10 M barium hydroxide," which of the following is correct?
a) $[OH^-] = 0.10$ M, $[Ba^{2+}] = 0.050$ M
b) $[OH^-] = 0.20$ M, $[Ba^{2+}] = 0.20$ M
c) $[OH^-] = 0.10$ M, $[Ba^{2+}] = 0.10$ M
d) $[OH^-] = 0.10$ M, $[Ba^{2+}] = 0.20$ M
e) $[OH^-] = 0.20$ M, $[Ba^{2+}] = 0.10$ M

15. When 500 mL of 0.126 M KOH(aq) is mixed with 500 mL of 0.120 M $Ca(OH)_2$(aq), the pH of the resulting solution is
a) 13.26 b) 14.00 c) 13.09 d) 13.39 e) 13.56

16. In a solution labeled "1.0 M HCl," which of the following is correct?
a) $[H_3O^+]$ 1.0 M, $[Cl^-] = 1.0$ M
b) $[HCl] = 0.50$ M, $[H_3O^+] = 0.50$ M, $[Cl^-] = 0.50$ M
c) $[H_3O^+] = 1.0$ M, $[OH^-] = 1.0 \times 10^{-7}$ M
d) $[HCl] = 1.0$ M
e) $[H_3O^+] = 0.90$ M, $[Cl^-] = 0.10$ M

17. Calculate the concentration of OH^- for an aqueous solution with a pH of 9.60.
a) 0.64 M b) 4.0×10^{-5} M c) 1.5×10^{-5} M d) 1.0×10^{-14} M e) 2.5×10^{-10} M

18. Which of the following is the strongest acid?
a) HCN ($pK_a = 9.31$)
b) CH_3COOH ($pK_a = 4.75$)
c) HIO_3 ($pK_a = 0.77$)
d) HF ($pK_a = 3.45$)
e) HNO_2 ($pK_a = 3.37$)

19. Which of the following produces the strongest conjugate base?
a) HClO ($pK_a = 7.53$)
b) HF ($pK_a = 3.45$)
c) HCOOH ($pK_a = 3.75$)
d) HIO ($pK_a = 10.64$)
e) CH_3COOH ($pK_a = 4.75$)

20. Given:

acetic acid, $pK_a = 4.75$

$HClO_2$, $pK_a = 2.00$

HF, $pK_a = 3.45$

The order of these acids from strongest to weakest is

a) HF > $HClO_2$ > acetic acid
b) acetic acid > HF > $HClO_2$
c) HF > acetic acid > $HClO_2$
d) $HClO_2$ > HF > acetic acid
e) $HClO_2$ > acetic acid > HF

21. If a small amount of HCl(aq) is added to 0.10 M $CH_3NH_2(aq)$,

a) no change occurs.
b) the equilibrium concentration of methylamine increases.
c) the equilibrium concentration of the methylammonium ion is decreased.
d) K_a becomes larger.
e) the equilibrium concentration of $CH_3NH_3^+(aq)$ is increased.

22. If a small amount of CsOH(aq) is added to 0.10 M $NH_3(aq)$,

a) K_b becomes larger.
b) the equilibrium concentration of $NH_3(aq)$ is decreased.
c) the equilibrium concentration of $NH_3(aq)$ is increased.
d) no change occurs.
e) K_b becomes smaller.

23. If a small amount of CsOH(aq) is added to 0.10 M HF(aq),

a) K_a becomes smaller.
b) the equilibrium concentration of $F^-(aq)$ is increased.
c) no change occurs.
d) the equilibrium concentration of HF(aq) increases.
e) the equilibrium concentration of $F^-(aq)$ is decreased.

24. The pK_b for NH_3 is 4.74. This expression refers to which of the following reactions?

a) $NH_4^+(aq) + OH^-(aq) \leftrightarrow NH_3(aq) + H_2O(aq)$
b) $NH_4^+(aq) + H_2O(l) \leftrightarrow NH_3(aq) + H_3O^+(aq)$
c) $H_3O^+(aq) + OH^-(aq) \leftrightarrow 2H_2O(l)$
d) $NH_3(aq) + H_3O^+(aq) \leftrightarrow NH_4^+(aq) + H_2O(l)$
e) $NH_3(aq) + H_2O(l) \leftrightarrow NH_4^+(aq) + OH^-(aq)$

25. The pK_a for HCN is 3.45. This expression refers to which of the following reactions?

a) $HCN(aq) + H_2O(l) \leftrightarrow CN^-(aq) + H_3O^+(aq)$
b) $H_3O^+(aq) + OH^-(aq) \leftrightarrow 2H_2O(l)$
c) $CN^-(aq) + H_3O^+(aq) \leftrightarrow HCN(aq) + H_2O(l)$
d) $CN^-(aq) + H_2O(l) \leftrightarrow HCN(aq) + OH^-(aq)$
e) $HCN(aq) + OH^-(aq) \leftrightarrow CN^-(aq) + H_2O(l)$

26. Consider the following reaction:

$$HCO_3^-(aq) + H_2O(l) \leftrightarrow H_2CO_3(aq) + OH^-(aq)$$

Which of the following statements is true?

a) HCO_3^- is not an acid and thus the equation is incorrect.
b) H_2O is the base and H_2CO_3 is its conjugate acid.
c) HCO_3^- is the acid and OH^- is its conjugate base.
d) The reaction of HCO_3^- with H_2O produces H_3O^+ rather than OH^-.
e) HCO_3^- is the base and H_2CO_3 is its conjugate acid.

27. Which of the following is a Bronsted base?
a) H_3O^+ b) H_2CO_3 c) NH_4^+ d) H_2S e) PO_4^{3-}

28. All of the following are strong bases except
a) $B(OH)_3$ b) RbOH c) LiOH d) KOH e) $Ba(OH)_2$

29. Which of the following 0.10 M aqueous solutions has the highest pH?
a) CsOH b) NH_2OH c) C_6H_5OH d) NH_3 e) $(C_2H_5)_3N$

30. All of the following are strong acids except
a) HClO b) $HClO_4$ c) HI d) HBr e) HNO_3

31. Given: HClO, $pK_a = 7.53$
HF, $pK_a = 3.45$
H_2CO_3, $pK_{a1} = 6.37$ and $pK_{a2} = 10.25$

List the conjugate bases, ClO^-, CH_2ClCOO^-, and CO_3^{2-}, in order of decreasing basicity.

a) $F^- > ClO^- > CO_3^{2-}$
b) $CO_3^{2-} > ClO^- > F^-$
c) $ClO^- > CO_3^{2-} > F^-$
d) $F^- > CO_3^{2-} > ClO^-$
e) $CO_3^{2-} > F^- > ClO^-$

32. Which of the following is the strongest base?
a) LiOH b) NH_3 c) NH_2OH d) $B(OH)_3$ e) NH_2NH_2

33. Which of the following is the strongest base?
a) morphine ($pK_b = 5.79$)
b) ammonia ($pK_b = 4.75$)
c) ethylamine ($pK_b = 3.19$)
d) pyridine ($pK_b = 8.75$)
e) urea ($pK_b = 13.90$)

34. Which of the following produces the strongest conjugate acid?
a) pyridine ($pK_b = 8.75$)
b) methylamine ($pK_b = 3.44$)
c) urea ($pK_b = 13.90$)
d) nicotine ($pK_b = 5.98$)
e) ammonia (pK_b =4.74)

35. If the value of K_a for HCN is 4.9×10^{-10}, calculate the equilibrium constant for

$$CN^-(aq) + H_2O(l) \leftrightarrow HCN(aq) + OH^-(aq)$$

a) 4.9×10^{-10} b) -4.9×10^{-10} c) 2.0×10^{-5} d) 2.0×10^{-5} e) 2.0×10^{9}

36. If the value of K_b for pyridine is 1.8×10^{-9}, calculate the equilibrium constant for

$$C_5H_5NH^+(aq) + H_2O(l) \leftrightarrow C_5H_5N(aq) + H_3O^+(aq)$$

a) -1.8×10^{-9} b) 1.8×10^{-16} c) 5.6×10^{-6} d) 5.6×10^{8} e) 1.8×10^{-9}

37. Which of the following is the strongest acid in aqueous solution?
a) HIO_3 b) $HClO_4$ c) HCN d) HF e) All of the acids listed have the same strength in water.

38. Which of the following is the strongest acid?
a) CCl_3COOH b) $CH_2ClCOOH$ c) CF_3COOH d) CH_2FCOOH e) $CH_2BrCOOH$

39. Which of the following is the strongest acid?
a) $HClO_2$ b) $HClO_3$ c) HClO d) HBrO e) HIO

40. Which of the following has the highest pK_a?
a) HBrO b) $HClO_2$ c) $HClO_3$ d) $HClO_4$ e) HIO

41. Which of the following aqueous solutions gives a pH greater than 7?
a) 10^{-8} M HCOOH b) 10^{-8} M HNO_3 c) 10^{-8} M C_5H_5N d) 10^{-8} M CH_3COOH e) 10^{-8} M HCl

42. Which of the following acids gives the anion that is the strongest base in aqueous solution?
a) C_6H_5COOH ($pK_a = 4.19$)
b) HIO ($pK_a = 10.64$)
c) HClO ($pK_a = 7.53$)
d) $CH_2ClCOOH$ ($pK_a = 2.85$)
e) HNO_2 ($pK_a = 3.37$)

43. Which of the following 0.10 M aqueous solutions has the lowest pH?
a) HCl b) $HClO_2$ c) HIO d) HBrO e) HClO

44. If the pK_a of HClO is 7.52, the pK_a of $HClO_2$ may be roughly estimated to be
a) also 7.53. b) less than zero. c) about 2.0. d) about 9.0. e) about 16.

45. Estimate the pH of 10^{-7} M HCl(aq).
a) 6.9 b) 8.0 c) 7.0 d) 1.0 e) 5.0

46. The pH of 0.10 M HCOOH(aq) is 2.37. What is the percent ionization of HCOOH?
a) 4.2% b) 2.4% c) 0.71% d) 7.7% e) 0%

47. The pH of 0.80 M benzenesulfonic acid is 0.51. What is the percent ionization of benzenesulfonic acid?
a) 39% b) 64% c) 25% d) 5.0% e) 51%

48. The pH of 0.10 M CH_3COOH(aq) is 2.87. What is the value of K_a for CH_3COOH?
 a) 1.3×10^{-3} b) 1.8×10^{-6} c) 1.8×10^{-5} d) 0.037 e) 2.7×10^{-6}

49. The pH of 0.800 M aqueous benzenesulfonic acid is 0.51. What is the value of K_a for benzenesulfonic acid?
 a) 0.19 b) 0.51 c) 0.44 d) 0.12 e) 0.90

50. What is the pH of 0.010 M HClO(aq) ($K_a = 3.0 \times 10^{-8}$)?
 a) 4.76 b) 3.77 c) 2.00 d) 9.52 e) 7.52

51. What is the pH of 0.12 M HClO(aq) ($K_a = 3.0 \times 10^{-8}$)?
 a) 4.22 b) 3.76 c) 5.60 d) 0.92 e) 7.52

52. What is the pH of 0.14 M HCN(aq) (pK_a = 9.31)?
 a) 0.85 b) 4.66 c) 4.23 d) 5.08 e) 10.16

53. What is the pH of 0.25 M HBrO(aq) (pK_a = 8.69)?
 a) 4.65 b) 8.10 c) 5.90 d) 0.60 e) 9.30

54. A weak acid with $K_a = 1.8 \times 10^{-4}$ is 4.2% ionized in a 0.10 M solution. In a 1.0 M solution, the percent ionization would be
 a) less than 4.2%. b) 4.2%. c) greater than 4.2%. d) zero. e) 10%.

55. For a 0.10 M solution of a weak acid, HA, with pK_a = 10, which of the following is true?
 a) [HA] = [A^-] b) [HA] does not equal [H_3O^+] c) [HA] = [H_3O^+] d) [HA] = K_a e) [HA] $\cong$ 0

56. If a small amount of HCl(aq) is added to 0.10 M HF(aq),
 a) K_a becomes larger.
 b) the equilibrium concentration of hydrofluoric acid decreases.
 c) the equilibrium concentration of F^-(aq) is decreased.
 d) the equilibrium concentration of the fluoride ion is increased.
 e) no change occurs.

57. If an aqueous solution of HCN is diluted with water,
 a) the percent ionization of HCN(aq) decreases.
 b) the value of K_a changes.
 c) the percent ionization of HCN(aq) increases.
 d) no change occurs.
 e) more HCN is produced by reaction of CN^-(aq) with H_2O(l).

58. Consider the follwing acids:

$HClO_2$, $pK_a = 2.00$

HCN, $pK_a = 9.31$

HF, $pK_a = 3.45$

HIO, $pK_a = 10.64$

HClO, $pK_a = 7.53$

Which acid will have the least percent of solute ionized in a 0.10 M aqueous solution?

a) HIO b) All solutes are ionized to the same degree. c) $HClO_2$ d) HCN e) HF

59. When 126 mg of an organic base of molar mass 45.1 $g \cdot mol^{-1}$ is dissolved in 25.0 mL of water, the pH is found to be 11.82. Calculate the pK_b of the base and the pK_a of its conjugate acid.

60. The pH of 0.20 M $CH_3NH_2(aq)$ is 11.93. What is the percent of CH_3NH_2 protonated?

a) 4.3% b) 85% c) 43% d) 0.43% e) 8.5%

61. The pH of 0.010 M aniline(aq) is 8.32. What is the percent of aniline protonated?

a) 0.021% b) 0.21% c) 2.1% d) 0.12% e) 0.69%

62. The pH of 0.20 M $CH_3NH_2(aq)$ is 11.93. What is the value of K_b for CH_3NH_2?

a) 8.5×10^{-3} b) 1.4×10^{-5} c) 7.2×10^{-5} d) 1.7×10^{-3} e) 3.6×10^{-4}

63. The pH of 0.50 M ethylamine(aq) is 12.26. What is the value of K_b for ethylamine?

a) 9.1×10^{-3} b) 6.6×10^{-4} c) 3.3×10^{-4} d) 1.1×10^{-12} e) 5.5×10^{-13}

64. What is the pH of 0.010 M pyridine(aq) (pK_b= 8.75)?

a) 8.63 b) 5.37 c) 8.75 d) 12.00 e) 10.62

65. What is the pH of 0.12 M morphine(aq) ($K_b = 1.6 \times 10^{-6}$)?

a) 10.64 b) 3.56 c) 8.20 d) 11.56 e) 11.10

66. What is the pH of 0.025 M $(CH_3)_3N(aq)$ ($K_b = 6.5 \times 10^{-5}$)?

a) 9.81 b) 11.11 c) 11.58 d) 12.71 e) 8.21

67. What is the pH of 0.42 M $NH_3(aq)$ ($pK_b = 4.74$)?

a) 11.82 b) 11.44 c) 9.26 d) 9.49 e) 8.88

68. What is the pH of 0.125 M $NH_2NH_2(aq)$ ($K_b = 1.7 \times 10^{-6}$)?

a) 10.66 b) 3.34 c) 13.10 d) 8.23 e) 11.57

69. The equation for which one could write a K_{a2} expression for phosphoric acid is
a) $H_2PO_4^-(aq) + H_2O(l) \leftrightarrow HPO_4^{2-}(aq) + H_3O^+(aq)$
b) $HPO_4^{2-}(aq) + H_2O(l) \leftrightarrow PO_4^{3-}(aq) + H_3O^+(aq)$
c) $H_3PO_4(aq) + H_2O(l) \leftrightarrow H_2PO_4^-(aq) + H_3O^+(aq)$
d) $H_3PO_4(aq) + 2H_2O(l) \leftrightarrow HPO_4^{2-}(aq) + 2H_3O^+(aq)$
e) $HPO_4^{2-}(aq) + H_2O(l) \leftrightarrow H_2PO_4^-(aq) + OH^-(aq)$

70. If the pK_{a2} of H_2SO_4 is 1.92, the pH of a solution labeled "0.010 M H_2SO_4(aq)," is
a) slightly less than 2.0. b) 2.0. c) 1.7. d) 2.9. e) 0.

71. For a solution labeled "0.10 M H_2SO_3(aq)," pK_{a1} = 1.81 and pK_{a2} = 6.91, which of the following is true?
a) $[H^+] = 0.2$ M.
b) The pH is 1.0.
c) The pH is about 4.4.
d) The pH is 0.70.
e) The pH is about 1.5.

72. Calculate $[H^+]$ for a solution labeled "0.100 M H_2SO_3(aq)" (pK_{a1} = 1.81, pK_{a2} = 6.91).
a) 0.032 M b) 0.015 M c) 0.050 M d) 0.100 M e) 0.068 M

73. Calculate the equilibrium concentration of sulfurous acid in a solution labeled "0.0500 M H_2SO_3(aq)" (pK_{a1} = 1.81, pK_{a2} = 6.91).
a) 0.025 M b) 0.029 M c) 0.015 M d) 0.050 M e) 0.021 M

74. Calculate the equilibrium concentration of oxalic acid in a solution labeled "0.100 M $(COOH)_2$(aq), oxalic acid" (pK_{a1} = 1.23, pK_{a2} = 4.19).
a) 0.050 M b) 0.053 M c) 0.047 M d) 0.100 M e) 0.200 M

75. Calculate the pH of 0.100 M H_2S(aq) (pK_{a1} = 6.88, pK_{a2} = 14.15).
a) 3.94 b) 6.88 c) 7.12 d) 2.94 e) 3.44

76. Which of the following 0.10 M aqueous solutions gives the lowest pH?
a) CH_3COOH (pK_a = 4.75)
b) HF (pK_a = 3.45)
c) H_3PO_4 (pK_{a1} = 2.12)
d) HClO (pK_a = 2.00)
e) Since all are acids, the pH is the same for all solutions.

77. If pK_{a1} and pK_{a2} for H_2S are 6.88 and 14.15, respectively, calculate the equilibrium constant for the reaction below:

$$H_2S(aq) + 2H_2O(l) \leftrightarrow 2H_3O^+(aq) + S^{2-}(aq)$$

a) 9.2×10^{-22} b) 1.3×10^{-7} c) 7.1×10^{-15} d) 7.7×10^{-8} e) 1.1×10^{-7}

78. The predominant acid responsible for the acidity of "acid rain" is H_2SO_3 (pK_{a1} = 1.81 and pK_{a2} = 6.91). If a sample of water from a lake has a pH of 3.85, calculate the stoichiometric concentration of H_2SO_3 in the lake.

79. Calculate the pH of 1.0 M H_2SO_4(aq) (K_a for HSO_4^- is 0.012).

80. Calculate [H^+], [OH^-], [$H_2PO_4^-$], [H_3PO_4], and the pH of 0.0500 M H_3PO_4(aq) (pK_{a1} = 2.12, pK_{a2} = 7.21, and pK_{a3} = 12.67).

Chapter 15: Salts in Water
Form A

1. Which one of the following salts gives an acidic aqueous solution?
a) $NaHCO_3$ b) Na_2CO_3 c) $NaSO_4$ d) Na_2HPO_4 e) $NaHSO_4$

2. Which one of the following salts gives an acidic aqueous solution?
a) NaH_2PO_4 b) KBr c) KCN d) KNO_3 e) Na_3PO_4

3. Which one of the following salts gives an acidic aqueous solution?
a) $Cr(ClO_4)_3$ b) $CaCl_2$ c) LiF d) $CsNO_3$ e) $NaCH_3CO_2$

4. Which one of the following salts gives an acidic aqueous solution?
a) $NaCH_3CO_2$ b) $Mg(ClO_4)_2$ c) $(CH_3)_3NH(ClO_4)$ d) NaF e) KCN

5. Which one of the following salts gives an acidic aqueous solution?
a) $Cu(NO_3)_2$ b) NaBr c) $NaC_6H_5CO_2$ d) $Ca(ClO_4)_2$ e) $NaNO_2$

6. Which one of the following salts gives an acidic aqueous solution?
a) NH_4F b) NH_4CN c) $(NH_4)_2S$ d) $(NH_4)_3PO_4$ e) $(NH_4)_2CO_3$

7. Which one of the following salts gives a basic aqueous solution?
a) KCN b) NaBr c) NH_4ClO_4 d) $CuCl_2$ e) $NaNO_3$

8. Which one of the following salts gives a basic aqueous solution?
a) $NaCH_3CO_2$ b) KI c) NH_4Cl d) $Fe(ClO_4)_2$ e) $RbNO_3$

9. Which one of the following salts gives a basic aqueous solution?
a) NaHS b) $NaHSO_4$ c) $NiCl_2$ d) NH_4ClO_4 e) $(CH_3)NH_3Cl$

10. Which one of the following salts gives a basic aqueous solution?
a) MgO b) $NaHSO_4$ c) $Mg(ClO_4)_2$ d) $(CH_3)_3NHCl$ e) $Al(NO_3)_3$

11. Which one of the following salts gives a basic aqueous solution?
a) K_3PO_4 b) RbCl c) $Ca(ClO_4)_2$ d) NH_4ClO_4 e) $CrCl_3$

12. Which one of the following salts gives a basic aqueous solution?
a) NH_4ClO b) NH_4F c) NH_4NO_2 d) NH_4Cl e) NH_4HCO_2

13. Which one of the following gives a neutral aqueous solution?
a) $NaNO_2$ b) $Cu(ClO_4)_2$ c) $Mg(ClO_4)_2$ d) KF e) $C_5H_5NH(ClO_4)$

14. Which one of the following gives a neutral aqueous solution?
a) $Cu(ClO_4)_2$ b) $LiClO_4$ c) $NaNO_2$ d) KF e) $C_5H_5NH(ClO_4)$

15. Which one of the following 0.10 M aqueous solutions has the highest pH?
a) HNO_2 b) $KHCO_2$ c) NaI d) NH_4I e) HI

16. Which one of the following 0.10 M aqueous solutions has the highest pH?
a) LiOH b) $KHCO_2$ c) NaI d) NH_4I e) HI

17. Which one of the following 0.10 M aqueous solutions has the lowest pH?
a) HI b) HF c) KNO_2 d) $Al(ClO_4)_3$ e) NH_4Br

18. Which one of the following 0.10 M aqueous solutions has the lowest pH?
a) HF b) $NaCH_3CO_2$ c) KNO_2 d) $Ca(ClO_4)_2$ e) $NaClO_4$

19. For an aqueous solution labeled "0.10 M sodium fluoride,"
a) the pH is greater than 7. b) the pH is less than 7. c) the pH = 1. d) the pH = 7. e) the pH = 13.

20. For an aqueous solution labeled "0.10 M potassium iodide,"
a) the pH is greater than 7. b) the pH = 13. c) the pH = 1. d) the pH is less than 7. e) the pH = 7.

21. Which of the following pairs of ions **cannot** exist in large concentrations **simultaneously** in aqueous solution?
a) K^+ and F^- b) H_3O^+ and ClO_4^- c) NH_4^+ and OH^- d) Na^+ and CN^- e) Li^+ and OH^-

22. Which of the following pairs of ions **can** exist in large concentrations **simultaneously** in aqueous solution?
a) NH_4^+ and OH^- b) $CH_3NH_3^+$ and OH^- c) H_3O^+ and $CH_3CO_2^-$ d) H_3O^+ and CN^- e) NH_4^+ and F^-

23. Calculate the pH of a 1.0 M solution of aqueous sodium acetate. The value of K_a for acetic acid is 1.8×10^{-5}.
a) 9.37 b) 4.63 c) 7.00 d) 2.37 e) 9.26

24. Calculate the pH of a 0.136 M solution of aqueous sodium fluoride. The value of K_a for HF is 3.5×10^{-4}.
a) 8.29 b) 5.71 c) 8.73 d) 2.16 e) 11.84

25. Calculate the pH of a 0.30 M solution of aqueous ammonium bromide. The value of K_b for NH_3 is 1.8×10^{-5}.
a) 4.89 b) 9.11 c) 2.63 d) 4.74 e) 7.00

26. Calculate the pH of a 0.10 M solution of aqueous $Fe(ClO_4)_2$. The value of K_a for $Fe(ClO_4)_2$ is 1.3×10^{-6}.
 a) 3.44 b) 10.56 c) 6.89 d) 2.94 e) 7.00

27. Calculate the pH of a 0.10 M solution of aqueous pyridinium chloride. The value of K_b for pyridine is 1.8×10^{-9}.
 a) 3.13 b) 10.87 c) 4.87 d) 9.13 e) 4.37

28. Consider a solution which is 0.0300 M Na_2S(aq). Which of the following is correct? For H_2S, $K_{a1} = 1.3 \times 10^{-7}$ and $K_{a2} = 7.1 \times 10^{-15}$.
 a) The S^{2-} concentration is close to 0.0300 M.
 b) The OH^- concentration is about 4.8×10^{-5} M.
 c) The OH^- concentration is close to 0.0300 M.
 d) The OH^- concentration is about 0.2 M.
 e) The S^{2-} concentration is about 0.2 M.

29. The pH of 0.18 M aqueous sodium hypochlorite is 10.38. What is the value of K_a for HClO?
 a) 3.2×10^{-7} b) 3.1×10^{-8} c) 5.8×10^{-8} d) 4.2×10^{-11} e) 2.4×10^{-4}

30. The pH of 0.015 M $FeCl_2$(aq) is 3.85. What is the value of K_a for $Fe(OH_2)_6^{2+}$?
 a) 7.7×10^{-9} b) 2.0×10^{-8} c) 5.0×10^{-7} d) 1.3×10^{-6} e) 7.1×10^{-11}

31. The pH of 0.10 M $(CH_3)_3NHCl$(aq) is 5.40. What is the value of K_b for $(CH_3)_3N$?
 a) 4.0×10^{-12} b) 6.3×10^{-5} c) 4.0×10^{-6} d) 6.3×10^{-4} e) 1.6×10^{-10}

32. Draw a titration curve for the titration of 0.10 M phenol(aq) with 0.10 M NaOH(aq). Compare this curve to that for the titration of 0.10 M acetic acid with 0.10 M NaOH(aq) and explain any differences. The stoichiometric point and the point representing the pK_a of phenol should be exact.

33. Arrange the following 0.10 M aqueous solutions in order of increasing pH, 1---->14.
 NaCN, C_5H_5NHCl, CsOH, NaOCl, KNO_3, $HClO_4$, HOCl

34. The curve below corresponds to the titration of

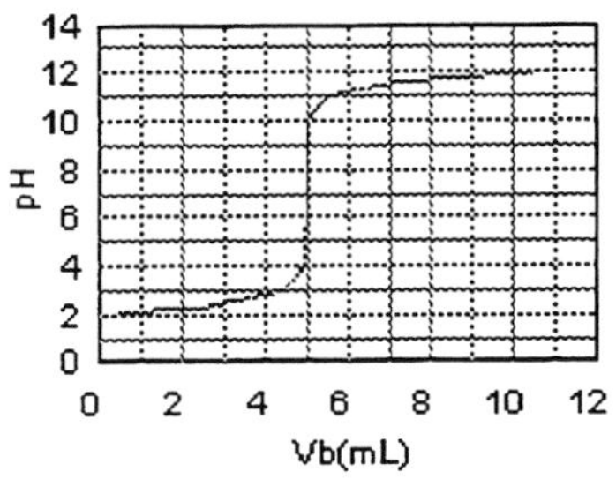

a) 0.010 M $HClO_4$(aq) with 0.10 M KOH(aq).
b) 0.10 M $HClO_2$(aq) ($pK_a = 2.0$) with 0.10 M KOH(aq).
c) 0.010 M HClO(aq) ($pK_a = 7.5$) with 0.10 M KOH(aq).
d) 0.10 M HCl(aq) with 0.10 M KOH(aq).
e) 0.010 M H_2SO_4(aq) ($pK_{a2} = 1.9$) with 0.10 M KOH(aq)

35. A buffer is prepared by mixing 0.25 moles of $NaNO_2$ and 0.75 moles of HNO_2 in 1.0 L of water. Calculate the pH of the solution after the addition of 0.20 moles of KOH. The K_a of HNO_2 is 4.3×10^{-4}.

36. For the titration of 25.0 mL of 0.100 M HClO(aq) ($pK_a = 7.5$) with 0.100 M NaOH(aq), the main species in solution after addition of 12.5 mL of base are
a) HClO(aq), ClO^-(aq), and H^+(aq).
b) HClO(aq), OH^-(aq), and Na^+(aq).
c) ClO^-(aq) and Na^+(aq).
d) HClO(aq), ClO^-(aq), and Na^+(aq).
e) ClO^-(aq), OH^-(aq), and Na^+(aq).

37. At the stoichiometric point in the titration of 0.130 M HCOOH(aq) with 0.130 M KOH(aq),
a) $[HCO_2^-] = 0.0650$ M
b) $[HCO_2^-] = 0.130$ M
c) the pH is less than 7.
d) the pH is 7.0.
e) [HCOOH] = 0.0650 M

38. At the stoichiometric point in the titration of 0.260 M CH_3NH_2(aq) with 0.260 M HCl(aq),
a) $[CH_3NH_2] = 0.130$ M.
b) the pH is greater than 7.
c) $[CH_3NH_3^+] = 0.260$ M
d) the pH is 7.0.
e) $[CH_3NH_3^+] = 0.130$ M

39. What is the pH at the stoichiometric point for the titration of 0.100 M CH_3COOH(aq) with 0.100 M KOH(aq)? The value of K_a for acetic acid is 1.8×10^{-5}.
a) 8.72 b) 5.28 c) 9.26 d) 8.89 e) 7.00

40. The curve for the titration of 50.0 mL of 0.0200 M HClO(aq) with 0.100 M NaOH(aq) is given below. Estimate the pK_a of HClO.

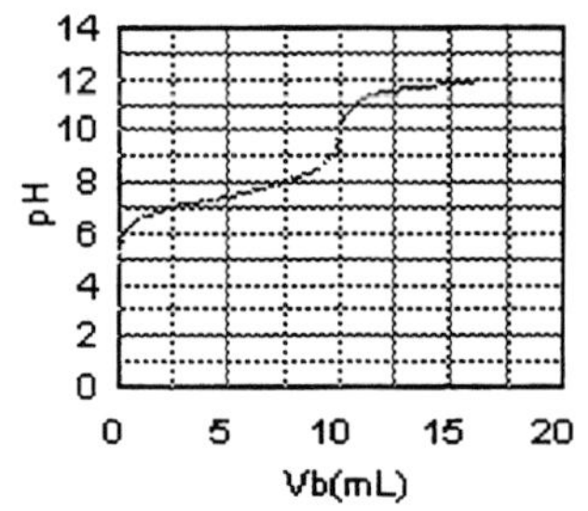

a) 7.5 b) 10.0 c) 5.0 d) 12.0 e) 7.0

41. The curve for the titration of 50.0 mL of 0.0200 M HClO(aq) with 0.100 M NaOH(aq) is given below. The main species in solution at the stoichiometric point are

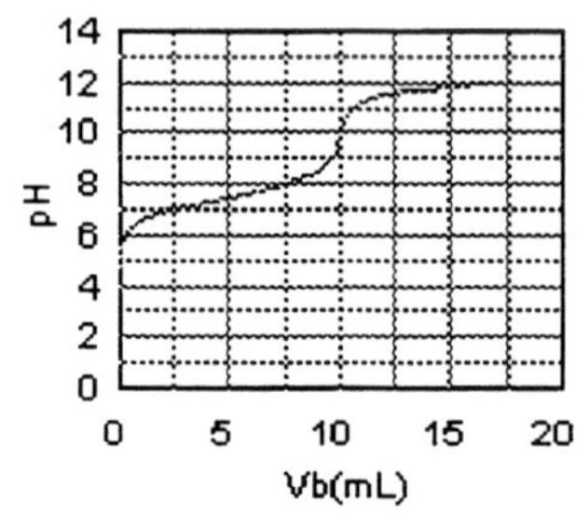

a) ClO^-(aq), OH^-(aq), Na^+(aq)
b) ClO^-(aq), Na^+(aq)
c) HClO(aq), ClO^-(aq), Na^+(aq)
d) HClO(aq), Na^+(aq), OH^-(aq)
e) HClO(aq)

42. What is the pH at the stoichiometric point for the titration of 0.26 M CH_3NH_2(aq) with 0.26 M $HClO_4$(aq)? For CH_3NH_2, $K_b = 3.6 \times 10^{-4}$.
a) 5.72 b) 2.16 c) 2.01 d) 5.57 e) 7.00

43. What is the pH at the half-stoichiometric point for the titration of 0.22 M HNO_2(aq) with 0.10 M KOH(aq)? For HNO_2, $K_a = 4.3 \times 10^{-4}$.
a) 3.37 b) 2.01 c) 2.16 d) 7.00 e) 2.31

44. What is the pH at the half-stoichiometric point for the titration of 0.20 M NH_3(aq) with 0.50 M HNO_3(aq)? For NH_3, $K_b = 1.8 \times 10^{-5}$.
a) 9.26 b) 4.74 c) 11.13 d) 8.56 e) 7.00

45. What is the pH at the half-stoichiometric point for the titration of 0.010 M morphine(aq) with 0.010 M HCl(aq)? For morphine, $K_b = 1.6 \times 10^{-6}$.

a) 8.20 b) 5.80 c) 10.10 d) 9.95 e) 7.00

46. What is the concentration of acetate ion at the stoichiometric point in the titration of 0.018 M CH_3COOH(aq) with 0.036 M NaOH(aq)? For acetic acid, $K_a = 1.8 \times 10^{-5}$.

a) 0.0090 M b) 0.018 M c) 0.012 M d) 0.024 M e) 0.036 M

47. For the titration of 50.0 mL of 0.020 M aqueous salicylic acid with 0.020 M KOH(aq), calculate the pH after the addition of 30.0 mL of KOH(aq). For salycylic acid, $pK_a = 2.97$.

a) 3.15 b) 2.12 c) 2.97 d) 2.33 e) 7.00

48. For the titration of 50.0 mL of 0.020 M aqueous salicylic acid with 0.020 M KOH(aq), calculate the pH after the addition of 55.0 mL of KOH(aq). For salycylic acid, $pK_a = 2.97$.

a) 10.98 b) 12.30 c) 11.26 d) 12.02 e) 7.00

49. The curve for the titration of 25.0 mL of 0.100 M C_5H_5N(aq), pyridine, with 0.100 M HCl(aq) is given below. Estimate the pK_b of pyridine.

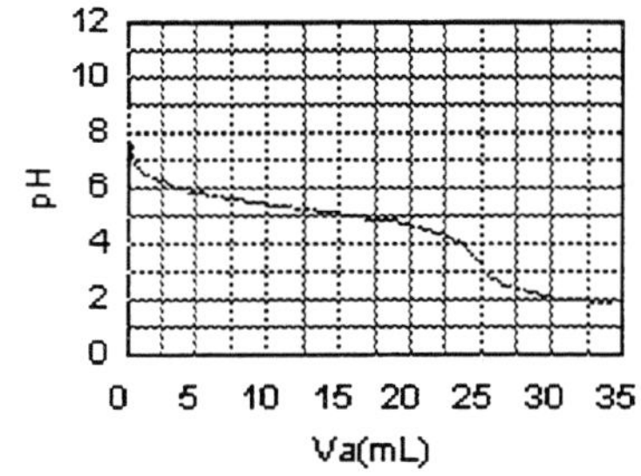

a) 8.8 b) 5.2 c) 3.0 d) 2.0 e) 11

50. The curve for the titration of 25.0 mL of 0.100 M C_5H_5N(aq), pyridine, with 0.100 M HCl is given below. Estimate the pH at the stoichiometric point.

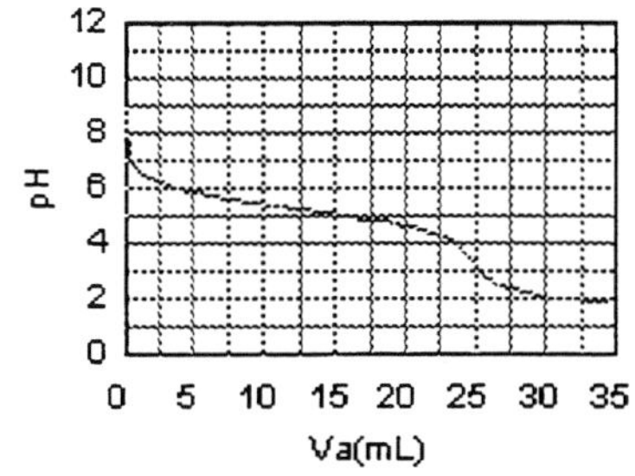

a) 3 b) 5 c) 8 d) 2 e) 7

51. For the titration of 25.0 mL of 0.100 M HClO(aq) (pK_a = 7.5) with 0.100 M NaOH(aq), the main species in solution after addition of 25.0 mL of base are
a) HClO(aq), ClO^-(aq), and Na^+(aq).
b) ClO^-(aq) and Na^+(aq).
c) HClO(aq), OH^-(aq), and Na^+(aq).
d) ClO^-(aq) and H^+(aq).
e) HClO(aq), H^+(aq), and ClO^-(aq).

52. Which of the following indicators would be most suitable for the titration of 0.10 M HBr(aq) with 0.10 M KOH(aq)?
a) bromothymol blue (pK_{In} = 7.1)
b) thymol blue (pK_{In} = 1.7)
c) bromophenol blue (pK_{In} = 3.9)
d) alizarin yellow (pK_{In} = 11.2)
e) methyl orange (pK_{In} = 3.4)

53. Which of the following indicators would be most suitable for the titration of 0.10 M $(CH_3)_3N$(aq) with 0.10 M $HClO_4$(aq)? For trimethyamine, pK_b = 4.19.
a) phenolphthalein (pK_{In} = 9.4)
b) alizarin yellow (pK_{In} = 11.2)
c) thymol blue (pK_{In} = 1.7)
d) bromocresol green (pK_{In} = 4.7)
e) bromothymol blue (pK_{In} = 7.1)

54. The curve for the titration of 50.0 mL of 0.020 M C_6H_5COOH(aq) with 0.100 M NaOH(aq) is given below. Which of the following indicators should be used for the titration?

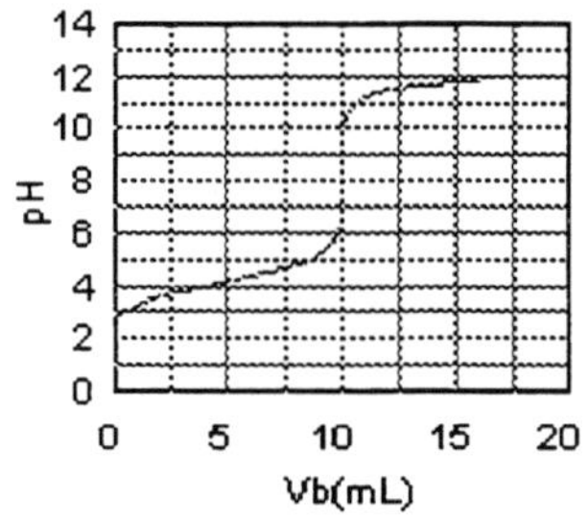

a) thymol blue (pK_{In} = 1.7)
b) bromocresol green (pK_{In} = 4.7)
c) phenol red (pK_{In} = 7.9)
d) methyl red (pK_{In} = 5.0)
e) alizarin (pK_{In} = 11.7)

55. If a small amount of a strong base is added to buffer made up of a weak acid, HA, and the sodium salt of its conjugate base, NaA, the pH of the buffer solution does not change appreciably because
a) the strong base reacts with A^- to give HA.
b) no reaction occurs.
c) the strong base reacts with HA to give AOH and H^+.
d) the K_a of HA is changed.
e) the strong base reacts with HA to give A^-.

56. The following aqueous solutions are available in the first year chemistry laboratory:
    0.10 M HCl(aq)
    0.10 M aniline(aq), $pK_b = 9.37$
    0.10 M HCN(aq), $pK_a = 9.31$
    0.10 M NaOH(aq)
    0.10 M methylamine(aq), $pK_b = 3.44$
    0.10 M HClO(aq), $pK_a = 7.53$

    (a) Describe how to prepare a buffer with a pH in the range 4–5.
    (b) Describe how to prepare a buffer with a pH in the range 7–8.

57. What is the pH of an aqueous solution that is 0.011 M HF ($K_a = 3.5 \times 10^{-4}$) and 0.015 M NaF?
a) 3.59 b) 3.33 c) 5.27 d) 1.95 e) 3.46

58. What is the pH of an aqueous solution that is 0.10 M HCOOH ($K_a = 1.8 \times 10^{-4}$) and 0.10 M $NaHCO_2$?
a) 3.74 b) 10.26 c) 5.62 d) 2.38 e) 5.74

59. What is the pH of an aqueous solution that is 0.12 M $C_6H_5NH_2$ ($K_b = 4.3 \times 10^{-10}$) and 0.018 M $C_6H_5NH_3Cl$?
a) 5.46 b) 8.54 c) 3.81 d) 10.19 e) 4.63

60. Calculate the $[OH^-]$ in an aqueous solution which is 0.125 M NH_3 and 0.300 M NH_4Cl. The value of K_b for NH_3 is 1.8×10^{-5}.
a) 4.3×10^{-5} M b) 7.5×10^{-6} M c) 1.8×10^{-5} M d) 0.425 M e) 0.125 M

61. Calculate the $[H^+]$ in an aqueous solution which is 0.0755 M HF and 0.100 M NaF. The value of K_a for HF is 3.5×10^{-4}.
a) 4.6×10^{-4} M b) 2.6×10^{-4} M c) 3.5×10^{-4} M d) 0.176 M e) 0.0755 M

62. If 100 mL of each of the following solutions is mixed, which one produces a buffer?
a) 1.0 M NH_3(aq) + 0.5 M KOH(aq)
b) 1.0 M NH_3(aq) + 1.0 M HCl(aq)
c) 1.0 M NH_3(aq) + 0.5 M HCl(aq)
d) 1.0 M NH_4Cl(aq) + 1.0 M KOH(aq)
e) 1.0 M NH_4Cl(aq) + 0.5 M HCl(aq)

63. Choose the effective pH range of an aniline-anilinium chloride buffer. The value of the K_b for aniline is 4.3×10^{-10}.
a) 3.6–5.6 b) 5.1–7.1 c) 10.1–12.1 d) 8.4–10.4 e) 1.1–3.1

64. Choose the effective pH range of a pyridine-pyridinium chloride buffer. For pyridine, the value of K_b is 1.8×10^{-9}.
a) 4.3–6.3 b) 7.7–9.7 c) 10.3–12.3 d) 1.4–3.4 e) 9.1–11.1

65. A buffer contains equal concentrations of a weak acid, HA, and its conjuate base, A^-. If the value of K_a for HA is 1.0×10^{-9}, what is the pH of the buffer?
a) 9.0 b) 5.0 c) 1.0 d) 13.0 e) 7.0

66. For NH_3, $pK_b = 4.74$. What is the pH of an aqueous buffer solution that is 0.050 M $NH_3(aq)$ and 0.20 M $NH_4Cl(aq)$?
a) 4.14 b) 8.66 c) 9.26 d) 9.86 e) 5.34

67. For HF, $pK_a = 3.45$. What is the pH of an aqueous buffer solution that is 0.100 M HF(aq) and 0.300 M KF(aq)?
a) 11.03 b) 2.97 c) 3.93 d) 10.07 e) 3.45

68. For pyridine, $pK_b = 8.75$. What is the pH of an aqueous buffer solution that is 0.300 M $C_5H_5N(aq)$ and 0.500 M $C_5H_5NHCl(aq)$?
a) 5.03 b) 8.53 c) 5.47 d) 8.97 e) 5.25

69. An aqueous solution of 0.050 M weak acid ($pK_a = 6.10$) is mixed with an equal volume of a 0.10 M aqueous solution of its sodium salt. What is the pH of the final mixture?
a) 6.40 b) 5.80 c) 8.20 d) 7.60 e) 6.10

70. Which one of the following aqueous solutions is a buffer with a pH greater than 7.0? For HCNO, $K_a = 2.2 \times 10^{-4}$ and for NH_3, $K_b = 1.8 \times 10^{-5}$.
a) 10 mL of 0.1 M $NH_3(aq)$ + 10 mL of 0.1 M HCl(aq)
b) 10 mL of 0.1 M HCNO(aq) + 5.0 mL of 0.1 M NaOH(aq)
c) 10 mL of 0.1 M HCNO(aq) + 10 mL of 0.1 M NaOH(aq)
d) 10 mL of 0.1 M $NH_3(aq)$ + 5.0 mL of 0.1 M HCl(aq)
e) 10 mL of 0.1 M $NH_3(aq)$ + 10 mL of 0.1 M HCNO(aq)

71. Which one of the following aqueous solutions is a buffer with a pH less than 7.0? For HCNO, $K_a = 2.2 \times 10^{-4}$ and for NH_3, $K_b = 1.8 \times 10^{-5}$.
a) 10 mL of 0.1 M HCNO(aq) + 10 mL of 0.1 M $NH_3(aq)$
b) 10 mL of 0.1 M $NH_3(aq)$ + 10 mL of 0.1 M HCl(aq)
c) 10 mL of 0.1 M $NH_3(aq)$ + 10 mL of 0.1 M $NH_4Cl(aq)$
d) 10 mL of 0.1 M HCNO(aq) + 10 mL of 0.1 M NaCNO(aq)
e) 10 mL of 0.1 M HCNO(aq) + 10 mL of 0.1 M NaOH(aq)

72. A commercial base solution is sold under the label "ammonium hydroxide". This label implies that large concentrations of $NH_4^+(aq)$ and $OH^-(aq)$ exist simultaneously in the solution.
(a) Write the net ionic equation for the reaction of ammonia with water and find the value of the equilibrium constant.
(b) Calculate the equilibrium constant for the reaction of $NH_4^+(aq)$ with $OH^-(aq)$.
(c) What would be a better label for this commercial base solution?

73. What is the K_{sp} expression for mercury(I) chloride?
a) $[Hg^{2+}][Cl^-]^2$ b) $[Hg^+][Cl^-]$ c) $[Hg_2^{2+}][Cl^-]^2$ d) $[Hg^+][Cl^-]/[HgCl]$ e) $[Hg_2^+][Cl^-]$

74. If the solubility of lead(II) bromide in water is 0.027 M at 25°C, calculate the solubility product.
a) 7.3×10^{-4} b) 2.0×10^{-5} c) 4.9×10^{-6} d) 2.9×10^{-3} e) 7.9×10^{-5}

75. If the value of the solubility product for AgBr is 5.0×10^{-13} at 25°C, calculate the solubility of AgBr(s) in water.
a) 5.0×10^{-7} b) 7.1×10^{-7} c) 2.5×10^{-13} d) 2.5×10^{-6} e) 1.4×10^{-6}

76. The relationship between the molar solubility in water, s, and K_{sp} for the ionic solid $Fe(OH)_2$ is
a) $K_{sp} = s^3$ b) $K_{sp} = 4s^3$ c) $K_{sp} = 2s^2$ d) $K_{sp} = s$ e) $K_{sp} = s^2$

77. If the value of K_{sp} for Ag_2SO_4 is 1.7×10^{-5} at 25°C, the solubility of Ag_2SO_4 in 2.0 M Na_2SO_4(aq) is
a) 2.0 M b) 2.1×10^{-6} M c) 4.1×10^{-3} M d) 1.7×10^{-5} M e) 1.5×10^{-3} M

78. Predict what will occur if 50.0 mL of 1.0 M $AgNO_3$(aq) is mixed with 50.0 mL of 0.010 M $NaBrO_3$(aq).The value of K_{sp} for $AgBrO_3$ is 5.8×10^{-5}.
a) The value of K_{sp} increases by a factor of 43.
b) No precipitation occurs.
c) The value of K_{sp} decreases by a factor of 43.
d) $AgBrO_3$(s) will precipitate spontaneously.
e) $NaNO_3$(s) will precipitate spontaneously.

79. If equal volumes of 0.1 M $Pb(NO_3)_2$(aq) and 0.2 M KI(aq) are mixed, what reaction, if any, occurs? The value of K_{sp} for PbI_2 is 1.4×10^{-8}.
a) K_{sp} changes to 5×10^{-4}
b) the solution turns purple due to formation of I_2
c) no reaction occurs
d) KNO_3(s) precipitates
e) PbI_2(s) precipitates

80. Which of the following water–insoluble salts is much more soluble in 1.0 M $HClO_4$(aq)?
a) PbI_2 b) Hg_2Br_2 c) $PbCO_3$ d) AgI e) AgCl

Chapter 15: Salts in Water
Form B

1. Which one of the following salts gives an acidic aqueous solution?
a) $Ca(ClO_4)_2$ b) $Ni(ClO_4)_2$ c) $Mg(ClO_4)_2$ d) $LiClO_4$ e) $NaClO_4$

2. Which one of the following salts gives an acidic aqueous solution?
a) $FeCl_3$ b) NaCN c) $Mg(ClO_4)_2$ d) $NaNO_3$ e) KNO_2

3. Which one of the following salts gives an acidic aqueous solution?
a) $AgClO_4$ b) $C_6H_5NH_3Cl$ c) KCN d) Na_2CO_3 e) Na_2HPO_4

4. Which one of the following salts gives an acidic aqueous solution?
a) C_5H_5NHCl b) KCN c) $AgClO_4$ d) $MgCl_2$ e) $NaHCO_2$

5. Which one of the following salts gives an acidic aqueous solution?
a) $NaHCO_3$ b) K_2S c) NH_4ClO_4 d) NaHS e) Na_2SO_4

6. Which one of the following salts gives an acidic aqueous solution?
a) $(NH_4)_2CO_3$ b) NH_4NO_2 c) NH_4CN d) $(NH_4)_3PO_4$ e) $(NH_4)_2S$

7. Which one of the following salts gives a basic aqueous solution?
a) $(CH_3)NH_3Cl$ b) NH_4ClO_4 c) $NaHSO_4$ d) $NiCl_2$ e) Na_2HPO_4

8. Which one of the following salts gives a basic aqueous solution?
a) NH_4ClO_4 b) $NiCl_2$ c) $(CH_3)NH_3Cl$ d) $NaHSO_4$ e) $NaNO_2$

9. Which one of the following salts gives a basic aqueous solution?
a) $NaHCO_3$ b) $Al(NO_3)_3$ c) $(CH_3)_3NHCl$ d) $Mg(ClO_4)_2$ e) $NaHSO_4$

10. Which one of the following salts gives a basic aqueous solution?
a) $NaC_6H_5CO_2$ b) $Mg(ClO_4)_2$ c) $Al(NO_3)_3$ d) $(CH_3)_3NHCl$ e) $NaHSO_4$

11. Which one of the following salts gives a basic aqueous solution?
a) KF b) RbCl c) $Ca(ClO_4)_2$ d) NH_4ClO_4 e) $CrCl_3$

12. Which one of the following salts gives a basic aqueous solution?
a) NH_4F b) NH_4HCO_2 c) NH_4Cl d) NH_4CN e) NH_4NO_2

13. Which one of the following gives a neutral aqueous solution?
a) $NaNO_3$ b) $NaNO_2$ c) $Cu(ClO_4)_2$ d) KF e) $C_5H_5NH(ClO_4)$

14. Which one of the following gives a neutral aqueous solution?
a) $Cu(ClO_4)_2$ b) $NaNO_2$ c) $AgNO_3$ d) KF e) $C_5H_5NH(ClO_4)$

15. Which one of the following 0.10 M aqueous solutions has the highest pH?
a) KF b) HNO_2 c) NaI d) NH_4I e) HI

16. Which one of the following 0.10 M aqueous solutions has the highest pH?
a) RbOH b) $KHCO_2$ c) NaI d) NH_4I e) HI

17. Which one of the following 0.10 M aqueous solutions has the lowest pH?
a) HBr b) HF c) KNO_2 d) $Al(ClO_4)_3$ e) NH_4Br

18. Which one of the following 0.10 M aqueous solutions has the lowest pH?
a) $Ca(ClO_4)_2$ b) KNO_2 c) $NaCH_3CO_2$ d) $Fe(ClO_4)_2$ e) $NaClO_4$

19. For an aqueous solution labeled " 0.10 M ammonium chloride,"
a) the pH is greater than 7. b) the pH = 13. c) the pH is less than 7. d) the pH = 7. e) the pH = 1.

20. For an aqueous solution labeled "0.10 M $Fe(ClO_4)_3$,"
a) the pH is less than 7. b) the pH = 13. c) the pH = 7. d) the pH = 1. e) the pH is greater than 7.

21. Which of the following pairs of ions **cannot** exist in large concentrations **simultaneously** in aqueous solution?
a) H_3O^+ and ClO_4^- b) Ba^+ and OH^- c) Mg^{2+} and $CH_3CO_2^-$ d) NH_4^+ and CN^- e) H_3O^+ and F^-

22. Which of the following pairs of ions **can** exist in large concentrations **simultaneously** in aqueous solution?
a) H_3O^+ and $CH_3CO_2^-$
b) Mg^{2+} and $CH_3CO_2^-$
c) H_3O^+ and CN^-
d) $CH_3NH_3^+$ and OH^-
e) NH_4^+ and OH^-

23. Calculate the pH of a 0.10 M solution of aqueous sodium acetate. The value of K_a for acetic acid is 1.8×10^{-5}.
a) 8.87 b) 5.13 c) 7.00 d) 2.87 e) 9.26

24. Calculate the pH of a 1.0 M solution of aqueous sodium fluoride. The value of K_a for HF is 3.5×10^{-4}.
a) 8.73 b) 5.27 c) 7.00 d) 1.73 e) 12.27

25. Calculate the pH of a 1.0 M solution of aqueous ammonium bromide. The value of K_b for NH_3 is 1.8×10^{-5}.
a) 4.63 b) 9.37 c) 9.26 d) 4.74 e) 7.00

26. Calculate the pH of a 0.10 M solution of aqueous $Ni(ClO_4)_2$. The value of K_a for $Fe(ClO_4)_2$ is 9.3×10^{-10}.
a) 5.02 b) 8.98 c) 5.48 d) 4.97 e) 7.00

27. Calculate the pH of a 1.0 M solution of aqueous pyridinium chloride. The value of K_b for pyridine is 1.8×10^{-9}.
a) 4.37 b) 9.63 c) 7.00 d) 5.25 e) 8.74

28. Consider a solution which is 0.0100 M $Na_2CO_3(aq)$. Which of the following is true? For H_2CO_3, $K_{a1} = 4.3 \times 10^{-7}$ and $K_{a2} = 5.6 \times 10^{-11}$.
a) To obtain the OH^- concentration, the quadratic equation must be solved because the amount of CO_3^{2-} that reacts cannot be neglected.
b) The pH of this solution is about 5.
c) The pH of this solution is about 7.
d) Because the equilibrium constants are so small, the amount of CO_3^{2-} that reacts can be neglected with respect to 0.0100 M.
e) The OH^- of this solution is about 0.0100 M.

29. The pH of 0.10 M aqueous sodium cyanide is 11.15. What is the value of K_a for HCN?
a) 5.0×10^{-9} b) 2.2×10^{-12} c) 2.0×10^{-5} d) 5.0×10^{-10} e) 2.0×10^{-6}

30. The pH of 0.020 M $NiCl_2(aq)$ is 5.37. What is the value of K_a for $Ni(OH_2)_6^{2+}$?
a) 1.8×10^{-11} b) 9.1×10^{-10} c) 1.1×10^{-5} d) 5.6×10^{-4} e) 4.6×10^{-10}

31. The pH of 0.30 M $NH_4Cl(aq)$ is 4.89. What is the value of K_b for ammonia?
a) 1.8×10^{-5} b) 1.3×10^{-5} c) 6.0×10^{-5} d) 1.7×10^{-10} e) 5.5×10^{-10}

32. Draw a titration curve for the titration of 0.10 M aniline(aq) with 0.10 M HCl(aq). Compare this curve to that for the titration of 0.10 M $NH_3(aq)$ with 0.10 M HCl(aq) and explain any differences. The stoichiometric point and the point representing the pK_a of aniline should be exact.

33. Arrange the following 0.10 M aqueous solutions in order of increasing pH, 1---->14.
RbOH, $(CH_3)_3NHCl$, KBr, HNO_3, KNO_2, HOCN, $NaHSO_4$

34. The curve below corresponds to the titration of

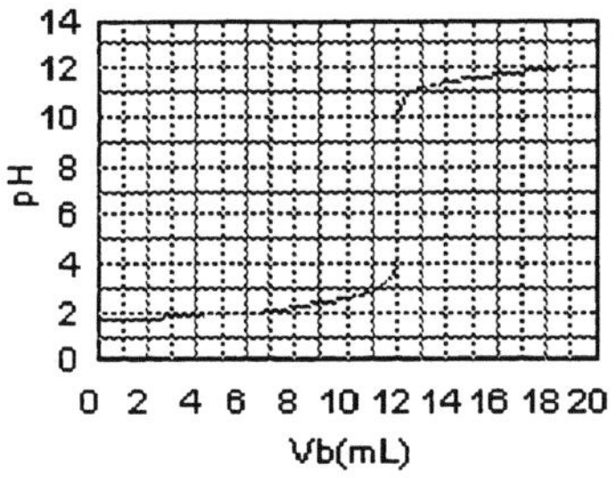

a) 0.10 M HCl(aq) with 0.10 M KOH(aq).
b) 0.10 M HClO(aq) (pK_a = 7.5) with 0.10 M KOH(aq).
c) 0.10 M H_2SO_4(aq) (pK_{a2} = 1.9) with 0.10 M KOH(aq).
d) 0.010 M $HClO_4$(aq) with 0.10 M KOH(aq).
e) 0.10 M $HClO_2$(aq) (pK_a = 2.0) with 0.10 M KOH(aq).

35. A buffer solution is 0.282 M $(CH_3)_3N$(aq) and 0.400 M $(CH_3)_3NHCl$(aq). Calculate the pH of the solution after the addition of 0.15 moles of NaOH. The $K_b = 6.5 \times 10^{-5}$.

36. For the titration of 25.0 mL of 0.100 M C_5H_5N(aq), pyridine, with 0.100 M HCl(aq), the main species in solution after the addition of 12.5 mL of acid are
a) C_5H_5N(aq), OH^-(aq), and Cl^-(aq).
b) $C_5H_5NH^+$(aq) and Cl^-(aq).
c) C_5H_5N(aq), $C_5H_5NH^+$(aq), and Cl^-(aq).
d) C_5H_5N(aq), H^+(aq), and Cl^-(aq).
e) C_5H_5N(aq).

37. At the stoichiometric point in the titration of 0.130 M HCOOH(aq) with 0.130 M KOH(aq),
a) the pH is less than 7.
b) the pH is greater than 7.
c) $[HCO_2^-]$ = 0.130 M.
d) the pH is 7.0.
e) [HCOOH] = 0.0650 M.

38. At the stoichiometric point in the titration of 0.260 M CH_3NH_2(aq) with 0.260 M HCl(aq),
a) the pH is less than 7.
b) $[CH_3NH_3^+]$ = 0.260 M
c) $[CH_3NH_2]$ = 0.130 M.
d) the pH is greater than 7.
e) the pH is 7.0.

39. What is the pH at the stoichiometric point for the titration of 0.400 M CH_3COOH(aq) with 0.400 M KOH(aq)? The value of K_a for acetic acid is 1.8×10^{-5}.
a) 9.02 b) 4.98 c) 9.26 d) 8.89 e) 7.00

40. The curve for the titration of 50.0 mL of 0.0200 M $C_6H_5COOH(aq)$ with 0.100 M NaOH(aq) is given below. Estimate the pK_a of benzoic acid.

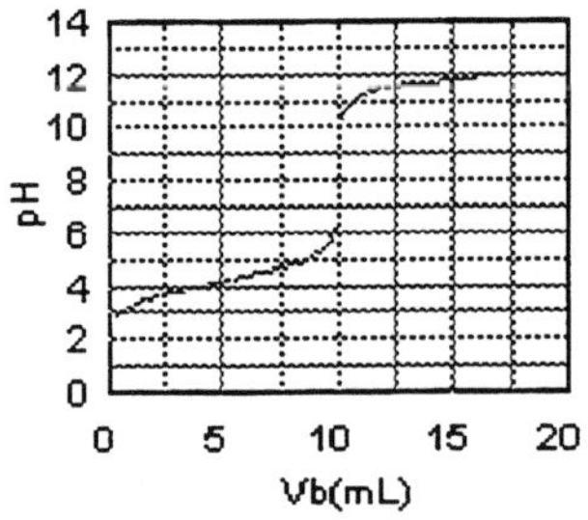

a) 4.2 b) 3.0 c) 8.0 d) 12.0 e) 3.8

41. The curve for the titration of 50.0 mL of 0.020 M $C_6H_5COOH(aq)$ with 0.100 M NaOH(aq) is given below. The main species in solution at the stoichiometric point are

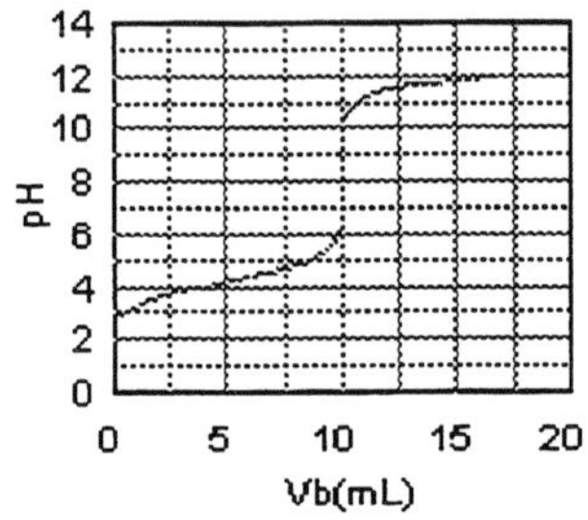

a) $C_6H_5COOH(aq)$
b) $C_6H_5CO_2^-(aq)$, $OH^-(aq)$, $Na^+(aq)$
c) $C_6H_5COOH(aq)$, $OH^-(aq)$, $Na^+(aq)$
d) $C_6H_5COOH(aq)$, $C_6H_5CO_2^-(aq)$, $Na^+(aq)$
e) $C_6H_5CO_2^-(aq)$, $Na^+(aq)$

42. What is the pH at the stoichiometric point for the titration of 0.10 M $CH_3NH_2(aq)$ with 0.10 M $HClO_4(aq)$? For CH_3NH_2, $K_b = 3.6 \times 10^{-4}$.
a) 5.93 b) 2.37 c) 2.22 d) 5.78 e) 7.00

43. What is the pH at the half-stoichiometric point for the titration of 0.88 M $HNO_2(aq)$ with 0.10 M KOH(aq)? For HNO_2, $K_a = 4.3 \times 10^{-4}$.
a) 3.37 b) 1.86 c) 1.71 d) 7.00 e) 2.01

44. What is the pH at the half-stoichiometric point for the titration of 0.022 M $NH_3(aq)$ with 0.10 M $HNO_3(aq)$? For NH_3, $K_b = 1.8 \times 10^{-5}$.
a) 9.26 b) 4.74 c) 10.65 d) 7.60 e) 7.00

45. What is the pH at the half-stoichiometric point for the titration of 0.10 M morphine(aq) with 0.15 M HCl(aq)? For morphine, $K_b = 1.6 \times 10^{-6}$.
a) 8.20 b) 5.80 c) 10.60 d) 10.45 e) 7.00

46. What is the concentration of acetate ion at the stoichiometric point in the titration of 0.018 M $CH_3COOH(aq)$ with 0.072 M NaOH(aq)? For acetic acid, $K_a = 1.8 \times 10^{-5}$.
a) 0.072 M b) 0.036 M c) 0.014 M d) 0.018 M e) 0.054 M

47. For the titration of 50.0 mL of 0.020 M aqueous salicylic acid with 0.020 M KOH(aq), calculate the pH after the addition of 20.0 mL of KOH(aq). For salycylic acid, $pK_a = 2.97$.
a) 2.79 b) 3.34 c) 2.42 d) 2.97 e) 7.00

48. For the titration of 50.0 mL of 0.020 M aqueous salicylic acid with 0.020 M KOH(aq), calculate the pH after the addition of 51.0 mL of KOH(aq). For salycylic acid, $pK_a = 2.97$.
a) 10.30 b) 12.31 c) 10.60 d) 12.02 e) 7.00

49. The curve for the titration of 25.0 mL of 0.100 M $C_5H_5N(aq)$, pyridine, with 0.100 M HCl(aq) is given below. Which indicator should be used for this titration?

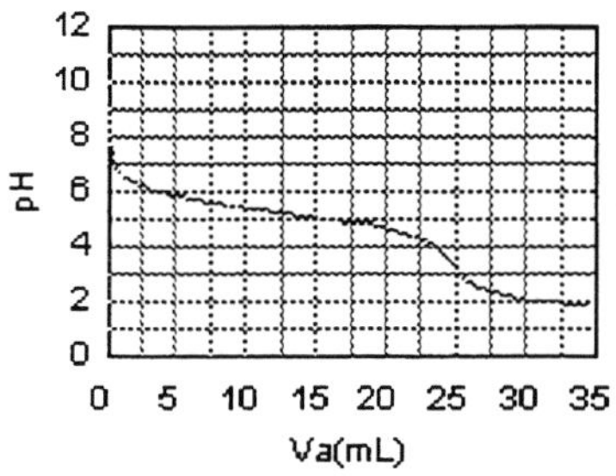

a) thymol blue ($pK_{In} = 1.7$)
b) phenolphthalein ($pK_{In} = 9.4$)
c) methyl red ($pK_{In} = 5.0$)
d) bromothymol blue ($pK_{In} = 7.1$)
e) methyl orange ($pK_{In} = 3.4$)

50. The curve for the titration of 25.0 mL of 0.100 M C_5H_5N(aq), pyridine, with 0.100 M HCl(aq) is given below. At 12.5 mL of added HCl(aq), the pH is approximately 5.2. This point corresponds to

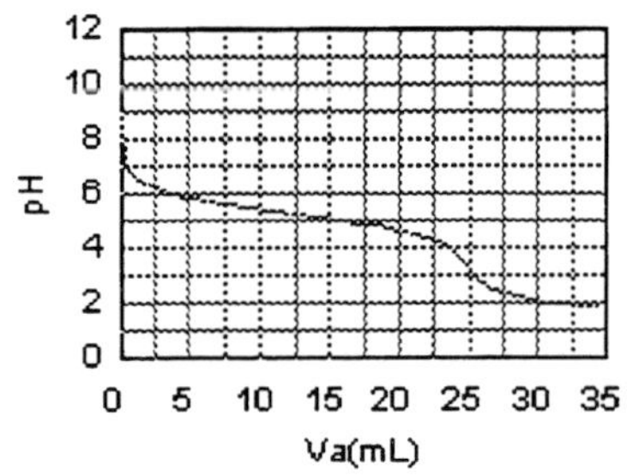

a) the value of the pK_a for the conjugate acid $C_5H_5NH^+$.
b) the value of the buffer capacity.
c) the end point.
d) the stoichiometric point.
e) the value of the pK_b for C_5H_5N.

51. For the titration of 25.0 mL of 0.100 M C_5H_5N(aq), pyridine, with 0.100 M HCl(aq), the main species in solution after the addition of 25.0 mL of acid are

a) C_5H_5N(aq), $C_5H_5NH^+$(aq) and Cl^-(aq).
b) $C_5H_5NH^+$(aq), and Cl^-(aq).
c) C_5H_5N(aq), OH^-(aq), and Cl^-(aq).
d) C_5H_5N(aq).
e) C_5H_5N(aq), H^+(aq), and Cl^-(aq).

52. Which of the following indicators would be most suitable for the titration of 0.10 M lactic acid with 0.10 M KOH(aq)? For lactic acid, pK_a = 3.08.

a) thymol blue (pK_{In} = 1.7)
b) phenol red (pK_{In} = 7.9)
c) alizarin yellow (pK_{In} = 11.2)
d) methyl orange (pK_{In} = 3.4)
e) bromophenol blue (pK_{In} = 3.9)

53. Which of the following indicators would be most suitable for the titration of 0.10 M C_5H_5NHCl(aq) with 0.10 M KOH(aq)?

a) methyl red (pK_{In} = 5.0)
b) thymol blue (pK_{In} = 1.7)
c) phenolphthalein (pK_{In} = 9.4)
d) bromothymol blue (pK_{In} = 7.1)
e) alizarin yellow (pK_{In} = 11.2)

54. The curve for the titration of 50.0 mL of 0.0200 M HClO(aq) with 0.100 M NaOH(aq) is given below. Which of the following indicators should be used for this titration?

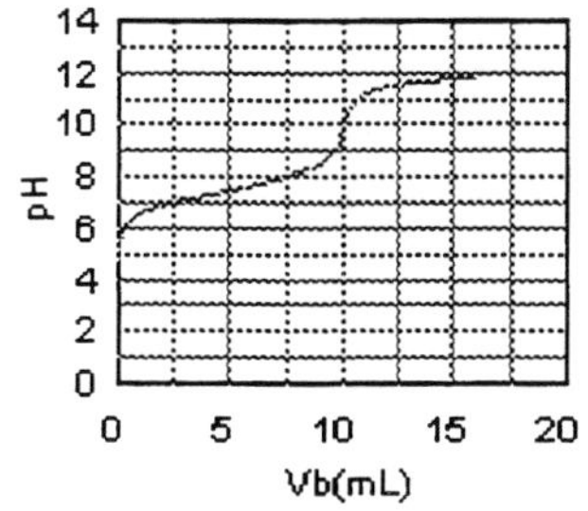

a) thymol blue (pK_{In} = 8.9)
b) litmus (pK_{In} = 6.5)
c) methyl orange (pK_{In} = 3.4)
d) thymolphthalein (pK_{In} = 10.0)
e) methyl red (pK_{In} = 5.0)

55. If a small amount of a strong acid is added to buffer made up of a weak acid, HA, and the sodium salt of its conjugate base, NaA, the pH of the buffer solution does not change appreciably because
a) the K_a of HA is changed.
b) the strong acid reacts with A^- to give H_2A^+.
c) the strong acid reacts with A^- to give HA.
d) the strong acid reacts with HA to give H_2A^+.
e) no reaction occurs.

56. The following 0.10 M aqueous solutions are available:
formic acid
sodium formate
perchloric acid
sodium hydroxide
Describe three ways to prepare a buffer, in each case using at least one different reagent, and write net ionic equations to show any reactions used.

57. What is the pH of an aqueous solution that is 0.015 M HF ($K_a = 3.5 \times 10^{-4}$) and 0.011 M NaF?
a) 3.32 b) 3.59 c) 5.27 d) 1.82 e) 3.46

58. What is the pH of an aqueous solution that is 0.20 M HNO_2 ($K_a = 4.3 \times 10^{-4}$) and 0.20 M $NaNO_2$?
a) 3.37 b) 10.63 c) 4.39 d) 2.37 e) 3.67

59. What is the pH of an aqueous solution that is 0.018 M $C_6H_5NH_2$ ($K_b = 4.3 \times 10^{-10}$) and 0.12 M $C_6H_5NH_3Cl$?
a) 3.81 b) 10.19 c) 5.46 d) 8.54 e) 4.63

60. Calculate the $[OH^-]$ in an aqueous solution which is 0.300 M NH_3 and 0.125 M NH_4Cl. The value of K_b for NH_3 is 1.8×10^{-5}.
a) 1.8×10^{-5} M b) 0.425 M c) 4.3×10^{-5} M d) 0.125 M e) 7.5×10^{-6} M

61. Calculate the $[H^+]$ in an aqueous solution which is 0.100 M HF and 0.0755 M NaF. The value of K_a for HF is 3.5×10^{-4}.
a) 4.6×10^{-4} M b) 3.5×10^{-4} M c) 2.6×10^{-4} M d) 0.0755 M e) 0.176 M

62. Which one of the following aqueous solutions is a buffer if the stoichiometric concentration of each component is 1.0 M?
a) HNO_3 + HCl b) HCl + NaCl c) HCN + NaCl d) NaCN + NaCl e) NaCN + HCN

63. Choose the effective pH range of a HF-NaF buffer. For HF, $K_a = 3.5 \times 10^{-4}$.
a) 2.5–4.5 b) 6.0–8.0 c) 5.0–7.0 d) 9.6–11.6 e) 0.7–2.7

64. Choose the effective pH range of a HNO_2-$NaNO_2$ buffer. For HNO_2, $K_a = 4.3 \times 10^{-4}$.
a) 2.4–4.4 b) 9.6–11.6 c) 0.7–2.7 d) 11.3–13.3 e) 6.0–7.0

65. A buffer contains equal concentrations of a weak base, B, and its conjugate acid, BH^+. If the value of K_b for B is 1.0×10^{-9}, what is the pH of the buffer?
a) 5.0 b) 9.0 c) 1.0 d) 13.0 e) 7.0

66. For NH_3, $pK_b = 4.74$. What is the pH of an aqueous buffer solution that is 0.50 M $NH_3(aq)$ and 0.20 M $NH_4Cl(aq)$?
a) 5.14 b) 9.66 c) 8.86 d) 4.34 e) 9.26

67. For HF, $pK_a = 3.45$. What is the pH of an aqueous buffer solution that is 0.400 M HF(aq) and 0.100 M KF(aq)?
a) 2.85 b) 4.05 c) 9.95 d) 11.15 e) 3.45

68. For pyridine, $pK_b = 8.75$. What is the pH of an aqueous buffer solution that is 0.500 M $C_5H_5N(aq)$ and 0.200 M $C_5H_5NHCl(aq)$?
a) 5.65 b) 9.15 c) 4.85 d) 8.35 e) 5.25

69. An aqueous solution of 0.10 M weak acid ($pK_a = 6.10$) is mixed with an equal volume of a 0.050 M aqueous solution of its sodium salt. What is the pH of the final mixture?
a) 5.80 b) 6.40 c) 8.20 d) 7.60 e) 6.10

70. Which one of the following aqueous solutions is a buffer with a pH greater than 7.0? For acetic acid, $K_a = 1.8 \times 10^{-5}$ and for NH_3, $K_b = 1.8 \times 10^{-5}$.
a) 10 mL of 0.1 M aqueous acetic acid + 10 mL of 0.1 M NaOH(aq)
b) 10 mL of 0.1 M aqueous acetic acid + 10 mL of 0.1 M aqueous sodium acetate
c) 10 mL of 0.1 M NH_3(aq) + 10 mL of 0.1 M HCl(aq)
d) 10 mL of 0.1 M NH_3(aq) + 10 mL of 0.1 M NH_4Cl(aq)
e) 10 mL of 0.1 M aqueous acetic acid + 10 mL of 0.1 M NH_3(aq)

71. Which one of the following aqueous solutions is a buffer with a pH less than 7.0? For acetic acid, $K_a = 1.8 \times 10^{-5}$ and for NH_3, $K_b = 1.8 \times 10^{-5}$.
a) 10 mL of 0.1 M NH_3(aq) + 10 mL of 0.1 M HCl(aq)
b) 10 mL of 0.1 M NH_3(aq) + 5.0 mL of 0.1 M HCl(aq)
c) 10 mL of 0.1 M aqueous acetic acid + 5.0 mL of 0.1 M NaOH(aq)
d) 10 mL of 0.1 M aqueous acetic acid + 10 mL of 0.1 M NH_3(aq)
e) 10 mL of 0.1 M aqueous acetic acid + 10 mL of 0.1 M NaOH(aq)

72. Write net ionic equations for the reactions that occur when $HClO_4$(aq) and KOH(aq) are added to the following solutions and calculate the equilibrium constant in each case.
(a) CH_3COOH-$NaCH_3CO_2$ buffer
(b) $C_2H_5NH_3Cl$-$C_2H_5NH_2$ buffer

73. What is the K_{sp} expression for iron(III) hydroxide?
a) $[Fe^{3+}][OH^-]^3$
b) $[Fe^{3+}][OH^-]^3/[Fe(OH)_3]$
c) $[Fe^{3+}][OH^-]/[Fe(OH)_3]$
d) $[Fe^{3+}]^3[OH^-]$
e) $3[Fe^{3+}][OH^-]$

74. If the solubility of calcium fluoride in water is 2.2×10^{-4} at 25°C, calculate the solubility product.
a) 4.8×10^{-8} b) 2.2×10^{-4} c) 2.1×10^{-14} d) 4.3×10^{-11} e) 1.1×10^{-11}

75. If the value of the solubility product for $Pb(IO_3)_2$ is 2.6×10^{-13} at 25°C, calculate the solubility of $Pb(IO_3)_2$(s) in water.
a) 4.0×10^{-5} b) 5.1×10^{-7} c) 6.0×10^{-7} d) 6.5×10^{-14} e) 5.1×10^{-6}

76. The relationship between the molar solubility in water, s, and K_{sp} for the ionic solid $Fe(OH)_3$ is
a) $K_{sp} = s$ b) $K_{sp} = 3s^4$ c) $K_{sp} = 27s^4$ d) $K_{sp} = 9s^4$ e) $K_{sp} = 3s^2$

77. If the value of K_{sp} for AgBr is 5.0×10^{-13} at 25°C, then the solubility of AgBr in 0.10 M KBr(aq) is
a) 0.10 M b) 5.0×10^{-12} M c) 2.2×10^{-6} M d) 5.0×10^{-13} M e) 7.0×10^{-7} M

78. Predict what will occur if 0.010 moles of the soluble salt $Cu(ClO_4)_2$ are added to a 0.0010 M $NaIO_3$(aq) solution. The value of K_{sp} for $Cu(IO_3)_2$ is 7.4×10^{-8}.

a) The value of K_{sp} increases by a factor of 0.14.
b) $NaClO_4$(s) will precipitate spontaneously.
c) $Cu(IO_3)_2$(s) will precipitate spontaneously.
d) The value of K_{sp} decreases by a factor of 0.14.
e) No precipitation will occur.

79. If equal volumes of 0.004 M $Pb(NO_3)_2$(aq) and 0.004 M KI(aq) are mixed, what reaction, if any, occurs? The value of K_{sp} for PbI_2 is 1.4×10^{-8}.

a) no reaction occurs
b) K_{sp} changes to 9×10^{-9}
c) the solution turns purple due to formation of I_2
d) KNO_3(s) precipitates
e) PbI_2(s) precipitates

80. Silver bromide is most soluble in

a) 0.10 M $AgNO_3$(aq)
b) dilute HNO_3(aq)
c) pure H_2O(l)
d) dilute NH_3(aq)
e) 0.10 M NaCl(aq)

Chapter 16: Energy in Transition: Thermodynamics
Form A

1. Calculate ΔU for a system that does 300 kJ of work on the surroundings when 150 kJ of heat are absorbed by the system.
a) +450 kJ b) 0 kJ c) –450 kJ d) –150 kJ e) +150 kJ

2. Calculate ΔU for a system that absorbs 325 kJ of heat and has 65 kJ of work done on the system.
a) –260 kJ b) +390 kJ c) 0 kJ d) –390 kJ e) +260 kJ

3. Calculate ΔU for a system that loses 225 kJ of heat and has 150 kJ of work done on the sysyem.
a) +375 kJ b) –375 kJ c) +75 kJ d) 0 kJ e) –75 kJ

4. Calculate ΔU for a system that loses 325 kJ of heat while doing 200 kJ of work on the surroundings.
a) +125 kJ b) –525 kJ c) 0 kJ d) –125 kJ e) +525 kJ

5. In a certain exothermic reaction at constant pressure, $\Delta H = -75$ kJ and 35 kJ of work was required to make room for products. What is ΔU?
a) –40 kJ b) 0 kJ c) –110 kJ d) +110 kJ e) +40 kJ

6. In a certain endothermic reaction at constant pressure, $\Delta H = +175$ kJ and 45 kJ of work was required to make room for products. What is ΔU?
a) +220 kJ b) –220 kJ c) –130 kJ d) 0 kJ e) +130 kJ

7. For a certain reaction at constant pressure, $\Delta U = -125$ kJ and 22 kJ of expansion work is done by the system. What is ΔH for this process?
a) –125 kJ b) +147 kJ c) +103 kJ d) –147 kJ e) –103 kJ

8. For a certain reaction at constant pressure, $\Delta U = +45$ kJ and 14 kJ of expansion work is done by the system. What is ΔH for this process?
a) –59 kJ b) +59 kJ c) +31 kJ d) +45 kJ e) –31 kJ

9. Calculate the work needed to make room for products in the combustion of 1 mole of CH_4(g) to carbon dioxide and water vapor at STP (1 L · atm = 101 J).
a) –4.52 kJ b) –2.26 kJ c) –6.79 kJ d) –11.3 kJ e) no work is needed

10. Calculate the work needed to make room for products in the combustion of 1 mole of C_3H_8(g) to carbon dioxide and water vapor where reactants and products are brought to STP (1 L · atm = 101 J).
a) no work is needed b) –15.8 kJ c) –13.6 kJ d) –4.52 kJ e) –2.26 kJ

11. Which of the following reactions leads to a decrease in entropy of the system?
a) $2H_2O(l) \rightarrow 2H_2(g) + O_2(g)$
b) $C(s) + H_2O(g) \rightarrow CO(g) + H_2(g)$
c) $PCl_5(s) \rightarrow PCl_3(l) + Cl_2(g)$
d) $C_6H_{12}O_6(s) + 6O_2(g) \rightarrow 6CO_2(g) + 6H_2O(l)$
e) $N_2(g) + 3H_2(g) \rightarrow 2NH_3(g)$

12. Which of the following reactions has the largest positive molar entropy change?
a) $6Li(s) + N_2(g) \rightarrow 2Li_3N(s)$
b) $2Na(l) + H_2(g) \rightarrow 2NaH(s)$
c) $NH_3(g) + HCl(g) \rightarrow NH_4Cl(s)$
d) $CaSO_4{\cdot}2H_2O(s) \rightarrow CaSO_4(s) + 2H_2O(g)$
e) $2Mg(s) + CO_2(g) \rightarrow 2MgO(s) + C(s)$

13. Which of the following has the largest entropy at 298 K?
a) $I_2(g)$ b) $Cl_2(g)$ c) $Br_2(l)$ d) $F_2(g)$ e) $Br_2(g)$

14. Which of the following has the largest entropy at 298 K?
a) Xe(g) b) Ne(g) c) Kr(g) d) He(g) e) Ar(g)

15. Which of the following has the largest entropy at 298 K?
a) $CH_3Cl(l)$ b) $CH_4(g)$ c) $CCl_4(l)$ d) $CCl_4(g)$ e) $CH_3Cl(g)$

16. All of the following changes give a positive ΔS_r° except
a) $H_2O(s) \rightarrow H_2O(l)$
b) $H_2O(l, 20°C) \rightarrow H_2O(l, 50°C)$
c) $N_2(g) + 3H_2(g) \rightarrow 2NH_3(g)$
d) $H_2O(l) \rightarrow H_2O(g)$
e) $Na(l) \rightarrow Na(g)$

17. Which of the following processes would have a positive ΔS_r°?
a) He(g, 1 atm) $\rightarrow$ He(g, 10 atm)
b) $2NO(g) \rightarrow N_2O_4(g)$
c) $H_2(g) + I_2(s) \rightarrow 2HI(g)$
d) $NH_3(g) \rightarrow NH_3(aq)$
e) $Pb(s) + Br_2(l) \rightarrow PbBr_2(s)$

18. When 1 mole of white tin changes to gray tin at 13.0°C, the change in entropy is $-7.5\ J \cdot K^{-1}$. This means that
a) the change from white to gray tin is predicted to be spontaneous on the basis of the entropy change.
b) gray tin has an entropy of $-7.5\ J \cdot K^{-1}$.
c) gray tin is more ordered than white tin.
d) gray tin has a higher molar entropy than white tin.
e) white tin is more ordered than gray tin.

19. Which of the following reactions has the largest positive molar entropy change?
a) $N_2(g) + 3H_2(g) \rightarrow 2NH_3(g)$
b) $KClO_4(s) + 4C(s) \rightarrow KCl(s) + 4CO(g)$
c) $CH_4(g) + 2O_2(g) \rightarrow CO_2(g) + 2H_2O(g)$
d) $PCl_5(g) \rightarrow PCl_3(g) + Cl_2(g)$
e) $H_2O(s) \rightarrow H_2O(g)$

20. Which of the following has the smallest molar entropy at 298 K?
a) S(s, rhombic) b) C(s, diamomd) c) Sn(s, gray) d) Sn(s, white) e) $CaCO_3(s)$

21. List the following in order of decreasing molar entropy at 298 K.
HCl(g), $Cl_2(g)$, HCl(aq), Cl(g)
a) HCl(aq) > Cl(g) > HCl(g) > $Cl_2(g)$
b) $Cl_2(g)$ > Cl(g) > HCl(g) > HCl(aq)
c) HCl(g) > HCl(aq) > $Cl_2(g)$ > Cl(g)
d) Cl(g) > $Cl_2(g)$ > HCl(g) > HCl(aq)
e) $Cl_2(g)$ > HCl(g) > Cl(g) > HCl(aq)

22. For the freezing of water at 0°C,
a) $\Delta S_{surroundings}$ is less than ΔS_{system}
b) $\Delta S_{surroundings}$ is greater than ΔS_{system}
c) $\Delta S_{surroundings} = \Delta S_{system}$
d) $(\Delta S_{surroundings} + \Delta S_{system})$ is less than zero
e) $\Delta S_{surroundings}$ is negative

23. If the enthalpy of vaporization of water at 100°C is 40.7 kJ · mol^{-1}, calculate $\Delta S°$ for vaporization of one mole of $H_2O(l)$ at 100°C.
a) 136 J · K^{-1} b) 109 J · K^{-1} c) 40.7 × 10^3 J · K^{-1} d) –109 J · K^{-1} e) –40.7 × 10^3 J · K^{-1}

24. If the enthalpy of fusion of water at its normal melting point is 6.00 kJ · mol^{-1}, calculate $\Delta S°$ for freezing 1 mole of water at this temperature.
a) –20.1 J · K^{-1} b) –6.00 × 10^3 J · K^{-1} c) +6.00 × 10^3 J · K^{-1} d) –22.0 J · K^{-1} e) 22.0 J · K^{-1}

25. The entropy of the universe
a) is always increasing.
b) has nothing to do with the universe.
c) remains constant.
d) is zero.
e) is also decreasing.

26. If $\Delta H° = -1202$ kJ and $\Delta S° = -217$ J · K^{-1} for the combustion of 2 moles of magnesium, the change in entropy of the surroundings at 298 K is
a) 217 J · K^{-1} b) –217 J K^{-1} c) 4.03 × 10^3 J · K^{-1} d) –4.03 × 10^3 J · K^{-1} e) 5.54 × 10^3 J · K^{-1}

27. The normal boiling point of benzene is 80°C. If the enthalpy of vaporization of benzene is 33.9 kJ · mol^{-1}, calculate the molar entropy of vaporization at its boiling point.
a) 96.0 J · K^{-1} b) 139 J · K^{-1} c) 235 J · K^{-1} d) 373 J · K^{-1} e) 423 J · K^{-1}

28. Estimate the normal boiling point of Br_2(l), given the molar enthalpy and molar entropy of vaporization of 30.7 kJ · mol^{-1} and 93.0 J · K^{-1} · mol^{-1}, respectively.
a) 57°C b) 303°C c) 100°C d) 30°C e) 330°C

29. Estimate the normal boiling point of ethanol, C_2H_5OH, given the molar enthalpy and entropy of vaporization of 38.7 kJ · mol^{-1} and 110 J · K^{-1} · mol^{-1}, respectively.
a) 100°C b) 78.8°C c) 11.0°C d) 352°C e) 284°C

30. The enthalpy of vaporization of ethanol, C_2H_5OH, is 38.7 kJ · mol^{-1} at its normal boiling point, 78°C. Calculate $\Delta S°$ for vaporization of 1 mole of ethanol at its normal boiling point.
a) 103 J · K^{-1} b) 292 J · K^{-1} c) 496 J · K^{-1} d) 110 J · K^{-1} e) 142 J · K^{-1}

31. Consider the following reaction

$$N_2H_4(l) + 2H_2O_2(l) \rightarrow N_2(g) + 4H_2O(g)$$

If $\Delta H_r° = -530$ kJ at 298 K, then
a) the reaction will only be spontaneous at very low temperatures.
b) the reaction is spontaneous at all temperatures.
c) the reaction is not spontaneous at any temperature.
d) the reaction will only be spontaneous at very high temperatures.
e) to determine if the reaction will be spontaneous, $\Delta S_r°$ must be given.

32. Which of the following conditions will result in a spontaneous reaction at all temperatures?
a) $\Delta H < 0, \Delta S < 0$ b) $\Delta H > 0, \Delta S < 0$ c) $\Delta H < 0, \Delta S > 0$ d) $\Delta H > 0, \Delta S > 0$ e) $\Delta H > 0, \Delta S = 0$

33. Consider the following reaction.

$$PCl_5(g) \rightarrow PCl_3(g) + Cl_2(g)$$

Which statement is true for this reaction?
a) $\Delta S > 0$ b) $\Delta S = 0$ c) $\Delta G = \Delta H + T\Delta S$ d) $S_m° = 0$ for Cl_2(g) e) $\Delta S < 0$

34. For the reaction

$$2SO_3(g) \rightarrow 2SO_2(g) + O_2(g)$$

$\Delta H_r° = 197.8$ kJ at 298 K. This reaction will
a) be driven by the enthalpy.
b) not be spontaneous at high temperatures.
c) not be spontaneous at any temperature.
d) be spontaneous at high temperatures.
e) be spontaneous at all temperatures.

35. Consider the following reaction.

$$CaCO_3(s) \rightarrow CaO(s) + CO_2(g)$$

If this reaction is endothermic, which one of the following statements is true at 298 K and 1 atm?
a) $\Delta G°_r$ is positive at all temperatures.
b) $\Delta G_r°$ is negative at low temperatures.
c) $\Delta G_r°$ is negative at high temperatures.
d) $\Delta S_r°$ is negative.
e) $\Delta H_r°$ is negative.

36. Which one of the following statements is true?
a) Spontaneous reactions always have $\Delta H_r° < 0$.
b) Reactions with positive values of $\Delta S_r°$ always become spontaneous at high temperatures.
c) All of these statements are false.
d) Spontaneous reactions always have $\Delta H_r° > 0$.
e) Spontaneous reactions always have $\Delta G_r° > 0$.

37. The vapor pressure of water at 25°C is 24 Torr. ΔG for the reaction

$$H_2O(l) \rightarrow H_2O(g,\ 24\ \text{Torr})$$

at 25°C is therefore
a) 0 b) 8.56 kJ c) –8.56 kJ d) 20.7 kJ e) –7.87 kJ

38. The vapor pressure of water at 25°C is 0.0316 atm. $\Delta G_r°$ for the reaction

$$H_2O(l) \rightarrow H_2O(g,\ 24\ \text{Torr})$$

at 25°C is therefore
a) 0
b) –(8.314)(298)ln(0.0316)
c) +(8.314)(298)ln(0.0316)
d) –(8.314)(25)ln(0.0316)
e) +(8.314)(25)ln(0.0316)

39. Calculate $\Delta G_r°$ for the decomposition of ammonium nitrate at 298 K.

$$NH_4NO_3(s) \rightarrow N_2O(g) + 2H_2O(l)$$

	$NH_4NO_3(s)$	$N_2O(g)$	$H_2O(l)$
$\Delta H_f°$, kJ · mol^{-1}	–365.56	82.05	–285.83
$S_m°$, J · K^{-1} · mol^{-1}	151.08	219.85	69.91

a) –186.21 kJ b) –332.64 kJ c) +99.62 kJ d) –124.05 kJ e) +120.45 kJ

40. Calculate $\Delta G_r°$ for the decomposition of mercury(II) oxide at 298 K.

$$2HgO(s) \rightarrow 2Hg(l) + O_2(g)$$

	$HgO(s)$	$Hg(l)$	$O_2(g)$
$\Delta H_f°$, kJ · mol^{-1}	–90.83		
$S_m°$, J · K^{-1} · mol^{-1}	70.29	76.02	205.14

a) +246.2 kJ b) –117.1 kJ c) –246.2 kJ d) –64.5 kJ e) +117.1 kJ

41. For a reaction, $\Delta H_r° = -92$ kJ and $\Delta S_r° = -65$ J · K^{-1}. Calculate the value of $\Delta G_r°$ for this reaction at 25°C.
a) +19,300 kJ b) –111 kJ c) –157 kJ d) –73 kJ e) –85 kJ

42. For the following reaction

$$CuO(s) + H_2(g) \rightarrow Cu(s) + H_2O(g)$$

at 298 K, $H_r° = -87.0$ kJ and $\Delta S_r° = 47.0$ J · K^{-1}. Calculate $\Delta G_r°$ at 400 K.
a) –106 kJ b) –18.9 kJ c) –68.2 kJ d) +18.9 kJ e) +106 kJ

43. For the following reaction

$C_6H_6(g) + 3H_2(g) \rightarrow C_6H_{12}(g)$

at 298 K, $\Delta H_r° = -206$ kJ and $\Delta S_r° = -363$ J · K^{-1}. Calculate $\Delta G_r°$ at 500°C.

a) −24.5 kJ b) −97.8 kJ c) +74.6 kJ d) −487 kJ e) −157 kJ

44. The standard free energy of formation of $CS_2(l)$, a common solvent, is 65.27 kJ · mol^{-1} at 298 K. This means that at 298 K

a) $CS_2(l)$ is thermodynamically unstable.
b) $CS_2(l)$ is thermodynamically stable.
c) no catalyst can be found to decompose $CS_2(l)$ into its elements.
d) $CS_2(l)$ has a negative entropy.
e) $CS_2(l)$ will not spontaneously form C(s) + 2S(s).

45. Consider the following compounds and their standard free energies of formation:

(1)	(2)	(3)	(4)
$C_6H_6(l)$	$Ccl_4(l)$	$CS_2(l)$	$PCl_3(g)$
124 kJ · mol^{-1}	−65 kJ · mol^{-1}	65 kJ · mol^{-1}	−268 kJ · mol^{-1}

Which of these compounds is/are thermodynamically unstable?

a) (1) and (3) b) (2) and (4) c) (4) d) (1) e) (2)

46. For the reaction

$2NH_3(g) \rightarrow 3H_2(g) + N_2(g)$

$\Delta H_r° = 92.22$ kJ and $\Delta S_r° = 198.75$ J · K^{-1} at 298 K. With all reactants and products in their standard states, this reaction will be spontaneous

a) at no temperature.
b) at all temperatures.
c) at temperatures above 464 K.
d) at 273 K.
e) at temperatures below 463 K.

47. For the reaction

$2H_2(g) + CO(g) \rightarrow CH_3OH(l)$

$\Delta H_r° = -128.13$ kJ and $\Delta S_r° = -332.23$ J · K^{-1} at 298. Calculate $\Delta G_r°$ at 100°C.

a) −128.46 kJ b) −252.05 kJ c) −4.21 kJ d) −218.8 kJ e) −29.13 kJ

48. Consider the reactions below.

1. $C_6H_6(g) + 3H_2(g) \rightarrow C_6H_{12}(g)$, $\Delta H_r° = -206$ kJ, $\Delta S_r° = -363$ J · K^{-1}
2. $H_2S(g) \rightarrow H_2(g) + S(s)$, $\Delta H_r° = 20.6$ kJ, $\Delta S_r° = -43.3$ J · K^{-1}
3. $CaCO_3(s) \rightarrow CaO(s) + CO_2(g)$, $\Delta H_r° = 179$ kJ, $\Delta S_r° = 161$ J · K^{-1}
4. $C_6H_6(l) \rightarrow 6C(s) + 3H_2(g)$, $\Delta H_r° = -49$ kJ, $\Delta S_r° = 254$ J · K^{-1}

For which of these reactions would it be feasible to search for a catalyst to speed up the reaction at 298 K?

a) 1. and 4. b) all the reactions c) 3. only d) 2. and 3. e) 2. only

49. If ΔH_f^o and S_m^o for $CCl_4(l)$ and $CCl_4(g)$ are –135.4 and –103.0 kJ · mol^{-1}, and 215.4 and 308.7 J · K^{-1} · mol^{-1}, respectively, calculate the boiling point of carbon tetrachloride.
a) 260 K b) 347°C c) 61.4°C d) 629 K e) 70.0°C

50. Is it possible to use the following reaction to produce fluorine, (a) at 25°C; (b) at 250°C?
$2NaF(s) + Cl_2(g) \rightarrow 2NaCl(s) + F_2(g)$

	$2NaF(s)$	$Cl_2(g)$	$2NaCl(s)$	$F_2(g)$
ΔH_f^o, kJ · mol^{-1}	–573.65		–411.15	
S_m^o, J · K^{-1} · mol^{-1}	+51.46	+223.07	+72.13	+202.78

51. If the standard molar free energy of formation of NO(g) is 86.69 kJ · mol^{-1} at 298 K, calculate ΔG_r^o for the reaction below.
$N_2(g) + O_2(g) \rightarrow 2NO(g)$
a) –173.4 kJ b) +173.4 kJ c) +86.69 kJ d) –86.69 kJ e) +42.35 kJ

52. Copper(II) sulfate can be obtained in the anhydrous form, $CuSO_4(s)$, and as the blue pentahydrate, $CuSO_4{\cdot}5H_2O(s)$. If the anhydrous form is being considered as a potential dehydrating agent, determine whether it will spontaneously pick up water from the atmosphere at a temperature of 10°C and a vapor pressure of water of 9.98 Torr. Assume the product is the pentahydrate.

Data:
$CuSO_4(s)$: ΔH_f^o = –771 kJ · mol^{-1}, S° = 109 J · K^{-1} · mol^{-1}
$CuSO_4{\cdot}5H_2O(s)$: ΔH_f^o = –2280 kJ · mol^{-1}, S° = 300 J · K^{-1} · mol^{-1}
$H_2O(g)$: ΔH_f^o = –242 kJ · mol^{-1}, S° = 189 J · K^{-1} · mol^{-1}

53. Calculate ΔG_r at 298 K for the reaction
$C_2H_5OH(l) \rightarrow C_2H_5OH(g, 0.0263\ atm)$
given ΔG_r^o = 6.2 kJ at 298 K.
a) –2.8 kJ b) 2.8 kJ c) –15 kJ d) 6.2 kJ e) 15 kJ

54. Consider the following reaction
$CuSO_4(s) \rightarrow CuO(s) + SO_3(g)$
If ΔG^o = –14.6 kJ at 950°C for this reaction, calculate ΔG_r for an $SO_3(g)$ pressure of 20 atm at this temperature.
a) 30.5 kJ b) 45.1 kJ c) –45.1 kJ d) –14.6 kJ e) 15.9 kJ

55. If ΔG_r^o = 27.1 kJ at 25°C for the reaction
$CH_3COOH(aq) + H_2O(l) \leftrightarrow CH_3COO^-(aq) + H_3O^+(aq)$
calculate K_a for this reaction at 298 K.
a) 1.78×10^{-5} b) 9.89×10^{-1} c) 5.63×10^{4} d) 1.15×10^{-11} e) 1.01

56. Consider the following reaction

$NO(g) + \leftrightarrow O_2(g) \rightarrow NO_2(g)$

If $\Delta H_r^o = -56.52$ kJ and $\Delta S^o = -72.60\ J \cdot K^{-1}$ at 298 K, calculate the equilibrium constant for the reaction at 298 K.

a) 7.63×10^{-7} b) 1.31×10^{6} c) 8.08×10^{9} d) 660 e) 1.22×10^{14}

57. Consider the reaction

$2SO_3(g) \rightarrow 2SO_2(g) + O_2(g)$

If $\Delta H_r^o = 196$ kJ and $\Delta S_r^o = 190\ J \cdot K^{-1}$ for this reaction at 298 K, the value of the equilibrium constant will be greater than 1

a) at no temperature.
b) at temperatures below 1032 K.
c) at temperatures below 759°C.
d) at temperatures above 1032 K.
e) at all temperatures.

58. Consider the reaction

$N_2(g) + O_2(g) \rightarrow 2NO(g)$

If the standard molar free energy of formation of NO(g) at 298 K is 86.69 kJ · mol^{-1}, calculate the value of the equilibrium constant for this reaction at 298 K.

a) 4.06×10^{-31} b) 9.35×10^{-31} c) 6.37×10^{-16} d) 1.47×10^{-15} e) 1.57×10^{-31}

59. Calculate the equilibrium constant at 298 K for the reaction below.

$NH_3(g) + HCl(g) \rightarrow NH_4Cl(s)$

ΔG_f^o, kJ · mol^{-1} −16.45 −95.30 −202.87

a) 3.88×10^{19} b) 9.39×10^{15} c) 1.42×10^{55} d) 6.00×10^{36} e) 9.11×10

60. Consider the reaction

$2NO(g) \rightarrow N_2(g) + O_2(g)$

If the standard molar free energy of formation of NO(g) at 298 K is 86.69 kJ · mol^{-1}, calculate the value of the equilibrium constant for this reaction at 298 K.

a) 6.80×10^{14} b) 1.57×10^{15} c) 1.07×10^{30} d) 2.47×10^{30} e) 6.37×10^{30}

61. The vapor pressure of water at 25°C is 24 Torr. ΔG_r^o for the reaction

$H_2O(l) \rightarrow H_2O(g)$

at 25°C is therefore

a) −7.87 kJ b) −8.56 kJ c) 8.56 kJ d) 0 e) 20.7 kJ

62. The equilibrium constant for the reaction

$Hg(l) \leftrightarrow Hg(g)$

is 3.6×10^{-4} at 100°C. Calculate ΔG_r^o for this reaction.

a) −24.6 kJ b) 24.6 kJ c) 0 d) 6.59 kJ e) −6.59 kJ

63. If the enthalpy of vaporization of ethanol is 38.7 kJ · mol^{-1} at its normal boiling point of 78°C, determine $\Delta G_r°$ for the reversible vaporization of 1 mole of ethanol at 78°C and 1 atm.
a) 0 b) 110 kJ c) −110 kJ d) 38.7 kJ e) −38.7 kJ

64. Calculate the standard free energy of formation of $CHCl_3$(l) at 25°C given that $\Delta G_f°$ for $CHCl_3$(g) is −70.12 kJ · mol^{-1} and the vapor pressure of chloroform at 25°C is 0.500 atm.
a) −70.12 kJ · mol^{-1} b) −71.84 kJ · mol^{-1} c) 0 d) −1.72 kJ · mol^{-1} e) 68.40 kJ · mol^{-1}

65. Consider the reaction

$$NH_3(aq) + H_2O(l) \leftrightarrow NH_4^+(aq) + OH^-(aq)$$

If $K_c = 1.75 \times 10^{-5}$ at 25°C, calculate $\Delta G_r°$ for this reaction.
a) −27.1 kJ b) 2.27 kJ c) −2.27 kJ d) 6.47 kJ e) 27.1 kJ

66. If K_p for the reaction

$$N_2O_4(g) \leftrightarrow 2NO_2(g)$$

is 0.15 at 25°C, calculate $\Delta G_r°$ for this reaction.
a) 4.7 kJ b) −2.0 kJ c) −0.4 kJ d) 2.0 kJ e) −4.7 kJ

67. If the vapor pressure of Br_2(l) at 298 K is 217 Torr, calculate $\Delta G_r°$ for vaporization of 1 mole of Br_2(l) at 298 K.
a) −8.33 kJ b) −3.11 kJ c) 8.33 kJ d) 3.11 kJ e) 0

68. Consider the reaction

$$2CuBr_2(s) \rightarrow 2CuBr(s) + Br_2(g)$$

If the equilibrium vapor pressure of Br_2(g) is 1.43×10^{-5} Torr at 298 K, calculate ΔG_r at this temperature when Br_2(g) is produced at a pressure of 7.50×10^{-7} Torr.
a) 39.9 kJ b) −7.31 kJ c) −3.17 kJ d) −4.15 kJ e) 7.31 kJ

69. If $\Delta G_r = -46.9$ kJ for the reaction of 5 mL of 0.20 M $AgNO_3$(aq) with 5 mL of 0.60 M NaCl(aq) at 298 K, calculate the solubility product of AgCl at this temperature.
a) 7.2×10^{10} b) 3.5×10^{-21} c) 7.2×10^{-10} d) 1.8×10^{-10} e) 1.8×10^{10}

70. If K_{sp} of AgBr is 7.7×10^{-13} at 298 K, calculate ΔG_r for the reaction of 5 mL of 0.20 M $AgNO_3$(aq) with 5 mL of 0.20 M NaBr(aq) at 298 K.
a) −25.1 kJ b) −57.7 kJ c) 61.1 kJ d) −61.1 kJ e) 57.7 kJ

71. Calculate ΔG_r for the following reaction at 298 K.

$$Ag(NH_3)_2^+(aq, 0.100\ M) \rightarrow Ag^+(aq, 0.00100\ M) + 2NH_3(aq, 0.00200\ M)$$

$\Delta G_r° = -41.0$ kJ for this reaction at 298 K.
a) −42.2 kJ b) −1.2 kJ c) 14.2 kJ d) 83.2 kJ e) −83.2 kJ

72. Consider the following reaction at 298 K.

$HClO(aq) + H_2O(l) \leftrightarrow ClO^-(aq) + H_3O^+(aq) \quad K_c = 3.0 \times 10^{-8}$

Calculate ΔG_r when $[ClO^-] = [H_3O^+] = 1.0 \times 10^{-6}$ M and $[HClO] = 0.10$ M.

a) −62.6 kJ b) −23.0 kJ c) −19.8 kJ d) 42.8 kJ e) −42.8 kJ

73. If the K_{sp} of AgBr is 7.7×10^{-13} at 25°C, calculate the standard free energy of reaction for

$Ag^+(aq) + Br^-(aq) \rightarrow AgBr(s)$

a) −69.1 kJ b) +69.1 kJ c) −5.80 kJ d) +30.0 kJ e) −30.0 kJ

74. For the following process

$C_6H_6(l) \rightarrow C_6H_6(g)$

$\Delta H^\circ = 33.90\ kJ \cdot mol^{-1}$ and $\Delta S^\circ = 96.4\ J \cdot K^{-1} \cdot mol^{-1}$. Calculate the temperature at which the vapor pressure of benzene is 35 Torr.

75. The equilibrium constant for the reaction

$NH_4Cl(s) \leftrightarrow NH_3(g) + HCl(g)$

is 1.1×10^{-16} at 25°C. If the equilibrium constant is 6.5×10^{-2} at 300°C, which statement is correct?

a) The reaction is exothermic.
b) The reaction is not spontaneous at any temperature.
c) This is an enthalpy-driven reaction.
d) The reaction is spontaneous at all temperatures.
e) The reaction is spontaneous at high temperatures.

76. At the normal boiling point of $Cl_2(l)$, 238.5 K, the value of the standard enthalpy of vaporization is 20.4 $kJ \cdot mol^{-1}$. Assuming that $Cl_2(g)$ behaves as an ideal gas and that the molar volume of $Cl_2(l)$ is negligible compared to that of $Cl_2(g)$, calculate the values of q, w, ΔU°, ΔS°, and ΔG° for the reversible vaporization of 1 mole of $Cl_2(l)$ at 238.5 K and 1 atm.

77. Estimate the temperature at which oxygen reacts with Hg(l).

$2Hg(l) + O_2(g) \rightarrow 2HgO(s)$

	$2Hg(l)$	$O_2(g)$	$2HgO(s)$
ΔH_f°, $kJ \cdot mol^{-1}$			−90.83
S°, $J \cdot K^{-1} \cdot mol^{-1}$	76.02	205.14	70.29

78. Estimate the temperature at which mercury(II) oxide can be expected to decompose.

$2HgO(s) \rightarrow 2Hg(l) + O_2(g)$

	$2HgO(s)$	$2Hg(l)$	$O_2(g)$
ΔH_f°, $kJ \cdot mol^{-1}$			−90.83
S_m°, $J \cdot K^{-1} \cdot mol^{-1}$	70.29	76.02	205.14

a) HgO(s) is unstable and decomposes at all temperatures
b) less than 839 K
c) HgO(s) does not decompose at any temperature
d) less than 566°C
e) greater than 566°C

79. For the production of formaldehyde,

$$H_2(g) + CO(g) \rightarrow H_2CO(g)$$

$\Delta H_r° = 2$ kJ and $\Delta G_r° = 35$ kJ at 298 K. Estimate the temperature at which the reaction is spontaneous.

a) The reaction is not spontaneous at any temperature.
b) The reaction is spontaneous at all temperatures.
c) above 291°C
d) below 291°C
e) above 18 K

80. The purification of nickel by the Mond process uses the reactions below.

Ni(s, impure) + 4CO(g) Æ $Ni(CO)_4(g)$ (1)

$Ni(CO)_4(g) \rightarrow$ Ni(s, pure) + 4CO(g) (2)

If reaction (1) occurs at 353 K and reaction (2) at 473 K, reactions (1) and (2) are

a) both have $\Delta G = 0$.
b) entropy- and enthalpy-driven, respectively.
c) both entropy-driven,
d) both entalpy-driven.
e) enthalpy- and entropy-driven, respectively.

Chapter 16: Energy in Transition: Thermodynamics Form B

1. Calculate ΔU for a system that does 100 kJ of work on the surroundings when 150 kJ of heat are absorbed by the system.
a) +350 kJ b) 0 kJ c) +50 kJ d) –50 kJ e) –350 kJ

2. Calculate ΔU for a system that absorbs 350 kJ of heat and has 625 kJ of work done on the system.
a) +975 kJ b) –975 kJ c) 0 kJ d) –275 kJ e) +275 kJ

3. Calculate ΔU for a system that loses 125 kJ of heat and has 250 kJ of work done on the system.
a) +125 kJ b) +375 kJ c) –125 kJ d) –375 kJ e) 0 kJ

4. Calculate ΔU for a system that loses 100 kJ of heat while doing 150 kJ of work on the surroundings.
a) –250 kJ b) –50 kJ c) 0 kJ d) +50 kJ e) +250 kJ

5. In a certain reaction at constant pressure, $\Delta H = -125$ kJ and 25 kJ of work was required to make room for products. What is ΔU?
a) +100 kJ b) –100 kJ c) –150 kJ d) +150 kJ e) 0 kJ

6. In a certain reaction at constant pressure, $\Delta H = +225$ kJ and 35 kJ of work was required to make room for products. What is ΔU?
a) –190 kJ b) +260 kJ c) –260 kJ d) +190 kJ e) 0 kJ

7. For a certain reaction at constant pressure, $\Delta U = -90$ kJ and 14 kJ of expansion work is done by the system. What is ΔH for this process?
a) +76 kJ b) +104 kJ c) –104 kJ d) –90 kJ e) –76 kJ

8. For a certain reaction at constant pressure, $\Delta U = +65$ kJ and 23 kJ of expansion work is done by the system. What is ΔH for this process?
a) +42 kJ b) –88 kJ c) –42 kJ d) +88 kJ e) +65 kJ

9. Calculate the work needed to make room for products in the combustion of 1 mole of C(s) at STP (1 L · atm = 101 kJ).
a) no work is needed b) –22.4 kJ c) –4.52 kJ d) –2.26 kJ e) –6.79 kJ

10. Calculate the work needed to make room for products in the combustion of 2 moles of $C_4H_{10}(g)$ to carbon dioxide and water vapor at STP (1 L · atm = 101 J).
a) –4.52 kJ b) –6.79 kJ c) –40.7 kJ d) –3.39 e) no work is required

11. Which of the following reactions leads to a decrease in entropy of the system?
a) $PCl_5(s) \rightarrow PCl_3(l) + Cl_2(g)$
b) $2H_2O(l) \rightarrow 2H_2(g) + O_2(g)$
c) $2NO_2(g) \rightarrow N_2O_4(g)$
d) $C_6H_{12}O_6(s) + 6O_2(g) \rightarrow 6CO_2(g) + 6H_2O(l)$
e) $C(s) + H_2O(g) \rightarrow CO(g) + H_2(g)$

12. Which of the following reactions has the largest positive molar entropy change?
a) $2Mg(s) + CO_2(g) \rightarrow 2MgO(s) + C(s)$
b) $6Li(s) + N_2(g) \rightarrow 2Li_3N(s)$
c) $BaSO_4{\cdot}2H_2O(s) \rightarrow BaSO_4(s) + 2H_2O(g)$
d) $NH_3(g) + HCl(g) \rightarrow NH_4Cl(s)$
e) $2Na(l) + H_2(g) \rightarrow 2NaH(s)$

13. Which of the following has the largest entropy at 298 K?
a) $Br_2(l)$ b) $F_2(g)$ c) $Cl_2(g)$ d) $I_2(s)$ e) $Br_2(g)$

14. Which of the following has the largest entropy at 298 K?
a) $F_2(g)$ b) Ne(g) c) $O_2(g)$ d) $N_2(g)$ e) $H_2(g)$

15. Which of the following has the largest entropy at 298 K?
a) $CaCO_3(s)$ b) C(diamond) c) C(graphite) d) Pb(s) e) CaO(s)

16. All of the following changes give a positive ΔS_r^o except
a) $N_2(g) + 3H_2(g) \rightarrow 2NH_3(g)$
b) $H_2O(s) \rightarrow H_2O(l)$
c) $Na(l) \rightarrow Na(g)$
d) $O_2(g) \rightarrow O_2(aq)$
e) $H_2O(l, 20^oC) \rightarrow H_2O(l, 50^oC)$

17. Which of the following processes would have a positive ΔS_r^o?
a) $2NO(g) \rightarrow N_2O_4(g)$
b) $He(g, 1\ atm) \rightarrow He(g, 10\ atm)$
c) $Na_2SO_4(s) \rightarrow 2Na^+(aq) + SO_4^{2-}(aq)$
d) $NH_3(g) \rightarrow NH_3(aq)$
e) $Pb(s) + Br_2(l) \rightarrow PbBr_2(s)$

18. When 1 mole of gray tin changes to white tin at 13.0°C, the change in entropy is $+7.5\ J \cdot K^{-1}$. This means that
a) white tin is more ordered than gray tin.
b) gray tin has a higher molar entropy than white tin.
c) white tin has a higher molar entropy than gray tin.
d) the change from gray to white tin is predicted to be nonspontaneous on the basis of the entropy change.
e) gray tin has an entropy of $-7.5\ J \cdot K^{-1}$.

19. Which of the following reactions has the largest positive molar entropy change?
a) $Br_2(g) \rightarrow Br_2(l)$
b) $CO_2(g) + CaO(s) \rightarrow CaCO_3(s)$
c) $C(s) + O_2(g) \rightarrow CO_2(g)$
d) $SF_6(g) \rightarrow S(s) + 3F_2(g)$
e) $H_2O(l) \rightarrow H_2O(s)$

20. Which of the following has the smallest molar entropy at 298 K?
a) C(s, graphite) b) Sn(s, gray) c) S(s, rhombic) d) $CaCO_3(s)$ e) Pb(s)

21. List the following in order of decreasing molar entropy at 298 K.
Ar(g), $H_2O(l)$, $CO_2(g)$, Ne(g)
a) Ar(g) > $CO_2(g)$ > Ne(g) > $H_2O(l)$
b) $CO_2(g)$ > Ne(g) > Ar(g) > $H_2O(l)$
c) $CO_2(g)$ > Ar(g) > Ne(g) > $H_2O(l)$
d) $H_2O(l)$ > Ne(g) > Ar(g) > $CO_2(g)$
e) Ar(g) > Ne(g) > $CO_2(g)$ > $H_2O(l)$

22. For the vaporization of water at 100°C,
a) $\Delta S_{surroundings}$ is less than ΔS_{system}
b) $\Delta S_{surroundings}$ is greater than ΔS_{system}
c) $\Delta S_{surroundings}$ is negative
d) $\Delta S_{surroundings} = \Delta S_{system}$
e) $(\Delta S_{surroundings} + \Delta S_{system})$ is less than zero

23. If the enthalpy of vaporization of water at 100°C is 40.7 kJ · mol^{-1}, calculate $\Delta S°$ for condensation of one mole of $H_2O(l)$ at 100°C.
a) 136 J · K^{-1} b) −109 J · K^{-1} c) 109 J · K^{-1} d) 40.7 × 10^3 J · K^{-1} e) −40.7 × 10^3 J · K^{-1}

24. If the enthalpy of fusion of water at its normal melting point is 6.00 kJ · mol^{-1}, calculate $\Delta S°$ for melting 1 mole of water at this temperature.
a) 22.0 J · K^{-1} b) +6.00 × 10^3 J · K^{-1} c) −6.00 × 10^3 J · K^{-1} d) −22.0 J · K^{-1} e) −20.1 J · K^{-1}

25. The second law of thermodynamics requires that a spontaneous change
a) can only be carried out at high temperatures.
b) is accompanied by an increase of ΔS of the change.
c) has a negative enthalpy.
d) has a positive enthalpy.
e) is accompanied by an increase in the total entropy.

26. If $\Delta H° = -890$ kJ and $\Delta S° = -243$ J · K^{-1} for combustion of 1 mole of $CH_4(g)$ to carbon dioxide and water at 298 K, the change in entropy of the surroundings is
a) −243 J · K^{-1} b) 2.74 × 10^3 J · K^{-1} c) −2.99 × 10^3 J · K^{-1} d) 2.99 × 10^3 J K^{-1} e) 243 J · K^{-1}

27. The normal boiling point of $Br_2(l)$ is 57°C and the molar enthalpy of vaporization is 30.7 kJ · mol^{-1}. Calculate the molar entropy of vaporization.
a) 539 J · K^{-1} b) 30.7 × 10^3 J · K^{-1} c) 93.0 J · K^{-1} d) 10.7 J · K^{-1} e) 0

28. Estimate the normal boiling point of benzene, C_6H_6(l), given the molar enthalpy and entropy of vaporization of 33.90 $kJ \cdot mol^{-1}$ and 96.4 $J \cdot K^{-1} \cdot mol^{-1}$, respectively.
a) 11.0°C b) 125°C c) 25.0°C d) 78.7°C e) 0°C

29. Estimate the normal boiling point of mercury, Hg(l), given the molar enthalpy and entropy of vaporization of 61.32 $kJ \cdot mol^{-1}$ and 98.94 $J \cdot K^{-1} \cdot mol^{-1}$, respectively.
a) 77°C b) 347°C c) 620°C d) 534°C e) 806°C

30. The enthalpy of vaporization of Hg(l) is 61.32 $kJ \cdot mol^{-1}$ at its normal boiling point, 347°C. Calculate $\Delta S°$ for vaporization of 1 mole of Hg(l) at its normal boiling point.
a) 829 $J \cdot K^{-1}$ b) 30.7×10^3 $J \cdot K^{-1}$ c) 98.9 $J \cdot K^{-1}$ d) 177 $J \cdot K^{-1}$ e) 21.3 $J \cdot K^{-1}$

31. The following reaction is exothermic.

$$2CuO(s) + C(s) \rightarrow 2Cu(s) + CO_2(g)$$

Which of the following statements is correct?
a) The reaction will only be spontaneous at very high temperatures.
b) It is impossible to determine whether the reaction is spontaneous without calculations.
c) The reaction is spontaneous at all temperatures.
d) The reaction is not spontaneous at low temperatures.
e) The reaction is not spontaneous at high temperatures.

32. Which of the following conditions will prevent a reaction from being spontaneous at all temperatures?
a) $\Delta H > 0, \Delta S < 0$ b) $\Delta H < 0, \Delta S = 0$ c) $\Delta H > 0, \Delta S > 0$ d) $\Delta H < 0, \Delta S > 0$ e) $\Delta H < 0, \Delta S < 0$

33. Consider the following reaction.

$$2SO_2(g) + O_2(g) \rightarrow 2SO_3(g)$$

Which statement is true for this reaction?
a) $\Delta G = \Delta H + T\Delta S$ b) $\Delta S < 0$ c) $\Delta S = 0$ d) $S_m° = 0$ for O_2(g) e) $\Delta S > 0$

34. Consider the following reaction.

$$2SO_2(g) + O_2(g) \rightarrow 2SO_3(g)$$

If the standard enthalpy and free energy changes are –192 and –142 kJ, respectively, at 298 K, which of the following is true?
a) The reaction is not spontaneous at very low remperatures.
b) The negative value of the free energy change means the reaction will be spontaneous at any temperature.
c) The reaction is favored by raising the temperature.
d) The reaction is spontaneous despite an unfavorable entropy change.
e) The reaction is entropy-driven.

35. Consider the following reaction.

$$2H_2(g) + CO(g) \rightarrow CH_3OH(g)$$

If $\Delta H_r^\circ = -90$ kJ for this reaction, which statement is true?

a) The reaction is spontaneous at all temperatures.
b) The reaction is spontaneous at low temperatures.
c) The reaction is not spontaneous at 298 K, but becomes spontaneous at high temperatures.
d) The reaction is not spontaneous at any temperature.
e) The reaction is entropy-driven.

36. Which one of the following statements is true?

a) Spontaneous reactions always have $\Delta H_r^\circ < 0$.
b) Spontaneous reactions always have $\Delta H_r^\circ > 0$.
c) Spontaneous reactions always have $\Delta G_r^\circ > 0$.
d) Reactions with negative values of ΔS_r° always become spontaneous at low temperatures, if $\Delta H_r^\circ < 0$.
e) All of these statements are false.

37. The vapor pressure of acetic acid at 25°C is 16 Torr. ΔG for the reaction

$$CH_3CO_2H(l) \rightarrow CH_3CO_2H(g, 16\text{ Torr})$$

at 25°C is therefore

a) 0 b) 9.60 kJ c) –9.60 kJ d) 1.85 kJ e) –1.85 kJ

38. The vapor pressure of acetic acid at 25°C is 16 Torr. ΔG_r° for the reaction

$$CH_3CO_2H(l) \rightarrow CH_3CO_2H(g, 16\text{ Torr})$$

at 25°C is therefore

a) 0
b) –(8.314)(298)ln(16/760)
c) +(8.314)(298)ln(16/760)
d) –(8.314)(298)ln(16)
e) +(8.314)(298)ln(16)

39. Calculate ΔG_r° for the production of methanol at 298 K.

	$CO(g)$	$+\ 2H_2(g)$	$\rightarrow CH_3OH(l)$
ΔH_f°, kJ · mol^{-1}	–110.5		–239.0
S_m°, J · K^{-1} · mol^{-1}	198.0	131.0	127.0

a) –68.3 kJ b) –227.7 kJ c) –29.3 kJ d) –461 kJ e) –205 kJ

40. Calculate ΔG_r° for the following reaction at 298 K.

	$HgS(s)$	$+\ O_2(g)$	$\rightarrow Hg(l)$	$+\ SO_2(g)$
ΔH_f°, kJ · mol^{-1}	–58.2			–296.8
S_m°, kJ · mol^{-1}	82.4	205.14	76.02	248.22

a) –227.7 kJ b) –249.5 kJ c) –365.9 kJ d) –344.1 kJ e) +227.7 kJ

41. For a reaction, $H_r° = -297$ kJ and $\Delta S_r° = 10.1\ \text{J} \cdot \text{K}^{-1}$. Calculate the value of $\Delta G_r°$ for this reaction at 25°C.
a) −300 kJ b) −294 kJ c) −549 kJ d) −44.5 kJ e) −3310 kJ

42. For the following reaction
$CuO(s) + H_2(g) \rightarrow Cu(s) + H_2O(g)$
at 298 K, $\Delta H_r° = -87.0$ kJ and $\Delta S_r° = 47.0\ \text{J} \cdot \text{K}^{-1}$. Calculate $\Delta G_r°$ at 150 K.
a) −80.0 kJ b) +94.0 kJ c) −7140 kJ d) +80.0 kJ e) −94.0 kJ

43. For the following reaction
$C_6H_6(g) + 3H_2(g) \rightarrow C_6H_{12}(g)$
at 298 K, $\Delta H_r° = -206$ kJ and $\Delta S_r° = -363\ \text{J} \cdot \text{K}^{-1}$. Calculate $\Delta G_r°$ at 100°C.
a) −170 kJ b) +70.6 kJ c) +157 kJ d) +242 kJ kJ e) −157 kJ

44. The standard free energy of formation of $C_6H_6(l)$, a common solvent, is 124.3 kJ · mol^{-1} at 298 K. This means that at 298 K
a) $C_6H_6(l)$ is thermodynamically stable.
b) $C_6H_6(l)$ has a negative entropy.
c) $C_6H_6(l)$ is thermodynamically unstable.
d) $C_6H_6(l)$ will not spontaneously form $6C(s) + 3H_2(g)$.
e) no catalyst can be found to decompose $C_6H_6(l)$ into its elements.

45. Consider the following compounds and their standard free energies of formation:

(1)	(2)	(3)	(4)
$C_6H_6(l)$	$Ccl_4(l)$	$CS_2(l)$	$PCl_3(g)$
124 kJ · mol^{-1}	−65 kJ · mol^{-1}	65 kJ · mol^{-1}	−268 kJ · mol^{-1}

Which of these compounds is/are thermodynamically stable?
a) (2) and (4) b) (1) and (3) c) (4) d) (1) e) (3)

46. For the reaction
$6C(s) + 3H_2(g) \rightarrow C_6H_6(l)$
$\Delta H_r° = 49.0$ kJ and $\Delta S_r° = -253.18\ \text{J} \cdot \text{K}^{-1}$ at 298 K. With all reactants and products in their standard states, this reaction will be spontaneous
a) at temperatures above 194 K.
b) at temperatures below 194 K.
c) at all temperatures.
d) at no temperature,
e) at 273 K.

47. For the reaction
$2H_2(g) + CO(g) \rightarrow CH_3OH(l)$
$\Delta H_r° = -128.13$ kJ and $\Delta S_r° = -332.23\ \text{J} \cdot \text{K}^{-1}$ at 298. Calculate $\Delta G_r°$ at 300°C.
a) −29.13 kJ b) −204.1 kJ kJ c) +62.24 kJ d) −218.8 kJ e) −28.46 kJ

48. Consider the reactions below.
   1. $MgCO_3(s) \rightarrow MgO(s) + CO_2(g)$, $\Delta H_r^o = 101$ kJ, $\Delta S_r^o = 175\ J \cdot K^{-1}$
   2. $2Fe_2O_3(s) + 3C(s) \rightarrow 4Fe(s) + 3CO_2(g)$, $\Delta H_r^o = 462$ kJ, $\Delta S_r^o = 558\ J \cdot K^{-1}$
   3. $Fe_2O_3(s) + 2Al(s) \rightarrow 2Fe(s) + Al_2O_3(g)$, $\Delta H_r^o = -852$ kJ, $\Delta S_r^o = -39\ J \cdot K^{-1}$
   4. $MgO(s) + C(s) \rightarrow Mg(s) + CO(g)$, $\Delta H_r^o = 490$ kJ, $\Delta S_r^o = 198\ J \cdot K^{-1}$

   For which of these reactions would it be feasible to search for a catalyst to speed up the reaction at 673 K?

a) 3 only b) 2 and 4 c) 1 and 2 d) 1 and 3 e) 3 and 4

49. If the standard enthalpy, entropy, and free energy changes for the reaction
   $HCOOH(l) \rightarrow HCOOH(g)$
   at 298 K are 46.60 kJ, 122 $J \cdot K^{-1}$, and 10.3 kJ, respectively, calculate the normal boiling point of HCOOH(l).

a) 109°C b) 262°C c) 382°C d) 84.4°C e) 84.4 K

50. Is it worthwhile to search for a catalyst for the synthesis of $CH_3SH(g)$ by the reaction

$CH_4(g) + H_2S(g) \rightarrow CH_3SH(g) + H_2(g)$				
ΔH_f^o, $kJ \cdot mol^{-1}$	−74.81	−20.63	−12.4	
S_m^o, $J.K^{-1}.mol^{-1}$	+186.26	+205.79	+254.8	+130.68

(a) at 25°C?
(b) at 300°C?

51. If the standard molar free energy of formation of $H_2O(g)$ is $-228.6\ kJ \cdot mol^{-1}$ at 298 K, calculate ΔG_r^o for the reaction below.
   $2H_2(g) + O_2(g) \rightarrow 2H_2O(l)$

a) −228.6 kJ b) +457.2 kJ c) +228.6 kJ d) −457.2 kJ e) −57.15 kJ

52. Determine whether the reaction below occurs spontaneously at 25°C.
   $PCl_5(g, 760\ Torr) \rightarrow PCl_3(g, 0.760\ Torr) + Cl_2(g, 0.760\ Torr)$
   The values of ΔG_f^o for $PCl_3(g)$ and $PCl_5(g)$ are −286.3 and $-324.6\ kJ \cdot mol^{-1}$, respectively.

53. Calculate ΔG_r at 298 K for the reaction
   $C_2H_5OH(l) \rightarrow C_2H_5OH(g, 0.0400\ atm)$
   given $\Delta G_r^o = 6.2$ kJ at 298 K.

a) 2.7 kJ b) −14 kJ c) −1.8 kJ d) 14 kJ e) 1.8 kJ

54. Consider the following reaction
   $CuSO_4(s) \rightarrow CuO(s) + SO_3(g)$
   If $\Delta G^o = -14.6$ kJ at 950°C for this reaction, calculate ΔG_r for an $SO_3(g)$ pressure of 50 atm at this temperature.

a) 25.2 kJ b) 2.68 kJ c) −54.4 kJ d) 54.4 kJ e) 16.3 kJ

55. If $\Delta G_r^o = -27.1$ kJ at 25°C for the reaction

$$CH_3COO^-(aq) + H_3O^+(l) \leftrightarrow CH_3COOH(aq) + H_2O(aq)$$

calculate the value of the equilibrium constant for this reaction at 298 K.

a) 9.89×10^{-1} b) 5.63×10^4 c) 1.15×10^{-11} d) 1.78×10^{-5} e) 1.01

56. Consider the following reaction

$$2Fe_2O_3(s) + 3C(s) \rightarrow 4Fe(s) + 3CO_2(g) \quad \Delta H_r^o = 462 \text{ kJ}, \Delta S_r^o = 558 \text{ J} \cdot \text{K}^{-1}$$

Calculate the equilibrium constant for this reaction at 525°C.

a) 2.18×10^{-2} b) 5.20×10^{-7} c) 1.9×10^6 d) 8.07×10^{-2} e) 3.04×10^{-3}

57. Consider the reaction

$$2Fe_2O_3(s) + 3C(s) \rightarrow 4Fe(s) + 3CO_2(g) \quad \Delta H_r^o = 462 \text{ kJ}, \Delta S_r^o = 558 \text{ J} \cdot \text{K}^{-1} \text{ at } 298 \text{ K.}$$

The value of the equilibrium constant for this reaction will be greater than 1

a) at all temperatures.
b) at temperatures below 829 K.
c) at temperatures above 829 K.
d) at temperatures below 556°C.
e) at no temperature.

58. Consider the reaction

$$N_2(g) + 3H_2(g) \rightarrow 2NH_3(g)$$

If the standard molar free energy of formation of $NH_3(g)$ at 298 K is –16.45 kJ · mol^{-1}, calculate the equilibrium constant for this reaction at 298 K.

a) 6.02×10^2 b) 4.36×10^6 c) 1.90×10^{13} d) 5.85×10^5 e) 27.7

59. Calculate the equilibrium constant at 298 K for the reaction below.

$$2NH_3(g) + 3CuO(s) \rightarrow N_2(g) + 3H_2O(g) + 3Cu(s)$$

ΔG_f^o· kJ · mol^{-1} –16.45 –129.7 –228.57

a) 1.68×10^{46}
b) 2.80×10^{14}
c) 1.85×10^{33}
d) 1.16×10^{40}
e) To do the calculation we need ΔG_f^o values for nitrogen and copper.

60. Consider the reaction

$$2NH_3(g) \rightarrow N_2(g) + 3H_2(g)$$

If the standard molar free energy of formation of $NH_3(g)$ at 298 K is –16.45 kJ · mol^{-1}, calculate the equilibrium constant for this reaction at 298 K.

a) 3.62×10^{-2} b) 1.66×10^{-3} c) 5.26×10^{-14} d) 1.71×10^{-6} e) 2.29×10^{-7}

61. The vapor pressure of acetic acid at 25°C is 16 Torr. $\Delta G_r°$ for the reaction

$$CH_3CO_2H(l) \rightarrow CH_3CO_2H(g)$$

at 25°C is therefore

a) 9.57 kJ b) −9.57 kJ c) 0 d) −1.85 kJ e) 1.85 kJ

62. The equilibrium constant for the reaction

$$Hg(l) \leftrightarrow Hg(g)$$

is 2.6×10^{-6} at 25°C. Calculate $\Delta G_r°$ for this reaction.

a) 31.9 kJ b) −31.9 kJ c) 0 d) 2.67 kJ e) −2.67 kJ

63. If the enthalpy of vaporization of methanol is 37.6 kJ · mol^{-1} at its normal boiling point of 65°C, determine $\Delta G_r°$ for reversible vaporization of 1 mole of methanol at 65°C and 1 atm.

a) 0 b) 37.6 kJ c) −37.6 kJ d) 111 kJ e) −111 kJ

64. Calculate the standard free energy of formation of $CCl_4(l)$ at 25°C given a $\Delta G_f°$ for $CCl_4(g)$ of 60.63 kJ · mol^{-1} and a vapor pressure for carbon tetrachloride at 25°C of 0.154 atm.

a) −65.27 kJ · mol^{-1} b) −4.64 kJ · mol^{-1} c) 0 d) 55.99 kJ · mol^{-1} e) −60.63 kJ · mol^{-1}

65. Consider the reaction

$$CaCO_3(s) \leftrightarrow Ca^{2+}(aq) + CO_3^{2-}(aq)$$

If $K_c = 2.8 \times 10^{-9}$ at 25°C, calculate $\Delta G_r°$ for this reaction.

a) 21.2 kJ b) −21.2 kJ c) −48.8 kJ d) 48.8 kJ e) 69.9 kJ

66. If K_p for a certain reaction is 0.233 Torr at 1073 K, calculate $\Delta G_r°$ for this reaction at this temperature.

a) −11.3 kJ b) −72.2 kJ c) −31.3 kJ d) 72.2 kJ e) 11.3 kJ

67. Consider the reaction

$$COCl_2(g) \leftrightarrow CO(g) + Cl_2(g)$$

If $K_c = 0.00463$ at 527°C, calculate $\Delta G_r°$ for the reaction at this temperature.

a) 71.4 kJ b) −35.8 kJ c) 35.8 kJ d) −15.5 kJ e) 15.5 kJ

68. Consider the reaction

$$2CuBr_2(s) \rightarrow 2CuBr(s) + Br_2(g)$$

If the equilibrium vapor pressure of $Br_2(g)$ is 1.43×10^{-5} Torr at 298 K, calculate ΔG_r at this temperature when $Br_2(g)$ is produced at a pressure of 7.50×10^{-8} Torr.

a) 5.65 kJ b) −3.42 kJ c) −5.65 kJ d) −13.0 kJ e) 13.0 kJ

69. If $\Delta G_r = -57.7$ kJ for the reaction of 5 mL of 0.20 M $AgNO_3(aq)$ with 5 mL of 0.20 M NaBr(aq) at 298 K, calculate the solubility product of AgBr at this temperature.

a) 3.1×10^{-12} b) 7.7×10^{-13} c) 5.1×10^{-26} d) 5.1×10^{26} e) 7.7×10^{13}

70. If K_{sp} of AgI is 1.5×10^{-16} at 298 K, calculate ΔG_r for the reaction of 5 mL of 0.20 M $AgNO_3(aq)$ with 5 mL of 0.20 M NaI(aq) at 298 K.
a) −82.3 kJ b) 82.3 kJ c) −78.9 kJ d) 78.9 kJ e) −34.2 kJ

71. Calculate ΔG_r for the following reaction at 298 K.

$$Ag(NH_3)_2^+(aq, 0.400\ M) \rightarrow Ag^+(aq, 0.100\ M) + 2NH_3(aq, 0.300\ M)$$

$\Delta G^o = -41.0$ kJ for this reaction at 298 K.
a) 31.6 kJ b) −9.4 kJ c) −50.4 kJ d) 50.4 kJ e) −45.1 kJ

72. Consider the following reaction at 298 K.

$$HClO(aq) + H_2O(l) \leftrightarrow ClO^-(aq) + H_3O^+(aq) \qquad K_c = 3.0 \times 10^{-8}$$

Calculate ΔG_r when $[ClO^-] = [H_3O^+] = 1.0 \times 10^{-3}$ M and $[HClO] = 0.10$ M.
a) −6.25 kJ b) 31.5 kJ c) 14.4 kJ d) −14.4 kJ e) 6.25 kJ

73. If the K_a of acetic acid is 1.8×10^{-5} at 25°C, calculate the standard free energy of reaction for

$$CH_3CO_2^-(aq) + H_3O^+(aq) \rightarrow CH_3COOH(aq) + H_2O(l)$$

a) −2.27 kJ b) +11.8 kJ c) −11.8 kJ d) −27.1 kJ e) +27.1 kJ

74. For the following process

$$C_6H_6(l) \rightarrow C_6H_6(g)$$

$\Delta H^o = 33.90\ kJ \cdot mol^{-1}$ and $\Delta S^o = 96.4\ J \cdot K^{-1} \cdot mol^{-1}$. Calculate the temperature at which the vapor pressure of benzene is 85 Torr.

75. The equilibrium constant for a certain reaction is 0.742 at 298 K and 0.321 at 400 K. Which statement is correct?
a) The equilibrium constant cannot become smaller if the temperature is increased.
b) The reaction is not spontaneous at any temperature if the standard entropy of reaction is negative.
c) The reaction is spontaneous at all temperatures if the standard entropy of reaction is positive.
d) The reaction is not spontaneous at any temperature if the standard entropy of reaction is positive.
e) The reaction is endothermic.

76. At the normal boiling point of $C_2H_5OH(l)$, 315.7 K, the value of the standard enthalpy of vaporization is $38.7\ kJ \cdot mol^{-1}$. Assuming that $C_2H_5OH(g)$ behaves as an ideal gas and that the molar volume of $C_2H_5OH(l)$ is negligible compared to that of $C_2H_5OH(g)$, calculate the values of q, w, ΔU^o, ΔS^o, and ΔG^o for the reversible vaporization of 1 mole of $C_2H_5OH(l)$ at 315.7 K and 1 atm.

77. Estimate the temperature at which HgS(s) reacts with oxygen.

	$HgS(s)$	$+\ O_2(g)$	$\rightarrow Hg(l)$	$+\ SO_2(g)$
ΔH_f^o, $kJ \cdot mol^{-1}$	−58.2			−296.8
S^o, $J \cdot K^{-1} \cdot mol^{-1}$	82.4	205.14	76.02	248.22

78. Estimate the temperature at which mercury(II) oxide can be expected to decompose.

	$2HgO(s)$	$\rightarrow 2Hg(l)$	$+ O_2(g)$
$\Delta H_f°$, $kJ \cdot mol^{-1}$			–90.83
$S_m°$, $J \cdot K^{-1} \cdot mol^{-1}$	70.29	76.02	205.14

a) HgO(s) does not decompose at any temperature
b) greater than 839 K
c) HgO(s) is unstable and decomposes at all temperatures
d) less than 839 K
e) less than 566°C

79. For the decomposition of $MgCO_3(s)$,

$$MgCO_3(s) \rightarrow MgO(s) + CO_2(g)$$

$\Delta H_r° = 100.59$ kJ and $\Delta S_r° = 174.98\ J \cdot K^{-1}$ at 298 K. Estimate the temperature at which the reaction is spontaneous.

a) greater than 302°C
b) less than 302°C
c) The reaction is not spontaneous at any temperature.
d) less than 575 K
e) The reaction is spontaneous at all temperatures.

80. For a certain reaction, $\Delta H_r° = -6.90$ kJ and $\Delta S_r° = -15.4\ J \cdot K^{-1}$ at 298 K. Under standard conditions, the reaction is

a) not spontaneous at any temperature.
b) spontaneous at all temperatures.
c) spontaneous at temperatures above about 448 K.
d) spontaneous at temperatures below about 175°C.
e) spontaneous at temperatures above about 175°C.

Chapter 17: Electrons in Transition: Electrochemistry
Form A

1. Given: $Cl^-(aq) \rightarrow ClO_3^-(aq)$, acidic solution.
   How many electrons appear in the balanced half-reaction?
   a) 6 b) 3 c) 9 d) 12 e) 1

2. Given: $Br^-(aq) \rightarrow BrO_3^-(aq)$, basic solution.
   How many electrons appear in the balanced half-reaction?
   a) 6 b) 3 c) 2 d) 8 e) 1

3. Given: $NO_3^-(aq) \rightarrow NO(g)$, acidic solution.
   How many electrons appear in the balanced half-reaction?
   a) 2 b) 6 c) 8 d) 4 e) 3

4. Given: $ClO^-(aq) \rightarrow Cl^-(aq)$, basic solution.
   How many electrons appear in the balanced half-reaction?
   a) 2 b) 1 c) 6 d) 4 e) 8

5. Given: $MnO_2(s) \rightarrow Mn^{2+}(aq)$, acidic solution.
   How many electrons appear in the balanced half-reaction?
   a) 2 b) 5 c) 7 d) 10 e) 4

6. Given: $Cr(OH)_3(s) \rightarrow CrO_4^{2-}(aq)$, basic solution.
   How many electrons appear in the balanced half-reaction?
   a) 3 b) 6 c) 4 d) 5 e) 7

7. Given: $MnO_4^-(aq) + H^+(aq) + Cl^-(aq) \rightarrow Mn^{2+}(aq) + Cl_2(g) + H_2O(l)$
   If the coefficient of MnO_4^- in the balanced equation is 2, what are the coefficients of H^+, Cl^- and Cl_2, respectively?
   a) 4, 8, 4 b) 16, 10, 5 c) 10, 10, 5 d) 8, 10, 5 e) 8, 5, 5

8. Given: $I^-(aq) + MnO_4^-(aq) + H_2O(l) \rightarrow I_2(s) + MnO_2(s) + OH^-(aq)$
   If the coefficient of MnO_4^- in the balanced equation is 2, what are the coefficients of I^- and OH^-, respectively?
   a) 3 and 12 b) 6 and 6 c) 3 and 4 d) 6 and 12 e) 6 and 8

9. Given: $Zn(s) + OH^-(aq) + H_2O(l) + NO_3^-(aq) \rightarrow Zn(OH)_4^{2-}(aq) + NH_3(g)$
   If the coefficient of NO_3^- in the balanced equation is 1, how many electrons are transferred in the reaction?
   a) 8 b) 6 c) 4 d) 10 e) 2

10. Given: $Zn(s) + OH^-(aq) + H_2O(l) + NO_3^-(aq) \rightarrow Zn(OH)_4^{2-}(aq) + NH_3(g)$
If 8 electrons are transferred in the balanced equation, what is the coefficient of OH^-?
a) 7 b) 8 c) 4 d) 6 e) 9

11. When concentrated $H_2SO_4(l)$ is added to KI(aq), $I_2(s)$ and $H_2S(g)$ are formed. When the equation for this reaction is balanced, what is the number of electrons transferred?
a) 2 b) 4 c) 8 d) 1 e) 6

12. Given: $Cr_2O_7^{2-}(aq) \rightarrow Cr^{3+}(aq)$, acidic solution.
When this half-reaction is balanced, both H^+ and H_2O appear in the equation. If the coefficient of Cr^{3+} is 2 in the balanced half-reaction, what is the coefficient of H^+?
a) 14 b) 8 c) 10 d) 12 e) 4

13. When the reaction below is balanced, how many electrons are transferred?
$O_2(g) + H_2O(l) + Fe(s) \rightarrow Fe(OH)_3(s)$
a) 12 b) 4 c) 3 d) 2 e) 6

14. When the Ag(s) | AgCl(s) | $Cl^-(aq)$ electrode acts as a cathode, the half-reaction is
a) $Ag^+(aq) + e^- \rightarrow Ag(s)$
b) $AgCl(s) + e^- \rightarrow Ag(s) + Cl^-(aq)$
c) $Ag(s) + Cl^-(aq) \rightarrow AgCl(s) + e^-$
d) $2AgCl(s) + 2e^- \rightarrow 2Ag^+(aq) + Cl_2(g)$
e) $Ag(s) \rightarrow Ag^+(aq) + e^-$

15. When the Ag(s) | AgBr(s) | $Br^-(aq)$ electrode acts as a cathode, the half-reaction is
a) $2AgBr(s) + 2e^- \rightarrow 2Ag^+(aq) + Br_2(l)$
b) $AgBr(s) + e^- \rightarrow Ag(s) + Br^-(aq)$
c) $Ag^+(aq) + e^- \rightarrow Ag(s)$
d) $Ag(s) \rightarrow Ag^+(aq) + e^-$
e) $Ag(s) + Br^-(aq) \rightarrow AgBr(s) + e^-$

16. For the cell diagram
Pt | $H_2(g)$ | $H^+(aq)$ | $Fe^{3+}(aq)$,$Fe^{2+}(aq)$ | Pt
which reaction occurs at the anode?
a) $2H^+(aq) + 2e^- \rightarrow H_2(g)$
b) $2Fe^{3+}(aq) + H_2(g) \rightarrow 2Fe^{2+}(aq) + 2H^+(aq)$
c) $H_2(g) \rightarrow 2H^+(aq) + 2e^-$
d) $Fe^{2+}(aq) \rightarrow Fe^{3+}(aq) + e^-$
e) $Fe^{3+}(aq) + e^- \rightarrow Fe^{2+}(aq)$

17. For the cell diagram
Pt | $Br_2(l)$ | $Br^-(aq)$ | $Cl^-(aq)$,$Cl_2(g)$ | Pt
which reaction occurs at the cathode?
a) $Cl_2(g) + 2Br^-(aq) \rightarrow 2Cl^-(aq) + Br_2(l)$
b) $2Br_2(l) + 2e^- \rightarrow 2Br^-(aq)$
c) $Cl_2(g) + 2e^- \rightarrow 2Cl^-(aq)$
d) $2Br^-(aq) \rightarrow Br_2(l) + 2e^-$
e) $2Cl^-(aq) \rightarrow Cl_2(g) + 2e^-$

18. For the cell diagram

$Pt \mid H_2(g), H^+(aq) \mid Cu^{2+}(aq) \mid Cu(s)$

Which reaction occurs at the anode?

a) $Cu(s) \rightarrow Cu^{2+}(aq) + 2e^-$
b) $2H^+(aq) + 2e^- \rightarrow H_2(g)$
c) $Cu^{2+}(aq) + 2e^- \rightarrow Cu(s)$
d) $H_2(g) \rightarrow 2H^+(aq) + 2e^-$
e) $2H^+(aq) + Cu(s) \rightarrow H_2(g) + Cu^{2+}(aq)$

19. The cell diagram for the reaction in which hygrogen gas reduces $Cu^{2+}(aq)$ is

a) $Cu(s) \mid Cu^{2+}(aq) \mid O_2(g) \mid H_2(g) \mid Pt$
b) $Pt \mid Cu^{2+}(aq), H_2(g) \mid H^+(aq) \mid Pt$
c) $Cu(s) \mid Cu^{2+}(aq) \mid H^+(aq) \mid H_2(g) \mid Pt$
d) $Pt \mid H_2(g) \mid H^+(aq) \mid Cu^{2+}(aq) \mid Cu(s)$
e) $Pt \mid Cu^{2+}(aq), H^+(aq) \mid H_2(g) \mid Pt$

20. Write the cell diagram for the reaction

$Cl_2(g) + 2Br^-(aq) \rightarrow 2Cl^-(aq) + Br_2(l)$

a) $Pt \mid Cl_2(g) \mid Br^-(aq) \mid Cl^-(aq) \mid Br_2(l) \mid Pt$
b) $Pt \mid Cl_2(g) \mid Cl^-(aq) \mid Br^-(aq) \mid Br_2(l) \mid Pt$
c) $Pt \mid Cl^-(aq), Br^-(aq) \mid Br_2(l) \mid Cl_2(g) \mid Pt$
d) $Pt \mid Br_2(l) \mid Cl_2(g) \mid Cl^-(aq) \mid Br^-(aq) \mid Pt$
e) $Pt \mid Br_2(l) \mid Br^-(aq) \mid Cl^-(aq) \mid Cl_2(g) \mid Pt$

21. Consider the following cell:

$Pb(s) \mid PbSO_4(s) \mid SO_4^{2-}(aq) \mid Pb^{2+}(aq) \mid Pb(s)$

The reaction utilized by this cell is

a) $PbSO_4(s) \rightarrow Pb^{2+}(aq) + SO_4^{2-}(aq)$
b) $SO_4^{2-}(aq) + H^+(aq) \rightarrow HSO_4^-(aq)$
c) $Pb^{2+}(aq) + SO_4^{2-}(aq) \rightarrow PbSO_4(s)$
d) $SO_4^{2-}(aq) + H_2O(l) \rightarrow HSO_4^-(aq) + OH^-(aq)$
e) $Pb(s) + 2H^+(aq) \rightarrow Pb^{2+}(aq) + H_2(g)$

22. The standard potential of the Ag^+/Ag electrode is +0.80 V and the standard potential of the cell

$Fe(s) \mid Fe^{3+}(aq) \mid Ag^+(aq) \mid Ag(s)$

is +0.84 V. What is the standard potential of the Fe^{3+}/Fe electrode?

a) +0.04 V b) +1.64 V c) –1.64 V d) –0.04 V e) –0.12 V

23. The standard potential of the Cu^{2+}/Cu electrode is +0.34 V and the standard potential of the cell

$Ag(s) \mid AgCl(s) \mid Cl^-(aq) \mid Cu^{2+}(aq) \mid Cu(s)$

is +0.12 V. What is the standard potential of the $AgCl/Ag, Cl^-$ electrode?

a) –0.22 V b) +0.46 V c) +0.24 V d) –0.46 V e) +0.22 V

24. The standard potential of the Pb^{2+}/Pb electrode is –0.13 V and the standard potential of the cell

$Zn(s) \mid Zn^{2+}(aq) \mid Pb^{2+}(aq) \mid Pb(s)$

is +0.63 V. What is the standard potential of the Zn^{2+}/Zn electrode?

a) –0.50 V b) –0.76 V c) –1.52 V d) +0.50 V e) +0.76 V

25. Consider the following reaction:

$$2Ag^{+}(aq) + Cu(s) \rightarrow Cu^{2+}(aq) + 2Ag(s)$$

If the standard potentials of Ag^{+} and Cu^{2+} are +0.80 V and +0.34 V, respectively, calculate the value of E^{o} for the given reaction.

a) +1.48 V b) −0.46 V c) +1.26 V d) +0.46 V e) −1.26 V

26. Calculate E^{o} for the following cell.

$$Zn(s)\,|\,Zn^{2+}(aq)\,|\,Cl^{-}(aq)\,|\,AgCl(s)\,|\,Ag(s)$$

a) −1.20 V b) +1.20 V c) +0.98 V d) −0.54 V e) +0.54 V

27. The standard potential of the cell

$$Tl(s)\,|\,Tl^{+}(aq)\,|\,Cl^{-}(aq)\,|\,AgCl(s)\,|\,Ag(s)$$

is +0.56 V at 25°C. If the standard potential of Tl^{+} is −0.34 V, calculate the standard potential of the $AgCl/Ag,Cl^{-}$ couple.

a) −0.22 V b) +0.22 V c) −0.90 V d) −0.56 V e) +0.90 V

28. The standard potential of the cell

$$Pb(s)\,|\,Pb^{2+}(aq)\,|\,Cl^{-}(aq)\,|\,AgCl(s)\,|\,Ag(s)$$

is +0.35 V at 25°C. If the standard potential of Pb^{2+} is −0.13 V, calculate the standard potential of the $AgCl/Ag,Cl^{-}$ couple.

a) −0.22 V b) +0.48 V c) −0.48 V d) +0.22 V e) −0.35 V

29. The standard potential of the cell

$$Tl(s)\,|\,Tl^{+}(aq)\,|\,Cl^{-}(aq)\,|\,AgCl(s)\,|\,Ag(s)$$

is +0.56 V at 25°C. If the standard potential of the $AgCl/Ag,Cl^{-}$ couple is +0.22 V, calculate the standard potential of the Tl^{+}/Tl couple.

a) +0.34 V b) −0.78 V c) −0.34 V d) −0.56 V e) +0.78 V

30. The standard potential of the cell

$$In(s)\,|\,In^{3+}(aq)\,|\,Cu^{2+}(aq)\,|\,Cu(s)$$

is +0.68 V at 25°C. If the standard potential of the Cu^{2+}/Cu couple is +0.34 V, calculate the standard potential of the In^{3+}/In couple.

a) +0.34 V b) −0.34 V c) −0.23 V d) +1.02 V e) −0.68 V

31. Calculate the standard potential for the following half-reaction:

$$Tl^{3+}(aq) + 3e^{-} \rightarrow Tl(s)$$

The standard potentials of the Tl^{3+}/Tl^{+} and Tl^{+}/Tl couples are +1.21 and −0.34 V, respectively.

32. Which species will oxidize V^{2+} but not Br^{-}?

a) O_2 b) Fe^{3+} c) Zn^{2+} d) Cr^{3+} e) Mn^{2+}

33. Which species will reduce Br_2 but not V^{3+}?

a) Ag b) Zn c) In^{+} d) Al e) Ce

34. Which species will oxidize Cr^{2+} but not Mn^{2+}?
a) V^{3+} b) Pb^{4+} c) Zn^{2+} d) Fe^{2+} e) O_3

35. Which species will reduce Mn^{3+} but not Cr^{3+}?
a) Pb b) Zn c) Fe d) Al e) In^{2+}

36. Which species will reduce Fe^{3+} but not Zn^{2+}?
a) Br^- b) MnO_4^- c) V^{3+} d) V^{2+} e) Mn^{2+}

37. Which pair of metals will dissolve in nitric acid?
a) Ag, Fe b) Ag, Au c) Pt, Fe d) Pt, Ag e) Pt, Au

38. If the standard potentials for the couples Cu^{2+}/Cu, Ag^+/Ag, and Fe^{2+}/Fe are +0.34, + 0.80, and –0.44 V, respectively, which is the strongest oxidizing agent?
a) Cu^{2+} b) Fe^{2+} c) Ag^+ d) Cu e) Ag

39. If the standard potentials for the couples Cu^{2+}/Cu, Ag^+/Ag, and Fe^{2+}/Fe are +0.34, + 0.80, and –0.44 V, respectively, which is the strongest reducing agent?
a) Fe b) Ag^+ c) Fe^{2+} d) Cu e) Ag

40. If the standard potentials for the couples Fe^{3+}/Fe^{2+}, $MnO_4^-,H^+/Mn^{2+}$, Zn^{2+}/Zn, V^{3+}/V^{2+}, and Br_2/Br^- are +0.77, +1.51, –0.76, –0.26, and +1.09 V, respectively, which is the strongest oxidizing agent?
a) MnO_4^- b) Br_2 c) Fe^{3+} d) Mn^{2+} e) Zn^{2+}

41. If the standard potentials for the couples Fe^{3+}/Fe^{2+}, $MnO_4^-,H^+/Mn^{2+}$, Zn^{2+}/Zn, V^{3+}/V^{2+}, and Br_2/Br^- are +0.77, +1.51, –0.76, –0.26, and +1.09 V, respectively, which is the weakest reducing agent?
a) Br^- b) V^{2+} c) Mn^{2+} d) Zn e) Fe^{2+}

42. Which of the following is the strongest reducing agent?
a) Fe^{2+} b) H_2 c) Co^{2+} d) F^- e) Cr^{2+}

43. Which of the following is the strongest oxidizing agent?
a) Ca^{2+} b) Fe^{2+} c) Cr^{3+} d) Cu^{2+} e) Co^{2+}

44. Given:

$$Ag^+(aq) + e^- \rightarrow Ag(s) \qquad E^o = 0.80 \text{ V}$$
$$Fe^{3+}(aq) + e^- \rightarrow Fe^{2+}(aq) \qquad E^o = 0.77 \text{ V}$$
$$Cu^{2+}(aq) + 2e^- \rightarrow Cu(s) \qquad E^o = 0.34 \text{ V}$$

Which is the strongest reducing agent?
a) Cu b) Ag^+ c) Ag d) Fe^{2+} e) Cu^{2+}

45. Which of the following will be reduced by Fe^{2+}?
a) Cu^{2+} b) Ag^+ c) AgCl d) Cu^+ e) H^+

46. Place the following in order of decreasing strength as a reducing agent.
H_2, Na, Zn, Ag
a) Na > H_2 > Ag > Zn
b) Ag > Zn > H_2 > Na
c) Na > Ag > H_2 > Zn
d) Na > Zn > H_2 > Ag
e) Ag > H_2 > Zn > Na

47. Which of the following occurs when HNO_3(aq), Cu(s), and Pt(s) are mixed under standard conditions?
a) Cu(s) dissolves and H_2(g) is formed
b) Cu(s) dissolves
c) no reaction takes place
d) Pt(s) dissolves and H_2(g) is formed
e) Pt(s) dissolves

48. An aqueous acid solution of the dichromate ion and NO_2^-(aq) are mixed and the solution changes color from orange to pale violet. Write a balanced net ionic equation for this reaction.

49. Predict whether lead metal will dissolve in 1.0 M HCl(aq) and write a net ionic equation for any reaction.

50. Consider the following cell at 25°C:
Pb(s) | Pb^{2+}(aq, saturated PbI_2(s)) | Pb^{2+}(aq, 0.500 M) | Pb(s)
If the voltage of this cell is 0.0774 V, calculate K_{sp} for PbI_2.

51. The standard potential of the cell
Ag(s) | Ag^+(aq) | Cl^-(aq) | AgCl(s) | Ag(s)
is –0.58 V at 25°C. Calculate the equilibrium constant for the cell reaction.
a) 6.3×10^9 b) 1.2×10^{-5} c) 5.7×10^{-18} d) 2.7×10^{-23} e) 1.6×10^{-10}

52. The standard potential of the cell
Pb(s) | $PbSO_4$(s) | SO_4^{2-}(aq) | Pb^{2+}(aq) | Pb(s)
is +0.23 V at 25°C. Calculate the equilibrium constant for the reaction of Pb^{2+}(aq) with SO_4^{2-}(aq).
a) 1.7×10^{-8} b) 7.7×10^3 c) 3.7×10^{16} d) 8.0×10^{17} e) 6.0×10^7

53. The equilibrium constant for the reaction
$Ni(s) + Hg_2Cl_2(s) \rightarrow 2Hg(l) + 2Cl^-(aq) + Ni^{2+}(aq)$
is 1.8×10^{19} at 25°C. Calculate the value of $E°$ for a cell utilizing this reaction.
a) +1.14 V b) +0.25 V c) +0.57 V d) –0.57 V e) –1.14 V

54. If $E°$ for the disproportionation of Cu^+(aq) to Cu^{2+}(aq) and Cu(s) is +0.37 V at 25°C, calculate the equilibrium constant for the reaction.
a) 3.3×10^{12} b) 1.8×10^6 c) 3.9×10^{74} d) 35.7 e) 2.5×10^{14}

55. The standard voltage of the cell

$Pt \mid H_2(g) \mid H^+(aq \mid Cl^-(aq) \mid AgCl(s) \mid Ag(s)$

is 0.22 V at 25°C. Calculate the equilibrium constant for the reaction below.

$2AgCl(s) + H_2(g) \rightarrow 2Ag(s) + 2H^+(aq) + 2Cl^-(aq)$

a) 2.7×10^7 b) 3.7 c) 7.4 d) 5.2×10^3 e) 1.7×10^3

56. If the standard free energy change for combustion of 1 mole of $CH_4(g)$ is –818 kJ/mol, calculate the standard voltage that could be obtained from a fuel cell using this reaction.

a) +1.06 V b) +0.53 V c) +4.24 V d) +8.48 V e) –1.06 V

57. Given: $2Cu^+(aq) \rightarrow Cu(s) + Cu^{2+}(aq)$

If $E°$ for this reaction is 0.18 V at 25°C, calculate ΔG°.

a) –17 kJ b) –95 kJ c) +35 kJ d) –35 kJ e) +17 kJ

58. A hydrogen electrode with a $H_2(g)$ pressure of 1 atm is combined with a standard calomel electrode whose half-reaction is given below.

$Hg_2Cl_2(s) + 2e^- \rightarrow 2Hg(l) + 2Cl^-(aq)$ $E° = +0.270$ V

If the cell voltage is +0.800 V when the calomel electrode is the cathode, calculate the pH of the solution around the hydrogen electrode.

59. Consider the following cell:

$Zn(s) \mid Zn^{2+}(aq, 0.00200\ M) \mid Cl^-(aq, 0.100\ M) \mid AgCl(s)Ag(s)$

If the standard potentials of the Zn^{2+}/Zn and $AgCl/Ag,Cl^-$ couples are –0.76 and 0.22 V, respectively, at 25°C, calculate the voltage of the cell.

a) +0.96 V b) +1.09 V c) +1.00 V d) +0.84 V e) +1.12 V

60. Consider the following cell:

$Ni(s) \mid Ni^{2+}(aq, 0.200\ M) \mid Cl^-(aq, 0.0200\ M) \mid Cl_2(g, 0.500\ atm) \mid Pt$

If the standard voltage for this cell is 1.59 V at 25°C, calculate the voltage of this cell.

a) +1.48 V b) +1.70 V c) +1.65 V d) +1.59 V e) +1.81 V

61. Consider the following cell:

$Pt \mid Sn^{2+}(aq, 0.40\ M), Sn^{4+}(aq, 0.10\ M) \mid Sn^{4+}(aq, 0.030\ M), Sn^{2+}(aq, 0.010\ M) \mid Pt$

If the standard potential for the Sn^{4+}/Sn^{2+} couple is 0.150 V, calculate the cell voltage at 25°C.

a) +0.118 V b) 0 V c) +0.214 V d) +0.032 V e) +0.182 V

62. Consider the following cell:

$Ag(s) \mid Ag^+(aq, 0.100\ M) \mid Ag^+(aq, 1.00\ M) \mid Ag(s)$

What is the voltage of this cell?

a) +0.0296 V b) –0.0296 V c) –0.0592 V d) +0.80 V e) +0.0592 V

63. Calculate E for the half-reaction below.

$2H^+(aq, 1.00 \times 10^{-7}\ M) + 2e^- \rightarrow H_2(g, 1.00\ atm)$

a) 0 V b) –0.414 V c) +0.414 V d) –0.829 V e) +0.829 V

64. Consider the following cell:

$Mn(s)\ |\ Mn^{2+}(aq, 0.100\ M)\ |\ H^+(aq, ?)\ |\ H_2(g, 0.500\ atm)\ |\ Pt$

If E = 0.97 V and $E°$ = 1.03 V at 25°C, calculate the concentration of H^+ in the cathode cell compartment.

a) 4.7×10^{-4} M b) 0.022 M c) 0.011 M d) 4.8×10^{-3} M e) 0.070 M

65. Consider the following cell:

$Pt\ |\ H_2(g, 1\ atm)\ |\ H^+(aq, ?\ M)\ |\ Ag^+(aq, 1.0\ M)\ |\ Ag(s)$

If the voltage of this cell is 1.04 V at 25°C and the standard potential of the Ag^+/Ag couple is +0.80 V, calculate the hydrogen ion concentration in the anode compartment.

a) 9.4×10^{-3} M b) 4.6×10^{-10} M c) 1.0 M d) 8.8×10^{-5} M e) 3.7×10^{-8} M

66. Consider the following cell:

$Zn(s)\ |\ Zn^{2+}(aq, 0.100\ M)\ |\ Cl^-(aq, ?\ M)\ |\ Cl_2(g, 0.500\ atm)\ |\ Pt$

For this cell, $E°$ = 2.12 V and E = 2.27 V at 25°C. Calculate the $Cl^-(aq)$ concentration in the cathode compartment.

a) 4.3×10^{-5} M b) 2.9×10^{-3} M c) 1.2×10^{-1} M d) 1.5×10^{-3} M e) 6.5×10^{-3} M

67. Consider the following cell:

$Zn(s)\ |\ Zn^{2+}(aq, 0.200\ M)\ |\ Cl^-(aq, 0.100\ M)\ |\ AgCl(s)\ |\ Ag(s)$

If $E°$ for the cell is 0.98 V at 25°C, write the Nernst equation for the cell at this temperature.

a) $E = -0.98 + 0.01285\ln[(0.100)^2(0.200)]$
b) $E = 0.98 - 0.02569\ln[(0.100)(0.200)]$
c) $E = 0.98 + 0.02569\ln[(0.100)^2(0.200)]$
d) $E = 0.98 - 0.02569\ln[(0.100)^2(0.200)]$
e) $E = 0.98 - 0.01285\ln[(0.100)^2(0.200)]$

68. When a lead–acid battery is charged,

a) $PbSO_4(s)$ is formed at the anode.
b) sulfuric acid is produced.
c) $PbSO_4(s)$ is formed at the cathode.
d) sulfuric acid is depleted.
e) $PbO_2(s)$ dissolves.

69. In the standard dry cell, the cathode reaction involves

a) reduction of $H_2O(l)$.
b) reduction of $Zn(NH_3)_4{}^{2+}(aq)$.
c) migration of Zn^{2+} ions towards the anode.
d) reduction of $Zn^{2+}(aq)$.
e) reduction of $MnO_2(s)$.

70. Which metal would be suitable to provide cathodic protection from corrosion for an iron bridge?

a) Mg b) Sn c) Pb d) Ni e) Cu

71. The products of the electrolysis of $CuF_2(aq)$ are
a) $Cu(s)$ and $F_2(g)$.
b) $Cu(s)$ and $H_2(g)$.
c) $Cu(s)$, $O_2(g)$, and $H^+(aq)$.
d) $H_2(g)$ and $F_2(g)$.
e) $H_2(g)$ and $O_2(g)$.

72. The products of the electrolysis of $Na_2SO_4(aq)$ are
a) $H_2(g)$ and $H_2SO_3(aq)$.
b) $H_2(g)$ and $OH^-(aq)$.
c) $H_2(g)$ and $O_2(g)$.
d) $Na(s)$ and $O_2(g)$.
e) $O_2(g)$ and $H^+(aq)$.

73. Magnesium is produced by electrolysis of molten magnesium chloride. What are the products at the anode and cathode, respectively?
a) $Cl_2(g)$ and $Mg(l)$
b) $Cl^-(aq)$ and $MgO(l)$
c) $Cl_2(g)$ and $MgO(l)$
d) $O_2(g)$ and $Mg(l)$
e) $Mg(l)$ and $O_2(g)$

74. The half-reaction that occurs at the anode during electrolysis of molten sodium chloride is
a) $Na(l) \rightarrow Na^+(l) + e^-$
b) $2H_2O(l) \rightarrow O_2(g) + 4H^+(aq) + 4e^-$
c) $Na^+(l) + e^- \rightarrow Na(l)$
d) $2Cl^-(l) \rightarrow Cl_2(g) + 2e^-$
e) $Cl_2(g) + 2e^- \rightarrow 2Cl^-(l)$

75. If 306 C of charge is passed through a solution of $Cu(NO_3)_2(aq)$, calculate the number of moles of copper deposited.
a) 0.00634 mol b) 0.00317 mol c) 0.00159 mol d) 1.00 mol e) 2.00 mol

76. How many moles of $O_2(g)$ are produced by electrolysis of $Na_2SO_4(aq)$ if 0.120 A is passed through the solution for 65.0 min?
a) 0.0000202 mol b) 0.0000808 mol c) 0.00121 mol d) 0.00485 mol e) 0.00242 mol

77. If 1020 C of charge is passed through a solution of $AgNO_3(aq)$, calculate the number of moles of silver deposited.
a) 0.0212 mol b) 0.0106 mol c) 0.00530 mol d) 0.0424 mol e) 1.00 mol

78. How long will it take to deposit 0.00470 moles of gold by electrolysis of $KAuCl_4(aq)$ using a current of 0.214 amperes?
a) 106 min b) 35.3 min c) 23.0 min d) 212 min e) 70.7 min

79. How long will it take to prepare 0.600 moles of $Cl_2(g)$ by the electrolysis of concentrated sodium chloride, if 0.500 A are passed through the solution? The equation for this process, the "chloralkali" process is

$$2NaCl(aq) + 2H_2O(l) \rightarrow 2NaOH(aq) + H_2(g) + Cl_2(g)$$

a) 965 min b) 1608 min c) 1930 min d) 3860 min e) 7720 min

80. How many moles of $Cl_2(g)$ are produced by the electrolysis of concentrated sodium chloride, if 2.00 A are passed through the solution for 8.00 hours? The equation for this process, the "chloralkali" process is

$$2NaCl(aq) + 2H_2O(l) \rightarrow 2NaOH(aq) + H_2(g) + Cl_2(g)$$

a) 0.00496 mol b) 0.596 mol c) 0.894 mol d) 0.149 mol e) 0.298 mol

Chapter 17: Electrons in Transition: Electrochemistry Form B

1. Given: $Cl^-(aq) \rightarrow Cl_2(g)$, acidic solution.
   How many electrons appear in the balanced half-reaction?
a) 2 b) 3 c) 1 d) 4 e) 5

2. Given: $NO_3^-(aq) \rightarrow NH_3(aq)$, basic solution.
   How many electrons appear in the balanced half-reaction?
a) 4 b) 10 c) 6 d) 8 e) 2

3. Given: $H_2SO_3(aq) \rightarrow HSO_4^-(aq)$, acidic solution.
   How many electrons appear in the balanced half-reaction?
a) 2 b) 3 c) 7 d) 6 e) 5

4. Given: $MnO_4^-(aq) \rightarrow MnO_2(s)$, basic solution.
   How many electrons appear in the balanced half-reaction?
a) 3 b) 2 c) 6 d) 7 e) 5

5. Given: $H_2C_2O_4(aq) \rightarrow CO_2(g)$, acidic solution.
   How many electrons appear in the balanced half-reaction?
a) 2 b) 3 c) 5 d) 10 e) 1

6. Given: $Zn(OH)_4^{2-}(aq) \rightarrow Zn(s)$, basic solution.
   How many electrons appear in the balanced half-reaction?
a) 2 b) 6 c) 4 d) 1 e) 0

7. Given: $MnO_4^-(aq) + H^+(aq) + Cl^-(aq) \rightarrow Mn^{2+}(aq) + Cl_2(g) + H_2O(l)$
   If the coefficient of Cl^- in the balanced equation is 10, what are the coefficients of MnO_4^-, H^+, and Cl_2, respectively?
a) 2, 10, 5 b) 8, 16, 5 c) 2, 16, 5 d) 1, 16, 5 e) 5, 8, 5

8. Given: $I^-(aq) + MnO_4^-(aq) + H_2O(l) \rightarrow I_2(s) + MnO_2(s) + OH^-(aq)$
   If the coefficient of I^- in the balanced equation is 6, what are the coefficients of MnO_4^- and OH^-, respectively?
a) 3 and 12 b) 2 and 8 c) 2 and 6 d) 1 and 8 e) 1 and 6

9. Given: $MnO_4^-(aq) + H^+(aq) + Cl^-(aq) \rightarrow Mn^{2+}(aq) + Cl_2(g) + H_2O(l)$
   If the coefficient of MnO_4^- in the balanced equation is 2, how many electrons are transferred in the reaction?
a) 10 b) 5 c) 2 d) 7 e) 8

10. Given: $MnO_4^-(aq) + H^+(aq) + Cl^-(aq) \rightarrow Mn^{2+}(aq) + Cl_2(g) + H_2O(l)$
If 10 electrons are transferred in the balanced equation, what is the coefficient of Cl^-?
a) 10 b) 5 c) 2 d) 7 e) 8

11. When alkaline $ClO^-(aq)$ is mixed with $Cr(OH)_3(s)$, $CrO_4^{2-}(aq)$ and $Cl^-(aq)$ are produced. When the equation for this reaction is balanced, what is the number of electrons transferred?
a) 6 b) 8 c) 4 d) 2 e) 3

12. Given: $SO_4^{2-}(aq) \rightarrow H_2SO_3(aq)$, acidic solution.
When this half-reaction is balanced, both H^+ and H_2O appear in the equation. If the coefficient of H_2SO_3 is 1 in the balanced half-reaction, what is the coefficient of H^+?
a) 4 b) 8 c) 12 d) 10 e) 2

13. When the reaction below is balanced, how many electrons are transferred?
$Cr^{3+}(aq) + OH^-(aq) + H_2O_2(aq) \rightarrow CrO_4^{2-}(aq) + H_2O(l)$
a) 6 b) 12 c) 3 d) 2 e) 8

14. When the $Ag(s) \mid AgCl(s) \mid Cl^-(aq)$ electrode acts as an anode, the half-reaction is
a) $2AgCl(s) + 2e^- \rightarrow 2Ag^+(aq) + Cl_2(g)$
b) $Ag^+(aq) + e^- \rightarrow Ag(s)$
c) $Ag(s) + Cl^-(aq) \rightarrow AgCl(s) + e^-$
d) $Ag(s) \rightarrow Ag^+(aq) + e^-$
e) $AgCl(s) + e^- \rightarrow Ag(s) + Cl^-(aq)$

15. When the $Ag(s) \mid AgBr(s) \mid Br^-(aq)$ electrode acts as an anode, the half-reaction is
a) $AgBr(s) + e^- \rightarrow Ag(s) + Br^-(aq)$
b) $Ag(s) \rightarrow Ag^+(aq) + e^-$
c) $Ag^+(aq) + e^- \rightarrow Ag(s)$
d) $2AgBr(s) + 2e^- \rightarrow 2Ag^+(aq) + Br_2(l)$
e) $Ag(s) + Br^-(aq) \rightarrow AgBr(s) + e^-$

16. For the cell diagram
$Pt \mid H_2(g) \mid H^+(aq) \mid Fe^{3+}(aq), Fe^{2+}(aq) \mid Pt$
which reaction occurs at the cathode?
a) $2H^+(aq) + 2e^- \rightarrow H_2(g)$
b) $Fe^{2+}(aq) \rightarrow Fe^{3+}(aq) + e^-$
c) $H_2(g) \rightarrow 2H^+(aq) + 2e^-$
d) $Fe^{3+}(aq) + e^- \rightarrow Fe^{2+}(aq)$
e) $2Fe^{3+}(aq) + H_2(g) \rightarrow 2Fe^{2+}(aq) + 2H^+(aq)$

17. For the cell diagram
$Pt \mid Br_2(l) \mid Br^-(aq) \mid Cl^-(aq), Cl_2(g) \mid Pt$
which reaction occurs at the anode?
a) $2Br_2(l) + 2e^- \rightarrow 2Br^-(aq)$
b) $Cl_2(g) + 2e^- \rightarrow 2Cl^-(aq)$
c) $2Cl^-(aq) \rightarrow Cl_2(g) + 2e^-$
d) $Cl_2(g) + 2Br^-(aq) \rightarrow 2Cl^-(aq) + Br_2(l)$
e) $2Br^-(aq) \rightarrow Br_2(l) + 2e^-$

18. For the cell diagram

$Pt | H_2(g), H^+(aq) | Cu^{2+}(aq) | Cu(s)$

Which reaction occurs at the cathode?

a) $Cu(s) \rightarrow Cu^{2+}(aq) + 2e^-$
b) $Cu^{2+}(aq) + 2e^- \rightarrow Cu(s)$
c) $2H^+(aq) + 2e^- \rightarrow H_2(g)$
d) $H_2(g) \rightarrow 2H^+(aq) + 2e^-$
e) $2H^+(aq) + Cu(s) \rightarrow H_2(g) + Cu^{2+}(aq)$

19. The cell diagram for the reaction in which hygrogen gas reduces $Ag^+(aq)$ is

a) $Pt | H_2(g) | H^+(aq) | Ag^+(aq) | Ag(s)$
b) $Pt | Ag^+(aq), H_2(g) | H^+(aq) | Pt$
c) $Pt | Ag^+(aq), H^+(aq) | H_2(g) | Pt$
d) $Ag(s) | Ag^+(aq) | H^+(aq) | H_2(g) | Pt$
e) $Ag(s) | Ag^+(aq) | O_2(g) | H_2(g) | Pt$

20. Write the cell diagram for the reaction

$2AgCl(s) + H_2(g) \rightarrow 2Ag(s) + 2H^+(aq) + 2Cl^-(aq)$

a) $Pt | H_2(g) | H^+(aq) | Cl^-(aq) | Ag(s) | Pt$
b) $Ag(s) | AgCl(s) | Cl^-(aq) | H^+(aq) | H_2(g) | Pt$
c) $Pt | H_2(g) | H^+(aq) | Cl^-(aq) | AgCl(s) | Ag(s)$
d) $Pt | Cl^-(aq) | H^+(aq) | H_2(g) | AgCl(s) | Ag(s)$
e) $Ag(s) | AgCl(s) | H^+(aq) | Cl^-(aq) | H_2(g) | Pt$

21. Consider the following cell diagram:

$Pt | Fe^{3+}(aq), Fe^{2+}(aq) | Cl^-(aq) | Cl_2(g) | Pt$

The reaction utilized by this cell is

a) $Fe^{2+}(aq) + 2Cl^-(aq) \rightarrow Fe(s) + Cl_2(g)$
b) $Fe(s) + Cl_2(g) \rightarrow Fe^{2+}(aq) + 2Cl^-(aq)$
c) $2Fe^{3+}(aq) + 2Cl^-(aq) \rightarrow 2Fe^{2+}(aq) + Cl_2(g)$
d) $2Fe^{2+}(aq) + Cl_2(g) \rightarrow 2Fe^{3+}(aq) + 2Cl^-(aq)$
e) $Fe^{3+}(aq) + Cl^-(aq) \rightarrow Fe^{2+}(aq) + 1/2Cl_2(g)$

22. The standard potential of the Ag^+/Ag electrode is +0.80 V and the standard potential of the cell

$Fe(s) | Fe^{2+}(aq) | Ag^+(aq) | Ag(s)$

is +1.24 V. What is the standard potential of the Fe^{2+}/Fe electrode?

a) +2.04 V b) –0.44 V c) –2.04 V d) +0.44 V e) –0.88 V

23. The standard potential of the Cu^{2+}/Cu electrode is +0.34 V and the standard potential of the cell

$Pb(s) | Pb^{2+}(aq) | Cu^{2+}(aq) | Cu(s)$

is +0.47 V. What is the standard potential of the Pb^{2+}/Pb electrode?

a) +0.13 V b) –0.13 V c) +0.81 V d) –0.81V e) –0.26 V

24. The standard potential of the Pb^{2+}/Pb electrode is –0.13 V and the standard potential of the cell

$Sn(s) | Sn^{2+}(aq) | Pb^{2+}(aq) | Pb(s)$

is +0.01 V. What is the standard potential of the Sn^{2+}/Sn electrode?

a) –0.12 V b) –0.14 V c) +0.14 V d) +0.12 V e) –0.28 V

25. Consider the following reaction:

$$2Cu^{+}(aq) \rightarrow Cu(s) + Cu^{2+}(aq)$$

If the standard potentials of Cu^{2+} and Cu^{+} are +0.34 and +0.52 V, respectively, calculate the value of E° for the given reaction.

a) –0.70 V b) –0.18 V c) +0.70 V d) +0.18 V e) +0.86 V

26. Calculate E° for the following cell.

$$Pt\,|\,Fe^{3+}(aq),Fe^{2+}(aq)\,|\,Cl^{-}(aq)\,|\,Cl_2(g)\,|\,Pt$$

a) +2.13 V b) +0.59 V c) +1.77 V d) +1.00 V e) +0.95 V

27. The standard potential of the cell

$$Ni(s)\,|\,Ni^{2+}(aq)\,|\,Cl^{-}(aq)\,|\,AgCl(s)\,|\,Ag(s)$$

is +0.45 V at 25°C. If the standard potential of Ni^{2+} is –0.23 V, calculate the standard potential of the AgCl/Ag,Cl^{-} couple.

a) –0.22 V b) –0.68 V c) –0.45 V d) +0.22 V e) +0.68 V

28. The standard potential of the cell

$$Cu(s)\,|\,Cu^{2+}(aq)\,|\,Cl^{-}(aq)\,|\,AgCl(s)\,|\,Ag(s)$$

is –0.12 V at 25°C. If the standard potential of Cu^{2+} is +0.34 V, calculate the standard potential of the AgCl/Ag,Cl^{-} couple.

a) –0.46 V b) –0.22 V c) +0.12 V d) +0.22 V e) +0.46 V

29. The standard potential of the cell

$$Ni(s)\,|\,Ni^{2+}(aq)\,|\,Cl^{-}(aq)\,|\,AgCl(s)\,|\,Ag(s)$$

is +0.45 V at 25°C. If the standard potential of the AgCl/Ag,Cl^{-} couple is +0.22 V, calculate the standard potential of the Ni^{2+}/Ni couple.

a) –0.45 V b) –0.67 V c) +0.23 V d) –0.23 V e) +0.67 V

30. The standard potential of the cell

$$Sn(s)\,|\,Sn^{2+}(aq)\,|\,Cl^{-}(aq)\,|\,AgCl(s)\,|\,Ag(s)$$

is +0.36 V at 25°C. If the standard potential of the AgCl/Ag,Cl^{-} couple is 0.22 V, calculate the standard potential of the Sn^{2+}/Sn couple.

a) +0.36 V b) +0.58 V c) –0.14 V d) +0.14 V e) –0.07 V

31. Calculate the standard potential for the following half-reaction:

$$Ti^{3+}(aq) + 3e^{-} \rightarrow Ti(s)$$

The standard potentials of the Ti^{3+}/Ti^{2+} and Ti^{2+}/Ti couples are –0.37 and –1.63 V, respectively.

32. Which species will oxidize V^{2+} but not Br^{-}?

a) Cu^{2+} b) Fe^{2+} c) Zn^{2+} d) Mn^{2+} e) O_2

33. Which species will reduce Br_2 but not V^{3+}?

a) Cu b) Cr^{2+} c) Zn d) Al e) Ce

34. Which species will oxidize Cr^{2+} but not Mn^{2+}?
a) Fe^{2+} b) O_3 c) Zn^{2+} d) Pb^{4+} e) Pb^{2+}

35. Which species will reduce Cr^{3+} but not Mn^{3+}?
a) Sn b) Zn c) Fe d) Al e) In^{2+}

36. Which species will reduce Ag^+ but not Fe^{2+}?
a) H_2 b) Au c) Cr d) V e) Pt

37. Which metal will dissolve in hydrochloric acid?
a) Fe b) Pt c) Ag d) Au e) All of the metals listed will dissolve.

38. If the standard potentials for the couples Cu^{2+}/Cu, Ag^+/Ag, and Fe^{2+}/Fe are +0.34, + 0.80, and –0.44 V, respectively, which is the weakest oxidizing agent?
a) Cu^{2+} b) Fe^{2+} c) Ag^+ d) Cu e) Ag

39. If the standard potentials for the couples Cu^{2+}/Cu, Ag^+/Ag, and Fe^{2+}/Fe are +0.34, + 0.80, and –0.44 V, respectively, which is the weakest reducing agent?
a) Ag b) Cu^{2+} c) Fe^{2+} d) Cu e) Fe

40. If the standard potentials for the couples Fe^{3+}/Fe^{2+}, $MnO_4^-,H^+/Mn^{2+}$, Zn^{2+}/Zn, V^{3+}/V^{2+}, and Br_2/Br^- are +0.77, +1.51, –0.76, –0.26, and +1.09 V, respectively, which is the strongest reducing agent?
a) Zn b) Br^- c) V^{2+} d) Zn^{2+} e) Fe^{2+}

41. If the standard potentials for the couples Fe^{3+}/Fe^{2+}, $MnO_4^-,H^+/Mn^{2+}$, Zn^{2+}/Zn, V^{3+}/V^{2+}, and Br_2/Br^- are +0.77, +1.51, –0.76, –0.26, and +1.09 V, respectively, which is the weakest oxidizing agent?
a) Fe^{3+} b) Br_2 c) Zn^{2+} d) MnO_4^- e) V^{3+}

42. Which of the following is the strongest reducing agent.
a) Zn b) Cu c) Hg d) Pf e) Cu^+

43. Which of the following is the strongest oxidizing agent?
a) Ag^+ b) Cu^{2+} c) Co^{3+} d) Al e) H_2

44. Given:

$$Ag^+(aq) + e^- \rightarrow Ag(s) \qquad E^\circ = 0.80\ V$$
$$Fe^{3+}(aq) + e^- \rightarrow Fe^{2+}(aq) \qquad E^\circ = 0.77\ V$$
$$Cu^{2+}(aq) + 2e^- \rightarrow Cu(s) \qquad E^\circ = 0.34\ V$$

Which is the strongest oxidizing agent?
a) Ag^+ b) Fe^{2+} c) Fe^{3+} d) Ag e) Cu^{2+}

45. Which of the following will be reduced by Cr^{2+}?
a) Pb^{2+} b) Al^{3+} c) Ca^{2+} d) Fe^{2+} e) Zn^{2+}

46. Place the following in order of decreasing strength as an oxidizing agent.
Fe^{3+}, Pb^{2+}, Co^{3+}, Cu^{+}
a) $Co^{3+} > Fe^{3+} > Cu^{+} > Pb^{2+}$
b) $Co^{3+} > Cu^{+} > Fe^{3+} > Pb^{2+}$
c) $Pb^{2+} > Cu^{+} > Fe^{3+} > Co^{3+}$
d) $Co^{3+} > Pb^{2+} > Fe^{3+} > Cu^{+}$
e) $Pb^{2+} > Fe^{3+} > Co^{3+} > Cu^{+}$

47. Which of the following occurs when HCl(aq), Cu(s), and Fe(s) are mixed under standard conditions?
a) Fe(s) dissolves
b) Cu(s) dissolves
c) O_2(g) is formed
d) no reaction takes place
e) Cl_2(g) is formed

48. When Cr^{3+}(aq) is treated with alkaline hydrogen peroxide, the yellow chromate ion is formed. Write a balanced net ionic equation for this reaction.

49. Predict what reaction, if any, will occur when Co^{2+}(aq), Co(s), and Fe^{2+}(aq) are mixed under standard conditions.

50. Consider the following cell at 25°C:
Ag(s) | AgCl(s) | Cl^{-}(aq) | Ag^{+}(aq) | Ag(s)
(a) Write a balanced net ionic equation for the overall cell reaction and calculate the value of $E°$.
(b) Calculate the value of K_{sp} for AgCl.

51. The standard potential of the cell
Ag(s) | Ag^{+}(aq) | Cl^{-}(aq) | AgCl(s) | Ag(s)
is –0.58 V at 25°C. Calculate the equilibrium constant for the reaction of 1 Ag^{+}(aq) with Cl^{-}(aq).
a) 1.2×10^{-5} b) 1.7×10^{17} c) 6.3×10^{9} d) 1.6×10^{-10} e) 3.7×10^{22}

52. The standard potential of the cell
Pb(s) | $PbSO_4$(s) | SO_4^{2-}(aq) | Pb^{2+}(aq) | Pb(s)
is +0.23 V at 25°C. Calculate the K_{sp} of $PbSO_4$.
a) 6.0×10^{7} b) 1.3×10^{-4} c) 1.7×10^{-8} d) 1.3×10^{-18} e) 2.7×10^{-17}

53. The equilibrium constant for the reaction
$2Hg(l) + 2Cl^{-}(aq) + Ni^{2+}(aq) \rightarrow Ni(s) + Hg_2Cl_2(s)$
is 5.6×10^{-20} at 25°C. Calculate the value of $E°$ for a cell utilizing this reaction.
a) +0.57 V b) –1.14 V c) –0.57 V d) –0.25 V e) +1.14 V

54. If $E°$ for the reaction
$Cr_2O_7^{2-}(aq) + 6Fe^{2+}(aq) + 14H^{+}(aq) \rightarrow 2Cr^{3+}(aq) + 6Fe^{3+}(aq) + 7H_2O(l)$
is +0.56 V at 25°C, calculate the equilibrium constant for the reaction.
a) 6.2×10^{56} b) 37.8 c) 2.9×10^{9} d) 2.5×10^{28} e) 1.4×10^{3}

55. The standard voltage of the cell

$Ag(s) \mid Ag^+(aq) \mid Br^-(aq) \mid AgBr(s) \mid Ag(s)$

is –0.73 V at 25°C. Calculate the K_{sp} for AgBr.

a) 2.2×10^{12} b) 4.6×10^{-13} c) 3.9×10^{-29} d) 2.0×10^{-15} e) 5.1×10^{14}

56. If the standard free energy change for the reaction

$2H_2(g) + O_2(g) \rightarrow 2H_2O(l)$

is –475 kJ, calculate the standard voltage that could be obtained from a fuel cell utilizing this reaction.

a) +4.92 V b) 0 V c) –1.23 V d) +1.23 V e) +2.46 V

57. Consider the cell below at standard conditions:

$Zn(s) \mid Zn^{2+}(aq) \mid Fe^{2+}(aq) \mid Fe(s)$

Calculate the value of $\Delta G°$ for the reaction that occurs when current is drawn from this cell.

a) –31 kJ b) +230 kJ c) +62 kJ d) –62 kJ e) –230 kJ

58. A standard silver electrode is connected to a hydrogen electrode in which the pressure of hydrogen gas is 1 atm and the hydrogen ion concentration is unknown. Calculate the pH if the cell voltage is +0.90 V, with silver the cathode.

59. Consider the following cell:

$Zn(s) \mid Zn^{2+}(aq, 0.200\ M) \mid Cl^-(aq, 0.100\ M) \mid AgCl(s)Ag(s)$

If the standard potentials of the Zn^{2+}/Zn and $AgCl/Ag,Cl^-$ couples are –0.76 and 0.22 V, respectively, at 25°C, calculate the voltage of the cell.

a) +0.62 V b) +0.90 V c) +1.06 V d) –0.62 V e) +1.01 V

60. Consider the following cell:

$Ni(s) \mid Ni^{2+}(aq, 0.100\ M) \mid Cl^-(aq, 0.200\ M) \mid Cl_2(g, 1.50\ atm) \mid Pt$

If the standard voltage for this cell is 1.59 V at 25°C, calculate the voltage of this cell.

a) +1.67 V b) +1.62 V c) +1.74 V d) +1.51 V e) +1.44 V

61. Consider the following cell:

$Zn(s) \mid Zn^{2+}(aq, 1.00 \times 10^{-4}\ M) \mid Br^-(aq, 0.100\ M) \mid AgBr(s) \mid Ag(s)$

If the standard potentials of the Zn^{2+}/Zn and $AgBr/Ag,Br^-$ couples are –0.76 and 0.07 V, respectively, calculate the cell voltage at 25°C.

a) +1.19 V b) +1.01 V c) +0.83 V d) +1.13 V e) +0.98 V

62. Consider the following cell:

$Ag(s) \mid Ag^+(aq, 1.00 \times 10^{-5}\ M) \mid Ag^+(aq, 1.00\ M) \mid Ag(s)$

What is the voltage of this cell?

a) +0.296 V b) +0.0118 V c) +0.0592 V d) –0.296 V e) +0.80 V

63. Calculate E for the half-reaction below.

$2H^+(aq, 1.00 \times 10^{-2}\ M) + 2e^- \rightarrow H_2(g, 1.00\ atm)$

a) +0.829 V b) –0.118 V c) +0.118 V d) –0.829 V e) 0 V

64. Consider the following cell:

$Zn(s) \mid Zn^{2+}(aq, 0.200\ M) \mid H^{+}(aq, ?) \mid H_2(g, 1.00\ atm) \mid Pt$

If $E = +0.66$ V and $E^{o} = +0.76$ V at 25°C, calculate the concentration of H^{+} in the cathode cell compartment.

a) 8.4×10^{-5} M b) 9.2×10^{-3} M c) 4.0×10^{-3} M d) 4.0×10^{-1} M e) 2.1×10^{-2} M

65. Consider the following cell:

$Pt \mid H_2(g, 1\ atm) \mid H^{+}(aq, ?\ M) \mid Ag^{+}(aq, 0.0100\ M) \mid Ag(s)$

If the voltage of this cell is 0.90 V at 25°C and the standard potential of the Ag^{+}/Ag couple is +0.80 V, calculate the hydrogen ion concentration in the anode compartment.

a) 2.0×10^{-6} M b) 4.2×10^{-4} M c) 1.3×10^{-6} M d) 2.0×10^{-4} M e) 1.0 M

66. Consider the following cell:

$Ni(s) \mid Ni^{2+}(aq, 0.100\ M) \mid Cl^{-}(aq, ?\ M) \mid Cl_2(g, 0.800\ atm) \mid Pt$

For this cell, $E^{o} = 1.59$ V and $E = 1.79$ V at 25°C. Calculate the $Cl^{-}(aq)$ concentration in the cathode compartment.

a) 4.2×10^{-4} M b) 1.2×10^{-3} M c) 3.4×10^{-3} M d) 5.8×10^{-2} M e) 6.5×10^{-2} M

67. Consider the following cell:

$Pb(s) \mid PbSO_4(s) \mid SO_4^{2-}(aq, 0.60\ M) \mid H_2(g, 192.5\ kPa) \mid H^{+}(aq, 0.70\ M) \mid Pt$

If E^{o} for the cell is 0.36 V at 25°C, write the Nernst equation for the cell at this temperature.

a) $E = 0.36 + 0.01285\ln[192.5/\{(0.70)^2(0.60)\}]$
b) $E = 0.36 - 0.02569\ln[1.90/\{(0.70)^2(0.60)\}]$
c) $E = 0.36 - 0.01285\ln[1.90/\{(0.70)^2(0.60)\}]$
d) $E = 0.36 + 0.01285\ln[1.90/\{(0.70)^2(0.60)\}]$
e) $E = 0.36 - 0.01285\ln[1.90/\{(0.70)(0.60)\}]$

68. When a lead–acid battery discharges,

a) Pb(s) is formed at the anode.
b) $PbO_2(s)$ is formed at the cathode.
c) sulfuric acid is produced.
d) sulfuric acid is consumed.
e) $H_2O(l)$ is consumed.

69. In the standard dry cell, the anode reaction is

a) $Zn(s) \rightarrow Zn^{2+}(aq) + 2e^{-}$
b) $MnO_2(s) + H_2O(l) + e^{-} \rightarrow MnO(OH)(s) + OH^{-}(aq)$
c) $Mn^{2+}(aq) \rightarrow Mn^{3+}(aq) + e^{-}$
d) $2H_2O(l) \rightarrow 4H^{+}(aq) + O_2(g) + 4e^{-}$
e) $Zn^{2+}(aq) + 2e^{-} \rightarrow Zn(s)$

70. Which metal would be suitable to provide cathodic protection from corrosion for an iron bridge?

a) None of the metals listed is suitable. b) Sn c) Pb d) Ni e) Cu

71. The products of the electrolysis of NaF(aq) are
a) Na(l) and $O_2(g)$.
b) $H_2(g)$ and $O_2(g)$.
c) $H_2(g)$, $OH^-(aq)$, and $F_2(g)$.
d) Na(l) and $F_2(g)$.
e) $H_2(g)$ and $F_2(g)$.

72. The products of the electrolysis of $CuSO_4(aq)$ are
a) $H_2(g)$ and $H_2SO_3(aq)$.
b) Cu(s) and $H_2SO_3(aq)$.
c) $H_2(g)$ and $O_2(g)$.
d) $H_2SO_3(aq)$ and $O_2(g)$.
e) Cu(s) and $O_2(g)$.

73. Sodium is produced by electrolysis of molten sodium chloride. What are the products at the anode and cathode, respectively?
a) $Cl^-(aq)$ and $Na_2O(l)$
b) $Cl_2(g)$ and Na(l)
c) Na(l) and $O_2(g)$
d) $Cl_2(g)$ and $Na_2O(l)$
e) $O_2(g)$ and Na(l)

74. The half-reaction that occurs at the cathode during electrolysis of molten magnesium chloride is
a) $Mg^{2+}(l) + 2e^- \rightarrow Mg(l)$
b) $2Cl^-(l) \rightarrow Cl_2(g) + 2e^-$
c) $Mg(l) \rightarrow Mg^{2+}(l) + 2e^-$
d) $Cl_2(g) + 2e^- \rightarrow 2Cl^-(l)$
e) $2H_2O(l) \rightarrow O_2(g) + 4H^+(aq) + 4e^-$

75. If 612 C of charge is passed through a solution of $Cu(NO_3)_2(aq)$, calculate the number of moles of copper deposited.
a) 1.00 mol b) 0.00317 mol c) 0.0127 mol d) 0.00634 mol e) 2.00 mol

76. How many moles of $O_2(g)$ are produced by electrolysis of $Na_2SO_4(aq)$ if 0.240 A is passed through the solution for 65.0 min?
a) 0.00484 mol b) 0.000162 mol c) 0.0000404 mol d) 0.00970 mol e) 0.00242 mol

77. If 8686 C of charge is passed through molten magnesium chloride, calculate the number of moles of Mg(l) produced.
a) 0.0225 mol b) 0.0450 mol c) 0.0110 mol d) 0.0900 mol e) 2.00 mol

78. How long will it take to deposit 0.00235 moles of gold by electrolysis of $KAuCl_4(aq)$ using a current of 0.214 amperes?
a) 26.5 min b) 70.7 min c) 53.0 min d) 17.7 min e) 106 min

79. How long will it take to prepare 0.300 moles of $Cl_2(g)$ by the electrolysis of concentrated sodium chloride, if 0.500 A are passed through the solution? The equation for this process, the "chloralkali" process is

$$2NaCl(aq) + 2H_2O(l) \rightarrow 2NaOH(aq) + H_2(g) + Cl_2(g)$$

a) 3860 min b) 483 min c) 804 min d) 1930 min e) 965 min

80. How many moles of $Cl_2(g)$ are produced by the electrolysis of concentrated sodium chloride, if 2.00 A are passed through the solution for 4.00 hours? The equation for this process, the "chloralkali" process is

$$2NaCl(aq) + 2H_2O(l) \rightarrow 2NaOH(aq) + H_2(g) + Cl_2(g)$$

a) 0.149 mol b) 0.447 mol c) 0.298 mol d) 0.0745 mol e) 0.00248 mol

Chapter 18: Kinetics: The Rates of Reactions
Form A

1. The rate of formation of oxygen in the reaction

$$2N_2O_5(g) \rightarrow 4NO_2(g) + O_2(g)$$

is 2.28 (mol O_2)·L^{-1}·s^{-1}. What is the rate at which N_2O_5 is used?

a) –9.12 (mol N_2O_5)·L^{-1}·s^{-1}
b) –1.14 (mol N_2O_5)·L^{-1}·s^{-1}
c) –4.56 (mol N_2O_5)·L^{-1}·s^{-1}
d) –0.57 (mol N_2O_5)·L^{-1}·s^{-1}
e) –2.28 (mol N_2O_5)·L^{-1}·s^{-1}

2. The rate of formation of oxygen in the reaction

$$2N_2O_5(g) \rightarrow 4NO_2(g) + O_2(g)$$

is 1.45 (mol O_2)·L^{-1}·s^{-1}. What is the rate at which N_2O_5 is used?

a) –1.45 (mol N_2O_5)·L^{-1}·s^{-1}
b) –0.725 (mol N_2O_5)·L^{-1}·s^{-1}
c) –2.90 (mol N_2O_5)·L^{-1}·s^{-1}
d) –5.80 (mol N_2O_5)·L^{-1}·s^{-1}
e) –1.45 (mol N_2O_5)·L^{-1}·s^{-1}

3. Given:

$$2NO_2(g) + F_2(g) \rightarrow 2NO_2F(g) \qquad \text{rate} = -\Delta[F_2]/\Delta t$$

The rate of the reaction can also be expressed as

a) $\Delta[NO_2F]/\Delta t$ b) $-\Delta[NO_2]/\Delta t$ c) $-2\Delta[NO_2]/\Delta t$ d) $1/2\Delta[NO_2F]/\Delta t$ e) $1/2\Delta[NO_2]/\Delta t$

4. For the reaction

$$2NO_2(g) \rightarrow 2NO(g) + O_2(g)$$

initial rate = $k[NO_2]_o^2$. For concentration in M and time in s, the units of the rate constant k are

a) $M^{-2}\cdot s^{-1}$ b) $M^{-1}\cdot s^{-1}$ c) s^{-1} d) $M^2\cdot s^{-1}$ e) $M\cdot s^{-1}$

5. Given:

$$2N_2O_5(g) \rightarrow 4NO_2(g) + O_2(g) \qquad \text{rate} = k[N_2O_5]$$

The overall order of the reaction is

a) 1 b) 0 c) 2 d) 3 e) 7

6. Given:

$$4Fe^{2+}(aq) + O_2(aq) + 2H_2O(l) \rightarrow 4Fe^{3+}(aq) + 4OH^-(aq) \qquad \text{rate} = k[Fe^{2+}][OH^-]^2[O_2]$$

The overall order of the reaction and the order with respect to O_2 are

a) 4 and 2. b) 4 and 1. c) 5 and 1. d) 3 and 1. e) 7 and 1.

7. Given:

$$2A(g) + B(g) \rightarrow C(g) + D(g)$$

When [A] = [B] = 0.10 M, the rate is 2.0 $M\cdot s^{-1}$; for [A] = [B] = 0.20 M, the rate is 8.0 $M\cdot s^{-1}$; and for [A] = 0.10 M, [B] = 0.20 M, the rate is 2.0 $M\cdot s^{-1}$. The order of the reaction is

a) 2 b) 4 c) 1 d) 0 e) 1.5

8. If the rate of a reaction increases by a factor of 64 when the concentration of reactant increases by a factor of 4, the order of the reaction with respect to this reactant is

a) 3 b) 4 c) 2 d) 16 e) 1

9. For the reaction

$$2A + B \rightarrow \text{products}$$

determine the overall order of the reaction given the following data:

Initial Concentration, M		Initial Rate, $M\cdot s^{-1}$
A	B	
0.10	0.10	2.0×10^{-2}
0.20	0.10	8.0×10^{-2}
0.30	0.10	1.8×10^{-1}
0.20	0.20	8.0×10^{-2}
0.30	0.30	1.8×10^{-1}

a) 2 b) 3 c) 0 d) 1 e) 1.5

10. If the rate of reaction increases by a factor of 2.3 when the concentration of reactant increases by a factor of 1.5, the order of the reaction with respect to this reactant is

a) 2 b) 1 c) 3 d) 4 e) 1.5

11. The reaction

$$2NO(g) + 2H_2(g) \rightarrow N_2(g) + 2H_2O(g)$$

is first-order in H_2 and second-order in NO. Starting with equal concentrations of H_2 and NO, the rate after 50% of the H_2 has reacted is what percent of the initial rate?

a) 12.5% b) 25.0% c) 37.5% d) 18.8% e) 50.0%

12. The reaction

$$2NO(g) + Br_2(g) \rightarrow 2NOBr(g)$$

is second-order in NO and first-order in Br_2. If the initial concentrations of NO and Br_2 are 0.0200 M and 0.0100 M, respectively, the rate after 50% of the Br_2 has reacted is what percent of the initial rate?

a) 12.5% b) 37.5% c) 18.8% d) 2.50% e) 50.0%

13. For the reaction

$$C_2H_6(g) \rightarrow 2CH_3(g) \qquad \text{rate} = k[C_2H_6]$$

If $k = 5.50 \times 10^{-4}\ s^{-1}$ and $[C_2H_6]_{initial} = 0.0200$ M, calculate the rate of reaction after 1 hour.

a) The rate of reaction is zero since the reaction is complete.
b) 1.52×10^{-6} M·s^{-1}
c) 5.50×10^{-4} M·s^{-1}
d) 2.76×10^{-3} M·s^{-1}
e) 1.10×10^{-5} M·s^{-1}

14. Consider the following:

$$2A \rightarrow A_2 \qquad \text{rate} = k[A]^2$$

If the rate constant is 1.43 M^{-1}·s^{-1} and the initial concentration of A is 0.0180 M, calculate the time for the rate of consumption of A to drop to 1.25×10^{-5} M·s^{-1}.

a) 80.0 s b) 197 s c) 236 s d) 159 s e) 99.0 s

15. Given: $4PH_3(g) \rightarrow P_4(g) + 6H_2(g)$ rate = $k[PH_3]$

If the rate constant is 0.0278 s^{-1}, calculate the percent of the original PH_3 which has reacted after 76.0 s.

a) 87.9% b) 12.1% c) 47.3% d) 97.9% e) 2.10%

16. Consider the following reaction:

$$2N_2O(g) \rightarrow 2N_2(g) + O_2(g) \qquad \text{rate} = k[N_2O]$$

For an initial concentration of N_2O of 0.50 M, calculate the concentration of N_2O remaining after 2.0 min if $k = 3.4 \times 10^{-3}\ s^{-1}$.

a) 0.17 M b) 0.66 M c) 0.33 M d) 0.55 M e) 0.50 M

17. Consider the following reaction:

$$2N_2O(g) \rightarrow 2N_2(g) + O_2(g) \qquad \text{rate} = k[N_2O]$$

For an initial concentration of N_2O of 0.50 M, calculate the concentration of N_2O remaining after 1.0 min if $k = 3.4 \times 10^{-3}\ s^{-1}$.

a) 0.50 M b) 0.63 M c) 0.82 M d) 0.41 M e) 0.31 M

18. Consider the following reaction:

$$2N_2O(g) \rightarrow 2N_2(g) + O_2(g) \qquad \text{rate} = k[N_2O]$$

For an initial concentration of N_2O of 0.50 M, calculate the concentration of N_2O remaining after 2 min if $k = 6.8 \times 10^{-3}\ s^{-1}$.

a) 0.30 M b) 0.49 M c) 0.44 M d) 0.22 M e) 0.15 M

19. A first-order reaction has a rate constant of 0.00300 s^{-1}. The time required for 75% reaction is

a) 231 s b) 201 s c) 41.7 s d) 462 s e) 95.9 s

20. A first-order reaction has a rate constant of 0.00300 s^{-1}. The time required for 85% reaction is

a) 23.5 s b) 316 s c) 632 s d) 54.2 s e) 275 s

21. For a first-order reaction, after 230 s, 10% of the reactants remain. Calculate the rate constant for the reaction.
a) 0.000458 s^{-1} b) 0.510 s^{-1} c) 0.0195 s^{-1} d) 100 s^{-1} e) 0.0100 s^{-1}

22. For a first-order reaction, after 2.00 min, 20% of the reactants remain. Calculate the rate constant for the reaction.
a) 0.00186 s^{-1} b) 0.00582 s^{-1} c) 0.0134 s^{-1} d) 0.000808 s^{-1} e) 74.6 s^{-1}

23. For a first-order reaction, after 5.00 s, 10% of the reactants remain. Calculate the rate constant for the reaction.
a) 2.17 s^{-1} b) 0.00915 s^{-1} c) 0.200 s^{-1} d) 0.461 s^{-1} e) 0.0211 s^{-1}

24. Consider the following reaction:
$2N_2O_5(g) \rightarrow 4NO_2(g) + O_2(g)$ rate = $k[N_2O_5]$
Calculate the time for the concentration of N_2O_5 to fall to one–fourth its initial value if the rate constant for the reaction is 5.20×10^{-3} s^{-1}.
a) 66.6 s b) 267 s c) 33.3 s d) 533 s e) 133 s

25. Consider the following reaction:
$2N_2O_5(g) \rightarrow 4NO_2(g) + O_2(g)$ rate = $k[N_2O_5]$
Calculate the time for the concentration of N_2O_5 to fall to one–sixth its initial value if the rate constant for the reaction is 5.20×10^{-3} s^{-1}.
a) 2070 s b) 800 s c) 22.2 s d) 345 s e) 57.4 s

26. Given: $SO_2Cl_2(g) \rightarrow SO_2(g) + Cl_2(g)$ rate = $k[SO_2Cl_2]$
If 70% of the SO_2Cl_2 has reacted after 80 s, calculate the rate constant.
a) 0.0053 s^{-1} b) 5.3 s^{-1} c) 0.0045 s^{-1} d) 0.015 s^{-1} e) 0.029 s^{-1}

27. The following data were obtained for the disappearance of A at 25°C:

[A], M	**Time, s**
0.090	0
0.069	5
0.054	10
0.042	15
0.032	20
0.025	25
0.019	30

What is the order of the reaction and the rate constant?

28. The following data were obtained for the disappearance of A at 44.3°C:

[A], M	Time, s
0.568	30
0.509	70
0.443	120
0.337	220
0.257	320
0.193	420
0.127	570

Determine the order of this reaction, and calculate the rate constant and half-life.

29. Given: $2H_2O_2(aq) \rightarrow 2H_2O(l) + O_2(g)$ rate = $k[H_2O_2]$
If 75% of the H_2O_2 has reacted after 40 min, calculate the rate constant.
a) 0.0072 s^{-1} b) 0.035 s^{-1} c) 0.069 s^{-1} d) 0.000012 s^{-1} e) 0.00058 s^{-1}

30. Given: A $\rightarrow$ P rate = k[A]
If 20% of A reacts in 5.12 min, calculate the time required for 60% of A to react.
a) 30.7 min b) 21.0 min c) 48.3 min d) 11.7 min e) 15.4 min

31. Consider the following reaction:
$2N_2O(g) \rightarrow 2NO(g) + O_2(g)$ rate = $k[N_2O]$
If 46% of N_2O reacts in 1.0 s, what is the rate constant?
a) 0.78 s^{-1} b) 0.34 s^{-1} c) 0.62 s^{-1} d) 3.8 s^{-1} e) 0.27 s^{-1}

32. For the reaction
cyclopropane $\rightarrow$ propene
a plot of ln[cyclopropane] versus time in seconds gives a straight line with slope -2.8×10^{-4} s^{-1} at 500°C. What is the rate constant for this reaction?
a) 5.6×10^{-4} s^{-1} b) 1.7×10^{-2} s^{-1} c) 1.2×10^{-4} s^{-1} d) 6.4×10^{-4} s^{-1} e) 2.8×10^{-4} s^{-1}

33. For the first-order reaction
$2N_2O(g) \rightarrow 2NO(g) + O_2(g)$
the rate constant is 3.4 s^{-1}. Calculate the percent N_2O remaining after 1 second.
a) 3.3% b) 30% c) 87% d) 20% e) 66%

34. For the first-order reaction
$2N_2O(g) \rightarrow 2NO(g) + O_2(g)$
the rate constant is 3.4 s^{-1}. The time required for 75% reaction is
a) 85 ms b) 0.41 s c) 37 ms d) 0.18 s e) 2.6 s

35. A first-order reaction has a rate constant of 26 s^{-1}. If the initial concentration of reactant is 0.082 M, how long will it take for the reactant to reach 0.0010 M?
a) 0.074 s b) 2.2 s c) 0.097 s d) 0.038 s e) 0.17 s

36. Consider the following reaction:

$2N_2O_5(g) \rightarrow 4NO_2(g) + O_2(g)$ rate = $k[N_2O_5]$

If the initial concentration of N_2O_5 is 0.80 M, the concentration after 2 half-lives is

a) 0.55 M b) 0.20 M c) 0.35 M d) 0.40 M e) 0.28 M

37. Consider the following reaction:

$2N_2O_5(g) \rightarrow 4NO_2(g) + O_2(g)$ rate = $k[N_2O_5]$

If the initial concentration of N_2O_5 is 0.80 M, the concentration after 4 half-lives is

a) 0.10 M b) 0.20 M c) 0.063 M d) 0.050 M e) 0.025 M

38. A compound decomposes with a half-life of 8.0 s and the half-life is independent of the concentration. How long does it take for the concentration to decrease to one–sixteenth of its initial value?

a) 24 s b) 32 s c) 40 s d) 130 s e) 64 s

39. For a first-order reaction, what fraction of the starting material will remain after 4 half-lives?

a) 1/16 b) 1/9 c) 1/4 d) 1/8 e) 1/3

40. A first-order reaction has a half-life of 1.10 s. If the initial concentration of reactant is 0.142 M, how long will it take for the reactant concentration to reach 0.00100 M?

a) 0.127 s b) 4.50 s c) 7.87 s d) 1.59 s e) 3.09 s

41. A first-order reaction has a rate constant of 0.021 s^{-1}. What is the half-life for this reaction?

a) 0.69 s b) 0.015 s c) 14 s d) 33 s e) 0.030 s

42. Given: $2A + B \rightarrow P$ rate = $k[A]$

Which of the following is true?

a) The overall order of the reaction is 3.
b) $t_{1/2} = 0.693/k$
c) $[A] = 1/(k \times t_{1/2})$
d) $1/[A] = kt$
e) $\ln[A] = k/t$

43. For a first-order reaction, a straight line is obtained from a plot of

a) $\ln(t)$ vs. [A] b) [A] vs. t c) 1/[A] vs. t d) $\ln(1/t)$ vs. [A] e) $\ln[A]$ vs. t

44. Given: $4PH_3(g) \rightarrow P_4(g) + 6H_2(g)$ rate = $k[PH_3]$

If the rate constant for this reaction is 0.0278 s^{-1}, what is the half-life?

a) 99.6 s b) 36.0 s c) 18.0 s d) 24.9 s e) 49.8 s

45. Consider the following reaction:

$2N_2O(g) \rightarrow 2NO(g) + O_2(g)$ rate = $k[N_2O]$

If the half–time for the reaction is 910 ms, the rate constant is

a) 0.69 s^{-1} b) 1.3 s^{-1} c) 0.32 s^{-1} d) 0.76 s^{-1} e) 1.1 s^{-1}

46. For a certain first-order reaction the rate constant is 0.92 s^{-1}. What percent of reactant remains after 5 half-lives?
 a) 3.1% b) 6.3% c) 25% d) 74% e) 2.5%

47. The rate law for the decomposition of hydrogen peroxide is
 rate = $k[H_2O_2]$
 If the half-life for this reaction is 7.0 min, what is the concentration of H_2O_2 after 10 minutes when the initial concentration of hydrogen peroxide is 0.20 M?
 a) 0.027 M b) 0.18 M c) 0.14 M d) 0.074 M e) 0.043 M

48. Iodine-131 can be used to measure the activity of the thyroid gland. If the rate of decay of a sample containing I-131 is 2.15×10^5 disintegrations per minute initially, and 6.43×10^4 disintegrations per minute after 2 weeks, calculate the half-life of I-131.
 a) 9.70 days b) 18.5 days c) 14.0 days d) 4.02 days e) 8.04 days

49. An "old rock" was found to have 395 C-14 disintegrations per hour and a contemporary rock 503 C-14 disintegrations per hour. If the half-life of C-14 is 5.73×10^3 years, estimate the age of the "old rock." The decay of C-14 is first-order.
 a) 10800 years b) 8260 years c) 1230 years d) 4500 years e) 2000 years

50. The isotope I-131 has a half-life of 8.05 days. What fraction of the initial concentration of iodine-131 remains after 3 weeks. The decay of I-131 is first-order.
 a) 0.164 b) 0.836 c) 0.0861 d) 0.124 e) 0.383

51. A second-order reaction has a rate constant of 1.25 $M^{-1} \cdot s^{-1}$. If the initial reactant concentration is 1.0 M, calculate the time required for 90% reaction.
 a) 0.13 s b) 17 s c) 7.2 s d) 0.89 s e) 1.3 s

52. Consider the dimerization reaction below:
 $2A \rightarrow A_2$ rate = $k[A]^2$
 When the initial concentration of A is 2.0 M, it requires 1 hour for 60% of A to react. Calculate the rate constant at these conditions.
 a) $1.6 \times 10^{-4}\ M^{-1} \cdot s^{-1}$
 b) $5.6 \times 10^{-4}\ M^{-1} \cdot s^{-1}$
 c) $9.3 \times 10^{-5}\ M^{-1} \cdot s^{-1}$
 d) $2.1 \times 10^{-4}\ M^{-1} \cdot s^{-1}$
 e) $2.5 \times 10^{-4}\ M^{-1} \cdot s^{-1}$

53. Consider the dimerization of butadiene:
 $2C_4H_6(g) \rightarrow C_8H_{12}(g)$ rate = $k[C_4H_6]^2$
 If the rate constant is 0.014 $M^{-1} \cdot s^{-1}$, calculate the time required for dimerization of 90% of the butadiene for an initial concentration of butadiene of 0.10 M.
 a) 6400 s b) 160 s c) 6.4 s d) 50 s e) 79 s

54. Consider the following reaction:

$2NO_2(g) \rightarrow 2NO(g) + O_2(g)$ rate = $k[NO_2]^2$

When the initial concentration of NO_2 is 100 mM, it takes 74 s for 80% of the NO_2 to react. Calculate the rate constant.

a) 0.54 $M^{-1}\cdot s^{-1}$
b) 0.022 $M^{-1}\cdot s^{-1}$
c) 3.0×10^{-3} $M^{-1}\cdot s^{-1}$
d) 3.4×10^{-5} $M^{-1}\cdot s^{-1}$
e) 5.4×10^{-4} $M^{-1}\cdot s^{-1}$

55. Consider the following:

$2A \rightarrow A_2$ rate = $k[A]^2$

If the initial concentration of A is 0.0200 M and 80% of A reacts in 1500 s, calculate the rate constant.

a) 0.00833 $M^{-1}\cdot s^{-1}$ b) 0.0333 $M^{-1}\cdot s^{-1}$ c) 0.133 $M^{-1}\cdot s^{-1}$ d) 0.166 $M^{-1}\cdot s^{-1}$ e) 0.0417 $M^{-1}\cdot s^{-1}$

56. Consider the following reaction:

$NOBr(g) \rightarrow NO(g) + 1/2Br_2(g)$

A plot of $[NOBr]^{-1}$ versus time gives a straight line with a slope of 2.00 $M^{-1}\cdot s^{-1}$. The order of the reaction and the rate constant, respectively, are

a) first-order and 0.241 s^{-1}
b) second-order and 16.6 $M^{-1}\cdot s^{-1}$
c) second-order and 0.500 $M^{-1}\cdot s^{-1}$
d) first-order and 2.00 s^{-1}
e) second-order and 2.00 $M^{-1}\cdot s^{-1}$

57. A reaction has k = 8.39 $M^{-1}\cdot s^{-1}$. How long does it take for the reactant concentration to drop from 0.0420 M to 0.0110 M?

a) 8.00 s b) 2.84 s c) 4.00 s d) 10.8 s e) 8.39 s

58. The activation energy of a reaction is given by

a) $-R \div$ (slope of a plot of ln k vs. $1/T$)
b) –(slope of a plot of ln k vs. $1/T$) $\div R$
c) –(slope of a plot of ln k vs. $1/T$) $\times R$
d) +(slope of a plot of ln k vs. $1/T$) $\div R$
e) +(slope of a plot of ln k vs. $1/T$) $\times R$

59. For a certain reaction the following data were collected:

***T*, °C**	***k*, s^{-1}**
25	5.8
33	16
38	27
43	47

Calculate the activation energy for this reaction.

60. For the reaction

$C_2H_5Br(aq) + OH^-(aq) \rightarrow C_2H_5OH(aq) + Br^-(aq)$

a plot of lnk versus 1/T gives a straight line with a slope equal to -1.07×10^4 K. What is the activation energy for the reaction?

a) 89.0 kJ/mol b) 205 kJ/mol c) 24.6 kJ/mol d) 1.29 kJ/mol e) 10.7 kJ/mol

61. A certain reaction has a rate constant of 0.0503 s^{-1} at 298 K and 6.71 s^{-1} at 333 K. What is the activation energy for this reaction?
a) 115 kJ/mol b) 89.4 kJ/mol c) 85.3 kJ/ml d) 49.9 kJ/mol e) 34.5 kJ/mol

62. Calculate the rate constant at 43°C for the reaction
$$(NH_3)_5CoNC\text{–}R^{3+}(aq) + OH^-(aq) \rightarrow (NH_3)_5CoNHCOR^{2+}(aq)$$
if $k = 5.8 \times 10^6\ M^{-1}\cdot s^{-1}$ at 25°C and $E_a = 92$ kJ/mol.
a) $6.8 \times 10^7\ M^{-1}\cdot s^{-1}$
b) $4.8 \times 10^7\ M^{-1}\cdot s^{-1}$
c) $5.3 \times 10^8\ M^{-1}\cdot s^{-1}$
d) $6.4 \times 10^7\ M^{-1}\cdot s^{-1}$
e) $2.0 \times 10^7\ M^{-1}\cdot s^{-1}$

63. The activation energy for the reaction
$$N_2O(g) \rightarrow N_2(g) + O(g)$$
is 250 kJ/mol. If the rate constant is 3.4 s^{-1} at 1050 K, at what temperature will the rate constant be one thousand times smaller?
a) 573 K b) 846 K c) 777 K d) 1384 K e) 1204 K

64. A reaction that has a very low activation energy
a) must be second-order.
b) gives a curved Arrhenius plot.
c) must be first-order.
d) has a rate that does not change much with temperature.
e) has a rate that is very sensitive to temperature.

65. When a catalyst is used in a reaction,
a) the enthalpy change for the reaction becomes more exothermic.
b) the activation energy for the forward reaction is not changed.
c) it does not affect the final amounts of reactants and products.
d) the activation energy of the reverse reaction is increased.
e) the forward reaction is increased while the reverse reaction is retarded.

66. An elementary process has an activation energy of 40 kJ/mol. If the enthalpy change for the reaction is 30 kJ/mol, what is the activation energy for the reverse reaction?
a) 50 kJ/mol b) 10 kJ/mol c) 30 kJ/mol d) –40 kJ/mol e) 40 kJ/mol

67. An elementary process has an activation energy of 40 kJ/mol. If the activation energy for the reverse reaction is 10 kJ/mol, what is the enthalpy change for the reaction?
a) 40 kJ/mol b) –30 kJ/mol c) –10 kJ/mol d) 30 kJ/mol e) –40 kJ/mol

68. An elementary process has an activation energy of 88 kJ/mol. If the enthalpy change for the reaction is –60 kJ/mol, what is the activation energy for the reverse reaction?
a) 148 kJ/mol b) –88 kJ/mol c) –60 kJ/mol d) 60 kJ/mol e) 28 kJ/mol

69. An elementary process has an activation energy of 32 kJ/mol. If the activation energy for the reverse reaction is 20 kJ/mol, what is the enthalpy change for the reaction?
a) –52 kJ/mol b) –12 kJ/mol c) –20 kJ/mol d) 52 kJ/mol e) 12 kJ/mol

70. An elementary process has an activation energy of 42 kJ/mol. If the activation energy for the reverse reaction is 54 kJ/mol, what is the enthalpy change for the reaction?
a) 12 kJ/mol b) –66 kJ/mol c) –12 kJ/mol d) 66 kJ/mol e) –54 kJ/mol

71. Given:

$$CH_4(g) + Cl_2(g) \rightarrow CH_3Cl(g) + HCl(g)$$

The rate law for this elementary process is
a) rate = $k[CH_4][Cl_2]$ b) rate = $k[CH_4]$ c) rate = $k[Cl_2]$ d) rate = $k[CH_4]^2$ e) $k[CH_3Cl][HCl]$

72. Consider the mechanism:

$$O_3(g) \leftrightarrow O_2(g) + O(g) \qquad \text{fast}$$

$$O(g) + O_3(g) \rightarrow 2O_2(g) \qquad \text{slow}$$

An intermediate in this reaction is
a) O(g) b) $O_2(g)$ c) $O_3(g)$ d) $O_2(g)$ and O(g) e) There is no intermediate in this reaction.

73. The rate law for the following mechanism is

$$NO_2(g) + F_2(g) \rightarrow NO_2F(g) + F(g) \qquad k_1\text{, slow}$$

$$F(g) + NO_2(g) \rightarrow NO_2F(g) \qquad k_2\text{, fast}$$

a) rate = $k_1k_2[NO_2]^2$
b) rate = $k_1[NO_2][F_2]$
c) rate = $k_2[NO_2]^2$
d) rate = $k_2[NO_2][F]$
e) rate = $k_1[NO_2F][F]$

74. The reaction

$$2NO_2(g) \rightarrow 2NO(g) + O_2(g)$$

is postulated to occur via the mechanism below:

$$NO_2(g) + NO_2(g) \rightarrow NO(g) + NO_3(g) \qquad \text{slow}$$

$$NO_3(g) \rightarrow NO(g) + O_2(g) \qquad \text{fast}$$

An intermediate in this reaction is
a) $O_2(g)$ b) $NO_3(g)$ c) ON–$NO_3(g)$ d) NO(g) e) $NO_2(g)$

75. The reaction

$$2NO_2(g) \rightarrow 2NO(g) + O_2(g)$$

is postulated to occur via the mechanism below:

$NO_2(g) + NO_2(g) \rightarrow NO(g) + NO_3(g)$ k_1, slow

$NO_3(g) \rightarrow NO(g) + O_2(g)$ k_2, fast

The rate law for this mechanism is

a) rate = $k_2[NO_3]$
b) rate = $k_1[NO_2]$
c) rate = $k_1[NO_2]^2$
d) rate = $k_1k_2[NO_2]^2[NO_3]$
e) rate = $k_1k_2[NO_3]^2$

76. The rate law for the mechanism below is

$Cl_2(g) \leftrightarrow 2Cl(g)$ K_1, fast

$Cl(g) + CO(g) \leftrightarrow COCl(g)$ K_2, fast

$COCl(g) + Cl_2(g) \rightarrow COCl_2(g) + Cl(g)$ k_3, slow

a) rate = $k_3[COCl][Cl_2]$
b) rate = $k_3 K_1 K_2[CO][Cl_2]$
c) rate = $k_3[COCl][Cl_2]^{1.5}$
d) rate = $k_3 K_1^{\ 0.5} K_2[CO][Cl_2]^{1.5}$
e) rate = $k_3 K_1^{\ 0.5} K_2[CO][Cl_2]^{0.5}$

77. For the following mechanism

$2NO(g) \leftrightarrow N_2O_2(g)$ K, fast

$N_2O_2(g) + O_2(g) \rightarrow 2NO_2(g)$ k, slow

the rate law is

rate = $kK[NO]^2[O_2]$.

A plot of potential energy versus progress of reaction, the reaction profile, shows

a) one maximum for the first step.
b) two maxima, the first maxima being highest.
c) two maxima, the second maxima being highest.
d) two minima representing the two reactions.
e) one maximum for the second step.

78. The following mechanism has been suggested for the reaction between chlorine and chloroform:

$Cl_2 \leftrightarrow 2Cl$ K, fast

$CHCl_3 + Cl \leftrightarrow CCl_3 + HCl$ k_2, slow

$CCl_3 + Cl \leftrightarrow CCl_4$ k_3, fast

(a) Determine the rate law for this mechanism.
(b) What is the order of the reaction?
(c) What, if any, are the intermediates in the reaction?

79. The following mechanisms have been proposed for the reaction of $NO_2(g)$ with carbon monoxide to give carbon dioxide and NO(g):

(a) $2NO_2 \leftrightarrow NO + NO_3$ — K, fast

$NO_3 + CO \rightarrow NO_2 + CO_2$ — k, slow

(b) $2NO_2 \rightarrow NO + NO_3$ — k_1, slow

$NO_3 + CO \rightarrow NO_2 + CO_2$ — k_2, fast

(c) $NO_2 + CO \rightarrow NO + CO_2$ — k, slow

(a) Which mechanism agrees with the experimental rate law, rate = $k[NO_2]^2$?

(b) Is an intermediate involved in the mechanism? If yes, what is it?

80. The HBr synthesis is thought to involve the following reactions:

$Br_2 \rightarrow 2Br\cdot$ (1)

$Br\cdot + H_2 \rightarrow HBr + H\cdot$ (2)

$H\cdot + Br_2 \rightarrow HBr + Br\cdot$ (3)

$2Br\cdot \rightarrow Br_2$ (4)

$2H\cdot \rightarrow H_2$ (5)

$H\cdot + Br\cdot \rightarrow HBr$ (6)

The chain termination reactions in this mechanism are reactions

a) 4, 5, and 6 b) 6 c) 3 and 4 d) 3, 4, and 5 e) 4 and 5

Chapter 18: Kinetics: The Rates of Reactions
Form B

1. The rate of formation of oxygen in the reaction
$$2N_2O_5(g) \rightarrow 4NO_2(g) + O_2(g)$$
is 2.28 (mol O_2)·L^{-1}·s^{-1}. What is the rate of formation of NO_2?

a) 9.12 (mol NO_2)·L^{-1}·s^{-1}
b) 2.28 (mol NO_2)·L^{-1}·s^{-1}
c) 1.14 (mol NO_2)·L^{-1}·s^{-1}
d) 0.57 (mol NO_2)·L^{-1}·s^{-1}
e) 4.56 (mol NO_2)·L^{-1}·s^{-1}

2. The rate of formation of $NO_2(g)$ in the reaction
$$2N_2O_5(g) \rightarrow 4NO_2(g) + O_2(g)$$
is 5.78 (mol NO_2)·L^{-1}·s^{-1}. What is the rate at which N_2O_5 is used?

a) −11.6 (mol N_2O_5)·L^{-1}·s^{-1}
b) −1.45 (mol N_2O_5)·L^{-1}·s^{-1}
c) −0.723 (mol N_2O_5)·L^{-1}·s^{-1}
d) −2.89 (mol N_2O_5)·L^{-1}·s^{-1}
e) −5.78 (mol N_2O_5)·L^{-1}·s^{-1}

3. Given:
$$2NO_2(g) + F_2(g) \rightarrow 2NO_2F(g) \qquad \text{rate} = -\Delta[F_2]/\Delta t$$
The rate of the reaction can also be expressed as

a) $-\Delta[NO_2]/\Delta t$ b) $-2\Delta[NO_2]/\Delta t$ c) $1/2\Delta[NO_2]/\Delta t$ d) $\Delta[NO_2F]/\Delta t$ e) $-1/2\Delta[NO_2]/\Delta t$

4. For the reaction
$$2N_2O(g) \rightarrow 2N_2(g) + O_2(g)$$
initial rate = $k[N_2O]_o$. For concentration in M and time in s, the units of the rate constant k are

a) M^{-2}·s^{-1} b) M^{2}·s^{-1} c) s^{-1} d) M·s^{-1} e) M^{-1}·s^{-1}

5. Given:
$$NO_2(g) + CO(g) \rightarrow NO(g) + CO_2(g) \qquad \text{rate} = k[NO_2]^2$$
The overall order of the reaction is

a) 2 b) 1 c) 3 d) 0 e) 4

6. Given:
$$2O_3(g) \rightarrow 3O_2(g) \qquad \text{rate} = k[O_3]^2[O_2]^{-1}$$
The overall order of the reaction is

a) 1 b) 2 c) 3 d) −1 e) 0

7. Given:

$$2A(g) + B(g) \rightarrow C(g) + D(g)$$

When [A] = [B] = 0.10 M, the rate is 2.0 M·s^{-1}; for [A] = [B] = 0.20 M, the rate is 8.0 M·s^{-1}; and for [A] = 0.10 M, [B] = 0.20 M, the rate is 2.0 M·s^{-1}. The rate law is

a) rate = k[A] b) rate = $k[A][B]^0$ c) rate = $k[B]^2$ d) rate = k[A][B] e) rate = $k[A]^2$

8. If the rate of a reaction increases by a factor of 9 when the concentration of reactant increases by a factor of 3, the order of the reaction with respect to this reactant is

a) 2 b) 3 c) 9 d) 4 e) 1

9. For the reaction

$$2A + B \rightarrow \text{products}$$

determine the rate law for the reaction given the following data:

Initial Concentration, M		Initial Rate, M·s^{-1}
A	B	
0.10	0.10	2.0×10^{-2}
0.20	0.10	8.0×10^{-2}
0.30	0.10	1.8×10^{-1}
0.20	0.20	8.0×10^{-2}
0.30	0.30	1.8×10^{-1}

a) rate = $k[B]^2$ b) rate = k[A] c) rate = k[A][B] d) rate = $k[A]^2$ e) rate = $k[A][B]^0$

10. If the rate of reaction increases by a factor of 9.6 when the concentration of reactant increases by a factor of 3.1, the order of the reaction with respect to this reactant is

a) 2 b) 1 c) 3 d) 4 e) 1.5

11. The reaction

$$2NO(g) + 2H_2(g) \rightarrow N_2(g) + 2H_2O(g)$$

is first-order in H_2 and second-order in NO. Starting with equal concentrations of H_2 and NO, the rate after 25% of the H_2 has reacted is what percent of the initial rate?

a) 42.2% b) 1.56% c) 56.3% d) 6.25% e) 75.0%

12. The reaction

$$2ClO_2(g) + F_2(g) \rightarrow 2FClO_2(g)$$

is first-order in both ClO_2 and F_2. When the initial concentratons of ClO_2 and F_2 are equal, the rate after 25% of the F_2 has reacted is what percent of the initial rate?

a) 37.5% b) 12.5% c) 18.8% d) 28.1% e) 75.0%

13. For the reaction

$C_2H_6(g) \rightarrow 2CH_3(g)$ rate = $k[C_2H_6]$

If $k = 5.50 \times 10^{-4}\ s^{-1}$ and $[C_2H_6]_{initial} = 0.0200$ M, calculate the rate of reaction after 30 min.

a) 4.09×10^{-6} M·s^{-1}
b) 2.75×10^{-2} M·s^{-1}
c) 1.10×10^{-5} M·s^{-1}
d) The rate of reaction is zero since the reaction is complete.
e) 2.1×10^{-6} M·s^{-1}

14. Consider the following:

$2A \rightarrow A_2$ rate = $k[A]^2$

If the rate constant is 1.43 M^{-1}·s^{-1} and the initial concentration of A is 0.0180 M, calculate the time for the rate of consumption of A to drop to 3.75×10^{-5} M·s^{-1}.

a) 54.3 s b) 97.7 s c) 140 s d) 1.26 s e) 137 s

15. Given: $4PH_3(g) \rightarrow P_4(g) + 6H_2(g)$ rate = $k[PH_3]$

If the rate constant is 0.0278 s^{-1}, calculate the percent of the original PH_3 which has reacted after 65.0 s.

a) 83.6% b) 16.4% c) 1.56% d) 98.4% e) 59.2%

16. Consider the following reaction:

$2N_2O(g) \rightarrow 2N_2(g) + O_2(g)$ rate = $k[N_2O]$

For an initial concentration of N_2O of 0.50 M, calculate the concentration of N_2O remaining after 1.5 min if $k = 3.4 \times 10^{-3}\ s^{-1}$.

a) 0.99 M b) 0.50 M c) 0.37 M d) 0.25 M e) 0.74 M

17. Consider the following reaction:

$2N_2O(g) \rightarrow 2N_2(g) + O_2(g)$ rate = $k[N_2O]$

For an initial concentration of N_2O of 0.50 M, calculate the concentration of N_2O remaining after 3.0 min if $k = 3.4 \times 10^{-3}\ s^{-1}$.

a) 0.54 M b) 0.50 M c) 0.12 M d) 0.27 M e) 0.99 M

18. Consider the following reaction:

$2N_2O(g) \rightarrow 2N_2(g) + O_2(g)$ rate = $k[N_2O]$

For an initial concentration of N_2O of 0.75 M, calculate the concentration of N_2O remaining after 2 min if $k = 6.8 \times 10^{-3}\ s^{-1}$.

a) 0.44 M b) 0.49 M c) 0.74 M d) 0.99 M e) 0.33 M

19. A first-order reaction has a rate constant of 0.00300 s^{-1}. The time required for 60% reaction is

a) 153 s b) 170 s c) 305 s d) 133 s e) 73.9 s

20. A first-order reaction has a rate constant of 0.00300 s^{-1}. The time required for 55% reaction is

a) 116 s b) 199 s c) 86.5 s d) 58.0 s e) 266 s

21. For a first-order reaction, after 230 s, 33% of the reactants remain. Calculate the rate constant for the reaction.
   a) 0.000756 s^{-1} b) 0.00482 s^{-1} c) 0.00174 s^{-1} d) 0.00209 s^{-1} e) 207 s^{-1}

22. For a first-order reaction, after 3.00 min, 5% of the reactants remain. Calculate the rate constant for the reaction.
   a) 0.00723 s^{-1} b) 0.0166 s^{-1} c) 0.000285 s^{-1} d) 0.999 s^{-1} e) 0.0000124 s^{-1}

23. For a first-order reaction, after 2.10 ms, 15% of the reactants remain. Calculate the rate constant for the reaction.
   a) 77.4 s^{-1} b) 0.903 s^{-1} c) 392 s^{-1} d) 33.6 s^{-1} e) 903 s^{-1}

24. Consider the following reaction:
   $2N_2O_5(g) \rightarrow 4NO_2(g) + O_2(g)$ rate = $k[N_2O_5]$
   Calculate the time for the concentration of N_2O_5 to fall to one–eighth its initial value if the rate constant for the reaction is 5.20×10^{-3} s^{-1}.
   a) 400 s b) 16.7 s c) 533 s d) 1070 s e) 33.3 s

25. Consider the following reaction:
   $2N_2O_5(g) \rightarrow 4NO_2(g) + O_2(g)$ rate = $k[N_2O_5]$
   Calculate the time for the concentration of N_2O_5 to fall to one–third its initial value if the rate constant for the reaction is 5.20×10^{-3} s^{-1}.
   a) 211 s b) 634 s c) 400 s d) 44.4 s e) 267 s

26. Given: $SO_2Cl_2(g) \rightarrow SO_2(g) + Cl_2(g)$ rate = $k[SO_2Cl_2]$
   If 95% of the SO_2Cl_2 has reacted after 12 s, calculate the rate constant.
   a) 0.60 s^{-1} b) 4.0 s^{-1} c) 0.48 s^{-1} d) 0.25 s^{-1} e) 0.0.0043 s^{-1}

27. The following data were obtained for the reaction of B at 35°C:

[B], M	**Time, s**
0.286	0
0.274	20
0.261	40
0.250	60
0.240	80
0.231	100
0.206	160
0.192	200

   What is the order of the reaction and the rate constant?

28. The following data were obtained for the disappearance of A at 25°C:

[A], M	Time, s
0.377	0
0.235	20
0.165	35
0.130	45
0.105	55
0.0801	65
0.0650	75
0.0501	85
0.0360	100

Determine the order of the reaction, and calculate the rate constant and half-life.

29. Given: $2H_2O_2(aq) \rightarrow 2H_2O(l) + O_2(g)$ rate = $k[H_2O_2]$
If 85% of the H_2O_2 has reacted after 4.0 min, calculate the rate constant.
a) 0.00068 s^{-1} b) 0.00029 s^{-1} c) 0.0034 s^{-1} d) 0.0079 s^{-1} e) 0.47 s^{-1}

30. Given: A $\rightarrow$ P rate = k[A]
If 20% of A reacts in 5.12 min, calculate the time required for 90% of A to react.
a) 52.8 min b) 2.42 min c) 3170 min d) 1.05 min e) 22.9 min

31. Consider the following reaction:
$2N_2O(g) \rightarrow 2NO(g) + O_2(g)$ rate = $k[N_2O]$
If 67% of N_2O reacts in 1.0 s, what is the rate constant?
a) 0.48 s^{-1} b) 0.17 s^{-1} c) 0.40 s^{-1} d) 0.33 s^{-1} e) 1.1 s^{-1}

32. For the reaction
cyclopropane $\rightarrow$ propene
a plot of ln[cyclopropane] versus time in seconds gives a straight line with slope -4.1×10^{-3} s^{-1} at 550°C. What is the rate constant for this reaction?
a) 1.8×10^{-3} s^{-1} b) 3.9×10^{-2} s^{-1} c) 4.1×10^{-3} s^{-1} d) 8.2×10^{-3} s^{-1} e) 2.1×10^{-3} s^{-1}

33. For the first-order reaction
$2N_2O(g) \rightarrow 2NO(g) + O_2(g)$
the rate constant is 0.34 s^{-1}. Calculate the percent N_2O remaining after 1 second.
a) 7.1% b) 4.6% c) 2.0% d) 1.0% e) 2.9%

34. For the first-order reaction
$2N_2O(g) \rightarrow 2NO(g) + O_2(g)$
the rate constant is 3.4 s^{-1}. The time required for 95% reaction is
a) 0.88 s b) 6.55 ms c) 15.1 ms d) 0.38 s e) 0.17 s

35. A first-order reaction has a rate constant of 26 s^{-1}. If the initial concentration of reactant is 0.820 M, how long will it take for the reactant to reach 0.0010 M?
a) 0.038 s b) 21 s c) 0.11 s d) 7.7 ms e) 0.26 s

36. Consider the following reaction:
$2N_2O_5(g) \rightarrow 4NO_2(g) + O_2(g)$ rate = $k[N_2O_5]$
If the initial concentration of N_2O_5 is 0.80 M, the concentration after 5 half-lives is
a) 0.050 M b) 0.16 M c) 0.11 M d) 0.025 M e) 0.032 M

37. Consider the following reaction:
$2N_2O_5(g) \rightarrow 4NO_2(g) + O_2(g)$ rate = $k[N_2O_5]$
If the initial concentration of N_2O_5 is 0.60 M, the concentration after 3 half-lives is
a) 0.075 M b) 0.20 M c) 0.067 M d) 0.10 M e) 0.14 M

38. A compound decomposes with a half-life of 8.0 s and the half-life is independent of the concentration. How long does it take for the concentration to decrease to one–ninth of its initial value?
a) 64 s b) 32 s c) 72 s d) 3.6 s e) 25 s

39. For a first-order reaction, what fraction of the starting material will remain after 5 half-lives?
a) 1/32 b) 1/16 c) 1/8 d) 1/4 e) 1/12

40. A first-order reaction has a half-life of 1.10 s. If the initial concentration of reactant is 0.384 M, how long will it take for the reactant concentration to reach 0.00100 M?
a) 0.106 s b) 4.10 s c) 0.244 s d) 9.45 s e) 1.52 s

41. A first-order reaction has a rate constant of 6.3 s^{-1}. What is the half-life for this reaction?
a) 0.11 s b) 0.048 s c) 0.69 s d) 9.1 s e) 21 s

42. Given: $2A + B \rightarrow P$ rate = $k[A]$
Which of the following is true?
a) $\ln[A] = k/t$
b) $[A] = 1/(k \times t_{1/2})$
c) $1/[A] = kt$
d) $k = 0.693/t_{1/2}$
e) The overall order of the reaction is 3.

43. For a second-order reaction, a straight line is obtained from a plot of
a) ln[A] vs. t b) [A] vs. t c) 1/[A] vs. t d) ln(t) vs. [A] e) ln(1/t) vs. [A]

44. Given: $4PH_3(g) \rightarrow P_4(g) + 6H_2(g)$ rate = $k[PH_3]$
If the rate constant for this reaction is 0.568 s^{-1}, what is the half-life?
a) 0.566 s b) 2.54 s c) 0.820 s d) 1.22 s e) 0.394 s

45. Consider the following reaction:

$2N_2O(g) \rightarrow 2NO(g) + O_2(g)$ rate = $k[N_2O]$

If the half–time for the reaction is 5.90 ms, the rate constant is

a) 234 s^{-1} b) 4.1 s^{-1} c) 17.0 s^{-1} d) 8.5 s^{-1} e) 117 s^{-1}

46. For a certain first-order reaction the rate constant is 2.9 s^{-1}. What percent of reactant remains after 5 half-lives?

a) 3.1% b) 6.3% c) 25% d) 74% e) 2.5%

47. The rate law for the decomposition of hydrogen peroxide is

rate = $k[H_2O_2]$

If the half-life for this reaction is 7.0 min, what is the concentration of H_2O_2 after 15 minutes when the initial concentration of hydrogen peroxide is 0.20 M?

a) 0.0065 M b) 0.0017 M c) 0.055 M d) 0.045 M e) 0.033 M

48. If the rate of decay of a sample containing C-14 was 920 disintegrations per hour initially, and 395 disintegrations per hour after 6.99 $\times 10^3$ years, calculate the half-life of C-14.

a) 1.32 $\times 10^3$ years
b) 6.99 $\times 10^3$ years
c) 3.00 $\times 10^3$ years
d) 2.86 $\times 10^3$ years
e) 5.73 $\times 10^3$ years

49. An "old rock" was found to have 450 C-14 disintegrations per hour and a contemporary rock 503 C–14 disintegrations per hour. If the half-life of C-14 is 5.73 $\times 10^3$ years, estimate the age of the "old rock." The decay of C-14 is first-order.

a) 400 years b) 5700 years c) 18,600 years d) 6600 years e) 920 years

50. The isotope I-131 has a half-life of 8.05 days. What fraction of the initial concentration of iodine-131 remains after 2 weeks. The decay of I-131 is first-order.

a) 0.300 b) 0.0623 c) 0.575 d) 0.176 e) 0.0182

51. A second-order reaction has a rate constant of 2.8 $M^{-1}\cdot s^{-1}$. If the initial reactant concentration is 1.0 M, calculate the time required for 90% reaction.

a) 0.036 s b) 0.040 s c) 3.6 s d) 0.40 s e) 3.2 s

52. Consider the dimerization reaction below:

$2A \rightarrow A_2$ rate = $k[A]^2$

When the initial concentration of A is 2.0 M, it requires 30 min for 60% of A to react. Calculate the rate constant at these conditions.

a) 1.9 $\times 10^{-4}$ $M^{-1}\cdot s^{-1}$
b) 1.1 $\times 10^{-3}$ $M^{-1}\cdot s^{-1}$
c) 5.0 $\times 10^{-4}$ $M^{-1}\cdot s^{-1}$
d) 4.2 $\times 10^{-4}$ $M^{-1}\cdot s^{-1}$
e) 3.2 $\times 10^{-4}$ $M^{-1}\cdot s^{-1}$

53. Consider the dimerization of butadiene:

$2C_4H_6(g) \rightarrow C_8H_{12}(g)$ rate = $k[C_4H_6]^2$

If the rate constant is 0.014 $M^{-1}\cdot s^{-1}$, calculate the time required for dimerization of 70% of the butadiene for an initial concentration of butadiene of 0.10 M.

a) 5.1 min b) 12 min c) 17 min d) 40 min e) 28 min

54. Consider the following reaction:

$2NO_2(g) \rightarrow 2NO(g) + O_2(g)$ rate = $k[NO_2]^2$

When the initial concentration of NO_2 is 100 mM, it takes 55 s for 90% of the NO_2 to react. Calculate the rate constant.

a) 1.6×10^{-3} $M^{-1}\cdot s^{-1}$ b) 0.13 $M^{-1}\cdot s^{-1}$ c) 1.6 $M^{-1}\cdot s^{-1}$ d) 0.042 $M^{-1}\cdot s^{-1}$ e) 1.8×10^{-5} $M^{-1}\cdot s^{-1}$

55. Consider the following:

$2A \rightarrow A_2$ rate = $k[A]^2$

If the initial concentration of A is 0.0200 M and 80% of A reacts in 975 s, calculate the rate constant.

a) 0.0513 $M^{-1}\cdot s^{-1}$ b) 0.0128 $M^{-1}\cdot s^{-1}$ c) 0.256 $M^{-1}\cdot s^{-1}$ d) 0.00165 $M^{-1}\cdot s^{-1}$ e) 0.205 $M^{-1}\cdot s^{-1}$

56. Consider the following reaction:

$NOBr(g) \rightarrow NO(g) + \leftrightarrow Br_2(g)$

A plot of $[NOBr]^{-1}$ versus time gives a straight line with a slope of 1.76 $M^{-1}\cdot s^{-1}$. The order of the reaction and the rate constant, respectively, are

a) second-order and 14.6 $M^{-1}\cdot s^{-1}$
b) second-order and 1.76 $M^{-1}\cdot s^{-1}$
c) second-order and 0.568 $M^{-1}\cdot s^{-1}$
d) first-order and 1.76 s^{-1}
e) first-order and 0.212 s^{-1}

57. A reaction has k = 8.39 $M^{-1}\cdot s^{-1}$. How long does it take for the reactant concentration to drop from 0.0840 M to 0.0220 M?

a) 1.42 s b) 2.00 s c) 8.39 s d) 5.42 s e) 4.00 s

58. According to collision theory, the increase in the rate constant with increasing temperature is due mostly to the fact that

a) the activation energy decreases with increasing temperature.
b) the fraction of the collisions having sufficient energy to react increases with increasing temperature.
c) the pressure of the reagents increases with increasing temperature.
d) the enthalpy change for most reactions is negative.
e) the fraction of the collisions having the proper orientation for reaction increases with increasing temperature.

59. For a certain reaction the following data were collected:

***T*, °C**	***k*, s^{-1}**
15	0.67
25	0.92
35	1.3
45	1.7

Calculate the activation energy for this reaction.

60. For the reaction

$$HO(g) + H_2(g) \rightarrow H_2O(g) + H(g)$$

a plot of lnk versus 1/T gives a straight line with a slope equal to -5.1×10^3 K. What is the activation energy for the reaction?

a) 0.61 kJ/mol b) 12 kJ/mol c) 5.1 kJ/mol d) 42 kJ/mol e) 98 kJ/mol

61. A certain reaction has a rate constant of 8.8 s^{-1} at 298 K and 140 s^{-1} at 323 K. What is the activation energy for this reaction?

a) 23 kJ/mol b) 120 kJ/mol c) 38 kJ/mol d) 1.2 kJ/mol e) 89 kJ/mol

62. Calculate the rate constant at 53°C for the reaction

$$(NH_3)_5CoNC{-}R^{3+}(aq) + OH^-(aq) \rightarrow (NH_3)_5CoNHCOR^{2+}(aq)$$

if $k = 5.8 \times 10^6$ M^{-1}·s^{-1} at 25°C and $E_a = 92$ kJ/mol.

a) 9.0×10^9 M^{-1}·s^{-1}
b) 1.4×10^8 M^{-1}·s^{-1}
c) 1.3×10^7 M^{-1}·s^{-1}
d) 2.5×10^7 M^{-1}·s^{-1}
e) 7.3×10^7 M^{-1}·s^{-1}

63. The activation energy for the reaction

$$N_2O(g) \rightarrow N_2(g) + O(g)$$

is 250 kJ/mol. If the rate constant is 3.4 s^{-1} at 1050 K, at what temperature will the rate constant be one thousand times larger?

a) 1384 K b) 1204 K c) 573 K d) 846 K e) 777 K

64. A reaction that has a very high activation energy
a) must be first-order.
b) gives a curved Arrhenius plot.
c) has a rate that does not change much with temperature.
d) must be second-order.
e) has a rate that is very sensitive to temperature.

65. A catalyst facilitates a reaction by
a) making the reaction more exothermic.
b) shifting the position of the equilibrium of the reaction.
c) lowering the activation energy of the reaction.
d) decreasing the temperature at which the reaction will proceed spontaneously.
e) increasing the activation energy for the reverse reaction.

66. An elementary process has an activation energy of 40 kJ/mol. If the enthalpy change for the reaction is 20 kJ/mol, what is the activation energy for the reverse reaction?
a) 20 kJ/mol b) –40 kJ/mol c) 40 kJ/mol d) 60 kJ/mol e) 20 kJ/mol

67. An elementary process has an activation energy of 40 kJ/mol. If the activation energy for the reverse reaction is 20 kJ/mol, what is the enthalpy change for the reaction?
a) 20 kJ/mol b) –20 kJ/mol c) 60 kJ/mol d) 40 kJ/mol e) –60 kJ/mol

68. An elementary process has an activation energy of 88 kJ/mol. If the enthalpy change for the reaction is –40 kJ/mol, what is the activation energy for the reverse reaction?
a) 40 kJ/mol b) 48 kJ/mol c) –40 kJ/mol d) 128 kJ/mol e) –88 kJ/mol

69. An elementary process has an activation energy of 32 kJ/mol. If the activation energy for the reverse reaction is 30 kJ/mol, what is the enthalpy change for the reaction?
a) –62 kJ/mol b) 2 kJ/mol c) –30 kJ/mol d) –2 kJ/mol e) 62 kJ/mol

70. An elementary process has an activation energy of 42 kJ/mol. If the activation energy for the reverse reaction is 64 kJ/mol, what is the enthalpy change for the reaction?
a) –86 kJ/mol b) –22 kJ/mol c) –64 kJ/mol d) 86 kJ/mol e) 22 kJ/mol

71. Given:

$$O(g) + O_3(g) \rightarrow 2O_2(g)$$

The rate law for this elementary process is
a) rate = $k[O_2]^2$ b) rate = $k[O][O_3]$ c) rate = $k[O]$ d) rate = $k[O_2]$ e) rate = $k[O_3]$

72. Consider the mechanism:

$NO_2(g) + F_2(g) \rightarrow NO_2F(g) + F(g)$ slow

$F(g) + NO_2(g) \rightarrow NO_2F(g)$ fast

An intermediate in this reaction is
a) $NO_2(g)$ b) $NO_2F(g)$ c) There is no intermediate in this reaction. d) $NO_2F(g)$ and F(g) e) F(g)

73. The rate law for the following mechanism is

$ClO^-(aq) + H_2O(l) \leftrightarrow HOCl(aq) + OH^-(aq)$ K, fast

$I^-(aq) + HOCl(aq) \rightarrow HOI(aq) + Cl^-(aq)$ k_1, slow

$HOI(aq) + OH^-(aq) \rightarrow OI^-(aq) + H_2O(l)$ k_2, fast

a) rate = $k_1 k_2 K[ClO^-][I^-]$
b) rate = $k_1 K[ClO^-][I^-][OH^-]^{-1}$
c) rate = $k_1 K[ClO^-][I^-][OH^-]$
d) rate = $k_1[I^-][HOCl]$
e) rate = $k_1 K[ClO^-][I^-]$

74. The reaction

$NO_2(g) + CO(g) \rightarrow CO_2(g) + NO(g)$

is postulated to occur via the mechanism below:

$NO_2(g) + NO_2(g) \rightarrow NO_3(g) + NO(g)$ slow

$NO_3(g) + CO(g) \rightarrow NO_2(g) + CO_2(g)$ fast

An intermediate in this reaction is

a) $NO_3(g)$ b) $NO(g)$ c) $CO(g)$ d) $NO_2(g)$ e) $ON{-}NO_3(g)$

75. The reaction between nitrogen dioxide and carbon monoxide is thought to occur by the following mechanism:

$2NO_2(g) \rightarrow NO_3(g) + NO(g)$ k_1, slow

$NO_3(g) + CO(g) \rightarrow NO_2(g) + CO_2(g)$ k_2, fast

The rate law for this mechanism is

a) rate = $k_1[NO_3][NO]$
b) rate = $k_2[NO_3][CO]$
c) rate = $(k_1/k_2)[NO_2]^2[CO]$
d) rate = $k_1[NO_2]^2$
e) rate = $k_1k_2[NO_2]^2[CO]$

76. The following mechanism has been suggested for the reaction between chlorine and chloroform:

$Cl_2(g) \leftrightarrow 2Cl(g)$ K, fast

$CHCl_3(g) + Cl(g) \rightarrow CCl_3(g) + HCl(g)$ k_2, slow

$CCl_3(g) + Cl(g) \rightarrow CCl_4(g)$ k_3, fast

The rate law for this mechanism is

a) rate = $k_2 K^{0.5} k_3[CHCl_3][Cl_2]^{0.5}[CCl_3]$
b) rate = $k_2 K^{0.5}[CHCl_3][Cl_2]^{0.5}$
c) rate = $k_3[CCl_3][Cl]$
d) rate = $k_2[CHCl_3][Cl]$
e) rate = $k_2[CHCl_3]$

77. Consider the following reaction:

$2NO_2(g) + F_2(g) \rightarrow 2NO_2F(g)$ rate = $k[NO_2][F_2]$

Which of the following is true?

a) The overall order of the reaction is 1.
b) The reaction occurs in one step as indicated by the equation.
c) The order of the reaction with respect to NO_2 is 2.
d) The overall order of the reaction is 3.
e) The reaction occurs in more than one elementary step.

78. The following mechanism has been suggested for the reduction of Fe^{3+} to Fe^{2+} by dithionite ion, $S_2O_4^{2-}$, in the oxygen-storage protein myoglobin.

$S_2O_4^{2-} \leftrightarrow 2SO_2^-$ K, fast

$Fe^{3+} + SO_2^- \rightarrow Fe^{2+} + S(IV)$ k, slow

Derive the rate law for this mechanism.

79. Two mechanisms have been postulated for the loss of ozone in the atmosphere:

(a) $O_3 \leftrightarrow O_2 + O$ — K, fast

$O_3 + O \rightarrow 2O_2$ — k, slow

(b) $O_3 + O_3 \rightarrow 3O_2$ — k, slow

Which mechanism is consistent with the experimental rate law, rate = $k[O_3]^2/[O_2]$?

80. The HBr synthesis is thought to involve the following reactions:

$Br_2 \rightarrow 2Br\cdot$ (1)

$Br\cdot + H_2 \rightarrow HBr + H\cdot$ (2)

$H\cdot + Br_2 \rightarrow HBr + Br\cdot$ (3)

$2Br\cdot \rightarrow Br_2$ (4)

$2H\cdot \rightarrow H_2$ (5)

$H\cdot + Br\cdot \rightarrow HBr$ (6)

The chain propagation reactions in this mechanism are reactions

a) 6 b) 2, 3, and 6 c) 1, 2, and 3 d) 1, 2, 3, and 6 e) 2 and 3

Chapter 19:
The Main-Group Elements: I. The First Four Families
Form A

1. Hydrogen can be made from fossil fuels in a series of two catalyzed reactions. One of these reactions, the **shift reaction**, is
a) $Zn(s) + 2HCl(aq) \rightarrow ZnCl_2(aq) + H_2(g)$
b) $Cu(s) + 2H^+(aq) \rightarrow Cu^{2+}(aq) + H_2(g)$
c) $CH_3OH(l) \rightarrow 2H_2(g) + CO(g)$
d) $CO(g) + H_2O(g) \rightarrow CO_2(g) + H_2(g)$
e) $2H_2O(l) \rightarrow 2H_2(g) + O_2(g)$

2. Which of the following can be extracted by hydrometallurgical reduction of their ions with hydrogen? Consult a table of standard potentials.
a) Ag b) Fe c) Pb d) Ni e) Co

3. Which of the following can be extracted by hydrometallurgical reduction of their ions with hydrogen? Consult a table of standard potentials.
a) Cu b) Fe c) Mn d) V e) Co

4. The equation for the hydrometallurgical extraction of silver is
a) $Ag^+(aq) + 2S_2O_3^{2-}(aq) \rightarrow Ag(S_2O_3)_2^{3-}(aq)$
b) $Ag^+(aq) + H_2(g) \rightarrow 2Ag(s) + 2H^+(aq)$
c) $Ag^+(aq) + Cl^-(aq) \rightarrow AgCl(s)$
d) $2Ag(s) + 2H^+(aq) \rightarrow Ag^+(aq) + H_2(g)$
e) $Ag^+(aq) + 2CN^-(aq) \rightarrow Ag(CN)_2^-(aq)$

5. Which of the following is "synthesis gas?"
a) $CH_4(g)$ b) $CH_4(g) + H_2O(g)$ c) $CO(g) + H_2O(g)$ d) $CO(g) + H_2(g)$ e) $C_2H_2(g)$

6. Small quantities of hydrogen can be produced by dissolving a metal in hydrochloric acid. Which of the following does not liberate hydrogen from HCl(aq)?Consult a table of standard potentials.
a) Cu(s) b) Zn(s) c) Pb(s) d) Fe(s) e) Cd(s)

7. The mass 3 isotope of hydrogen is called
a) tritium b) deuterium c) heavy hydrogen d) radioactive hydrogen e) atomic hydrogen

8. The symbol of the radioactive isotope of hydrogen is
a) ^3H b) ^2D c) ^3T d) ^2T e) ^2H

9. Hydrogen is used in all of the following industrial processes except
a) the Haber process.
b) petroleum refining.
c) the production of HCl.
d) the reduction of metals from their oxides.
e) hydrogenation of vegetable oils.

10. Which of the following is likely to form a saline hydride?
a) Sr b) Ti c) W d) P e) As

11. All of the following hydrides are possible candidates for storing and transporting hydrogen except
a) $SiH_4(s)$ b) $CuH(s)$ c) $NaH(s)$ d) $CaH_2(s)$ e) $WH_3(s)$

12. Which of the following is likely to form a saline hydride?
a) Cs b) Fe c) Al d) Be e) C

13. When saline hydrides dissolve in water they produce
a) Saline hydrides are insoluble in water. b) $H^+(aq)$ c) $O_2(g)$ d) $H_2(g)$ e) $H_2O(l)$

14. Which of the following is a saline hydride?
a) BaH_2 b) NH_2^- c) RbOH d) HCl e) CuH

15. Which equation represents the reaction that occurs, if any, when potassium hydride is added to water?
a) $2KH(s) + H_2O(l) \rightarrow K_2O(aq) + 4H^+(aq)$
b) $KH_2(s) + H_2O(l) \rightarrow KO(s) + 2H_2(g)$
c) no reaction occurs
d) $2KH(s) + H_2O(l) \rightarrow K_2O(aq) + 2H_2(g)$
e) $KH(s) + H_2O(l) \rightarrow KOH(aq) + H_2(g)$

16. In the thermite reaction with $Fe_2O_3(s)$, how many moles of iron are produced from 1.00 kg of $Fe_2O_3(s)$?

17. Which of the following has the lowest melting point?
a) Cs b) Rb c) K d) Na e) Li

18. The alkali metals all react with water. Which is the least reactive metal?
a) Li b) Na c) K d) Rb e) Cs

19. Which of the following metals react with nitrogen at room temperature?
a) Li b) K c) Na d) Rb e) Cs

20. Lithium and sodium react with oxygen to produce
a) $Li_2O_2(s)$ and $NaO_2(s)$
b) $LiO_2(s)$ and $Na_2O_2(s)$
c) $Li_2O(s)$ and $Na_2O(s)$
d) $Li_2O(s)$ and $Na_2O_2(s)$
e) $LiO_2(s)$ and $Na_2O(s)$

21. The main product of the reaction of potassium with oxygen is
a) $K_2O(s)$ b) $KO_2(s)$ c) $K_2O_2(s)$ d) $KOH(s)$ e) $K_2O_3(s)$

22. Which group of elements all produce superoxides when burned in oxygen?
a) Li and Na b) Na, K, and Rb c) K, Rb, and Cs d) Li, Na, and K e) Na, Rb, and Cs

23. The reaction of potassium superoxide with water vapor produces
a) $H_2(g)$ b) $K_2O(s)$ c) $H_2O_2(l)$ d) $O_2(g)$ e) $KHCO_3(s)$

24. Sodium sulfate, "salt cake," is produced by the reaction of
   a) $HCl(g)$ and $NaHSO_4(s)$.
   b) $MnSO_4(s)$ and $NaCl(s)$.
   c) $H_2SO_4(l)$ and $NaCl(s)$.
   d) $NaHSO_4(s)$ and $H_2O(l)$.
   e) $SO_2(g)$ and $H_2O(l)$.

25. Calcium reacts with hydrogen to form A. When A is dissolved in water, B and C are formed. What are the compounds A, B, and C, respectively?
   a) $CaH_2(s)$, $CaO(s)$, and $H_2(g)$
   b) $CaH_2(s)$, $CaO_2(s)$, and $H_2(g)$
   c) $CaH_2(s)$, $Ca(OH)_2(s)$, and $H_2(g)$
   d) $CaH_2(s)$, $CaO(s)$, and $O_2(g)$
   e) $CaH_2(s)$, $CaO_2(s)$, and $O_2(g)$

26. Which of the substances below can be used to extinguish a magnesium fire?
   a) All of the substances listed react with burning magnesium.
   b) carbon dioxide
   c) water
   d) nitrogen
   e) sand

27. All of the following react with water to form hydrogen and a hydroxide except
   a) Be b) Ca c) Na d) K e) Mg

28. Which element from another group will resemble Al in its physical and chemical properties?
   a) Be b) Li c) Mg d) Ca e) Ga

29. An example of a diagonal relationship between elements in the periodic table is the similar properties of
   a) Li and Mg. b) Mg and B. c) Li and Be. d) Na and Ca. e) Mg and K.

30. The products of the reaction of $BaO(s)$ with $Al(s)$ are
   a) $Ba(s)$ and $Al_2O_3(s)$.
   b) $Ba(O_2)_2(s)$ and $Al_2O_3(s)$.
   c) $Ba(O_2)_2(s)$ and $Al_2(O_2)_3(s)$.
   d) $Ba(s)$ and $Al(O_2)_3(s)$.
   e) $BaO_2(s)$ and $Al_2O_3(s)$.

31. Write the balanced equation for the reaction which occurs, if any, when $Ba(s)$ is added to water.
   a) no reaction occurs
   b) $Ba(s) + H_2O(l) + CO_2(g) \rightarrow BaCO_3(s) + H_2(g)$
   c) $Ba(s) + 2H_2O(l) \rightarrow Ba(OH)_2(aq) + H_2(g)$
   d) $Ba(s) + H_2O(l) \rightarrow BaO(s) + H_2(g)$
   e) $Ba(s) + 4H_2O(l) \rightarrow Ba(OH_2)_4^{2+}(aq)$

32. All of the following react vigorously with dilute acids except
   a) beryllium b) magnesium c) radium d) barium e) strontium

33. Lithium has several anomalous properties with respect to the remaining alkali metals. This is attributed to
 a) the small ionization energy of Li.
 b) the large ionization energy of Li.
 c) the low electronegativity of Li.
 d) the small hydration energy of Li^+.
 e) the small ionic radius of Li^+.

34. Which of the following is a basic oxide?
 a) $MgO(s)$ b) $SiO_2(s)$ c) $SO_2(g)$ d) $CO_2(g)$ e) $CO(g)$

35. Which of the following is a property of a metal?
 a) low–melting halides b) basic oxides c) high electronegativity d) insulators e) acidic oxides

36. Which of the following has the lowest melting point?
 a) Ba b) Sr c) Ca d) Mg e) Be

37. Most carbonates decompose to the oxide and carbon dioxide when heated:

$$CO_3^{2-}(s) \rightarrow CO_2(g) + O^{2-}(s)$$

Carbonates also react with acid to produce carbon dioxide and water:

$$CO_3^{2-}(aq) + 2H^+(aq) \rightarrow CO_2(g) + H_2O(l)$$

(a) In the above equations indicate which species are Lewis acids and bases.
(b) Write the equation for the reaction which occurs when sodium carbonate is heated.
(c) Write the equation for the reaction which occurs when calcium carbonate is used to neutralize a sulfuric acid spill.

38. Write the equation for the thermite reaction for the reduction of $Cr_2O_3(s)$ and $BaO(s)$.

39. What is the formula of "baking soda"?
 a) Na_2CO_3 b) $NaHCO_3$ c) Na_2SO_4 d) $NaHSO_4$ e) $Na_2CO_3 \cdot 10H_2O$

40. What is the formula of anhydrous "Epsom salts"?
 a) $CaCl_2$ b) $MgCl_2$ c) $CaSO_4$ d) $MgSO_4$ e) $Mg(OH)_2$

41. The products of the reaction of sodium hydride with water are
 a) $Na_2O(s)$ and $H_2(g)$.
 b) $NaOH(aq)$ and $O_2(g)$.
 c) $Na_2O_2(s)$ and $O_2(g)$.
 d) $NaO_2(s)$ and $O_2(g)$.
 e) $NaOH(aq)$ and $H_2(g)$.

42. Write the balanced equation for the reaction which occurs, if any, when beryllium is added to NaOH(aq).
 a) no reaction occurs
 b) $Be(s) + 4OH^-(aq) \rightarrow Be(OH)_4^{2-}(aq)$
 c) $Be(s) + 2H_2O(l) \rightarrow Be(OH)_2(s) + H_2(g)$
 d) $Be(s) + 2OH^-(aq) + 2H_2O(l) \rightarrow Be(OH)_4^{2-}(aq) + H_2(g)$
 e) $Be(s) + 4OH^-(aq) \rightarrow Be(OH)_2(s) + H_2(g) + O_2(g)$

43. Which of the following has the highest melting point?
a) B b) Al c) Ga d) In e) Tl

44. Which of the following is amphoteric?
a) Be b) Ca c) Mg d) Sr e) Ba

45. Aluminum dissolves in aqueous base. What are the products of this reaction?
a) $Al(OH_2)_6^{3+}(aq)$ and $H_2(g)$
b) $Al(OH_2)_6^{3+}(aq)$ and $O_2(g)$
c) $Al(OH)_4^-(aq)$ and $O_2(g)$
d) $Al(OH)_4^-(aq)$ and $H_2(g)$
e) $Al_2O_3(s)$ and $H_2(g)$

46. Group 13 consists of the elements B, Al, Ga, In, and Tl. Identify the nonmetals.
a) B b) Tl and In c) Ga, In, and Tl d) B and Tl e) B and In

47. Which of the group 13 elements are amphoteric?
a) Al and Ga b) Ga and In c) B and Al d) B and In e) In and Tl

48. Which of the following group 13 elements tends to form compounds with oxidation number +3 and +1?
a) Tl and In b) B and Al c) Al and Ga d) Ga and B e) In and Al

49. Aluminum metal is produced by
a) the thermite reaction.
b) electrolysis of a molten mixture of alumina, Al_2O_3, and cryolite, Na_3AlF_6.
c) electrolysis of brine containing $Al_2(SO_4)_3$.
d) reduction of Al_2O_3 with carbon.
e) treatment of Al_2O_3 with sodium hydroxide.

50. Boron and aluminum
a) have basic and acidic oxides, respectively.
b) both have basic oxides.
c) both have acidic oxides.
d) both have amphoteric oxides.
e) have acidic and amphoteric oxides, respectively.

51. $Al_2O_3(s)$ dissolves in aqueous base to produce
a) $Al(OH)_4^-(aq)$ and $O_2(g)$
b) $Al(OH_2)_6^{3+}(aq)$
c) $Al(OH)_4^-(aq)$ and $H_2(g)$
d) $Al(OH)_4^-(aq)$
e) $Al(OH_2)_6^{3+}(aq)$ and $H_2(g)$

52. The formula of boric acid written as a Lewis acid is
a) $HB(OH)_2$ b) $HB(OH)_4$ c) $B(OH)_3$ d) $HB(OH)_3$ e) $B(OH)_4^-$

53. Which of the following is a Lewis acid?
a) none of the choices is a Lewis acid b) $B(OH)_4^-$ c) $B(OH)_3$ d) BH_4^- e) $B(OH)_3OH_2$

54. Boric acid is a weak monoprotic acid. When boric acid reacts with water, what are the products at equilibrium?
a) $B(OH)_2O^-$(aq) and H_3O^+(aq)
b) HBO_3^-(aq) and H_3O^+(aq)
c) $HB(OH)_2^-$(aq) and H_3O^+(aq)
d) $B(OH)_3$(aq) and H_3O^+(aq)
e) $B(OH)_4^-$(aq) and H_3O^+(aq)

55. The formula of lithium aluminum hydride is
a) $LiAlH_4$. b) LiAl(OH). c) $LiAl(OH)_4$. d) LiAlH. e) $LiAlH_3$.

56. The molecule Al_2Cl_6 is formed from two $AlCl_3$ molecules. This reaction is
a) a Lewis acid–base complex formation reaction.
b) a reduction reaction.
c) an oxidation reaction.
d) a redox dimerization reaction.
e) a precipitation reaction.

57. The molecule Al_2Br_6 is formed from two $AlBr_3$ molecules. This reaction is
a) a Lewis acid–base complex formation reaction.
b) a precipitation reaction.
c) an oxidation reaction.
d) a reduction reaction.
e) a redox dimerization reaction.

58. The formula of sodium borohydride is
a) $Na_2B_4O_5(OH)_4$ b) $Na_4B_2O_5$ c) NaB_2H_5 d) $NaB(OH)_4$ e) $NaBH_4$

59. Sodium borohydride is produced from the reaction of
a) NaH(s) and B_2O_3(s).
b) NaH(s) and BCl_3(l).
c) NaH(s) and $B(OH)_3$(aq).
d) NaH(s) and B(s).
e) $B(OH)_3$(aq) and BCl_3(l).

60. Diborane is produced by the reaction of
a) BH_3(g) and B(s)
b) B(s) and H_2(g).
c) BCl_3(g) and NaH(s)
d) B(s) and NH_3(g)
e) BF_3(g) and $NaBH_4$(s).

61. What is the approximate shape around each boron atom in diborane?
a) tetrahedral b) square planar c) trigonal bipyramidal d) see-saw e) angular

62. Write the equation for the reaction which occurs when anhydrous aluminum chloride is exposed to moist air.

63. Diborane has
 a) 2 bridging hydrogens and 4 terminal hydrogens.
 b) 4 bridging hydrogens and 2 terminal hydrogens.
 c) 6 terminal hydrogens and 1 boron–boron bond.
 d) 1 BF_3^- anion and 1 BF_3^+ cation.
 e) 1 bridging hydrogen, 4 terminal hydrogens, and an ionic hydrogen.

64. Group 14 consists of the elements C, Si, Ge, Sn, and Pb. Which elements are metals?
 a) Sn and Pb b) Ge, Sn, and Pb c) Ge and Pb d) Si and Ge e) Pb

65. The nonmetals in groups 13 and 14 are
 a) B, C, Si, and Ge b) B, C, and Ga c) B, C, and Si d) B, C, Si, and In e) B, C, Si, and Ga

66. Allotropes are
 a) gases with similar properties.
 b) different crystalline forms of the same element.
 c) molecules with the same number of electrons.
 d) different molecules which produce oxygen gas.
 e) different forms of the same element in the same physical state that have different physical or chemical properties.

67. Which of the following is amphoteric?
 a) $SnO(s)$ b) $SiO_2(s)$ c) $CO_2(g)$ d) $GeO_2(s)$ e) $SiO_4^{4-}(aq)$

68. All of the following are amphoteric except
 a) $GeO_2(s)$ b) $SnO(s)$ c) $PbO(s)$ d) $Sn(OH)_2(s)$ e) $SnO_2(s)$

69. Crude silicon is prepared by
 a) the reaction of $SiO_2(g)$ with $Cl_2(g)$.
 b) the reduction of $SiO_2(s)$ with high–purity carbon.
 c) the reaction of $SiO_2(s)$ with sodium hydroxide.
 d) zone refining.
 e) the reaction of $SiCl_4(l)$ with high–purity carbon.

70. The equation for the "roasting" of the ore galena, PbS, is
 a) $PbS(s) + 2NaOH(aq) \rightarrow Pb(OH)_2(s) + Na_2S(s)$
 b) $PbS(s) + C(s) \rightarrow Pb(s) + CS(g)$
 c) $2PbS(s) + 3O_2(g) \rightarrow 2PbO(s) + 2SO_2(g)$
 d) $PbS(s) + CO(g) \rightarrow Pb(s) + COS(g)$
 e) $PbS(s) + 2HCl(aq) \rightarrow PbCl_2(s) + H_2S(g)$

71. Carbon monoxide is produced commercially by the reaction
 a) $CO_2(g) + Ba(s) \rightarrow BaO(s) + CO(g)$
 b) $SO_2(g) + 2C(s) \rightarrow 2CO(g) + S(s)$
 c) $CH_4(g) + H_2O(g) \rightarrow CO(g) + 3H_2(g)$
 d) $2CH_4(g) + 3O_2(g) \rightarrow 2CO(g) + 4H_2O(g)$
 e) $2C(s) + O_2(g) \rightarrow 2CO(g)$

72. Carbon monoxide is the formal anhydride of
a) HCOOH b) CO_2 c) H_2CO_3 d) H_2CO e) HCO

73. An important silicate is SiO_4^{4-}. What is its shape?
a) tetrahedral b) see-saw c) square planar d) T-shaped e) trigonal bipyramidal

74. All of the following contain silicates in various forms except
a) soda lime b) talc c) mica d) asbestos e) sand

75. Which of the following is an acidic oxide?
a) $SiO_2(s)$ b) MgO(s) c) CaO(s) d) $Bi_2O_3(s)$ e) $Na_2O(s)$

76. Write the equation for the reaction of the Lewis base F^- with silica.

77. Which of the following is a saline carbide?
a) Al_4C_3 b) SiC c) W_2C d) Fe_3C e) Mo_2C

78. Which of the following is a covalent carbide?
a) SiC b) CaC_2 c) W_2C d) Fe_3C e) SrC_2

79. Saline carbides contain the anion
a) CH_2^{2-} b) CO_3^{2-} c) C_2^{2-} d) HC_2^- e) CN^-

80. The products of the reaction of $Al_4C_3(s)$ with water are
a) $Al(OH)_3(s)$ and $C_2H_2(g)$
b) $Al_2O_3(s)$ and $CH_4(g)$
c) $Al_2O_3(s)$ and $C_2H_2(g)$
d) $Al(OH)_3(s)$ and $CH_4(g)$
e) $Al(OH)_3(s)$ and $C_2H_4(g)$

Chapter 19:
The Main-Group Elements: I. The First Four Families
Form B

1. Hydrogen can be made from fossil fuels in a series of two catalyzed reactions. One of these reactions, the **re-forming reaction**, is
a) $Cu(s) + 2H^+(aq) \rightarrow Cu^{2+}(aq) + H_2(g)$
b) $Zn(s) + 2HCl(aq) \rightarrow ZnCl_2(aq) + H_2(g)$
c) $CH_4(g) + H_2O(g) \rightarrow CO(g) + 3H_2(g)$
d) $2H_2O(l) \rightarrow 2H_2(g) + O_2(g)$
e) $CH_3OH(l) \rightarrow 2H_2(g) + CO(g)$

2. Which of the following can be extracted by hydrometallurgical reduction of their ions with hydrogen? Consult a table of standard potentials.
a) Au b) Fe c) Pb d) Zn e) Cr

3. Which of the following can be extracted by hydrometallurgical reduction of their ions with hydrogen? Consult a table of standard potentials.
a) Pt b) Sn c) Pb d) Ni e) Cr

4. The equation for the hydrometallurgical extraction of copper is
a) $Cu^{2+}(aq) + 4CN^-(aq) \rightarrow Cu(CN)_4{}^{2-}(aq)$
b) $Cu^{2+}(aq) + CO_3{}^{2-}(aq) \rightarrow CuCO_3(s)$
c) $Cu^{2+}(aq) + H_2(g) \rightarrow Cu(s) + 2H^+(aq)$
d) $Cu(s) + 2H^+(aq) \rightarrow Cu^{2+}(aq) + H_2(g)$
e) $Cu^{2+}(aq) + 4Cl^-(aq) \rightarrow CuCl_4{}^{2-}(aq)$

5. The stable form of hydrogen at 25°C and 1 atm is
a) $H_2(g)$ b) $H(g)$ c) $H(l)$ d) $H(s)$ e) $H_2O(l)$

6. Small quantities of hydrogen can be produced by dissolving a metal in hydrochloric acid. Which of the following does not liberate hydrogen from HCl(aq)?Consult a table of standard potentials.
a) Ag(s) b) Zn(s) c) Pb(s) d) Fe(s) e) Cd(s)

7. The mass 2 isotope of hydrogen is called
a) deuterium b) tritiu c) heavy hydrogen d) radioactive hydrogen e) atomic hydrogen

8. The symbol of deuterium is
a) ^2D b) ^3D c) ^2T d) ^3H e) ^2H

9. Hydrogen is used in metallurgy to
a) produce metal amides.
b) produce coal.
c) reduce oxides to metals.
d) produce hydrides.
e) produce hydroxides.

10. Which of the following is likely to form a saline hydride?
a) Rb b) Zr c) V d) Sb e) As

11. When metallic hydrides are heated, they
a) produce hydrogen ions.
b) decompose explosively to yield metal oxides.
c) liberate hydrogen.
d) do not change because they are extremely stable.
e) liberate oxygen and hydrogen.

12. Which of the following is likely to form a saline hydride?
a) Ca b) Fe c) Al d) Be e) C

13. When potassium hydride dissolves in water, the products are
a) $KOH(aq)$ and $H_2(g)$ b) $O_2(g)$ and $KOH(aq)$ c) $H^+(aq)$ and $K_2O(aq)$ d) $KOH(aq)$ e) $K_2O(aq)$

14. Which of the following is a saline hydride?
a) SrH_2 b) NH_2^- c) RbOH d) HI e) CuH

15. When the hydride ion reacts with water, it acts as a
a) spectator ion. b) oxidizing agent. c) Bronsted acid. d) complexing agent. e) reducing agent.

16. In the thermite reaction with $Fe_2O_3(s)$, how many moles of Al are required to reduce 1.00 kg of $Fe_2O_3(s)$?

17. Which of the following has the highest boiling point?
a) Li b) Na c) K d) Rb e) Cs

18. The alkali metals all react with water. Which is the most reactive metal?
a) Cs b) Na c) K d) Rb e) Li

19. The alkali metals are extremely reactive. They react with
a) all nonmetals.
b) all the nonmetals but not the noble gases.
c) hydrogen and the noble gases.
d) most nonmetals but not hydrogen.
e) the group 16, 17, and 18 nonmetals.

20. Potassium, rubidium, and cesium react with oxygen to form mainly
a) $K_2O(s)$, $Rb_2O_2(s)$, and $CsO_2(s)$.
b) $K_2O(s)$, $Rb_2O(s)$, and $Cs_2O(s)$.
c) $KO_2(s)$, $Rb_2O(s)$, and $Cs_2O_2(s)$
d) $KO_2(s)$, $RbO_2(s)$, and $CsO_2(s)$.
e) $K_2O_2(s)$, $Rb_2O_2(s)$, and $Cs_2O_2(s)$.

21. The main product of the reaction of sodium with oxygen is
a) $NaO_2(s)$ b) $Na_2O_2(s)$ c) $Na_2O(s)$ d) $NaOH(s)$ e) $Na_2O_3(s)$

22. Which of the following forms a peroxide when burned in oxygen?
a) Na b) Li c) K d) Rb e) Cs

23. The reaction of sodium peroxide with water vapor produces
a) $O_2(g)$ b) $Na_2O(s)$ c) $NaHCO_3(s)$ d) $H_2O_2(l)$ e) $H_2(g)$

24. Calcium hydroxide, "slaked lime," is produced by the reaction of
a) $CaO(s)$ and $H_2(g)$.
b) $CaH_2(s)$ and $CO_2(g)$.
c) $CaO(s)$ and $H_2O(l)$.
d) $CaH_2(s)$ and $O_2(g)$.
e) $CaCO_3(s)$ and $H_2O(l)$.

25. Consider the following unbalanced equations:

$$Mg(s) + N_2(g) \rightarrow A$$

$$A + H_2O(l) \rightarrow Mg(OH)_2(aq) + B$$

Write the formulas for A and B, respectively.
a) $MgN_2(s)$ and $NO(g)$
b) $Mg_3N_2(s)$ and $N_2(g)$
c) $MgN_2(s)$ and $NH_3(g)$
d) $Mg(s)$ does not react with $N_2(g)$.
e) $Mg_3N_2(s)$ and $NH_3(g)$

26. A magnesium fire cannot be extinguished by covering with sand because of the reaction
a) $2Mg(s) + CO_2(g) \rightarrow 2MgO(s) + C(s)$
b) Burning magnesium is extinguished by covering with sand.
c) $2Mg(l) + SiO_2(s) \rightarrow 2MgO(s) + Si(s)$
d) $2Mg(s) + O_2(g) \rightarrow 2MgO(s)$
e) $Mg(s) + 2C(s) \rightarrow MgC_2(s)$

27. All of the following react with oxygen to form the metal oxide except
a) K b) Ca c) Be d) Mg e) Li

28. Beryllium has chemical and physical properties most similar to
a) Al b) Mg c) Na d) B e) K

29. An example of a diagonal relationship between elements in the periodic table is the similar properties of
a) Be and Al. b) Mg and B. c) Li and Be. d) Na and Ca. e) Mg and K.

30. The products of the reaction of calcium hydride with water are
a) $CaO(s)$ and $H_2(g)$.
b) $Ca(OH)_2(s)$ and $O_2(g)$.
c) $Ca^{2+}(aq)$ and $OH^-(aq)$.
d) $CaO(s)$ and $O_2(g)$.
e) $Ca(OH)_2(s)$ and $H_2(g)$.

31. Write the balanced equation for the reaction which occurs, if any, when Ca(s) is added to water.
a) $Ca(s) + 4H_2O(l) \rightarrow Ca(OH_2)_4{}^{2+}(aq)$
b) $Ca(s) + H_2O(l) \rightarrow CaO(s) + H_2(g)$
c) $Ca(s) + H_2O(l) + CO_2(g) \rightarrow CaCO_3(s) + H_2(g)$
d) no reaction occurs
e) $Ca(s) + 2H_2O(l) \rightarrow Ca(OH)_2(aq) + H_2(g)$

32. All of the following react with water except
a) aluminum b) potassium c) magnesium d) potassium hydride e) potassium superoxide

33. Beryllium has several anomalous properties with respect to the remaining alkaline earth metals. This is attributed to
a) the large negative standard potential of the Be^{2+}/Be couple.
b) the small electrobegativity of Be.
c) the large ionization energy of Be.
d) the small ionization energy of Be^{+}.
e) the small size and large charge of Be^{2+}.

34. Which of the following is a basic oxide?
a) $CaO(s)$ b) $SiO_2(s)$ c) $SO_2(g)$ d) $CO_2(g)$ e) $CO(g)$

35. Which of the following is a property of a nonmetal?
a) low electronegativity
b) high electrical conductance
c) basic oxides
d) nonvolatile halides
e) acidic oxides

36. Which of the following has the highest boiling point?
a) Be b) Ca c) Ba d) Sr e) Mg

37. (a) Write the equation for the thermite reaction.
(b) Explain why the reaction is so highly exothermic.

38. How much heat is produced in the thermite reaction if 1.00 kg of $Fe_2O_3(s)$ is reduced? The ΔH^{o} for the reaction is –853 kJ.

39. What is the formula of the main constituent of "milk of magnesia"?
a) $CaCO_3$ b) $MgSO_4$ c) $CaSO_4$ d) $MgCl_2$ e) $Mg(OH)_2$

40. What is the formula of "quicklime"?
a) CaO b) $Ca(OH)_2$ c) $CaCO_3$ d) $CaSO_4$ e) $MgSO_4$

41. The products of the reaction when $CaCO_3(s)$ is heated are
a) $CaO(s)$, $C(s)$ and $O_2(g)$.
b) $Ca(s)$ and $CO(g)$.
c) $CaO(s)$ and $CO_2(g)$.
d) $CaO(s)$ and $CO(g)$.
e) $Ca(s)$ and $CO_2(g)$.

42. Write the balanced equation for the reaction which occurs, if any, when beryllium is added to water
  a) $Be(s) + 4OH^-(aq) \rightarrow Be(OH)_2(s) + H_2(g) + O_2(g)$
  b) $Be(s) + H_2O(l) + CO_2(g) \rightarrow BeCO_3(s) + H_2(g)$
  c) $Be(s) + 2OH^-(aq) + 2H_2O(l) \rightarrow Be(OH)_4^{2-}(aq) + H_2(g)$
  d) $Be(s) + H_2O(l) \rightarrow BeO(aq) + H_2(g)$
  e) no reaction occurs.

43. Which of the following has the lowest boiling point?
  a) Tl b) In c) Ga d) Al e) B

44. Which of the following is amphoteric?
  a) Al b) B c) Mg d) Ca e) Ba

45. Aluminum dissolves in aqueous acid. What are the products of this reaction?
  a) $Al_2O_3(s)$ and $H_2(g)$
  b) $Al(OH_2)_6^{3+}(aq)$ and $H_2(g)$
  c) $Al(OH)_4^-(aq)$ and $O_2(g)$
  d) $Al(OH)_4^-(aq)$ and $H_2(g)$
  e) $Al(OH_2)_6^{3+}(aq)$ and $O_2(g)$

46. Group 13 consists of the elements B, Al, Ga, In, and Tl. Identify the metals.
  a) B and Al b) Al, Ga, and In c) Al, Ga, In, and Tl d) Al and Tl e) Al and In

47. The oxides of In and Tl
  a) are basic.
  b) are unstable and cannot be isolated.
  c) react with water.
  d) react with base.
  e) are amphoteric.

48. Which of the following statements is true regarding the group 13 elements?
  a) Thallium forms only Tl(I) compounds.
  b) All compounds of these elements have an oxidation number of +3.
  c) Indium forms only In(III) compounds.
  d) The oxidation number +3 becomes increasingly important on going down the group.
  e) The oxidation number +1 becomes increasingly important on going down the group.

49. Boron is obtained by
  a) reduction of the mineral kernite, $Na_2B_4O_4 \cdot 4H_2O(s)$ with Mg(s).
  b) treatment of $B_2O_3(s)$ with acid.
  c) reduction of $B_2O_3(s)$ with Mg(s).
  d) reduction of the mineral borax, $Na_2B_4O_4 \cdot 10H_2O(s)$ with Mg(s).
  e) heating $B_2O_3(s)$.

50. Aluminum and gallium
a) both have acidic oxides.
b) have acidic and amphoteric oxides.
c) both have amphoteric oxides.
d) both have basic oxides.
e) have amphoteric and acidic oxides.

51. $Al_2O_3(s)$ dissolves in aqueous acid to produce
a) $Al(OH_2)_6^{3+}(aq)$ and $H_2(g)$
b) $Al(OH_2)_6^{3+}(aq)$
c) $Al(OH)_4^-(aq)$ and $O_2(g)$
d) $Al(OH)_4^-(aq)$ and $H_2(g)$
e) $Al(OH)_4^-(aq)$

52. The formula of boric acid written as a Bronsted acid is
a) $HB(OH)_3$ b) $B(OH_2)_3$ c) $HB(OH)_2$ d) $B(OH)_3OH_2$ e) $B(OH)_4^-$

53. Which of the following is a Bronsted acid?
a) $Al(OH)_4^-$ b) $B(OH)_3$ c) none of the choices is a Bronsted acid d) $B(OH)_4^-$ e) $B(OH)_3OH_2$

54. The equation which corresponds to the K_a for boric acid is
a) $B(OH)_3(aq) + 2H_2O(l) \leftrightarrow B(OH)_4^-(aq) + H_3O^+(aq)$
b) $B(OH)_4^-(aq) + H^+(aq) \leftrightarrow B(OH)_3(aq) + H_2O(l)$
c) $B_4O_5(OH)_4^{2-}(aq) + 5H_2O(l) \leftrightarrow 4H_3BO_3(aq) + 2OH^-(aq)$
d) $B_4O_7^{2-}(aq) + H_2O(l) \leftrightarrow B_4O_8H^{3-}(aq) + H^+(aq)$
e) $H_3BO_3(aq) + H_2O(l) \leftrightarrow HB(OH)_3^+(aq) + OH^-(aq)$

55. The formula of boron nitride is
a) BN b) BN_3 c) B_2N_2 d) BN_2 e) B_2N_4

56. The molecule Al_2Cl_6 is formed from two $AlCl_3$ molecules. In the molecule Al_2Cl_6,
a) the aluminum atoms are the Lewis acids and a Cl atom on the neighboring Al atom is the Lewis base.
b) there is an Al–Al bond.
c) there is an electrostatic interaction between Cl^- ions and Al^{3+} ions.
d) 1 Al atom is a Lewis acid and the other is a Lewis base.
e) 1 $AlCl_3$ unit has a negative charge and the other has a positive charge.

57. The molecule Al_2Br_6 is formed from two $AlBr_3$ molecules. In the molecule Al_2Cl_6,
a) the aluminum atoms are the Lewis acids and a Br atom on the neighboring Al atom is the Lewis base.
b) there is an Al–Al bond.
c) there is an electrostatic interaction between Br^- ions and Al^{3+} ions.
d) 1 Al atom is a Lewis acid and the other is a Lewis base.
e) 1 $AlBr_3$ unit has a negative charge and the other has a positive charge.

58. The formula of diborane is
a) BH_3 b) B_2H c) B_2H_6 d) B_2H_4 e) B_2H_2

59. Boron nitride is produced from the reaction of
a) $B_2H_6(g)$ and $N_2(g)$.
b) $B_2O_3(s)$ and $NH_3(g)$
c) B(s) and $N_2(g)$.
d) B(s) and $NH_3(g)$.
e) B(s) and HCN(g).

60. The products of the reaction of diborane with water are
a) $B(OH)_4^-(aq)$ and $H_2(g)$.
b) $B(OH)_3(aq)$ and $O_2(g)$.
c) $BH_3(g)$ and $B(OH)_3(aq)$.
d) $B(OH)_3(aq)$ and $H_2(g)$.
e) $B_2H_5^-(aq)$ and $H_3O^+(aq)$.

61. What is the approximate shape around each aluminum atom in Al_2Cl_6?
a) tetrahedral b) square planar c) trigonal bipyramidal d) see-saw e) angular

62. Explain why boron forms electron-deficient compounds like diborane.

63. What is the shape of boron trifluoride?
a) trigonal planar b) trigonal pyramidal c) see-saw d) T-shaped e) trigonal bipyramidal

64. Group 14 consists of the elements C, Si, Ge, Sn, and Pb. Which elements are metalloids?
a) Si and Ge b) Ge and Sn c) Si d) Ge e) C and Ge

65. The metals in groups 13 and 14 are
a) Al and Pb
b) Al, Ga, In, Tl, Sn, and Pb
c) Al, In, Tl, and Pb
d) Al, Sn, and Pb
e) Al, In, Tl, and Pb

66. Which of the following are allotropes?
a) silicon carbide, diamond, and C_{60}
b) graphite, diamond, and C_{60}
c) carbon monoxide and carbon dioxide
d) silicon, carbon, and C_{60}
e) boron carbide and carbon

67. Which of the following is amphoteric?
a) PbO(s) b) $SiO_2(s)$ c) $CO_2(g)$ d) $GeO_2(s)$ e) $SiO_4^{4-}(aq)$

68. All of the following are amphoteric except
a) $SiO_2(s)$ b) SnO(s) c) PbO(s) d) $Sn(OH)_2(s)$ e) $SnO_2(s)$

69. Ultrapure silicon is obtained by
a) repeated distillation of $SiCl_4(l)$ and subsequent reaction with Mg(s).
b) decomposition of $SiH_4(g)$ to Si(l) and $H_2(g)$.
c) reaction of $SiO_2(s)$ with strong acids.
d) electrolysis of molten SiO_2.
e) zone refining.

70. Lead is obtained from the oxide by
  a) reduction with carbon.
  b) reduction with aluminum.
  c) reduction with $CO(g)$.
  d) heating to 1600°C.
  e) treatment with concentrated sulfuric acid.

71. All of the following reactions involve carbon monoxide acting as a reducing agent except
  a) $CO(g) + H_2O(g) \rightarrow CO_2(g) + H_2(g)$
  b) $Ni(s) + 4CO(g) \rightarrow Ni(CO)_4(g)$
  c) $Fe_2O_3(s) + 3CO(g) \rightarrow 2Fe(s) + 3CO_2(g)$
  d) $2CO(g) + O_2(g) \rightarrow 2CO_2(g)$
  e) $PbO(s) + CO(g) \rightarrow Pb(s) + CO_2(g)$

72. Carbon dioxide is the acid anhydride of
  a) H_2CO_3 b) HCOOH c) HCO d) H_2CO e) CH_3COOH

73. The mineral spondumene contains long, straight chain silicates involving the SiO_3^{2-} unit. What is the shape of SiO_3^{2-}?
  a) trigonal planar b) tetrahedral c) trigonal pyramidal d) see-saw e) T-shaped

74. All of the following contain silicates in various forms except
  a) alum b) talc c) mica d) asbestos e) sand or quartz

75. Which of the following is an acidic oxide?
  a) $CO_2(g)$ b) $MgO(s)$ c) $CaO(s)$ d) $Bi_2O_3(s)$ e) $Na_2O(s)$

76. Write the equation for the reaction used to obtain tin from the mineral cassiterite.

77. Which of the following is a saline carbide?
  a) CaC_2 b) SiC c) W_2C d) Fe_3C e) Mo_2C

78. Which of the following is an interstitial carbide?
  a) Al_4C_3 b) CaC_2 c) K_2C_2 d) SrC_2 e) W_2C

79. Saline carbides contain the anion
  a) C^{4-} b) CN^- c) HC_2^- d) CO_3^{2-} e) CH_2^{2-}

80. The products of the reaction of calcium carbide with water are
  a) $Ca(OH)_2(s)$ and $C_2H_2(g)$
  b) $Ca(OH)_2(s)$ and $CH_4(g)$
  c) $CaH_2(s)$ and $CH_4(g)$
  d) $CaO(s)$ and $C_2H_2(g)$
  e) $CaO(s)$ and $CH_4(g)$

Chapter 20:
The Main-Group Elements: II. The Last Four Families
Form A

1. Which of the following is a metalloid?
a) Sb b) Bi c) Pxw d) Se e) S

2. Which of the following is a metalloid?
a) As b) Bi c) P d) Se e) S

3. Which of the following is a metal?
a) Bi b) Sb c) Po d) Te e) As

4. Which of the following elements form oxides which are acidic?
a) Sb and As b) Bi and Sb c) N and P d) Bi e) N and Bi

5. Lithium nitride reacts directly with water to produce
a) ammonia. b) hydrazine. c) nitric acid. d) nitrogen. e) lithium nitrate.

6. Hydrazine is produced by the reaction of
a) ammonia with alkaline hypochlorite solution.
b) nitrogen dioxide with water.
c) nitrogen with water vapor at high pressure.
d) nitric acid with hygrogen.
e) ammonia with hydrogen.

7. Write the name and formulas of compounds in which nitrogen has oxidation numbers from +5 to –3.

8. The formula of sodium nitrite is
a) Na_3N. b) $NaNO_2$. c) $NaNO_3$. d) NaNO. e) Na_2NO_2.

9. In the Ostwald process, the total change in oxidation number of nitrogen from reactant to nitric acid is
a) 8. b) 7. c) 6. d) 5. e) 3.

10. Which of the following compounds contains nitrogen with an oxidation number of +5?
a) HNO_3 b) NO c) N_2O d) NH_3 e) NO_2

11. In order to produce nitric acid starting with ammonia gas,
a) ammonia must be oxidized.
b) ammonia is first converted to $N_2(g)$ and $H_2(g)$.
c) ammonia must be reduced.
d) Ammonia is not used to produce nitric acid.
e) ammonia must be liquefied.

12. Nonmetal chlorides react with water to give
a) the nonmetal oxide and $Cl_2(g)$.
b) oxoacids without a change in oxidation number.
c) oxochloro acids in a redox reaction.
d) the nonmetal oxide and hypochlorous acid.
e) the nonmetal and HCl(aq).

13. The main product of the reaction of dinitrogen trioxide and water is
a) $N_2H_5^+(aq)$. b) $HN_3(aq)$. c) $HNO_3(aq)$. d) $HNO_2(aq)$. e) $H_2N_2O_4(aq)$.

14. Dinitrogen trioxide can be prepared by
a) reacting hydrazine with water.
b) reacting N_2O with O_2.
c) dehydrating nitric acid.
d) dehydrating nitrous acid.
e) reacting ammonia with water.

15. Dinitrogen pentoxide is the anhydride of
a) nitric acid. b) nitrous acid. c) hydrazoic acid. d) hydrazine. e) dinitrogen tetroxide.

16. The anhydride of phosphoric acid is produced by the reaction
a) $2H_3PO_4(l) \rightarrow H_4P_2O_7(l) + H_2O(l)$
b) $P_4(s) + 5O_2(g) \rightarrow P_4O_{10}(s)$
c) $P_4(s) + 3O_2(g) \rightarrow P_4O_6(s)$
d) $P_4(g) + 3OH^-(aq) + 3H_2O(l) \rightarrow 3H_2PO_2^-(aq) + PH_3(g)$
e) $H_3PO_4(aq) \rightarrow HPO_3(aq) + H_2O(l)$

17. Phosphorus(III) oxide reacts with water to produce
a) $H_3PO_2^-$. b) H_3PO_2. c) H_3PO_3. d) H_3PO_4. e) HPO_3.

18. The formula of phosphine is
a) P_2H_6. b) PH_4Cl. c) PH_3. d) POH_3. e) KPH_2.

19. White phosphorus
a) is very unreactive.
b) consists of large, random aggregates of phosphorus atoms.
c) consists of tetrahedral P_4 molecules.
d) is an amorphous substance.
e) is not very toxic.

20. The geometry of PCl_3 is
a) trigonal pyramidal. b) tetrahedral. c) T-shaped. d) seesaw. e) trigonal bipyramidal.

21. When the drying agent $P_4O_{10}(s)$ reacts with water vapor, the product is
a) $H_2PO_3(aq)$. b) $H_5P_3O_{10}(aq)$. c) $H_3PO_4(aq)$. d) $PH_4^+(aq)$. e) $PH_3(g)$.

22. Phosphoric and phosphorous acids are
a) both triprotic.
b) diprotic and triprotic, respectively.
c) triprotic and diprotic, respectively.
d) triprotic and monoprotic, respectively.
e) both diprotic.

23. The formula of phosphorous acid is
a) $H_4P_2O_7$. b) H_3PO_3. c) H_3PO_4. d) H_2PO_3. e) HPO_2.

24. Phosphoric acid is sold for laboratory use as 85% H_3PO_4 by mass. If the concentration of the acid is 15 M, calculate the density of this solution.

25. Which of the following pairs are allotropes?
a) $O_2(g)$ and $O_3(g)$
b) $S_8(s)$ and $H_2S(g)$
c) $O_2(g)$ and $H_2O_2(l)$
d) C(s) and $CO_2(g)$
e) S(s) and $SO_2(g)$

26. Native sulfur is mined from underground deposits by the
a) reaction of sulfur dioxide with hydrogen sulfide.
b) Frasch process.
c) reaction of oxygen with hydrogen sulfide.
d) roasting of sulfide ores.
e) contact process.

27. The term *catenation* refers to
a) the fact that sulfur can exist as monatomic S(g).
b) the fact that ozone is an allotrope of oxygen.
c) the ability of sulfur to bond four oxygen atoms.
d) the synthesis of sulfuric acid.
e) the ability of sulfur to form chains of atoms.

28. Water can act as
a) All of the responses are correct.
b) a Lewis base.
c) a Bronsted acid.
d) a Bronsted base.
e) an oxidizing agent.

29. Which of the following is paramagnetic?
a) N_2O b) O_3 c) O_2^- d) O_2^{2-} e) N_2

30. Hydrogen peroxide is sold for laboratory use as 30% H_2O_2 by mass. If the density of this solution is about 1.11 g, mL^{-1}, calculate the molarity.

31. Which of the following can only be an oxidizing agent?
a) H_2SO_3 b) H_2SO_4 c) SO_2 d) H_2S e) SO_3^{2-}

32. In the contact process for the production of sulfuric acid, sulfur is first burned in oxygen to produce $SO_2(g)$. The $SO_2(g)$ is then
a) oxidized to $SO_3(g)$.
b) dissolved in water to form oleum.
c) dissolved in water to form $H_2SO_4(aq)$.
d) dissolved in water to form $H_2SO_3(aq)$.
e) reduced to $H_2S(g)$.

33. What is the formula of oleum?
a) $H_2S_2O_4$ b) $H_4S_2O_7$ c) $H_2S_2O_8$ d) $H_2S_2O_7$ e) $H_2SO_4.H_2O$

34. Oleum is produced from
a) $S(s)$ and $H_2SO_4(l)$.
b) $SO_3(g)$ and $H_2SO_4(l)$.
c) $SO_2(g)$ and $H_2SO_4(l)$.
d) $SO_3(g)$ and $H_2S(g)$.
e) $H_2S(g)$ and $H_2SO_4(l)$.

35. The contact process for the production of sulfuric acid is

$$S(g) + O_2(g) \rightarrow SO_2(g)$$
$$2SO_2(g) + O_2(g) \rightarrow 2SO_3(g)$$
$$SO_3(g) + H_2SO_4(l) \rightarrow H_2S_2O_7(l)$$
$$H_2S_2O_7(l) + H_2O(l) \rightarrow 2H_2SO_4(l)$$

How many moles of "new" sulfuric acid are produced from 75 mol S(g)?
a) 225 mol b) 300 mol c) 75 mol d) 150 mol e) 37.5 mol

36. Which of the following is not a property of sulfuric acid?
a) Sulfuric acid is a Bronsted acid.
b) Sulfuric acid is an oxidizing agent.
c) The reaction $H_2SO_4(aq) + H_2O(l) \rightarrow HSO_4^-(aq) + H_3O^+(aq)$ is complete.
d) Sulfuric acid is a dehydrating agent.
e) The reaction $H_2SO_4(aq) + 2H_2O(l) \rightarrow SO_4^{2-}(aq) + 2H_3O^+(aq)$ goes to completion.

37. What reaction, if any, is predicted between iron and concentrated sulfuric acid?
a) $Fe(s) + H_2SO_4(aq) \rightarrow FeSO_4(aq) + H_2(g)$
b) $Fe(s) + 2H_2SO_4(aq) \rightarrow FeSO_4(aq) + SO_2(g) + 2H_2O(l)$
c) $Fe(s) + 2H_2SO_4(aq) \rightarrow FeSO_4(aq) + H_2SO_3(aq) + H_2O(l)$
d) $Fe(s) + 2H_2SO_4(aq) \rightarrow FeS(aq) + SO_2(g) + 2H_2O(l) + 2O_2(g)$
e) No reaction occurs.

38. Salts of sulfuric and sulfurous acid, respectively, are called
a) sulfates and persulfates.
b) sulfates and sulfites.
c) sulfites and persulfates.
d) sulfates and persulfites.
e) sulfates and thiosulfites.

39. The shape of SO_3 is
a) trigonal planar. b) trigonal pyramidal. c) tetrahedral. d) seesaw. e) trigonal bipyramidal.

40. All of the following form clathrates with water except
a) carbon dioxide. b) hydrogen sulfide. c) sulfur dioxide. d) methane. e) argon.

41. Sulfur does not react directly with
a) iodine. b) fluorine. c) chlorine. d) chlorine and bromine. e) fluorine and chlorine.

42. The shape of sulfur dichloride is
a) angular. b) tetrahedral. c) trigonal planar. d) linear. e) trigonal pyramidal.

43. The total number of lone pairs on both sulfur atoms in sulfur dichloride is
a) 4. b) 2. c) 6. d) 0. e) 3.

44. Which element has the lowest melting point?
a) F b) Cl c) Br d) I e) At

45. Which element has the smallest atomic radius?
a) F b) Cl c) Br d) I e) At

46. Which halogen is the strongest oxidizing agent?
a) F_2 b) I_2 c) Cl_2 d) Br_2 e) At_2

47. The stable forms of chlorine, bromine, and iodine at 25°C and 1 atm are
a) $Cl_2(g)$, $Br_2(l)$, and $I_2(l)$.
b) $Cl_2(g)$, $Br_2(l)$, and $I_2(s)$.
c) $Cl_2(g)$, $Br(l)$, and $I_2(g)$.
d) $Cl_2(g)$, $Br_2(g)$, and $I_2(g)$.
e) $Cl_2(g)$, $Br(l)$, and $I(s)$.

48. Which of the following is true?
a) Fluorine can be produced by oxidation of fluoride ion with chlorine gas.
b) Bromine can be produced by oxidation of bromide ion with chlorine gas.
c) Fluorosubstituted hydrocarbons react with strong oxidizing agents.
d) When chlorine reacts with iron, the product is iron(II) chloride.
e) Chlorine is a better oxidizing agent than fluorine.

49. The lattice enthalpies of ionic compounds of fluorine tend to be very high because
a) fluorine gas is very reactive.
b) fluorine is a strong oxidant.
c) the fluoride ion is small.
d) fluoride ion has an oxidation state of –1.
e) fluorine has a high electronegativity.

50. All of the following reactions occur except
a) $2I^-(aq) + Cl_2(g) \rightarrow I_2(aq) + 2Cl^-(aq)$
b) $2Br^-(aq) + Cl_2(g) \rightarrow Br_2(l) + 2Cl^-(aq)$
c) $2F^-(aq) + Cl_2(g) \rightarrow F_2(g) + 2Cl^-(aq)$
d) $2F_2(g) + H_2O(g) \rightarrow OF_2(g) + 2HF(g)$
e) $F_2(g) + H_2(g) \rightarrow 2HF(g)$

51. What is the shape of the I_3^- ion?
a) linear b) trigonal bipyramidal c) angular d) tetrahedral e) trigonal planar

52. What is the shape of BrF_5?
a) square pyramidal b) octahedral c) trigonal bipyramidal d) square planar e) seesaw

53. What is the shape of ClF_5?
a) seesaw b) octahedral c) trigonal bipyramidal d) square planar e) square pyramidal

54. Which of the following is the weakest acid?
a) HF b) HCl c) HBr d) HI e) The acids all have the same strength.

55. Which of the following is unstable?
a) HFO b) HClO c) HBrO d) HF e) HI

56. What is the oxidation number of Br in $HBrO_4$?
a) +7 b) +5 c) +3 d) +8 e) +6

57. What is the oxidation number of Cl in $HClO_2$?
a) +4 b) +3 c) +5 d) +2 e) +1

58. When sulfuric acid and sodium chloride are mixed, a product is
a) S(s). b) Cl_2(g). c) H_2S(g). d) SO_2(g). e) HCl(g).

59. When phosphoric acid and potassium iodide are mixed, a product is
a) HI(g). b) I_2(g). c) PI_2(g). d) P_4(s). e) HIO(g).

60. Which of the following is the weakest acid?
a) HBrO b) $HBrO_4$ c) $HClO_4$ d) $HBrO_3$ e) HClO

61. What is the formula of dry bleach, calcium hypochlorite?
a) $Ca(ClO_4)_2$ b) $Ca(ClO)_2$ c) $Ca(ClO_3)_2$ d) $CaCl(ClO_3)$ e) CaCl(ClO)

62. When potassium chlorate is heated in the presence of a catalyst, a product is
a) KO_2(s). b) Cl_2(g). c) OCl_2(g). d) ClO_2(g). e) O_2(g).

63. Which of the following can only act as an oxidizing agent?
a) $NaClO_3$ b) ClO_2 c) $HClO_4$ d) Cl_2 e) NaClO

64. Which of the following is hypobromous acid?
a) HBrO b) $HBrO_2$ c) $HBrO_3$ d) $HBrO_4$ e) $HBrO_5$

65. The halogens form a series of oxoacids in which the oxidation number of the halogen atom can be all of the following except
a) −1. b) +1. c) +3. d) +5. e) +7.

66. What is the shape of the chlorite ion?
a) angular b) linear c) seesaw d) trigonal pyramidal e) tetrahedral

67. Write a balanced equation for the disproportionation of chlorine gas in water.

68. The phase diagram for helium indicates that
a) helium exists as He(g) and He(l) at about 2 K.
b) helium has four phases.
c) helium exists only as a gas.
d) helium can be liquefied at ordinary pressures.
e) helium has three phases.

69. Which of the Group 18 elements has the largest enthalpy of vaporization?
a) Xe b) Kr c) Ar d) Ne e) He

70. The Lunar Lander used the following reaction for propulsion:

$$NH_2N(CH_3)_2(l) + 2N_2O_4(l) \rightarrow 3N_2(g) + 2CO_2(g) + 4H_2O(g)$$

Calculate $-G^o$ and K_{eq} for this reaction. Do your calculations give you confidence that this reaction works? Explain. The standard free energies of formation of $NH_2N(CH_3)_2(l)$, $N_2O_4(l)$, $CO_2(g)$, and $H_2O(g)$ are +210 (estimated), +97.54, –394.36, and –228.57 kJ, mol^{-1}, respectively.

71. Which of the following noble gases is likely to have the smallest ionization energy?
a) Xe b) Kr c) Ar d) Ne e) He

72. Xenon tetrafluoride is a powerful oxidizing agent. A product in its reactions is likely to be
a) Xe. b) XeF_6. c) XeO_3. d) XeO_4^{2-}. e) XeO_4^{4-}.

73. If xenon is heated with excess fluorine under high pressure, the products are
a) XeF_3(s) and XeF_5^+(g).
b) Xe_2F_4(g) and XeF_4^{2-}(g).
c) XeF_2(s) and Xe_2F_2(g).
d) XeF_2(s), XeF_4(s), and XeF_6(s).
e) No reaction occurs.

74. Xenon tetroxide is prepared by the reaction of
a) XeO_3(s) with O_3(g).
b) Ba_2XeO_6(s) with sulfuric acid.
c) XeO_3(s) with O_2(g).
d) XeF_6(s) with H_2O(l).
e) XeF_4(s) with H_2O(l).

75. The formula of barium perxenate is
a) $BaXe_2O_6$. b) $BaXeO_3$. c) Ba_2XeO_4. d) Ba_2XeO_6. e) $BaXeO_4$.

76. What is the oxidation number of Xe in XeO_2F_2?
a) +6 b) +8 c) +4 d) +2 e) 0

77. What is the oxidation number of Xe in Ba_2XeO_6?
a) +8 b) +6 c) +4 d) +10 e) 0

78. Xenon can exhibit all of the following oxidation numbers in its compounds.
a) +2, +4, +6, +8 b) +2, +4, +5 c) +2, +4, +6, +7 d) +2, +3, +4, +5, +6 e) 0, +2, +3

79. What is the shape of XeO_3?
a) tetrahedral b) T-shaped c) trigonal pyramidal d) linear e) trigonal planar

80. What is the shape of XeO_2F_2?
a) seesaw b) T-shaped c) tetrahedral d) square planar e) trigonal bipyramidal

Chapter 20:
The Main-Group Elements: II. The Last Four Families
Form B

1. Which of the following is a metalloid?
a) Te b) Bi c) P d) Se e) S

2. Which of the following is a metalloid?
a) Po b) Bi c) P d) Se e) S

3. All of the following are metalloids except
a) Bi. b) As. c) Sb. d) Te. e) Po.

4. The most stable form of phosphorus at 25°C and 1 atm is
a) $P_2O_3(s)$ b) $P_4(g)$. c) $P_4O_{10}(s)$ d) P(s, red). e) P(s, white).

5. Magnesium nitride dissolves in water to produce
a) $NH_3(g)$ and $Mg(OH)_2(s)$.
b) $N_2H_4(l)$ and $Mg(OH)_2(s)$.
c) $N_2(g)$ and MgO(s).
d) $Mg(NO_3)_2(aq)$.
e) $HNO_3(aq)$ and MgO(s).

6. Dinitrogen oxide or "laughing gas" is used as a propellant in canned "whipped cream" products. It is produced by
a) heating $Mg(NO_3)_2(s)$.
b) heating $NH_4HCO_3(s)$.
c) reaction of Cu(s) with $HNO_3(aq)$.
d) heating $NH_4NO_3(s)$.
e) reaction of NO(g) with $NO_2(g)$.

7. Write balanced equations for the following.
(a) The preparation of hydrazine.
(b) The hydrolysis of PCl_5.
(c) The reaction of magnesium nitride with water.

8. The formula of hydrazine is
a) NH_2NH_2. b) N_2H_2. c) NH_2OH. d) HNO_2. e) HN_3.

9. The production of nitric acid in the Ostwald process starts with
a) $NH_3(g)$. b) $N_2(g)$. c) NO(g). d) $N_2O_3(g)$. e) $N_2H_4(aq)$.

10. Which of the following can only be an oxidizing agent?
a) HNO_3 b) NH_3 c) N_2 d) NO e) NO_2

11. The first reaction in the Ostwald process for the synthesis of nitric acid is the
a) production of $NH_3(l)$.
b) oxidation of ammonia to nitrogen oxide.
c) reduction of ammonia with copper.
d) oxidation of ammonia to nitrogen dioxide.
e) decomposition of ammonia to nitrogen and hydrogen.

12. Phosphorus pentachloride reacts with water to give
a) $H_3PO_4(l)$ and $HCl(g)$.
b) $H_3PO_4(l)$ and $HClO(aq)$.
c) $H_3PO_3(s)$ and $Cl_2(g)$.
d) $H_3PO_4(l)$ and $Cl_2(g)$.
e) $H_3PO_3(s)$ and $HCl(g)$.

13. The main product of the reaction of dinitrogen pentoxide and water is
a) $H_2N_2O_4(aq)$. b) $HNO_3(aq)$. c) $HNO_2(aq)$. d) $N_2H_5^+(aq)$. e) $HN_3(aq)$.

14. Dinitrogen pentoxide can be prepared by
a) dehydrating nitrous acid.
b) reacting ammonia with NO.
c) reacting hydrazine with water.
d) dehydrating nitric acid.
e) reacting N_2O with O_2.

15. Dinitrogen trioxide is the anhydride of
a) hydrazoic acid. b) nitrous acid. c) nitric acid. d) hydrazine. e) dinitrogen tetroxide.

16. The anhydride of phosphorous acid is produced by the reaction
a) $P_4(g) + 3OH^-(aq) + 3H_2O(l) \rightarrow 3H_2PO_2^-(aq) + PH_3(g)$
b) $H_3PO_4(aq) \rightarrow HPO_3(aq) + H_2O(l)$
c) $P_4(s) + 3O_2(g) \rightarrow P_4O_6(s)$
d) $P_4(s) + 5O_2(g) \rightarrow P_4O_{10}(s)$
e) $2H_3PO_4(l) \rightarrow H_4P_2O_7(l) + H_2O(l)$

17. Phosphorus(V) oxide reacts with water to produce
a) HPO_3. b) H_3PO_4. c) $H_3PO_2^-$. d) H_3PO_2. e) H_3PO_3.

18. The formula of phosphorus(V) oxide is
a) P_4O_{10}. b) P_2O_3. c) P_4O_6. d) PO_2. e) P_2O_4.

19. White phosphorus reacts directly with excess oxygen to produce
a) $PO_3(g)$. b) $P_4O_{10}(s)$. c) $PO_2(g)$. d) $P_4O_6(s)$. e) $P_2O_3(s)$.

20. The geometry of PCl_4^+ is
a) tetrahedral. b) trigonal pyramidal. c) trigonal bipyramidal. d) T-shaped. e) seesaw.

21. What are the products of the reaction of $PBr_5(s)$ with water?
a) $H_3PO_2(aq)$ and $HBr(l)$
b) $H_3PO_4(l)$ and $HBr(l)$
c) $H_3PO_4(l)$ and $Br_2(l)$
d) $H_3PO_3(s)$ and $HBr(l)$
e) $H_3PO_3(s)$ and $Br_2(l)$

22. Phosphorous and phosphoric acids have
a) 2 and 0 hydrogens bonded to the phosphorus atom.
b) 1 and 0 hydrogens bonded to the phosphorus atom, respectively.
c) 1 hydrogen bonded to the phosphorus atom.
d) 2 hydrogens bonded to the phosphorus atom.
e) 3 hydrogens bonded to oxygen atoms.

23. When phosphate rock is treated with phosphoric acid, *triple superphosphate* is produced. What is the formula of *triple superphosphate*?
a) $Ca_3(PO_4)_2$ b) Ca_3P_2 c) $Ca(H_2PO_4)_2$ d) $Ca(HPO_4)$ e) $Ca_2(P_2O_7)$

24. Nitric acid is sold for laboratory use as 70% HNO_3 by mass. If the concentration of the acid is 16 M, calculate the density of this solution.

25. Which of the following statements is true?
a) Oxygen has an unusual ability to catenate.
b) Sulfur does not catenate.
c) Oxygen is diamagnetic.
d) Oxygen forms chains and rings with itself.
e) Sulfur forms chains and rings with itself.

26. The allotropic forms of solid sulfur are
a) S_8 rings and limear S_8 chains.
b) galena and rhombic.
c) plastic and monoclinic.
d) S_8 and S_6.
e) rhombic and monoclinic.

27. When sulfur is produced by the Claus process, the starting material is
a) $H_2SO_4(l)$. b) $SO_2(g)$. c) $H_2S(g)$. d) $Na_2SO_3(s)$. e) $SO_3(g)$.

28. Hydrogen peroxide is
a) a very weak acid with a pK_{a1} of 11.75.
b) All of the responses are correct.
c) an oxidizing agent in basic solution.
d) a reducing agent in basic solution.
e) a reducing agent in acidic solution.

29. Which of the following is paramagnetic?
a) NO b) O_2^{2-} c) N_2 d) O_3 e) N_2O

30. Calculate the pH of 2.0 M hydrogen peroxide ($pK_{a1} = 11.75$).

31. Which of the following can only be a reducing agent?
a) SO_3 b) SO_2 c) H_2SO_4 d) H_2S e) SO_3^{2-}

32. The starting material for the production of sulfuric acid by the contact process is
a) S(g). b) H_2S(g). c) SO_2(g). d) SO_3(g). e) $H_2S_2O_7$(l).

33. What is the formula of sulfurous acid?
a) H_2SO_3 b) $H_2S_2O_7$ c) H_2SO_2 d) $H_2S_2O_4$ e) H_2SO_4

34. In the last step of the contact process for the production of sulfuric acid,
a) SO_2(g) is added to water.
b) H_2SO_3(l) is added to water.
c) SO_3(g) is added to water.
d) oleum is added to water.
e) H_2S(g) is added to $H_2S_2O_7$.

35. The contact process for the production of sulfuric acid is

$$S(g) + O_2(g) \rightarrow SO_2(g)$$
$$2SO_2(g) + O_2(g) \rightarrow 2SO_3(g)$$
$$SO_3(g) + H_2SO_4(l) \rightarrow H_2S_2O_7(l)$$
$$H_2S_2O_7(l) + H_2O(l) \rightarrow 2H_2SO_4(l)$$

How many moles of "new" sulfuric acid are produced from 150 mol S(g)?
a) 225 mol b) 450 mol c) 75 mol d) 300 mol e) 150 mol

36. Which of the following is not true regarding sulfuric acid?
a) Sulfuric acid is produced by a process called the contact process.
b) Sulfuric acid is used to produce sulfur.
c) Dry HCl(g) is produced from the reaction of sulfuric acid with sodium chloride.
d) Sulfuric acid is used in the manufacture of ammonium sulfate and phosphate fertilizers.
e) Concentrated sulfuric acid is a powerful dehydrating agent.

37. What reaction, if any, is predicted between silver and concentrated sulfuric acid?
a) $2Ag(s) + 2H_2SO_4(aq) \rightarrow Ag_2SO_4(aq) + H_2SO_3(aq) + H_2O(l)$
b) $2Ag(s) + H_2SO_4(aq) \rightarrow Ag_2SO_4(aq) + H_2(g)$
c) $2Ag(s) + 2H_2SO_4(aq) \rightarrow Ag_2SO_4(aq) + SO_2(g) + 2H_2O(l)$
d) $2Ag(s) + 2H_2SO_4(aq) \rightarrow Ag_2S(aq) + SO_2(g) + 2H_2O(l) + 2O_2(g)$
e) No reaction occurs.

38. The formula of sodium sulfite is
a) $Na_2S_2O_4$. b) Na_2SO_4. c) $Na_2S_2O_6$. d) Na_2SO_3. e) $Na_2S_2O_3$.

39. The total number of lone pairs on the oxygen atoms of sulfur trioxide is
a) 8. b) 9. c) 6. d) 10. e) 3.

40. All of the following form clathrates with water except
a) nitrogen.. b) carbon dioxide. c) sulfur dioxide. d) methane. e) argon.

41. Sulfur dichloride reacts with ethene to give
a) $S(CH_2CH_2Cl)_2$ b) $ClCH_2CH_2Cl$ c) $SCl(CH_2CH_3)$ d) $SCl(CH_2CH_2Cl)$ e) S_2Cl_2

42. The geometry around each sulfur atom in sulfur dichloride is
a) angular.
b) tetrahedral.
c) linear.
d) angular around one sulfur atom and linear around the other sulfur atom.
e) angular around one sulfur atom and terahedral arond the other sulfur atom.

43. When sulfur burns in fluorine, the product is
a) SF_4. b) SF_2. c) SF_6. d) S_2F_2. e) SOF_4.

44. Which element has the highest boiling point?
a) At b) I c) Br d) Cl e) F

45. Which element has the largest electronegativity?
a) F b) Cl c) Br d) I e) At

46. The following standard reduction voltages refer to the X_2/X^- couple of the halogens: +2.87, +1.36, +1.09, +0.54, and +0.10 V. Which halogen is likely to have the value +1.09 V?
a) bromine b) fluorine c) chlorine d) iodine e) astatine

47. The stable forms of bromine and iodine at 25°C and 1 atm are
a) $Br_2(s)$ and $I_2(s)$.
b) $Br_2(g)$ and $I_2(l)$.
c) $Br_2(l)$ and $I_2(s)$.
d) $Br_2(g)$ and $I_2(g)$.
e) $Br_2(l)$ and $I_2(g)$.

48. Which of the following is true?
a) Fluorine can be produced by oxidation of fluoride ion with chlorine gas.
b) Fluorosubstituted hydrocarbons react with strong oxidizing agents.
c) When chlorine reacts with iron, the product is iron(II) chloride.
d) Iodine can be produced by oxidation of iodide ion with chlorine gas.
e) Chlorine is a better oxidizing agent than fluorine.

49. The lower solubility of most fluorides results because
a) fluoride ion forms long chains.
b) fluorine gas is very reactive.
c) fluorine is a strong oxidant.
d) fluorine has a high electronegativity.
e) the lattice enthalpies of fluorides tend to be high.

50. All of the following are characteristic of fluorine or its compounds except
a) fluorine has an oxidation number of −1.
b) fluorine has a high electronegativity.
c) fluorine has a very small size.
d) fluorine brings about low oxidation numbers in other elements.
e) fluorine does not have available *d*-orbitals for bonding.

51. What is the shape of IF_3?
a) T-shaped b) trigonal planar c) trigonal pyramidal d) tetrahedral e) trigonal bipyramidal

52. What is the shape of ClF_3?
a) T-shaped b) trigonal pyramidal c) trigonal planar d) tetrahedral e) seesaw

53. What is the shape of BrF_3?
a) T-shaped b) trigonal pyramidal c) trigonal planar d) tetrahedral e) seesaw

54. Which of the following has the largest bond enthalpy?
a) HI b) HCl c) HBr d) HF e) All the bond enthalpies are the same.

55. All of the following acids have pK_as smaller than 2.0 except
a) HIO b) $HClO_4$ c) HIO_4 d) $HClO_3$ e) $HBrO_3$

56. What is the oxidation number of I in HIO_3?
a) +5 b) +6 c) +3 d) +4 e) +7

57. What is the oxidation number of I in HIO_4?
a) +5 b) +7 c) +3 d) +6 e) +8

58. When sulfuric acid and calcium fluoride are mixed, a product is
a) $F_2(g)$. b) $HF(g)$. c) $SF_2(g)$. d) $H_2S(g)$. e) $SO_2(g)$.

59. When phosphoric acid and potassium bromide are mixed, a product is
a) $HBr(g)$. b) $Br_2(g)$. c) $PBr_2(g)$. d) $P_4(s)$. e) $HBrO(g)$.

60. Which of the following is the weakest acid?
a) HIO b) $HBrO_4$ c) $HClO_4$ d) $HBrO_3$ e) HClO

61. Calcium hypochlorite is produced by the reaction of
a) hydrochloric acid with calcium oxide.
b) hydrochloric acid with calcium carbonate.
c) chlorine gas with calcium oxide.
d) perchloric acid with calcium oxide.
e) chlorine gas with calcium chloride.

62. When sodium chlorate is reduced by sulfur dioxide, a product is
a) $ClO_2(g)$. b) $Cl_2(g)$. c) $NaCl(s)$. d) $Na_2O_2(s)$. e) $SO_3(g)$.

63. The oxoacids of Group 17 have the general formula HXO_n where X is a halogen and n = 1 to 4. Which of the following is true?
a) The oxoacids with oxidation number +1 are reducing agents.
b) As the oxidation number of X increases, the oxidizing strength of the acid increases.
c) As the oxidation number of X increases, the strength of the acid decreases.
d) Only the oxoacids with oxidation number +1 are strong acids.
e) The oxoacids with oxidation number +7 are weak oxidizing agents.

64. Which of the following is chloric acid?
a) $HClO_2$ b) $HClO_4$ c) $HClO_3$ d) HClO e) $HClO_5$

65. Which of the following is sodium hypobromite?
a) NaBrO b) NaHBr c) $NaBrO_4$ d) $NaBrO_3$ e) $NaBrO_2$

66. What is the shape of the chlorate ion?
a) trigonal pyramidal b) tetrahedral c) trigonal planar d) trigonal bipyramidal e) seesaw

67. Write an equation for the condensation of phosphoric acid.

68. The phase diagram for helium
a) has two triple points.
b) indicates that helium exists as He(g) and He(l) at about 2 K.
c) indicates that helium can be liquefied at ordinary pressures.
d) indicates that helium has three phases.
e) indicates that helium exists only as a gas.

69. Which of the Group 18 elements has the smallest boiling point?
a) He b) Ne c) Ar d) Kr e) Xe

70. In the phase diagram for carbon dioxide, at 5.1 atm and –56°C, the solid, liquid, and gaseous states coexist. At what pressure and temperature, if any, do the solid, liquid, and gas states of helium coexist. Explain.

71. Which of the following noble gases is least likely to react with fluorine?
a) Ne b) Ar c) Kr d) Xe e) Rn

72. When XeF_4 acts as a fluorinating agent,
a) Xe is reduced.
b) XeF_3^+ is produced and is the reactive species.
c) Xe is oxidized.
d) XeF_4 disproportionates.
e) F_2 is produced.

73. The compound $XeF_2(s)$
a) disproportionates in water to give $XeF^+(aq)$ and $XeF_3^-(aq)$.
b) is a powerful reducing agent.
c) is relatively unreactive.
d) dissolves in water to give $Xe(OH_2)_4F_2(aq)$.
e) is a powerful fluorinating agent.

74. The reaction of $XeF_4(s)$ with $Pt(s)$ produces
a) $Xe(g)$ and $PtF_4(s)$.
b) $XeF_2(s)$ and $PtF_2(s)$.
c) $XeF_2(s)$ and $PtF_4(s)$.
d) dimeric $XeF_2PtF_4(s)$.
e) $Xe(g)$ and $PtF_6(s)$.

75. The formula of xenic acid is
a) H_2XeO_6. b) H_2XeO_3. c) $HXeO_4$. d) $HXeO_2$. e) H_2XeO_4.

76. What is the oxidation number of Xe in $XeOF_4$?
a) +6 b) +8 c) +4 d) +5 e) 0

77. What is the oxidation number of Xe in XeO_3?
a) +2 b) +3 c) +4 d) +6 e) 0

78. What is the oxidation number of Xe in H_2XeO_4?
a) +6 b) +8 c) +10 d) 0 e) +4

79. What is the shape of $XeOF_4$?
a) square pyramidal b) octahedral c) seesaw d) trigonal bipyramidal e) trigonal pyramidal

80. What is the shape of XeF_2?
a) linear b) angular c) tetrahedral d) trigonal pyramidal e) trigonal bipyramidal

Chapter 21: The *d*-Block: Metals in Transition Form A

1. The *d*-metals can be mixed together to form a wide range of alloys because
a) the nucleus is well shielded by the *d*-electrons.
b) the *d*-metals have low melting points.
c) the *d*-metals have a wide range of metal radii.
d) the range of *d*-metal radii is not very great.
e) the *d*-electrons interact strongly with each other.

2. Which of the following has the largest atomic radius?
a) Ru b) Sc c) Mn d) Ni e) Co

3. In the *d*-block elements, the third-row metallic radii are about the same as the second-row radii because of
a) The third-row metallic radii are greater than the second-row radii.
b) the lanthanide contraction.
c) greater shielding of *f*-electrons.
d) greater shielding of *d*-electrons.
e) the increased number of *d*-electrons.

4. The properties of the transition metals vary greatly but they all have
a) the same physical properties.
b) high densities and high melting points.
c) low boiling points.
d) the same strength as their alloys.
e) oxides which are used to produce alloys.

5. Which of the following metals has the greatest density?
a) Ni b) Fe c) Mn d) V e) Sc

6. Which of the following has the lowest density?
a) Sc b) Co c) Ni d) Ti e) Cr

7. Which of the following pairs has about the same atomic radius?
a) Mn and Tc b) Co and Rh c) Fe and Ru d) Ir and Rh e) Cu and Ag

8. Which metal is the least reactive?
a) Pt b) Ag c) Rh d) Cu e) Pd

9. Gold is much less reactive than silver because of
a) the higher density of gold.
b) the poor shielding of the *d*-electrons.
c) gold is more reactive than silver.
d) the lanthanide contraction.
e) the larger atomic size of gold.

10. Which of the following is ferromagnetic?
a) Co b) Sc c) Ti d) Zn e) Cu

11. Which pair has the electron configuration $[Ar]3d^3$?
a) V and Cr^{2+} b) V^{2+} and Cr^{3+} c) Mn^{3+} and Cr^{3+} d) V^{2+} and Mn^{3+} e) Ti and Cr^{2+}

12. Which pair has a d^{10} electron configuration?
a) Co and Ni b) Zn and Cu^{2+} c) Cu and Zn d) Cu^{+} and Zn^{2+} e) Cu^{2+} and Ni^{2+}

13. What are the electron configurations of Fe^{2+} and Co^{3+}, respectively?
a) both d^6 b) both d^5 c) d^6 and d^8 d) both d^7 e) d^7 and d^8

14. Ruthenium is directly below iron in the periodic table. The Ru^{2+} ion is a
a) d^6 ion. b) d^7 ion. c) d^5 ion. d) d^9 ion. e) d^8 ion.

15. Which of the following complexes contains a d^3 metal ion?
a) $[Cr(OH_2)_5Cl]Cl_2$ b) $[Ni(NH_3)_6]Cl_2$ c) $K_2[Ni(CN)_4]$ d) $K_4[Fe(CN)_6]$ e) $[Co(OH_2)_6](ClO_4)_2$

16. Predict the number of unpaired electrons in $[Cr(CN)_6]^{4-}$ and $[Cr(OH_2)_6]^{2+}$, respectively.
a) 0 and 2 b) 0 and 4 c) 4 and 2 d) 2 and 2 e) 2 and 4

17. Predict the number of unpaired electrons in the tetrahedral complex ion $[CuCl_4]^{2-}$.
a) 1 b) 2 c) 0 d) 3 e) 4

18. Which of the following is likely to be a good oxidizing agent?
a) CrO_4^{2-} b) Fe^{2+} c) Mn^{2+} d) V^{2+} e) Cr^{2+}

19. The maximum oxidation state of titanium is
a) +4. b) +3. c) +2. d) +1. e) 0.

20. The maximum oxidation state of manganese is
a) +7. b) +5. c) +3. d) +6. e) 0.

21. The oxidation state of zinc is
a) +2. b) +3. c) +1. d) +4. e) 0.

22. For the oxides CrO, Cr_2O_3, and CrO_3, which of the following is true?
a) CrO and Cr_2O_3 are acidic and CrO_3 is basic.
b) All the oxides are basic.
c) All the oxides are acidic.
d) CrO is acidic, and Cr_2O_3 and CrO_3 are basic.
e) CrO is basic, Cr_2O_3 is amphoteric, and CrO_3 is acidic.

23. The product of the first step in the manufacture of titanium is
a) $TiO_2(s)$. b) $Ti_2O_3(s)$. c) impure Ti(s). d) $TiCl_4(g)$. e) $Ti_2(SO_4)_3(s)$.

24. Vanadium metal is produced by the
a) roasting of $VO_2(s)$.
b) reduction of vanadium(V) oxide with calcium.
c) reduction of $VOCl_2(s)$ with magnesium.
d) decomposition of $VCl_4(g)$ at high temperature.
e) reaction of $V_2O_5(s)$ with coke.

25. What is the equation for the production of Cr by the thermite reaction?
a) $CrO_3(s) + 3CO(g) \rightarrow Cr(s) + 3CO_2(g)$
b) $Cr_2O_3(s) + 3C(s) \rightarrow 2Cr(s) + 3CO(g)$
c) $Cr_2O_3(s) + 2Al(s) \rightarrow Al_2O_3(s) + 2Cr(l)$
d) $FeCr_2O_4(s) + 4C(s) \rightarrow Fe(s) + 2Cr(s) + 4CO(g)$
e) $CrO_3(s) + 3C(s) \rightarrow Cr(s) + 3CO(g)$

26. What is the formula of sodium chromate?
a) $NaCrO_4$ b) $Na_2Cr_2O_7$ c) $NaCrO_3$ d) Na_2CrO_4 e) Na_2CrO_5

27. Manganese is obtained by the
a) reaction of $KMnO_4(aq)$ with coke.
b) purification of manganese nodules recovered from the sea.
c) reduction of $KMnO_4(aq)$.
d) decomposition of $MnCl_3(g)$ at high temperature.
e) reduction of $MnO_2(s)$ with Al(s).

28. The principal ores of iron are hematite and magnetite. What is the formula of magnetite?
a) FeO_2 b) Fe_3O_4 c) FeS_2 d) Fe_2O_3 e) FeO

29. In a blast furnace, which of the following reduces $Fe_2O_3(s)$ to produce iron?
a) CO(g) b) C(s) c) $SiO_2(s)$ d) $CO_2(g)$ e) $Al_2O_3(s)$

30. Impure nickel is purified by
a) reduction of the ore with CO(g).
b) first alloying with iron and then treating with carbon dioxide to give the metal.
c) producing pure $Ni(CO)_4(g)$ which can be decomposed to the metal.
d) reaction with $H_2S(aq)$.
e) reduction of the ore with C(s).

31. Steel is produced by
a) forcing oxygen and powdered limestone through molten pig iron.
b) adding chromium and nickel to molten pig iron.
c) simply melting $Fe_2O_3(s)$.
d) adding carbon to molten pig iron.
e) mixing carbon and phosphorus with molten pig iron.

32. If you were an antique dealer searching for platinum and gold jewelry, how could you differentiate between platinum (very valuable) and silver (relatively cheap), and gold and "fool's gold" (FeS_2), using only the strong acids. Explain.

33. Given the standard reduction voltages:

$Cu^{2+}(aq) + 2e^- \rightarrow Cu(s)$ $E° = +0.34$ V

$2H^+(aq) + 2e^- \rightarrow H_2(g)$ $E° = 0$

$NO_3^-(aq) + 4H^+(aq) + 3e^- \rightarrow NO(g) + 2H_2O(l)$ $E° = +0.96$ V

$Cl_2(g) + 2e^- \rightarrow 2Cl^-(aq)$ $E° = +1.36$ V

Which of the following statements regarding the reactivity of copper is true?

a) Copper reacts with hydrogen gas.
b) Copper dissolves in nitric acid.
c) Copper dissolves in hydrochloric acid.
d) Copper does not dissolve in nitric acid.
e) Copper dissolves in all acids.

34. Bronze is an alloy of

a) copper with tin.
b) nickel with zinc.
c) copper with iron.
d) copper with zinc.
e) copper with nickel.

35. Given the standard reduction voltages:

$Ag^+(aq) + e^- \rightarrow Ag(s)$ $E° = +0.80$ V

$2H^+(aq) + 2e^- \rightarrow H_2(g)$ $E° = 0$

$NO_3^-(aq) + 4H^+(aq) + 3e^- \rightarrow NO(g) + 2H_2O(l)$ $E° = +0.96$ V

$Cl_2(g) + 2e^- \rightarrow 2Cl^-(aq)$ $E° = +1.36$ V

Which of the following statements regarding the reactivity of silver is true?

a) Silver reacts with hydrogen gas.
b) Silver dissolves in all acids.
c) Silver does not dissolve in nitric acid.
d) Silver dissolves in nitric acid.
e) Silver dissolves in hydrochloric acid.

36. What is the percent gold in 10-carat gold?

a) 42% b) 10% c) 71% d) 100% e) 59%

37. What are the oxidation numbers of zinc and cadmium in their compounds?

a) Both zinc and cadmium have oxidation numbers of +2 and +3.
b) Both zinc and cadmium have an oxidation number of +2 in all their compounds.
c) Zinc has an oxidation number of +2 and cadmium +3.
d) Zinc has an oxidation number of +2 and cadmium +2 and +3.
e) Zinc has oxidation numbers of +2 and +3 while cadmium has an oxidation number of +2.

38. A ligand is
a) necessary in a complex to balance the charge on the metal ion.
b) a Lewis acid.
c) used to force octahedral geometry on a metal ion.
d) used to precipitate metal ions from aqueous solution.
e) a Lewis base.

39. All of the following ligands are monodentate except
a) ethylenediamine. b) aqua. c) cyano. d) carbonyl. e) ammine.

40. What is the coordination number of iron in $K_4[Fe(CN)_6]$?
a) 6 b) 10 c) 2 d) 4 e) 3

41. What is the oxidation number of cobalt in $[CoCl(NH_3)_5]Cl_2$?
a) +3 b) +2 c) +1 d) +6 e) +4

42. What is the name of the complex $[Fe(en)_2(OH_2)_2](ClO_4)_2$?
a) bis(aquaethylenediamine)iron(II) perchlorate
b) diaquadiethylenediamineiron(II) perchlorate
c) diaquabis(ethylenediamine)ferrate(II) perchlorate
d) diaquabis(ethylenediamine)iron(II) perchlorate
e) di(ethylenediamine)bis(aqua)iron(II) perchlorate

43. What is the name of the complex $K_3[Cr(ox)_2(CN)_2]$?
a) potassium dicyanodioxalatochromate(II)
b) potassium dioxalatedicyanochromate(II)
c) potassium dicyanobis(oxalato)chromate(III)
d) potassium dioxalatedicyanochromate
e) potassium oxalatecyanochromium(III)

44. What is the name of the complex $Na[Co(NH_3)_3Cl_3]$?
a) sodium trichlorotriamminecobalt(III)
b) sodium triamminetrichlorocobaltate(II)
c) sodium tris(aminechloro)cobaltate(II)
d) sodium triamminetrichlorocobalt(II)
e) sodium triamminetrichlorocobalt(III)

45. Which of the following is the formula of pentaamminechlorocobalt(III) chloride?
a) $[Co(NH_3)_5Cl]Cl_2$
b) $[Co(NH_3)_5Cl]Cl_4$
c) $[Co(NH_3)_5Cl]Cl$
d) $[Co(NH_3)_5Cl]$
e) $[Co(NH_3)_5Cl]Cl_3$

46. One mole of pentaamminechloroplatinum(IV) chloride dissolves in water to give
a) 4 moles of ions. b) 1 mole of ions. c) 3 moles of ions. d) 2 moles of ions. e) no ions.

47. How many unpaired electrons are predicted for $[Co(OH_2)_6]^{2+}$?
a) 3 b) 2 c) 1 d) 5 e) 0

48. How many unpaired electrons are predicted for $[V(OH_2)_6]^{2+}$?
a) 3 b) 2 c) 1 d) 4 e) 0

49. If an iron(III) complex is tetrahedral, how many unpaired electrons are predicted?
a) 4 b) 5 c) 1 d) Five or one depending on the ligands. e) Zero or four depending on the ligands.

50. Which of the following are ionization isomers?
a) $[CoBr(NH_3)_5]SO_4$ and $[CoSO_4(NH_3)_5]Br$
b) $[CoNCS(NH_3)_5]Cl_2$ and $[CoSCN(NH_3)_5]Cl_2$
c) *cis*-$[CoCl_2(en)_2]Cl$ and *trans*-$[CoCl_2(en)_2]Cl$
d) $[Cr(NH_3)_6][Fe(CN)_6]$ and $[Fe(NH_3)_6][Cr(CN)_6]$
e) $[CrCl(OH_2)_5]Cl_2.H_2O$ and $[CrCl_2(OH_2)_4]Cl.2H_2O$

51. How many geometric isomers are possible for the complex ion $[Co(en)_2Cl_2]^{+}$?
a) 2 b) 0 c) 3 d) 4 e) 5

52. Which one of the following complexes has geometric isomers?
a) $[Co(NH_3)_5Cl]^{2+}$
b) $[Co(en)_2Cl_2]^{+}$
c) $[Cr(OH_2)_5Cl]^{2+}$
d) $[Pt(NH_3)_3Cl]^{+}$ (square planar)
e) $[Co(NCS)_2Cl_2]^{2-}$ (tetrahedral)

53. How many geometric isomers are possible for the complex $[Co(en)(OH_2)Cl_3]$?
a) 2 b) 3 c) 4 d) 0 e) 5

54. A complex with the empirical formula $Co(NH_3)_4Cl_2Br$ dissolves in water to give two ions. How many possible structures can be drawn for this complex?
a) 4 b) 2 c) 3 d) 5 e) 7

55. How many geometric isomers are possible for the square planar complex $[Pt(NH_3)(OH_2)(Cl)(Br)]$?
a) 3 b) 2 c) 4 d) 5 e) 6

56. How many different isomers of all types are possible for the complex ion $[Co(NO_2)_2(en)_2]^{+}$?
a) 3 b) 6 c) 4 d) 2 e) 9

57. How many different isomers of all types are possible for the complex ion $[Cr(NH_3)_3(OH_2)(Cl)(CN)]^{+}$?
a) 5 b) 4 c) 3 d) 2 e) 6

58. How many different isomers of all types are possible for the complex $[Ir(CO)_2(Cl)_2(OH_2)Br]$?
a) 8 b) 4 c) 3 d) 5 e) 6

59. Which of the following complexes is chiral?

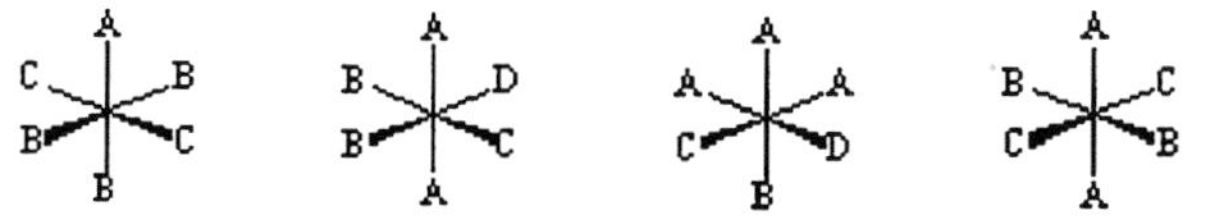

a) 3 b) 2 c) 4 d) 1 and 2 e) 1 and 3

60. Which of the following complexes is chiral?

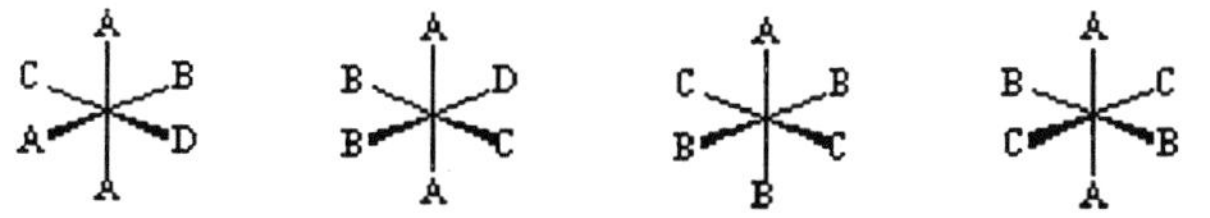

a) None of the complexes is chiral. b) 1 c) 2 d) 3 e) 4

61. Which of the following complexes is chiral?

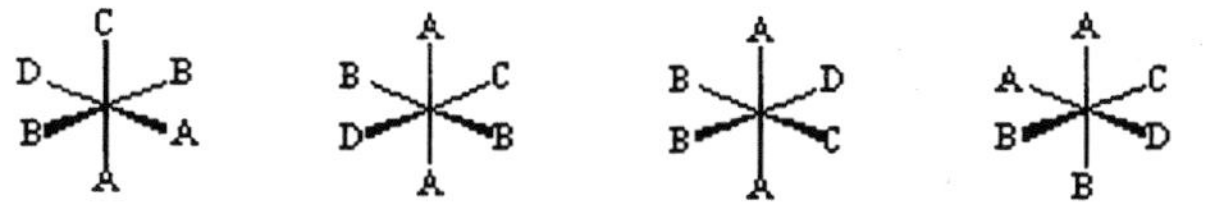

a) 3 b) 1 c) 2 d) 4 e) 1 and 3

62. How many isomers of all types are possible for the complex ion $[Co(en)(NH_3)_2(OH_2)Cl]^{2+}$?

a) 6 b) 4 c) 2 d) 3 e) 5

63. Which of the following complexes is chiral?

a) *cis*-$[Cr(en)_2Cl_2]^+$
b) *trans*-$[Cr(en)_2(NH_3)Cl]^{2+}$
c) *cis*-$[Cr(NH_3)_4Cl_2]^+$
d) $[Co(OH_2)_3(NH_3)_3]^{3+}$
e) *trans*-$[Cr(en)_2Cl_2]^+$

64. The complex ion $[Co(NH_3)_4Cl_2]^+$

a) has no geometric isomers.
b) is chiral.
c) has 2 geometric isomers neither of which is chiral.
d) has 2 geometric isomers both of which are chiral.
e) has 2 geometric isomers one of which is chiral.

65. Consider a complex with the formula $Co(NH_3)_2(OH_2)_2Cl_2$.
(a) Indicate how to determine if the structure of the complex is octahedral, tetrahedral, or square planar.
(b) For each structure, predict the number of geometric isomers.

66. Explain why there are no geometric isomers of the complex $[Ni(NH_3)_2Cl_2]$.

67. Which of the following complex ions has the greatest *d*-orbital splitting energy?

a) $[Co(OH_2)_6]^{3+}$ b) $[Co(NH_3)_6]^{3+}$ c) $[Co(ox)_3]^{3-}$ d) $[Co(CN)_6]^{3-}$ e) $[CoCl_6]^{3-}$

68. Which of the following complex ions has the lowest *d*-orbital splitting energy?
a) $[Fe(NH_3)_6]^{2+}$ b) $[FeCl_4]^{2-}$ c) $[FeCl_6]^{4-}$ d) $[Fe(OH_2)_6]^{2+}$ e) $[Fe(CN)_6]^{4-}$

69. Which of the following complex ions is likely to absorb at 600–700 nm?
a) $[Co(CN)_6]^{4-}$ b) $[Co(OH_2)_6]^{2+}$ c) $[CoCl_4]^{2-}$ d) $[Co(NH_3)_6]^{2+}$ e) $[Co(NO_2)_6]^{4-}$

70. Which of the following complex ions is colorless?
a) $[Ni(OH_2)_6]^{2+}$ b) $[Cr(OH_2)_6]^{3+}$ c) $[V(OH_2)_6]^{2+}$ d) $[Ti(OH_2)_6]^{3+}$ e) $[Zn(OH_2)_6]^{2+}$

71. The complex ion $[Ni(OH_2)_6]^{2+}$ absorbs at about 700 nm. The complex $[Co(CN)_6]^{3-}$ is likely to absorb at about
a) 700 nm b) 450 nm c) 800 nm d) 550 nm e) 600 nm

72. The complex $Co[CoCl_4](s)$ is blue. On humid days, the blue color fades to a light pink which is almost invisible. When the air becomes dryer, the faint pink color changes back to blue. Explain these observations using an equation. Hint: octahedral complexes usually have less intense colors with respect to tetrahedral complexes.

73. For which one of the following would it not be possible to distinguish between high-spin and low-spin complexes in octahedral geometry?
a) Ni(II) b) Co(III) c) Fe(II) d) Co(II) e) Cr(II)

74. The complex ion $[Cr(OH_2)_6]^{2+}$ has four unpaired electrons. This means that
a) the complex is low-spin.
b) Δ_o is very large.
c) the water ligands are difficult to remove.
d) the complex is high-spin.
e) the complex is diamagnetic.

75. Rhodium lies below cobalt in the periodic table. What is the *d*-electron configuration of $[Rh(CN)_6]^{3-}$?
a) t^5 b) t^4e^2 c) t^4e^1 d) t^6 e) t^5e^2

76. Predict the total number of *d*-electrons in a complex having one unpaired electron in a strong octahedral field and three unpaired electrons in a weak octahedral field.
a) 7 b) 5 c) 6 d) 8 e) 4

77. Both $[TiCl_6]^{3-}$ and $[Ti(CN)_6]^{3-}$ have one absorption band in the visible region of the spectrum. Predict the color of each complex and the wavelength of their absorption bands.

78. Which of the following is diamagnetic?
a) $[Cr(NH_3)_6]^{3+}$ b) $[Co(NH_3)_6]^{3+}$ c) $[Co(OH_2)_6]^{2+}$ d) $[Fe(OH_2)_6]^{2+}$ e) $[Fe(CN)_6]^{3-}$

79. What change in magnetic properties can be expected when Cl^- ligands in an octahedral complex are replaced by CN^- ligands in a d^6 complex?

a) The complex becomes diamagnetic.
b) No change occurs.
c) The complex becomes high-spin.
d) The complex becomes paramagnetic.
e) The complex becomes tetrahedral.

80. Comparing $[Co(CN)_6]^{3-}$ with $[CoCl_6]^{4-}$, which of the following statements is true?

a) $[Co(CN)_6]^{3-}$has more *d*-electrons than $[CoCl_6]^{4-}$.
b) $[Co(CN)_6]^{3-}$ is diamagnetic while $[CoCl_6]^{4-}$ is paramagnetic.
c) $[Co(CN)_6]^{3-}$ is paramagnetic while $[CoCl_6]^{4-}$ is diamagnetic.
d) $[Co(CN)_6]^{3-}$ has the same number of *d*-electrons as $[CoCl_6]^{4-}$.
e) Both complexes are paramagnetic.

Chapter 21: The *d*-Block: Metals in Transition Form B

1. Which of the following has the largest atomic radius?
a) Sc b) Zn c) Mn d) Fe e) Cr

2. Which of the following has the smallest atomic radius?
a) Zn b) Cd c) Pd d) Ru e) Mo

3. The lanthanide contraction results because
a) *f*-electrons interact more strongly with each other than *d*-electrons.
b) the *d*-electrons shield the *f*-electrons from the nucleus.
c) the shielding effect of the *d*-electrons is greater than that of the *f*-electrons.
d) the shielding effect of the *d*-electrons is less than that of the *f*-electrons.
e) of the poor shielding effect of the *f*-electrons and the higher nuclear charge.

4. All of the following are transition metals except
a) In b) V c) W d) Fe e) Cu

5. Which of the following has the greatest density?
a) Au b) Ag c) Cu d) Zn e) Hg

6. Which of the following has the lowest density?
a) Zn b) Au c) Cd d) Cu e) Hg

7. Which of the following pairs has about the same atomic radius?
a) Mn and Tc b) Fe and Ru c) Co and Rh d) Cu and Ag e) Ru and Os

8. Which metal is least reactive?
a) Au b) Pd c) Ag d) Cu e) Fe

9. Platinum is less reactive than palladium because of
a) the poor shielding of the *d*-electrons.
b) the lanthanide contraction.
c) the higher density of platinum.
d) the larger atomic size of platinum.
e) the electronic configuration of platinum.

10. Which of the following is ferromagnetic?
a) Fe b) Sc c) Ti d) Zn e) Cu

11. Which pair has a d^6 electron configuration?
a) Zn^{2+} and Ni^{2+} b) Co^{2+} and Ni^{2+} c) Fe^{2+} and Co^{3+} d) Cr^{3+} and Mn^{2+} e) Mn^{2+} and Fe^{3+}

12. What are the electron configurations of V^{2+} and Cr^{3+}, respectively?
a) both d^5 b) d^4 and d^5 c) both d^4 d) d^3 and d^4 e) both d^3

13. What are the electron configurations of Mn^{2+} and Ni^{2+}, respectively?
a) both d^5 b) d^5 and d^8 c) both d^8 d) d^5 and d^7 e) d^6 and d^9

14. Palladium and platinum are below nickel in the periodic table. The Pd^{2+} and Pt^{4+} ions, respectively, are
a) d^6 and d^7 ions. b) d^8 and d^6 ions. c) both d^8 ions. d) d^8 and d^7 ions. e) d^{10} and d^8 ions.

15. Which of the following complexes contains a d^7 metal ion?
a) $[Co(OH_2)_6](ClO_4)_2$
b) $K_4[Fe(CN)_6]$
c) $[Cr(OH_2)_5Cl]Cl_2$
d) $[Ni(NH_3)_6]Cl_2$
e) $[Fe(OH_2)_6](NO_3)_3$

16. Predict the number of unpaired electrons in $[Fe(CN)_6]^{3-}$ and $[Fe(OH_2)_6]^{3+}$, respectively.
a) 5 and 5 b) 1 and 5 c) 5 and 1 d) 0 and 4 e) 1 and 1

17. Predict the number of unpaired electrons in the tetrahedral complex ion $[CoCl_4]^{2-}$.
a) 3 b) 2 c) 0 d) 1 e) 4

18. Which of the following is likely to be a good reducing agent?
a) Cr_2O_3 b) $Cr_2O_7^{2-}$ c) Cr^{2+} d) MnO_4^- e) MnO_2

19. The oxidation state of scandium is
a) +3. b) +4. c) +2. d) +1. e) 0.

20. The maximum oxidation states for chromium and manganese are
a) +5 and +6, respectively.
b) +3 and +4, respectively.
c) both +7.
d) +6 and +7, respectively.
e) +7 and +8, respectively.

21. The most common oxidation states of copper are
a) +1 and +3. b) +2 and +4. c) +2 and +3. d) +1 and +4. e) +1 and +2.

22. Which of the following oxides is likely to be acidic?
a) Mn_2O_3 b) MnO_2 c) Mn_2O_7 d) MnO e) Mn_2O_7 and MnO

23. The formula of barium titanate is
a) $BaTiO_3$. b) Ba TiO_4. c) BaTiO. d) $BaTiO_2$. e) $Ba_2Ti_2O_7$.

24. The formula of vanadium(V) oxide is
a) VO_3. b) V_3O_6. c) V_2O_5. d) VO_2. e) V_2O_4.

25. What is the formula of chromite?
a) $FeCr_2O_4$. b) Cr_2O_5. c) Na_2CrO_3. d) CrO_3. e) Cr_2O_3.

26. What is the formula of the compound formed when sodium chromate is dissolved in acid?
a) Na_2CrO_4 b) $Na_2Cr_2O_7$ c) H_2CrO_3 d) H_2CrO_4 e) $NaCrO_2$

27. What is the formula of the permanganate ion?
a) MnO_2 b) MnO_4^{2-} c) MnO_4^- d) MnO^+ e) $Mn_2O_7^{2-}$

28. When iron is produced in a blast furnace, each kg of iron produced requires about 1.75 kg of ore, 0.75 kg of coke, and 0.25 kg of limestone. The purpose of the limestone is
a) to react with the ore to produce carbon dioxide.
b) to remove acidic anhydride and amphoteric oxides such as $SiO_2(s)$ and $Al_2O_3(s)$.
c) to produce calcium hydroxide to dissolve the ore.
d) to lower the melting point of the ore.
e) to produce carbon monoxide.

29. In a blast furnace, carbon monoxide is
a) complexes with iron to form $Fe(CO)_5(g)$.
b) removes amphoteric impurities from the ore.
c) the principal reducing agent.
d) removes acidic anhydride impurities from the ore.
e) used to produce carbon dioxide which is needed to reduce hematite.

30. What is the oxidation state of $Ni(CO)_4$?
a) 0 b) +4 c) +3 d) +2 e) −2

31. Stainless steel
a) contains less then 0.2% carbon.
b) is an alloy steel which contains about 15% chromium by mass.
c) contains about 4 or 5 percent carbon.
d) contains between 0.8 and 1.5 percent carbon.
e) is another description for pig iron.

32. Calculate ΔG° for the reactions that occur in Zones C and D in a blast furnace. Comment.

33. Brass is an alloy of
a) nickel with zinc.
b) copper with zinc.
c) copper with tin.
d) copper with nickel.
e) copper with iron.

34. Copper is protected by a hard pale green film of basic carbonate. What is the formula of this compound?
a) $CuCO_3$ b) Cu_2CO_3 c) $Cu(HCO_3)_2$ d) $Cu_2(OH)_2CO_3$ e) CuO

35. Given the standard reduction voltages:

$Au^{3+}(aq) + 3e^- \rightarrow Au(s)$ $E° = +1.40$ V

$2H^+(aq) + 2e^- \rightarrow H_2(g)$ $E° = 0$

$NO_3^-(aq) + 4H^+(aq) + 3e^- \rightarrow NO(g) + 2H_2O(l)$ $E° = +0.96$ V

$Cl_2(g) + 2e^- \rightarrow 2Cl^-(aq)$ $E° = +1.36$ V

Which of the following statements regarding the reactivity of gold is true?

a) Gold dissolves in all acids.
b) Gold reacts with hydrogen gas.
c) Gold does not dissolves in nitric acid.
d) Gold dissolves in hydrochloric acid.
e) Gold dissolves in nitric acid.

36. What is the percent gold in 14-carat gold?
a) 58% b) 82% c) 100% d) 14% e) 74%

37. The formulas of mercury(I) and mercury(II) chloride are
a) Hg_2Cl_3 and $HgCl_2$, respectively.
b) Hg_2Cl and $HgCl_2$, respectively.
c) Hg_2Cl_2 and $HgCl_2$, respectively.
d) HgCl and $HgCl_2$, respectively.
e) Hg_2Cl_2 and HgCl, respectively.

38. In the complex $Ni(CO)_4$,
a) Ni is a Lewis base and CO is a Lewis acid.
b) Ni is a Lewis acid and CO is a Lewis base.
c) Ni has an oxidation number of +2.
d) CO is a Lewis acid.
e) CO has a charge of −1 in the complex.

39. All of the following ligands are monodentate except
a) oxalato. b) aqua. c) cyano. d) carbonyl. e) ammine.

40. What is the coordination number of chromium in $[Cr(OH_2)_6](ClO_4)_2$?
a) 6 b) 8 c) 4 d) 3 e) 10

41. What is the oxidation number of iron in $K_4[Fe(CN)_6]$?
a) +2 b) +3 c) +6 d) +4 e) +10

42. What is the name of the complex $K_3[Fe(ox)_3]$?
a) tripotassium trioxalatocobaltate(III)
b) potassium trioxalatocobalt(III)
c) potassium tris(oxalate)cobaltate(III)
d) potassium tris(oxalato)cobalt(III)
e) potassium tris(oxalato)cobaltate(III)

43. What is the name of the complex $[Cr(en)_2(OH_2)Cl]Cl_2$?
a) aquachlorobis(ethylenediamine)chromium(III) chloride
b) aquachlorobis(ethylenediamine)chromium(II) dichloride
c) diethylenediaminebis(aquachloro)chromium(III) chloride
d) chloroaquabis(ethylenediamine)chromate(III) chloride
e) aquachloridebis(ethylenediamine)chromium(III) chloride

44. What is the name of the complex $[Co(NH_3)_5I]Cl_2$?
a) pentaammineiodocobalt(III) chloride
b) iodopentaamminecobalttate(II) chloride
c) pentaammineiodocobalt(II) dichloride
d) pentaammineiodocobaltate(II) dichloride
e) pentaammineiodocobaltate(III) chloride

45. Which of the following is the formula of potassium hexacyanoferrate(II)?
a) $K_4[Fe(CN)_6]$ b) $K_3[Fe(CN)_6]$ c) $[Fe(CN)_6]$ d) $K[Fe(CN)_6]$ e) $K_2[Fe(CN)_6]$

46. One mole of sodium hexacyanoferrate(III) dissolves in water to give
a) 2 moles of ions. b) 3 moles of ions. c) 4 moles of ions. d) 8 moles of ions. e) 5 moles of ions.

47. How many unpaired electrons are predicted for $[Mn(CN)_6]^{4-}$?
a) 1 b) 2 c) 3 d) 5 e) 0

48. How many unpaired electrons are predicted for $[Cr(OH_2)_6]^{2+}$?
a) 4 b) 2 c) 1 d) 3 e) 0

49. If an iron(II) complex is tetrahedral, how many unpaired electrons are predicted?
a) 4 b) 5 c) 1 d) Five or one depending on the ligands. e) Zero or four depending on the ligands.

50. Which of the following are hydrate isomers?
a) $[CoNCS(NH_3)_5]Cl_2$ and $[CoSCN(NH_3)_5]Cl_2$
b) $[CrCl(OH_2)_5]Cl_2.H_2O$ and $[CrCl_2(OH_2)_4]Cl.2H_2O$
c) *cis*-$[CoCl_2(en)_2]Cl$ and *trans*-$[CoCl_2(en)_2]Cl$
d) $[CoBr(NH_3)_5]SO_4$ and $[CoSO_4(NH_3)_5]Br$
e) $[Cr(NH_3)_6][Fe(CN)_6]$ and $[Fe(NH_3)_6][Cr(CN)_6]$

51. How many geometric isomers are possible for the complex ion $[Co(NH_3)_3(OH_2)_3]^{3+}$?
a) 2 b) 1 c) 3 d) 4 e) 0

52. Which one of the following complexes has geometric isomers?
a) $[Co(C_2O_4)_3]^{3-}$
b) $[CoBr_2Cl_2]^{2-}$ (tetrahedral)
c) $[Co(NH_3)_6]^{3+}$
d) $[Pt(NH_3)_5Cl]^{3+}$
e) $[Pt(NH_3)_2Cl_2]$ (square planar)

53. How many geometric isomers are possible for the complex $[Ru(OH_2)_3Cl_3]$?
a) 2 b) 3 c) 4 d) 0 e) 5

54. A complex with the empirical formula $Co(NH_3)_5Cl_2Br$ dissolves in water to give three ions. How many possible structures can be drawn for this complex?
a) 2 b) 4 c) 3 d) 5 e) 1

55. How many different isomers of all types are possible for the complex ion $[Co(NCS)_2(NH_3)_4]^+$?
a) 6 b) 4 c) 2 d) 9 e) 3

56. How many different isomers of all types are possible for the complex ion $[CoCl_2(en)_2]^+$?
a) 3 b) 1 c) 4 d) 2 e) 5

57. How many different isomers of all types are possible for the complex ion $[Cr(NH_3)_3(OH_2)_2Cl]^{2+}$?
a) 5 b) 4 c) 3 d) 2 e) 6

58. How many different isomers of all types are possible for the complex ion $[Pt(NH_3)_3I_3]^+$?
a) 4 b) 2 c) 3 d) 1 e) 5

59. Which of the following complexes is chiral?

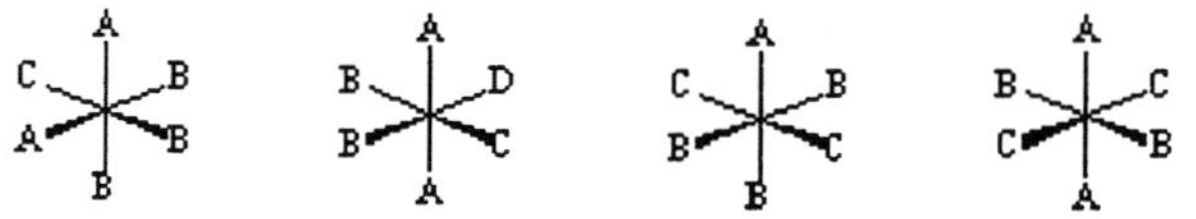

a) None of the complexes is chiral. b) 2 c) 1 d) 2 and 3 e) 4

60. Which of the following complexes is chiral?

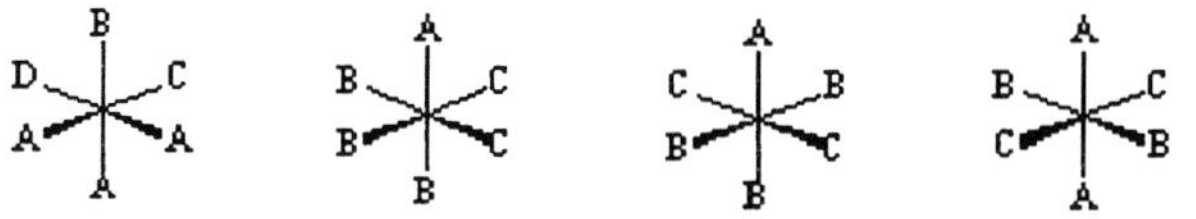

a) 1 b) 2 c) 3 d) 4 e) 1 an 2

61. Which of the following complexes is chiral?

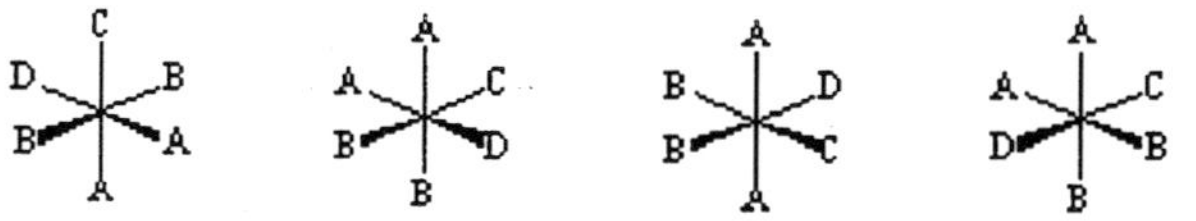

a) 2 b) 1 c) 3 d) 4 e) 1 and 3

62. For the complex ion $[Co(en)(NH_3)_2(OH_2)Cl]^{2+}$, there are four possible geometric isomers. How many of these geometric isomers are optically active?
a) 2 b) 1 c) None d) All of them e) 3

63. Which of the following complexes is chiral?
a) $[Ru(OH_2)_3Cl_3]$
b) $[Co(ox)_3]^{3-}$
c) *trans*-$[Co(ox)_2(OH_2)(NH_3)]^{3+}$
d) *trans*-$[Co(ox)_2(OH_2)_2]^-$
e) *cis*-$[Co(NH_3)_4(OH_2)_2]^{3+}$

64. The complex $[Ru(OH_2)_3Cl_3]$
a) has 2 geometric isomers both of which are chiral.
b) is chiral.
c) has 2 geometric isomers neither of which is chiral.
d) has no geometric isomers.
e) has 2 geometric isomers one of which is chiral.

65. Draw all possible geometric isomers of the complex ion $[Cr(en)(NH_3)_2(OH_2)Cl]^{2+}$ and indicate which are chiral.

66. Draw all possible geometric isomers of the complex ion $[Co(gly)_2(OH_2)_2]^{+}$ and indicate which are chiral. (gly^- is the bidentate chelate $NH_2CH_2CO_2^-$)

67. Which of the following complex ions has the greatest *d*-orbital splitting energy?
a) $[CrCl_6]^{3-}$ b) $[CrF_6]^{3-}$ c) $[Cr(ox)_3]^{3-}$ d) $[Cr(en)_3]^{3+}$ e) $[Cr(OH_2)_6]^{3+}$

68. Which of the following complex ions has the lowest *d*-orbital splitting energy?
a) $[Co(CN)_6]^{4-}$ b) $[Co(NH_3)_6]^{2+}$ c) $[CoCl_4]^{2-}$ d) $[Co(OH_2)_6]^{2+}$ e) $[CoCl_6]^{4-}$

69. Which of the following complex ions is likely to absorb at about 400 nm?
a) $[CrCl_6]^{3-}$ b) $[Cr(OH_2)_6]^{3+}$ c) $[Cr(ox)_3]^{3-}$ d) $[Cr(CN)_6]^{3-}$ e) $[CrF_6]^{3-}$

70. Which of the following complex ions is colorless?
a) $[Ti(OH_2)_6]^{3+}$ b) $[Ni(OH_2)_6]^{2+}$ c) $[Cr(OH_2)_6]^{3+}$ d) $[V(OH_2)_6]^{2+}$ e) $[Sc(OH_2)_6]^{3+}$

71. The complex ions $[Co(CN)_6]^{4-}$ and $[Co(OH_2)_6]^{2+}$
a) absorb at about 520 nm and 420 nm, respectively.
b) absorb at about 420 nm and 520 nm, respectively.
c) absorb at about 420 nm and 700 nm, respectively.
d) absorb at about 700 nm and 420 nm, respectively.
e) both absorb at the same wavelength because they are both d^7 ions.

72. (a) The complex ion $[CrCl_6]^{3-}$ is green. Predict the wavelength of the absorption band for this complex.
(b) If the chloride ligands are replaced with NH_3 ligands, predict the shift in the absorption band and the approximate wavelength of the band.
(c) Draw the *d*-orbital splitting diagram for each complex to the same scale.
(d) Predict whether the complexes are diamagnetic or paramagnetic.

73. For which one of the following would it not be possible to distinguish between high-spin and low-spin complexes in octahedral geometry?
a) Cr(III) b) Co(III) c) Fe(II) d) Co(II) e) Cr(II)

74. The complex ion $[Cr(OH_2)_6]^{2+}$ has four unpaired electrons. This means that

a) the complex is low-spin.
b) Δ_o is very large.
c) the complex is diamagnetic.
d) the water ligands are difficult to remove.
e) Δ_o is very small.

75. What is the *d*-electron configuration of the tetrahedral complex ion $[FeCl_4]^-$?

a) e^5 b) e^1t^4 c) e^2t^3 d) e^4t^1 e) e^3t^2

76. Predict the number of unpaired electrons in $[Fe(CN)_6]^{4-}$ and $[Fe(OH_2)_6]^{2+}$, respectively.

a) 0 and 2 b) 0 and 4 c) 2 and 4 d) 4 and 0 e) 0 and 3

77. Explain why there are no known low-spin tetrahedral complexes.

78. Which of the following is paramagnetic?

a) $[Cu(NCS)_4]^{3-}$ b) $[Zn(NH_3)_4]^{2+}$ c) $[Fe(CN)_6]^{4-}$ d) $[Co(NH_3)_6]^{3+}$ e) $[V(OH_2)_6]^{2+}$

79. What change in magnetic properties can be expected when NH_3 ligands in an octahedral complex are replaced by F^- ligands in a d^5 complex?

a) The complex changes from having 5 unpaired electrons to 1 unpaired electron.
b) The complex becomes tetrahedral.
c) The complex becomes high-spin.
d) No change occurs.
e) The complex changes from having 1 unpaired electron to 5 unpaired electrons.

80. If there are no geometric isomers for the complex $[MA_2B_2]$,

a) the complex is paramagnetic.
b) the complex is optically active.
c) the complex has a large Δ_o.
d) the complex is square planar.
e) the complex is tetrahedral.

Chapter 22: Nuclear Chemistry
Form A

1. How many protons, neutrons, and nucleons, respectively, are there in ^{235}U?
a) 235, 92, 327 b) 92, 235, 327 c) 143, 92, 235 d) 92, 327, 419 e) 92, 143, 235

2. How many protons, neutrons, and nucleons, respectively, are there in ^{24}Na?
a) 24, 11, 35 b) 11, 35, 46 c) 11, 13, 24 d) 13, 11, 24 e) 11, 24, 35

3. How many protons, neutrons, and nucleons, respectively, are there in ^{90}Sr?
a) 38, 128, 166 b) 38, 52, 90 c) 52, 38, 90 d) 90, 38, 128 e) 38, 90, 128

4. How many nucleons are there in ^{99}Mo?
a) 99 b) 42 c) 57 d) 141 e) 156

5. An α particle corresponds to
a) $_1^1H^+$. b) $_2^4He^{2+}$. c) $_{-1}^0e$. d) $_2^4He$. e) $_1^0e$.

6. When an α particle is emitted, the mass number
a) decreases by 4. b) decreases by 1. c) increases by 2. d) increases by 4. e) decreases by 2.

7. When a β particle is emitted, the atomic number changes by
a) +1. b) –2. c) +2. d) –1. e) 0.

8. When a positron is emitted, the mass number changes by
a) 0. b) +1. c) –1. d) +2. e) –2.

9. The decay of a proton to a neutron gives
a) a positron. b) an electron. c) a β particle. d) a deuterium atom. e) an α particle.

10. A positron has a mass number and charge, respectively, of
a) +1 and +1. b) 0 and +1. c) 0 and –1. d) 1 and –1. e) +4 and +2.

11. The nuclear equation for the disintegration of U-238 produces Th-234. What other nuclide is produced?
a) ^3He b) β c) ^4He d) positron e) neutron

12. If In-116 undergoes β emission, what other nuclide is produced?
a) Sn-116 b) Cd-116 c) Ag-112 d) Cd-115 e) Sb-120

13. A nuclide undergoes α decay and forms ^{110}I. What is the nuclide?
a) ^{114}Cs b) ^{114}I c) ^{110}Te d) ^{110}Xe e) ^{112}Cs

14. A nuclide undergoes positron emission to form ^{22}Ne. What is the nuclide?
a) ^{23}Na b) ^{23}Mg c) ^{22}Na d) ^{18}O e) ^{22}F

15. What nuclide is formed when tritium undergoes β decay?
a) β^- b) β^+ c) $^{1}H^+$ d) $^{4}He^{2+}$ e) ^{3}He

16. What nuclide is formed when ^{226}Ra (radium) undergoes α decay?
a) ^{222}Ra b) ^{222}Rn c) ^{222}Po d) ^{222}At e) ^{218}Po

17. What nuclide is formed when ^{22}Na undergoes positron emission
a) ^{22}Ne b) ^{18}O c) ^{23}Ne d) ^{22}F e) ^{21}Na

18. Bombarding ^{54}Fe with a neutron results in emission of a proton and formation of
a) ^{49}Ti. b) ^{54}Cr. c) ^{55}Fe. d) ^{54}Co. e) ^{54}Mn.

19. What nuclide is formed when ^{241}Am undergoes α decay?
a) ^{239}Np b) ^{237}Bk c) ^{239}Pa d) ^{237}Am e) ^{237}Np

20. What nuclide is formed when ^{234}Th undergoes β decay?
a) ^{232}Pu b) ^{230}Rn c) ^{234}Pa d) ^{233}Th e) ^{234}U

21. What type of particle is emitted in the transformation below?

$$^{201}Pt \rightarrow {}^{201}Au$$

a) β particle
b) positron
c) α particle
d) γ particle
e) No particle is emitted because electron capture occurs.

22. What is the element produced when ^{81}Rb undergoes positron emission?
a) Kr b) Se c) Y d) Br e) Sr

23. What is the element produced when ^{208}Fr undergoes α particle emission?
a) At b) Rn c) Po d) Ra e) Ac

24. What is the element produced when ^{37}Ar undergoes electron capture?
a) Cl b) S c) K d) Ca e) Sc

25. What process occurs when ^{210}Po forms the daughter nuclide ^{206}Pb?
a) α decay b) β decay c) neutron bombardment d) electron capture e) positron emisssion

26. When ^{116}Xe forms the daughter nuclide ^{116}I,
a) proton emission occurs.
b) either electron capture or positron emission occurs.
c) β emission occurs.
d) the reactant is bombarded by a neutron.
e) α emission occurs.

27. The overall nucleosynthesis of ^{60}Co results from bombardment of a certain element with 2 neutrons along with β emission. Identify the starting element.
a) ^{59}Fe b) ^{59}Co c) ^{58}Ni d) ^{60}Fe e) ^{58}Fe

28. What is the product nuclide when mercury-197 captures an electron?
a) Au-197 b) Au-193 c) Tl-197 d) Pt-193 e) Pb-201

29. Complete the following equations.
(a) $^{14}C \rightarrow$ __ + β
(b) $^{129}Hg \rightarrow$ __ + β^+
(c) ^{55}Fe + β $\rightarrow$ __

30. In the ^{232}Th decay series, the product is ^{208}Pb. How many α and β particles must be emitted?

31. The nuclide ^{178}Pt lies above the *band of stability*. This nuclide is likely to undergo
a) α emission. b) β emission. c) neutron emission. d) proton emission. e) positron emission.

32. Nuclei with atomic numbers greater than 84
a) are stable. b) do not emit. c) are radioactive. d) undergo fusion. e) have Z = N.

33. Predict which of the following is most stable.
a) Ne-20 b) Po-210 c) Fr-223 d) Kr-87 e) Sc-43

34. Which of the following is expected to be stable?
a) ^{24}Na b) ^{87}Kr c) ^{122}Sb d) ^{208}Pb e) ^{7}Be

35. Which of the following is most likely to undergo β decay?
a) ^{235}U b) ^{11}Li c) ^{226}Ra d) ^{114}Cs e) ^{222}Rn

36. Which of the following is most likely to undergo β decay?
a) ^{226}Ra b) ^{24}Na c) ^{222}Rn d) ^{114}Cs e) ^{235}U

37. Which of the following is most likely to undergo α decay?
a) ^{14}C b) ^{24}Na c) ^{47}K d) ^{3}H e) ^{238}U

38. Elements with an odd number of protons and neutrons
a) have a magic number of protons or neutrons.
b) never undergo α emission.
c) always have a proton to neutron ratio of about one.
d) are usually stable.
e) are usually unstable.

39. Nuclides with Z greater than 83 disintegrate mainly by
a) α particle emission.
b) positron emission.
c) proton emission.
d) β particle emission.
e) neutron emission.

40. If a nuclide lies below the *band of stability*, it should decay by
a) α particle emission.
b) β particle emission.
c) They do not decay.
d) neutron emission.
e) positron emission.

41. Which of the following nuclei is stable?
a) ^{118}Sb b) ^{16}O c) ^{15}O d) ^{8}B e) ^{40}K

42. In the ^{235}U decay series, there are seven α particles and four β particles lost starting with α emission from ^{235}U. What is the final product?
a) ^{210}Po b) ^{207}Pb c) ^{208}Pb d) ^{206}Pb e) ^{210}Bi

43. In the ^{238}U decay series, there are eight α particles and six β particles lost starting with α emission from ^{238}U. What is the final product?
a) ^{210}Bi b) ^{206}Pb c) ^{210}Po d) ^{207}Pb e) ^{208}Pb

44. The principal natural source of radiation in the body is potassium-40. What is its likely decay product?
a) Ca-40 b) Cl-36 c) Sc-44 d) Sc-40 e) Cl-37

45. Cobalt-60, used in the radiation treatment of cancer, is produced by absorption of two neutrons and emission of a β particle. What is the starting nuclide?
a) ^{59}Co b) ^{58}Fe c) ^{59}Fe d) ^{58}Mn e) ^{58}Ni

46. Rate the penetrating power of the following particles from most penetrating to least penetrating: α particle, photon, β particle.
a) α particle > photon > β particle
b) photon > β particle > α particle
c) β particle > α particle > photon
d) photon > α particle > β particle
e) β particle > photon > α particle

47. Which of the following is most penetrating?
a) n b) p c) α d) β e) β^+

48. Technetium-99 in an excited state is used in brain, liver, and spleen imaging. In this application the ^{99m}Tc (The *m* stands for metastable or excited state.) acts as a γ emitter. Write a balanced equation for the reaction.

49. Potassium-40 undergoes electron capture to give argon-40 with a half-life of 1.3×10^9 y. Estimate the age of a sedimentary rock with an argon-40 to potassium-40 ratio of 0.33.

50. Radioactive decay is a
a) second-order process.
b) zero-order process.
c) non-spontaneous process.
d) temperature-dependent process.
e) first-order process.

51. A 9.9-g sample of iodine-131 is stored for exactly 2 weeks. If the decay constant is $0.0861 \cdot day^{-1}$, what mass of the isotope remains?
a) 6.7 g b) 0.060 g c) 3.0 g d) 0.62 g e) 8.3 g

52. The decay constant for strontium-90 is $0.0247 \cdot y^{-1}$. How many grams of strontium-90 remain if 10.0 g decay for exactly 60 years?
a) 0.23 g b) 0.33 g c) 2.27 g d) 1.48 g e) 7.71 g

53. The half-life of strontium-90 is 28.1 years. What is the decay constant of strontium-90?
a) $3.34 \cdot y^{-1}$ b) $28.1 \cdot y^{-1}$ c) $0.0356 \cdot y^{-1}$ d) $1.04 \cdot y^{-1}$ e) $0.0247 \cdot y^{-1}$

54. The half-life of cobalt-60 is 5.27 years. Calculate the percent of a cobalt sample left after exactly 1 year.
a) 87.7% b) 19.0% c) 76.9% d) 73.9% e) 2.59%

55. The half-life of strontium-90 is 28.1 years. Calculate the percent of a strontium sample left after 100 years.
a) 8.5% b) 0.34% c) 63% d) 82% e) 76%

56. The decay constant for sodium-24 is $1.28 \times 10^{-5}\ s^{-1}$. How many mg of sodium-24 remain if 1.40 mg decay for 3.00 h?
a) 1.02 mg b) 1.40 mg c) 0.707 mg d) 0.124 mg e) 1.22 mg

57. The decay constant for sodium-24 is $1.28 \times 10^{-5}\ s^{-1}$. What fraction of sodium-24 remains after 3.00 h?
a) 0.871 b) 0.727 c) 1.00 d) 0.333 e) 0.290

58. The decay constant for lithium-8 is $0.825\ s^{-1}$. How many mg of litium-8 remain if 5.00 μg decay for 4.00 s?
a) 0.0460 μg b) 1.25 μg c) 0.00251 μg d) 0.184 μg e) 2.19 μg

59. The decay constant for lithium-8 is $0.825\ s^{-1}$. What fraction of Li-8 remains after 3.00 s?
a) 0.0841 b) 0.0.275 c) 0.00335 d) 0.0641 e) 0.251

60. The decay constant for cobalt-60 is $4.18 \times 10^{-9}\ s^{-1}$. What fraction of Co-60 remains after exactly 90 days?
a) 0.968 b) 0.923 c) 0.999 d) 0.0325 e) 0.151

61. The decay constant for cobalt-60 is $4.18 \times 10^{-9}\ s^{-1}$. How many g cobalt-60 remain if 2.50 g decay for 3 years (1095 days)?
a) 0.820 g b) 0.989 g c) 1.01 g d) 2.32 g e) 1.68 g

62. When a 2.50-g sample of cobalt-60 is stored for 10 years, 0.669 g remain. What is the decay constant of cobalt-60?
a) 0.0573 y^{-1} b) 0.268 y^{-1} c) 0.132 y^{-1} d) 0.0605 y^{-1} e) 0.0263 y^{-1}

63. When a 1.40-mg sample of sodium-24 is stored for 12 hours, 0.805 mg remain. What is the decay constant of sodium-24?
a) 0.0280 h^{-1} b) 0.0461 h^{-1} c) 7.14 h^{-1} d) 0.0200 h^{-1} e) 0.0122 h^{-1}

64. When a 5.00-μg sample of lithium-8 decays, 2.19 μg remain after 1.00 s. What is the decay constant of lithium-8?
a) 0.359 s^{-1} b) 0.438 s^{-1} c) 1.03 s^{-1} d) 0.449 s^{-1} e) 0.826 s^{-1}

65. What is the time needed for the activity of a radium-226 source to change from 1.5 Ci to 0.010 Ci? The half-life of radium-226 is 1.60×10^3 y.
a) 8.02×10^3 y b) 5.02×10^3 y c) 1.16×10^4 y d) 1.60×10^3 y e) 9.36×10^2 y

66. What is the time needed for the activity of a gold-198 source to change from 25 Ci to 2.0 Ci? The half-life of gold-198 is 64.6 h.
a) 236 h b) 103 h c) 301 h d) 0.0391 h e) 808 h

67. What is the time needed for the activity of a carbon-15 source to change from 0.52 mCi to 0.15 mCi? The half-life of carbon-15 is 2.4 s.
a) 2.26 s b) 4.30 s c) 1.87 s d) 0.518 s e) 0.692 s

68. What is the time needed for the activity of a sodium-25 source to change from 4.4 mCi to 1.2 mCi? The decay constant of sodium-25 is 0.0116 s^{-1}.
a) 112 s b) 55.5 s c) 48.6 s d) 23.5 s e) 128 s

69. Estimate the activity of a 1.5-Ci radium-226 source, $t_{1/2} = 1.60 \times 10^3$ y, after 2000 years have passed.
a) 0.20 Ci b) 1.5 Ci c) 0.45 Ci d) 0.63 Ci e) 1.2 Ci

70. Use the law of radioactive decay to determine the activity of a 2.0-μg sample of Tc-99, $t_{1/2} = 2.2 \times 10^5$ y.
a) 1.0 Ci b) 3.3×10^{-8} Ci c) 1.0×10^{-6} Ci d) 1.5 Ci e) 3.3×10^{-2} Ci

71. A sample of a wood carving from an archaeological dig gave 20,400 disintegrations per gram of carbon per day. A 1.00-g sample of carbon from a modern source gave 22,080 disintegrations per day. If the decay constant of carbon-14 is $1.2 \times 10^{-4} \cdot y^{-1}$, what is the age of the wood?
a) 660 y b) 1700 y c) 8300 y d) 6.2×10^4 y e) 290 y

72. A sample of charcoal gave 10.2 disintegrations per gram of carbon per minute. A 1.00-g sample of carbon from a modern source gave 15.3 disintegrations per minute. If the decay constant of carbon-14 is 1.2×10^{-4} y^{-1}, what is the age of the charcoal?
a) 5.8×10^3 y b) 3.4×10^3 y c) 5.9×10^3 y d) 1.3×10^4 y e) 1.5×10^3 y

73. The nuclear binding energy for lithium-7 is the energy released in the nuclear reaction
a) $3\,^1H + 4n \rightarrow {}^7Li$ b) $3\,^1H + 4\beta \rightarrow {}^7Li$ c) $3\,^1H + 7n \rightarrow {}^7Li$ d) $7\,^1H \rightarrow {}^7Li$ e) $^6Li + n \rightarrow {}^7Li$

74. The binding energy of a nucleus is
a) related to the mass lost in a chemical reaction.
b) equal to the energy required to break up a nucleus into its constituent protons and neutrons.
c) related to the number of electrons in a nucleus.
d) the energy evolved when a nucleus breaks up into fundamental particles.
e) the energy required to eject one neutron from the nucleus.

75. Calculate the nuclear binding energy of 1 mole of iron-56.
a) 7.90×10^{-11} kJ b) 1.59×10^{7} kJ c) 1.59×10^{1} kJ d) 7.90×10^{-14} kJ e) 4.76×10^{10} kJ

76. The nuclear binding energy for neon-20 is the energy released when
a) 20 protons form neon-20.
b) 10 protons and 10 neutrons form neon-20.
c) 10 protons and 10 electrons form neon-20.
d) neon-19 and 1 neutron form neon-20.
e) 20 neutrons form neon-20.

77. Calculate the binding energy per nucleon for oxygen-18.

78. The nuclear binding energy for cobalt-60 is the energy released in the nuclear reaction
a) $33p + 27n \rightarrow {}^{60}Co$
b) $^{59}Co + n \rightarrow {}^{60}Co$
c) $27p + 27n \rightarrow {}^{60}Co$
d) $^{59}Fe + n \rightarrow {}^{60}Co$
e) $27p + 33n \rightarrow {}^{60}Co$

79. Calculate the energy change when one ^{235}U nucleus undergoes the fission reaction

$$^{235}U + n \rightarrow {}^{142}Ba + {}^{92}Kr + 2n$$

The masses needed are ^{235}U, 235.04 u; ^{142}Ba, 141.92 u; ^{92}Kr, 91.92 u; n, 1.0087 u.
a) -3.2×10^{-28} J b) $+1.8 \times 10^{-10}$ J c) -2.9×10^{-11} J d) -1.7×10^{-10} J e) $+2.9 \times 10^{-11}$ J

80. The equation

$$4\,^1H \rightarrow {}^4He + 2\beta^+$$

a) fusion. b) fission. c) the binding energy of He. d) the decay of a proton. e) electron capture.

Chapter 22: Nuclear Chemistry
Form B

1. How many protons, neutrons, and nucleons, respectively, are there in ^{238}U?
a) 146, 92, 238 b) 238, 92, 330 c) 92, 146, 238 d) 92, 330, 422 e) 92, 238, 330

2. How many protons, neutrons, and nucleons, respectively, are there in ^{18}O?
a) 8, 26, 34 b) 8, 10, 18 c) 18, 8, 26 d) 8, 18, 26 e) 10, 8, 18

3. How many protons, neutrons, and nucleons, respectively, are there in ^{128}I?
a) 53, 128, 181 b) 53, 181, 234 c) 53, 75, 128 d) 128, 53, 181 e) 75, 53, 128

4. How many nucleons are there in ^{241}Am?
a) 241 b) 95 c) 146 d) 336 e) 431

5. A β particle corresponds to
a) ${}_{1}^{0}e$. b) ${}_{2}^{4}He^{2+}$. c) ${}_{0}^{1}n$. d) ${}_{1}^{0}H^{+}$. e) ${}_{-1}^{0}e$.

6. When an α particle is emitted, the atomic number
a) increases by 2. b) decreases by 2. c) increases by 4. d) decreases by 1. e) decreases by 4.

7. When a β particle is emitted, the mass number changes by
a) 0. b) −2. c) +2. d) −1. e) +1.

8. When a positron is emitted, the atomic number changes by
a) −1. b) +1. c) 0. d) +2. e) −2.

9. When ^{131}I emits a β particle, what nuclide is produced?
a) ^{131}Te b) ^{130}Te c) ^{127}Sb d) ^{131}Xe e) ^{130}I

10. A β particle has a mass number and charge, respectively, of
a) +1 and +1. b) 0 and +1. c) 0 and −1. d) +4 and +2. e) +1 and 0.

11. If U-238 undergoes α emission, what other nuclide is produced?
a) Th-234 b) Ac-236 c) U-234 d) Pa-234 e) Np-236

12. The nuclear equation for the disintegration of In-116 produces Sn-116 and
a) a positron. b) an electron. c) helium-4. d) a neutron. e) gamma rays.

13. A nuclide undergoes α decay and forms ^{222}Rn. What is the nuclide?
a) ^{224}Ra b) ^{226}Ra c) ^{218}At d) ^{222}Fr e) ^{224}Th

14. A nuclide undergoes proton emission to form ^{52}Fe. What is the nuclide?
a) ^{56}Ni b) ^{53}Co c) ^{52}Mn d) ^{53}Fe e) ^{52}Co

15. What nuclide is formed when ^{222}Rn (radon) undergoes α decay?
a) ^{231}Rn b) ^{222}Fr c) ^{222}At d) ^{218}Po e) ^{218}Ra

16. What nuclide is formed when ^{90}Sr undergoes β decay?
a) ^{89}Sr b) ^{90}Y c) ^{86}Kr d) ^{94}Zr e) ^{90}Rb

17. What nuclide is formed when ^{97}Tc undergoes electron capture?
a) ^{98}Tc b) ^{97}Mo c) ^{93}Y d) ^{97}Ru e) ^{97}Nb

18. What nuclide is formed when ^{237}U undergoes β decay?
a) ^{238}U b) ^{233}Pu c) ^{235}U d) ^{237}Pa e) ^{237}Np

19. What nuclide is formed when ^{207}Po undergoes positron emission?
a) ^{206}Bi b) ^{206}At c) ^{203}Pb d) ^{207}At e) ^{207}Bi

20. What nuclide is formed when ^{210}Po undergoes α decay?
a) ^{206}Rn b) ^{206}Ra c) ^{208}Pb d) ^{206}Pb e) ^{208}Rn

21. What type of particle is captured and what type is emitted in the nucleosynthesis below?

$$^{54}\text{Fe} \rightarrow {}^{54}\text{Mn}$$

a) Neutron capture and proton emission occurs.
b) Neutron capture and β particle emission.
c) Only neutron capture occurs.
d) Neutron capture and positron emission occurs.
e) Electron capture and β particle emission occurs.

22. What is the element produced when ^{43}K undergoes β particle emission?
a) Ca b) Cl c) Ar d) Sc e) S

23. What is the element produced when ^{83}Sr undergoes positron emission?
a) Rb b) Kr c) Y d) Br e) Zr

24. What is the element produced when ^{44}Ti undergoes electron capture?
a) Sc b) Ca c) K d) V e) Ar

25. What process occurs when ^{54}Fe is bombarded with a neutron and forms ^{54}Mn?
a) proton emission b) γ emission c) positron emisssion d) β emission e) α emission

26. What process occurs when ^{222}Rn forms the daughter nuclide ^{218}Po?
a) neutron emission b) electron capture c) β emission d) α emission e) positron emission

27. When ^{14}C decays by β emission, the product is
a) ^{13}C. b) ^{14}N. c) ^{16}O. d) ^{4}He. e) CO_2.

28. What is the product nuclide when boron-10 captures an electron?
a) B-11 b) C-12 c) C-10 d) Be-10 e) B-6

29. Complete the following equations.
(a) $^{238}U \rightarrow {}^{234}Th + \_\_$
(b) $^{234}Th \rightarrow \_\_ + \beta$
(c) $^{99}Tc \rightarrow \_\_ + \beta$

30. In the ^{238}U decay series, the product is ^{206}Pb. How many α and β particles must be emitted?

31. The nuclide ^{176}Au lies below the *band of stability*. This nuclide is likely to undergo
a) photon emission.
b) neutron emission.
c) α emission.
d) electron capture or positron emission.
e) β emission.

32. The nuclide ^{32}S
a) undergoes α emission.
b) undergoes positron emission.
c) undergoes β emission.
d) is stable.
e) undergoes proton emission.

33. All of the following are predicted to be stable except
a) N-16 b) He-4 c) Ca-40 d) O-16 e) Ne-20

34. All of the following are expected to be stable except
a) ^{118}Sn b) ^{122}Sb c) ^{40}Ca d) ^{206}Pb e) ^{208}Pb

35. Which of the following is most likely to undergo α decay?
a) ^{47}K b) ^{3}H c) ^{24}Na d) ^{232}Th e) ^{14}C

36. Which of the following is most likely to undergo β decay?
a) ^{226}Ra b) ^{9}Li c) ^{114}Cs d) ^{222}Rn e) ^{235}U

37. Which of the following is most likely to undergo α decay?
a) ^{3}H b) ^{228}Ra c) ^{24}Na d) ^{47}K e) ^{14}C

38. Elements with an even number of protons and neutrons
a) never undergo α emission.
b) are usually stable.
c) have a magic number of protons or neutrons.
d) always have a proton to neutron ratio of about one.
e) are usually unstable.

39. Nuclides with Z less than 83 disintegrate mainly by
a) proton emission.
b) positron emission.
c) α particle emission.
d) neutron emission.
e) β particle emission.

40. If a nuclide lies above the *band of stability*, it should decay by
a) neutron emission.
b) They do not decay.
c) α particle emission.
d) positron emission.
e) β particle emission.

41. Which of the following nuclei is stable?
a) ^{213}Fr b) ^{38}Cl c) ^{120}Sn d) ^{15}O e) ^{24}Na

42. In the ^{232}Th decay series, there are six α particles and four β particles lost. What is the final product?
a) ^{207}Pb b) ^{208}Pb c) ^{210}Po d) ^{206}Pb e) ^{210}Bi

43. When ^{214}Pb decays as part of the ^{238}U decay series, there are two α particles and four β particles lost starting with β emission from ^{214}Pb. What is the final product?
a) ^{210}Pb b) ^{210}Po c) ^{207}Pb d) ^{206}Pb e) ^{210}Bi

44. What is the product nuclide when potassium-38 emits a positron?
a) Ar-38 b) Ca-40 c) Sc-40 d) Cl-37 e) Sc-44

45. Technetium-97 can be produced by absorption of ^{2}H and emission of two neutrons. What is the starting nuclide?
a) ^{97}Mo b) ^{98}Ru c) ^{97}Ru d) ^{96}Nb e) ^{96}Mo

46. Rate the penetrating power of the following particles from most penetrating to least penetrating: α particle, neutron, β particle.
a) β particle > neutron > α particle
b) β particle > α particle > neutron
c) neutron > β particle > α particle
d) α particle > neutron > β particle
e) neutron > α particle > β particle

47. Which of the following is most penetrating?
a) γ b) p c) α d) β e) β^+

48. (a) Write a balanced equation for the reaction that occurs when ^{197}Hg captures an electron.
(b) Write a balanced equation for the reaction that occurs when ^{253}Es captures an α particle and emits a neutron.

49. A 1.00-g sample of antique wood shows 14,100 disintegrations in 24 hours. If a current sample of wood shows 15.3 disintegrations per minute per gram, what is the age of the *antique* wood?

50. Radioactive nuclei decay
a) by β emission only.
b) at a rate characteristic of the particular nuclei.
c) with rates that are highly dependent of temperature.
d) at the same rate.
e) to produce protons and electrons exclusively.

51. A 9.9-g sample of iodine-131 is stored for exactly 3 weeks. If the decay constant is $0.0861 \cdot day^{-1}$, what mass of the isotope remains?
a) 5.5 g b) 2.6 g c) 1.6 g d) 7.6 g e) 0.15 g

52. The decay constant for strontium-90 is $0.0247 \cdot y^{-1}$. How many grams of strontium-90 remain if 10.0 g decay for exactly 75 years?
a) 1.85 g b) 0.20 g c) 1.57 g d) 0.00329 g e) 0.14 g

53. The half-life of carbon-15 is 2.4 s. What is the decay constant of carbon-15?
a) 1.7 s^{-1} b) 2.4 s^{-1} c) 3.5 s^{-1} d) 0.42 s^{-1} e) 0.29 s^{-1}

54. The half-life of iodine-131 is 8.04 days. Calculate the percent of an iodine sample left after exactly 2 days.
a) 84.2% b) 82.0% c) 67.2% d) 6.17% e) 57.5%

55. The half-life of plutonium-239 is 24,100 years. Calculate the percent of a plutonium sample left after 100 years.
a) 99.7% b) 96.4% c) 90.0% d) 4.15% e) 50.0%

56. The decay constant for sodium-24 is 1.28×10^{-5} s^{-1}. How many mg of sodium-24 remain if 1.40 mg decay for 12.0 h?
a) 0.392 mg b) 0.124 mg c) 0.829 mg d) 1.40 mg e) 0.805 mg

57. The decay constant for sodium-24 is 1.28×10^{-5} s^{-1}. What fraction of sodium-24 remains after 12.0 h?
a) 0.575 b) 0.280 c) 1.00 d) 0.0833 e) 0.0479

58. The decay constant for lithium-8 is 0.825 s^{-1}. How many mg of litium-8 remain if 5.00 μg decay for 1.00 s?
a) 4.13 μg b) 0.748 μg c) 0.418 μg d) 2.19 μg e) 0.962 μg

59. The decay constant for lithium-8 is 0.825 s^{-1}. What fraction of Li-8 remains after 0.500 s?
a) 0.662 b) 0.387 c) 0.413 d) 0.192 e) 0.0224

60. The decay constant for cobalt-60 is 4.18×10^{-9} s^{-1}. What fraction of Co-60 remains after exactly 365 days?
a) 0.876 b) 0.738 c) 0.494 d) 0.00549 e) 0.442

61. The decay constant for cobalt-60 is $4.18 \times 10^{-9}\ s^{-1}$. How many g cobalt-60 remain if 2.50 g decay for 10 years?
a) 2.38 g b) 1.83 g c) 0.669 g d) 0.120 g e) 0.137 g

62. When a 2.50-g sample of cobalt-60 is stored for 10 years, 0.669 g remain. What is the half-life of cobalt-60?
a) 26.3 y b) 12.1 y c) 11.5 y d) 2.59 y e) 5.25 y

63. When a 1.40-mg sample of sodium-24 is stored for 12 hours, 0.805 mg remain. What is the half-life of sodium-24?
a) 34.7 h b) 0.0971 h c) 24.8 h d) 15.0 h e) 56.8 h

64. When a 5.00-μg sample of lithium-8 decays, 2.19 μg remain after 1.00 s. What is the half-life of lithium-8?
a) 0.673 s b) 0.839 s c) 1.93 s d) 1.58 s e) 1.54 s

65. What is the time needed for the activity of a radium-226 source to change from 1.5 Ci to 0.50 Ci? The half-life of radium-226 is 1.60×10^3 y.
a) 5.33×10^2 y b) 1.10×10^3 y c) 1.60×10^3 y d) 2.54×10^3 y e) 9.36×10^2 y

66. What is the time needed for the activity of a gold-198 source to change from 25 Ci to 2.0 Ci? The half-life of gold-198 is 2.69 d.
a) 5.42 d b) 4.25 d c) 12.5 d d) 0.939 d e) 9.79 d

67. What is the time needed for the activity of a carbon-15 source to change from 1.52 μCi to 0.20 μCi? The half-life of carbon-15 is 2.4 s.
a) 1.45 s b) 0.845 s c) 7.02 s d) 0.316 s e) 3.05 s

68. What is the time needed for the activity of a sodium-25 source to change from 9.2 mCi to 8.4 μCi? The decay constant of sodium-25 is $0.0116\ s^{-1}$.
a) 603 s b) 0.0787 s c) 12.7 s d) 262 s e) 191 s

69. Estimate the activity of a 1.5-Ci radium-226 source, $t_{1/2} = 1.60 \times 10^3$ y, after 5000 years have passed.
a) 0.33 Ci b) 0.047 Ci c) 0.48 Ci d) 0.010 Ci e) 0.17 Ci

70. Use the law of radioactive decay to determine the activity of a 2.0-mg sample of Tc-99, $t_{1/2} = 2.2 \times 10^5$ y.
a) 1.0×10^3 Ci b) 3.3×10^{-5} Ci c) 33 Ci d) 1.5×10^3 Ci e) 1.0×10^{-3} Ci

71. A sample of a dish from an archaeological dig gave 18, 500 disintegrations per gram of carbon per day. A 1.00-g sample of carbon from a modern source gave 22,080 disintegrations per day. If the decay constant of carbon-14 is $1.2 \times 10^{-4} \cdot y^{-1}$, what is the age of the dish?
a) 640 y b) 8300 y c) 68000 y d) 30000 y e) 1500 y

72. A sample of bone gave 12.6 disintegrations per gram of carbon per minute. A 1.00-g sample of carbon from a modern source gave 15.3 disintegrations per minute. If the decay constant of carbon-14 is $1.2 \times 10^{-4}\ y^{-1}$, what is the age of the bone?
a) 1.6×10^{3} y b) 8.2×10^{4} y c) 5.8×10^{3} y d) 700 y e) 3.6×10^{3} y

73. The nuclear binding energy for iron-56 is the energy released in the nuclear reaction
a) $^{55}Fe + n \rightarrow {}^{56}Fe$
b) $26\,^{1}H + 56n \rightarrow {}^{56}Fe$
c) $56\,^{1}H \rightarrow {}^{56}Fe$
d) $26\,^{1}H + 30n \rightarrow {}^{56}Fe$
e) $26\,^{1}H + 30\beta \rightarrow {}^{56}Fe$

74. The binding energy of a nucleus is
a) the energy released when Z protons and A - Z neutrons come together to form a nucleus.
b) related to the mass lost in a chemical reaction.
c) related to the number of electrons in a nucleus.
d) the energy evolved when a nucleus breaks up into fundamental particles.
e) the energy required to eject one neutron from the nucleus.

75. Calculate the nuclear binding energy of 1 mol lithium-7.
a) 6.31×10^{-15} kJ b) 1.14×10^{18} kJ c) 1.14×10^{15} kJ d) 3.80×10^{9} kJ e) 6.31×10^{-12} kJ

76. The nuclear binding energy for manganese-55 is the energy released when
a) 25 protons and 25 neutrons form manganese-55.
b) manganese-54 and 1 neutron form manganese-55.
c) 25 protons and 30 neutrons form manganese-55.
d) 30 protons and 25 neutrons form manganese-55.
e) chromium-54 and 1 proton form manganese-55.

77. Calculate the binding energy per nucleon for cerium-146.

78. Calculate the energy change for the synthesis of 1 mol ^{60}Co:

$$^{58}Fe + 2n \rightarrow {}^{60}Co + \beta$$

a) 2.53×10^{-15} kJ
b) 8.43×10^{-24} kJ
c) 2.53×10^{-12} kJ
d) The calculation cannot be done without the mass of β.
e) 1.52×10^{9} kJ

79. Calculate the energy change when one ^{238}U nucleus undergoes the reaction

$$^{238}U + n \rightarrow {}^{239}Pu + 2\beta$$

The masses needed are ^{238}U, 238.0508 u; ^{239}Pu, 239.0522 u; n, 1.0087 u.
a) -1.1×10^{-12} J b) -1.5×10^{-10} J c) -4.5×10^{-10} J d) $+1.5 \times 10^{-12}$ J e) $+3.0 \times 10^{-10}$ J

80. The most important uranium mineral is
a) UF_4 b) UF_6 c) UO_2 d) $UO_2(NO_3)_2$ e) UO_3

Chapter 1A Answers

1.	d	Section:	1.2
2.	a	Section:	1.2
3.	a	Section:	1.2
4.	a	Section:	1.3
5.	d	Section:	1.3
6.	e	Section:	1.4
7.	e	Section:	1.4
8.	c	Section:	1.4
9.	b	Section:	1.4
10.	e	Section:	1.4
11.	d	Section:	1.4
12.	e	Section:	1.4
13.	*(No answer.)*	Section:	1.4
14.	*(No answer.)*	Section:	1.4
15.	*(No answer.)*	Section:	1.4
16.	a	Section:	1.6
17.	a	Section:	1.6
18.	a	Section:	1.6
19.	b	Section:	1.6
20.	a	Section:	1.6
21.	a	Section:	1.7
22.	d	Section:	1.6
23.	a	Section:	1.6
24.	a	Section:	1.6
25.	a	Section:	1.6
26.	a	Section:	1.9
27.	b	Section:	1.10
28.	a	Section:	1.10
29.	d	Section:	1.10
30.	c	Section:	1.10
31.	a	Section:	1.10
32.	a	Section:	1.10
33.	a	Section:	1.10
34.	b	Section:	1.10
35.	a	Section:	1.10
36.	a	Section:	1.10
37.	c	Section:	1.10
38.	a	Section:	1.10
39.	a	Section:	1.10
40.	a	Section:	1.11
41.	b	Section:	1.11
42.	a	Section:	1.11
43.	c	Section:	1.11
44.	a	Section:	1.12
45.	d	Section:	1.12
46.	e	Section:	1.12
47.	a	Section:	1.12
48.	a	Section:	1.12
49.	e	Section:	1.12
50.	b	Section:	1.12
51.	d	Section:	1.12
52.	a	Section:	1.12
53.	a	Section:	1.12
54.	*(No answer.)*	Section:	1.12
55.	*(No answer.)*	Section:	1.12
56.	b	Section:	1.13
57.	d	Section:	1.13
58.	e	Section:	1.13
59.	e	Section:	1.13
60.	c	Section:	1.14
61.	c	Section:	1.14
62.	c	Section:	1.14
63.	b	Section:	1.14
64.	d	Section:	1.14
65.	b	Section:	1.14
66.	a	Section:	1.14
67.	e	Section:	1.14
68.	e	Section:	1.14
69.	a	Section:	1.14
70.	a	Section:	1.15
71.	c	Section:	1.15
72.	d	Section:	1.15
73.	d	Section:	1.15
74.	d	Section:	1.15
75.	c	Section:	1.15
76.	b	Section:	1.15
77.	d	Section:	1.16
78.	a	Section:	1.16
79.	b	Section:	1.16
80.	c	Section:	1.16

Chapter 1B Answers

1.	a	Section: 1.2
2.	a	Section: 1.2
3.	a	Section: 1.2
4.	c	Section: 1.3
5.	e	Section: 1.3
6.	c	Section: 1.4
7.	e	Section: 1.4
8.	c	Section: 1.4
9.	a	Section: 1.4
10.	b	Section: 1.4
11.	c	Section: 1.4
12.	b	Section: 1.4
13.	*(No answer.)*	Section: 1.4
14.	*(No answer.)*	Section: 1.4
15.	*(No answer.)*	Section: 1.4
16.	a	Section: 1.6
17.	a	Section: 1.6
18.	a	Section: 1.6
19.	a	Section: 1.6
20.	e	Section: 1.7
21.	a	Section: 1.6
22.	a	Section: 1.6
23.	d	Section: 1.6
24.	a	Section: 1.6
25.	a	Section: 1.6
26.	b	Section: 1.9
27.	d	Section: 1.10
28.	a	Section: 1.10
29.	a	Section: 1.10
30.	a	Section: 1.10
31.	b	Section: 1.10
32.	a	Section: 1.10
33.	c	Section: 1.10
34.	a	Section: 1.10
35.	a	Section: 1.10
36.	e	Section: 1.10
37.	b	Section: 1.10
38.	a	Section: 1.10
39.	a	Section: 1.10
40.	a	Section: 1.11
41.	a	Section: 1.11
42.	b	Section: 1.11
43.	a	Section: 1.11
44.	b	Section: 1.12
45.	c	Section: 1.12
46.	a	Section: 1.12
47.	a	Section: 1.12
48.	d	Section: 1.12
49.	b	Section: 1.12
50.	a	Section: 1.12
51.	d	Section: 1.12
52.	d	Section: 1.12
53.	e	Section: 1.12
54.	*(No answer.)*	Section: 1.12
55.	*(No answer.)*	Section: 1.12
56.	a	Section: 1.13
57.	a	Section: 1.13
58.	d	Section: 1.13
59.	d	Section: 1.13
60.	b	Section: 1.14
61.	d	Section: 1.14
62.	c	Section: 1.14
63.	c	Section: 1.14
64.	b	Section: 1.14
65.	a	Section: 1.14
66.	a	Section: 1.14
67.	b	Section: 1.14
68.	d	Section: 1.14
69.	a	Section: 1.14
70.	a	Section: 1.15
71.	a	Section: 1.15
72.	b	Section: 1.15
73.	c	Section: 1.15
74.	b	Section: 1.15
75.	b	Section: 1.15
76.	a	Section: 1.15
77.	b	Section: 1.16
78.	a	Section: 1.16
79.	c	Section: 1.16
80.	e	Section: 1.16

1.	a	Section: 2.2
2.	a	Section: 2.2
3.	b	Section: 2.2
4.	a	Section: 2.2
5.	e	Section: 2.2
6.	a	Section: 2.2
7.	b	Section: 2.2
8.	a	Section: 2.3
9.	b	Section: 2.3
10.	a	Section: 2.3
11.	c	Section: 2.3
12.	e	Section: 2.3
13.	a	Section: 2.3
14.	c	Section: 2.3
15.	b	Section: 2.3
16.	e	Section: 2.4
17.	a	Section: 2.4
18.	a	Section: 2.4
19.	b	Section: 2.4
20.	d	Section: 2.5
21.	b	Section: 2.5
22.	a	Section: 2.6
23.	a	Section: 2.6
24.	c	Section: 2.6
25.	d	Section: 2.6
26.	a	Section: 2.6
27.	a	Section: 2.6
28.	e	Section: 2.9
29.	c	Section: 2.9
30.	c	Section: 2.9
31.	c	Section: 2.9
32.	*(No answer.)*	Section: 2.9
33.	*(No answer.)*	Section: 2.9
34.	c	Section: 2.10
35.	b	Section: 2.10
36.	e	Section: 2.10
37.	c	Section: 2.10
38.	d	Section: 2.10
39.	a	Section: 2.11
40.	c	Section: 2.11
41.	a	Section: 2.11
42.	a	Section: 2.11
43.	c	Section: 2.11
44.	e	Section: 2.11
45.	c	Section: 2.11
46.	a	Section: 2.12
47.	a	Section: 2.12
48.	e	Section: 2.12
49.	a	Section: 2.12
50.	c	Section: 2.12
51.	e	Section: 2.12
52.	a	Section: 2.13
53.	d	Section: 2.13
54.	b	Section: 2.13
55.	*(No answer.)*	Section: 2.13
56.	*(No answer.)*	Section: 2.13
57.	d	Section: 2.14
58.	b	Section: 2.14
59.	e	Section: 2.14
60.	a	Section: 2.14
61.	a	Section: 2.14
62.	c	Section: 2.14
63.	d	Section: 2.14
64.	c	Section: 2.14
65.	b	Section: 2.14
66.	b	Section: 2.14
67.	c	Section: 2.14
68.	d	Section: 2.14
69.	e	Section: 2.14
70.	*(No answer.)*	Section: 2.14
71.	a	Section: 2.15
72.	c	Section: 2.15
73.	b	Section: 2.15
74.	c	Section: 2.15
75.	c	Section: 2.15
76.	b	Section: 2.15
77.	a	Section: 2.15
78.	b	Section: 2.15
79.	b	Section: 2.15
80.	a	Section: 2.15

Chapter 2B Answers

1.	a	Section: 2.2
2.	a	Section: 2.2
3.	d	Section: 2.2
4.	c	Section: 2.2
5.	e	Section: 2.2
6.	e	Section: 2.2
7.	c	Section: 2.2
8.	a	Section: 2.3
9.	a	Section: 2.3
10.	a	Section: 2.3
11.	c	Section: 2.3
12.	e	Section: 2.3
13.	d	Section: 2.3
14.	b	Section: 2.3
15.	a	Section: 2.3
16.	c	Section: 2.4
17.	d	Section: 2.4
18.	a	Section: 2.4
19.	c	Section: 2.4
20.	d	Section: 2.5
21.	d	Section: 2.5
22.	a	Section: 2.6
23.	a	Section: 2.6
24.	a	Section: 2.6
25.	d	Section: 2.6
26.	a	Section: 2.6
27.	a	Section: 2.6
28.	b	Section: 2.9
29.	d	Section: 2.9
30.	a	Section: 2.9
31.	e	Section: 2.9
32.	*(No answer.)*	Section: 2.9
33.	*(No answer.)*	Section: 2.9
34.	b	Section: 2.10
35.	b	Section: 2.10
36.	e	Section: 2.10
37.	a	Section: 2.10
38.	b	Section: 2.10
39.	a	Section: 2.11
40.	c	Section: 2.11
41.	a	Section: 2.11
42.	e	Section: 2.11
43.	e	Section: 2.11
44.	d	Section: 2.11
45.	b	Section: 2.11
46.	b	Section: 2.12
47.	a	Section: 2.12
48.	a	Section: 2.12
49.	c	Section: 2.12
50.	a	Section: 2.12
51.	d	Section: 2.12
52.	b	Section: 2.13
53.	d	Section: 2.13
54.	a	Section: 2.13
55.	*(No answer.)*	Section: 2.13
56.	*(No answer.)*	Section: 2.13
57.	b	Section: 2.14
58.	a	Section: 2.14
59.	c	Section: 2.14
60.	e	Section: 2.14
61.	d	Section: 2.14
62.	e	Section: 2.14
63.	a	Section: 2.14
64.	a	Section: 2.14
65.	c	Section: 2.14
66.	a	Section: 2.14
67.	b	Section: 2.14
68.	e	Section: 2.14
69.	e	Section: 2.14
70.	*(No answer.)*	Section: 2.14
71.	d	Section: 2.15
72.	c	Section: 2.15
73.	e	Section: 2.15
74.	c	Section: 2.15
75.	a	Section: 2.15
76.	a	Section: 2.15
77.	b	Section: 2.15
78.	a	Section: 2.15
79.	c	Section: 2.15
80.	b	Section: 2.15

Chapter 3A Answers

1.	d	Section: 3.1
2.	d	Section: 3.2
3.	a	Section: 3.2
4.	a	Section: 3.2
5.	b	Section: 3.2
6.	a	Section: 3.2
7.	a	Section: 3.2
8.	e	Section: 3.3
9.	a	Section: 3.3
10.	b	Section: 3.5
11.	d	Section: 3.5
12.	e	Section: 3.6
13.	a	Section: 3.6
14.	a	Section: 3.6
15.	c	Section: 3.6
16.	c	Section: 3.6
17.	a	Section: 3.6
18.	c	Section: 3.6
19.	c	Section: 3.6
20.	*(No answer.)*	Section: 3.6
21.	*(No answer.)*	Section: 3.6
22.	*(No answer.)*	Section: 3.6
23.	b	Section: 3.7
24.	a	Section: 3.8
25.	a	Section: 3.8
26.	a	Section: 3.8
27.	b	Section: 3.8
28.	a	Section: 3.9
29.	a	Section: 3.9
30.	d	Section: 3.9
31.	a	Section: 3.9
32.	a	Section: 3.9
33.	d	Section: 3.9
34.	c	Section: 3.10
35.	a	Section: 3.10
36.	b	Section: 3.11
37.	a	Section: 3.13
38.	a	Section: 3.13
39.	a	Section: 3.13
40.	a	Section: 3.13
41.	*(No answer.)*	Section: 3. 13
42.	b	Section: 3.14
43.	d	Section: 3.14
44.	b	Section: 3.14
45.	d	Section: 3.14
46.	c	Section: 3.14
47.	c	Section: 3.14
48.	a	Section: 3.14
49.	d	Section: 3.14
50.	b	Section: 3.14
51.	a	Section: 3.14
52.	a	Section: 3.14
53.	a	Section: 3.14
54.	a	Section: 3.14
55.	a	Section: 3.14
56.	a	Section: 3.14
57.	b	Section: 3.14
58.	a	Section: 3.14
59.	a	Section: 3.14
60.	a	Section: 3.14
61.	a	Section: 3.14
62.	d	Section: 3.14
63.	a	Section: 3.14
64.	a	Section: 3.14
65.	a	Section: 3.14
66.	c	Section: 3.14
67.	e	Section: 3.14
68.	b	Section: 3.14
69.	a	Section: 3.14
70.	d	Section: 3.14
71.	a	Section: 3.14
72.	d	Section: 3.14
73.	b	Section: 3.14
74.	a	Section: 3.14
75.	a	Section: 3.14
76.	b	Section: 3.14
77.	b	Section: 3.14
78.	a	Section: 3.14
79.	a	Section: 3.14
80.	*(No answer.)*	Section: 3.14

Chapter 3B Answers

1.	b	Section: 3.1
2.	a	Section: 3.2
3.	a	Section: 3.2
4.	a	Section: 3.2
5.	a	Section: 3.2
6.	a	Section: 3.2
7.	a	Section: 3.2
8.	a	Section: 3.3
9.	b	Section: 3.3
10.	e	Section: 3.5
11.	b	Section: 3.5
12.	b	Section: 3.6
13.	d	Section: 3.6
14.	a	Section: 3.6
15.	c	Section: 3.6
16.	d	Section: 3.6
17.	b	Section: 3.6
18.	b	Section: 3.6
19.	d	Section: 3.6
20.	*(No answer.)*	Section: 3.6
21.	*(No answer.)*	Section: 3.6
22.	*(No answer.)*	Section: 3.6
23.	b	Section: 3.7
24.	c	Section: 3.8
25.	a	Section: 3.8
26.	a	Section: 3.8
27.	e	Section: 3.8
28.	d	Section: 3.9
29.	a	Section: 3.9
30.	a	Section: 3.9
31.	a	Section: 3.9
32.	a	Section: 3.9
33.	c	Section: 3.9
34.	d	Section: 3.10
35.	a	Section: 3.10
36.	b	Section: 3.11
37.	a	Section: 3.13
38.	a	Section: 3.13
39.	a	Section: 3.13
40.	a	Section: 3.13
41.	*(No answer.)*	Section: 3.13
42.	c	Section: 3.14
43.	d	Section: 3.14
44.	c	Section: 3.14
45.	c	Section: 3.14
46.	e	Section: 3.14
47.	a	Section: 3.14
48.	d	Section: 3.14
49.	e	Section: 3.14
50.	e	Section: 3.14
51.	a	Section: 3.14
52.	a	Section: 3.14
53.	a	Section: 3.14
54.	a	Section: 3.14
55.	a	Section: 3.14
56.	a	Section: 3.14
57.	d	Section: 3.14
58.	a	Section: 3.14
59.	a	Section: 3.14
60.	a	Section: 3.14
61.	a	Section: 3.14
62.	b	Section: 3.14
63.	b	Section: 3.14
64.	a	Section: 3.14
65.	a	Section: 3.14
66.	e	Section: 3.14
67.	e	Section: 3.14
68.	a	Section: 3.14
69.	b	Section: 3.14
70.	b	Section: 3.14
71.	a	Section: 3.14
72.	c	Section: 3.14
73.	a	Section: 3.14
74.	e	Section: 3.14
75.	e	Section: 3.14
76.	a	Section: 3.14
77.	b	Section: 3.14
78.	a	Section: 3.14
79.	e	Section: 3.14
80.	*(No answer.)*	Section: 3.14

1.	c	Section: 4.1
2.	d	Section: 4.1
3.	b	Section: 4.1
4.	e	Section: 4.1
5.	b	Section: 4.1
6.	e	Section: 4.1
7.	a	Section: 4.1
8.	a	Section: 4.1
9.	b	Section: 4.1
10.	e	Section: 4.1
11.	b	Section: 4.1
12.	e	Section: 4.1
13.	d	Section: 4.1
14.	c	Section: 4.1
15.	c	Section: 4.1
16.	a	Section: 4.1
17.	b	Section: 4.1
18.	c	Section: 4.1
19.	b	Section: 4.1
20.	e	Section: 4.1
21.	b	Section: 4.1
22.	e	Section: 4.1
23.	c	Section: 4.1
24.	e	Section: 4.1
25.	a	Section: 4.1
26.	b	Section: 4.1
27.	a	Section: 4.1
28.	d	Section: 4.1
29.	b	Section: 4.2
30.	d	Section: 4.2
31.	c	Section: 4.3
32.	d	Section: 4.3
33.	c	Section: 4.3
34.	d	Section: 4.3
35.	a	Section: 4.3
36.	c	Section: 4.3
37.	d	Section: 4.3
38.	a	Section: 4.3
39.	a	Section: 4.3
40.	c	Section: 4.3
41.	*(No answer.)*	Section: 4.3
42.	*(No answer.)*	Section: 4.4
43.	a	Section: 4.4
44.	a	Section: 4.4
45.	a	Section: 4.4
46.	c	Section: 4.4
47.	b	Section: 4.4
48.	a	Section: 4.4
49.	a	Section: 4.4
50.	e	Section: 4.4
51.	a	Section: 4.5
52.	e	Section: 4.5
53.	c	Section: 4.5
54.	b	Section: 4.5
55.	d	Section: 4.5
56.	b	Section: 4.5
57.	b	Section: 4.5
58.	c	Section: 4.5
59.	e	Section: 4.5
60.	e	Section: 4.5
61.	d	Section: 4.5
62.	a	Section: 4.5
63.	e	Section: 4.5
64.	e	Section: 4.5
65.	b	Section: 4.5
66.	d	Section: 4.5
67.	e	Section: 4.5
68.	a	Section: 4.5
69.	a	Section: 4.5
70.	b	Section: 4.5
71.	d	Section: 4.5
72.	b	Section: 4.5
73.	b	Section: 4.5
74.	e	Section: 4.5
75.	*(No answer.)*	Section: 4.5
76.	a	Section: 4.6
77.	c	Section: 4.6
78.	b	Section: 4.6
79.	a	Section: 4.6
80.	*(No answer.)*	Section: 4.6

Chapter 4B Answers

1.	b	Section: 4.1
2.	b	Section: 4.1
3.	c	Section: 4.1
4.	c	Section: 4.1
5.	d	Section: 4.1
6.	b	Section: 4.1
7.	d	Section: 4.1
8.	c	Section: 4.1
9.	a	Section: 4.1
10.	a	Section: 4.1
11.	b	Section: 4.1
12.	a	Section: 4.1
13.	d	Section: 4.1
14.	e	Section: 4.1
15.	d	Section: 4.1
16.	e	Section: 4.1
17.	d	Section: 4.1
18.	e	Section: 4.1
19.	c	Section: 4.1
20.	c	Section: 4.1
21.	a	Section: 4.1
22.	c	Section: 4.1
23.	a	Section: 4.1
24.	a	Section: 4.1
25.	c	Section: 4.1
26.	b	Section: 4.1
27.	d	Section: 4.1
28.	a	Section: 4.1
29.	b	Section: 4.2
30.	a	Section: 4.2
31.	a	Section: 4.3
32.	c	Section: 4.3
33.	d	Section: 4.3
34.	c	Section: 4.3
35.	a	Section: 4.3
36.	c	Section: 4.3
37.	d	Section: 4.3
38.	c	Section: 4.3
39.	c	Section: 4.3
40.	d	Section: 4.3
41.	*(No answer.)*	Section: 4.3
42.	*(No answer.)*	Section: 4.4
43.	a	Section: 4.4
44.	a	Section: 4.4
45.	e	Section: 4.4
46.	a	Section: 4.4
47.	c	Section: 4.4
48.	a	Section: 4.4
49.	a	Section: 4.4
50.	a	Section: 4.4
51.	e	Section: 4.5
52.	e	Section: 4.5
53.	a	Section: 4.5
54.	a	Section: 4.5
55.	d	Section: 4.5
56.	e	Section: 4.5
57.	d	Section: 4.5
58.	c	Section: 4.5
59.	b	Section: 4.5
60.	e	Section: 4.5
61.	c	Section: 4.5
62.	a	Section: 4.5
63.	c	Section: 4.5
64.	e	Section: 4.5
65.	c	Section: 4.5
66.	b	Section: 4.5
67.	b	Section: 4.5
68.	e	Section: 4.5
69.	a	Section: 4.5
70.	b	Section: 4.5
71.	d	Section: 4.5
72.	c	Section: 4.5
73.	c	Section: 4.5
74.	d	Section: 4.5
75.	*(No answer.)*	Section: 4.5
76.	a	Section: 4.6
77.	a	Section: 4.6
78.	c	Section: 4.6
79.	a	Section: 4.6
80.	*(No answer.)*	Section: 4.6

1.	c	Section: 5.4
2.	c	Section: 5.4
3.	b	Section: 5.5
4.	e	Section: 5.6
5.	a	Section: 5.6
6.	e	Section: 5.7
7.	d	Section: 5.7
8.	a	Section: 5.9
9.	b	Section: *(None)*
10.	d	Section: 5.10
11.	c	Section: 5.10
12.	d	Section: 5.10
13.	e	Section: 5.10
14.	e	Section: 5.10
15.	e	Section: 5.10
16.	b	Section: 5.10
17.	a	Section: 5.10
18.	a	Section: 5.10
19.	a	Section: 5.10
20.	d	Section: 5.10
21.	e	Section: 5.10
22.	e	Section: 5.10
23.	e	Section: 5.10
24.	a	Section: 5.10
25.	a	Section: 5.10
26.	*(No answer.)*	Section: 5.10
27.	*(No answer.)*	Section: 5.10
28.	*(No answer.)*	Section: 5.10
29.	*(No answer.)*	Section: 5.10
30.	b	Section: 5.11
31.	e	Section: 5.11
32.	b	Section: 5.11
33.	a	Section: 5.11
34.	b	Section: 5.11
35.	a	Section: 5.11
36.	b	Section: 5.11
37.	d	Section: 5.11
38.	d	Section: 5.11
39.	a	Section: 5.11
40.	b	Section: 5.11
41.	b	Section: 5.11
42.	d	Section: 5.11
43.	b	Section: 5.11
44.	b	Section: 5.11
45.	c	Section: 5.11
46.	a	Section: 5.12
47.	d	Section: 5.12
48.	e	Section: 5.12
49.	a	Section: 5.13
50.	a	Section: 5.13
51.	c	Section: 5.13
52.	d	Section: 5.13
53.	c	Section: 5.13
54.	c	Section: 5.13
55.	a	Section: 5.13
56.	d	Section: 5.13
57.	a	Section: 5.13
58.	a	Section: 5.13
59.	b	Section: 5.13
60.	a	Section: 5.13
61.	b	Section: 5.14
62.	a	Section: 5.14
63.	d	Section: 5.14
64.	e	Section: 5.14
65.	d	Section: 5.14
66.	c	Section: 5.14
67.	a	Section: 5.14
68.	c	Section: 5.14
69.	a	Section: 5.14
70.	b	Section: 5.14
71.	b	Section: 5.14
72.	a	Section: 5.14
73.	a	Section: 5.15
74.	d	Section: 5.15
75.	a	Section: 5.15
76.	d	Section: 5.15
77.	b	Section: 5.16
78.	b	Section: 5.16
79.	e	Section: 5.16
80.	*(No answer.)*	Section: 5.17

Chapter 5B Answers

1.	e	Section: 5.4
2.	c	Section: 5.4
3.	b	Section: 5.5
4.	e	Section: 5.6
5.	e	Section: 5.6
6.	d	Section: 5.7
7.	a	Section: 5.7
8.	a	Section: 5.9
9.	a	Section: 5.9
10.	b	Section: 5.10
11.	e	Section: 5.10
12.	b	Section: 5.10
13.	e	Section: 5.10
14.	b	Section: 5.10
15.	b	Section: 5.10
16.	a	Section: 5.10
17.	c	Section: 5.10
18.	a	Section: 5.10
19.	a	Section: 5.10
20.	e	Section: 5.10
21.	d	Section: 5.10
22.	c	Section: 5.10
23.	b	Section: 5.10
24.	c	Section: 5.10
25.	e	Section: 5.10
26.	*(No answer.)*	Section: 5.10
27.	*(No answer.)*	Section: 5.10
28.	*(No answer.)*	Section: 5.10
29.	*(No answer.)*	Section: 5.10
30.	d	Section: 5.11
31.	c	Section: 5.11
32.	e	Section: 5.11
33.	e	Section: 5.11
34.	e	Section: 5.11
35.	b	Section: 5.11
36.	a	Section: 5.11
37.	e	Section: 5.11
38.	b	Section: 5.11
39.	e	Section: 5.11
40.	d	Section: 5.11
41.	a	Section: 5.11
42.	d	Section: 5.11
43.	a	Section: 5.11
44.	a	Section: 5.11
45.	d	Section: 5.11
46.	a	Section: 5.12
47.	e	Section: 5.12
48.	d	Section: 5.12
49.	a	Section: 5.13
50.	a	Section: 5.13
51.	b	Section: 5.13
52.	e	Section: 5.13
53.	a	Section: 5.13
54.	d	Section: 5.13
55.	b	Section: 5.13
56.	b	Section: 5.13
57.	c	Section: 5.13
58.	a	Section: 5.13
59.	a	Section: 5.13
60.	c	Section: 5.13
61.	e	Section: 5.14
62.	a	Section: 5.14
63.	d	Section: 5.14
64.	a	Section: 5.14
65.	a	Section: 5.14
66.	a	Section: 5.14
67.	d	Section: 5.14
68.	e	Section: 5.14
69.	c	Section: 5.14
70.	d	Section: 5.14
71.	a	Section: 5.14
72.	b	Section: 5.14
73.	a	Section: 5.15
74.	a	Section: 5.15
75.	a	Section: 5.15
76.	a	Section: 5.15
77.	a	Section: 5.16
78.	a	Section: 5.16
79.	a	Section: 5.16
80.	*(No answer.)*	Section: 5.17

1.	c	Section: 6.1
2.	e	Section: 6.3
3.	b	Section: 6.3
4.	e	Section: 6.3
5.	e	Section: 6.3
6.	d	Section: 6.3
7.	c	Section: 6.3
8.	d	Section: 6.3
9.	c	Section: 6.3
10.	d	Section: 6.3
11.	b	Section: 6.3
12.	d	Section: 6.4
13.	a	Section: 6.5
14.	c	Section: 6.5
15.	a	Section: 6.6
16.	a	Section: 6.6
17.	d	Section: 6.6
18.	a	Section: 6.7
19.	e	Section: 6.7
20.	e	Section: 6.8
21.	e	Section: 6.8
22.	b	Section: 6.8
23.	d	Section: 6.8
24.	c	Section: 6.8
25.	*(No answer.)*	Section: 6.8
26.	a	Section: 6.9
27.	c	Section: 6.9
28.	a	Section: 6.9
29.	a	Section: 6.10
30.	d	Section: 6.10
31.	c	Section: 6.10
32.	a	Section: 6.10
33.	b	Section: 6.10
34.	*(No answer.)*	Section: 6.11
35.	*(No answer.)*	Section: 6.11
36.	*(No answer.)*	Section: 6.11
37.	d	Section: 6.11
38.	c	Section: 6.11
39.	c	Section: 6.11
40.	a	Section: 6.11
41.	d	Section: 6.11
42.	c	Section: 6.11
43.	a	Section: 6.11
44.	d	Section: 6.11
45.	b	Section: 6.11
46.	b	Section: 6.12
47.	c	Section: 6.12
48.	a	Section: 6.12
49.	e	Section: 6.12
50.	a	Section: 6.12
51.	a	Section: 6.12
52.	a	Section: 6.12
53.	a	Section: 6.12
54.	b	Section: 6.12
55.	a	Section: 6.12
56.	b	Section: 6.12
57.	c	Section: 6.12
58.	c	Section: 6.12
59.	c	Section: 6.12
60.	a	Section: 6.12
61.	d	Section: 6.12
62.	d	Section: 6.12
63.	c	Section: 6.12
64.	c	Section: 6.12
65.	c	Section: 6.12
66.	c	Section: 6.12
67.	b	Section: 6.12
68.	b	Section: 6.12
69.	b	Section: 6.12
70.	b	Section: 6.12
71.	b	Section: 6.12
72.	b	Section: 6.12
73.	e	Section: 6.12
74.	d	Section: 6.12
75.	d	Section: 6.12
76.	b	Section: 6.12
77.	c	Section: 6.12
78.	b	Section: 6.12
79.	c	Section: 6.12
80.	*(No answer.)*	Section: 6.12

1.	d	Section: 6.1
2.	e	Section: 6.3
3.	c	Section: 6.3
4.	a	Section: 6.3
5.	b	Section: 6.3
6.	b	Section: 6.3
7.	c	Section: 6.3
8.	a	Section: 6.3
9.	d	Section: 6.3
10.	d	Section: 6.3
11.	c	Section: 6.3
12.	e	Section: 6.4
13.	c	Section: 6.5
14.	c	Section: 6.5
15.	a	Section: 6.6
16.	c	Section: 6.6
17.	e	Section: 6.6
18.	c	Section: 6.7
19.	a	Section: 6.7
20.	a	Section: 6.8
21.	e	Section: 6.8
22.	b	Section: 6.8
23.	e	Section: 6.8
24.	a	Section: 6.8
25.	*(No answer.)*	Section: 6.8
26.	c	Section: 6.9
27.	a	Section: 6.9
28.	e	Section: 6.9
29.	e	Section: 6.10
30.	d	Section: 6.10
31.	c	Section: 6.10
32.	a	Section: 6.10
33.	a	Section: 6.10
34.	*(No answer.)*	Section: 6.11
35.	*(No answer.)*	Section: 6.11
36.	*(No answer.)*	Section: 6.11
37.	a	Section: 6.11
38.	b	Section: 6.11
39.	b	Section: 6.11
40.	a	Section: 6.11
41.	b	Section: 6.11
42.	c	Section: 6.11
43.	d	Section: 6.11
44.	d	Section: 6.11
45.	a	Section: 6.11
46.	d	Section: 6.12
47.	c	Section: 6.12
48.	e	Section: 6.12
49.	e	Section: 6.12
50.	a	Section: 6.12
51.	a	Section: 6.12
52.	e	Section: 6.12
53.	c	Section: 6.12
54.	e	Section: 6.12
55.	e	Section: 6.12
56.	a	Section: 6.12
57.	b	Section: 6.12
58.	e	Section: 6.12
59.	b	Section: 6.12
60.	c	Section: 6.12
61.	e	Section: 6.12
62.	e	Section: 6.12
63.	b	Section: 6.12
64.	c	Section: 6.12
65.	d	Section: 6.12
66.	b	Section: 6.12
67.	d	Section: 6.12
68.	b	Section: 6.12
69.	b	Section: 6.12
70.	e	Section: 6.12
71.	b	Section: 6.12
72.	e	Section: 6.12
73.	a	Section: 6.12
74.	e	Section: 6.12
75.	b	Section: 6.12
76.	a	Section: 6.12
77.	d	Section: 6.12
78.	b	Section: 6.12
79.	c	Section: 6.12
80.	*(No answer.)*	Section: 6.12

1.	e	Section: 7.1
2.	a	Section: 7.1
3.	a	Section: 7.1
4.	e	Section: 7.1
5.	b	Section: 7.1
6.	e	Section: 7.1
7.	a	Section: 7.1
8.	c	Section: 7.1
9.	d	Section: 7.1
10.	d	Section: 7.1
11.	e	Section: 7.1
12.	b	Section: 7.2
13.	e	Section: 7.2
14.	a	Section: 7.2
15.	e	Section: 7.2
16.	d	Section: 7.2
17.	d	Section: 7.2
18.	c	Section: 7.2
19.	e	Section: 7.2
20.	a	Section: 7.2
21.	a	Section: 7.2
22.	a	Section: 7.2
23.	c	Section: 7.2
24.	d	Section: 7.2
25.	*(No answer.)*	Section: 7.2
26.	*(No answer.)*	Section: 7.2
27.	*(No answer.)*	Section: 7.2
28.	*(No answer.)*	Section: 7.4
29.	a	Section: 7.5
30.	a	Section: 7.6
31.	b	Section: 7.6
32.	a	Section: 7.6
33.	d	Section: 7.6
34.	e	Section: 7.6
35.	a	Section: 7.6
36.	a	Section: 7.6
37.	d	Section: 7.6
38.	b	Section: 7.6
39.	e	Section: 7.6
40.	e	Section: 7.6
41.	e	Section: 7.6
42.	a	Section: 7.6
43.	d	Section: 7.6
44.	a	Section: 7.6
45.	e	Section: 7.9
46.	e	Section: 7.11
47.	b	Section: 7.11
48.	b	Section: 7.11
49.	a	Section: 7.11
50.	a	Section: 7.11
51.	a	Section: 7.11
52.	a	Section: 7.12
53.	e	Section: 7.12
54.	a	Section: 7.12
55.	c	Section: 7.12
56.	e	Section: 7.12
57.	a	Section: 7.12
58.	c	Section: 7.13
59.	a	Section: 7.13
60.	c	Section: 7.13
61.	a	Section: 7.14
62.	a	Section: 7.15
63.	b	Section: 7.15
64.	e	Section: 7.15
65.	a	Section: 7.16
66.	e	Section: 7.16
67.	c	Section: 7.17
68.	a	Section: 7.17
69.	a	Section: 7.17
70.	a	Section: 7.17
71.	a	Section: 7.17
72.	c	Section: 7.17
73.	a	Section: 7.18
74.	d	Section: 7.20
75.	a	Section: 7.20
76.	a	Section: 7.20
77.	a	Section: 7.22
78.	a	Section: 7.22

1.	e	Section: 7.1
2.	b	Section: 7.1
3.	e	Section: 7.1
4.	a	Section: 7.1
5.	c	Section: 7.1
6.	c	Section: 7.1
7.	d	Section: 7.1
8.	a	Section: 7.1
9.	a	Section: 7.1
10.	e	Section: 7.1
11.	e	Section: 7.1
12.	a	Section: 7.2
13.	b	Section: 7.2
14.	c	Section: 7.2
15.	c	Section: 7.2
16.	d	Section: 7.2
17.	b	Section: 7.2
18.	b	Section: 7.2
19.	a	Section: 7.2
20.	b	Section: 7.2
21.	b	Section: 7.2
22.	b	Section: 7.2
23.	c	Section: 7.2
24.	e	Section: 7.2
25.	*(No answer.)*	Section: 7.2
26.	*(No answer.)*	Section: 7.2
27.	*(No answer.)*	Section: 7.2
28.	*(No answer.)*	Section: 7.4
29.	b	Section: 7.5
30.	c	Section: 7.6
31.	e	Section: 7.6
32.	a	Section: 7.6
33.	b	Section: 7.6
34.	e	Section: 7.6
35.	a	Section: 7.6
36.	a	Section: 7.6
37.	b	Section: 7.6
38.	d	Section: 7.6
39.	a	Section: 7.6
40.	e	Section: 7.6
41.	c	Section: 7.6
42.	a	Section: 7.6
43.	a	Section: 7.6
44.	c	Section: 7.6
45.	a	Section: 7.9
46.	a	Section: 7.11
47.	e	Section: 7.11
48.	a	Section: 7.11
49.	a	Section: 7.11
50.	a	Section: 7.11
51.	a	Section: 7.11
52.	a	Section: 7.12
53.	b	Section: 7.12
54.	a	Section: 7.12
55.	d	Section: 7.12
56.	c	Section: 7.12
57.	a	Section: 7.12
58.	a	Section: 7.13
59.	a	Section: 7.13
60.	b	Section: 7.13
61.	a	Section: 7.14
62.	a	Section: 7.15
63.	a	Section: 7.15
64.	a	Section: 7.15
65.	a	Section: 7.16
66.	b	Section: 7.16
67.	c	Section: 7.17
68.	a	Section: 7.17
69.	a	Section: 7.17
70.	a	Section: 7.17
71.	a	Section: 7.17
72.	a	Section: 7.17
73.	c	Section: 7.18
74.	d	Section: 7.20
75.	a	Section: 7.20
76.	a	Section: 7.20
77.	a	Section: 7.22
78.	a	Section: 7.22

1.	a	Section: 8.1
2.	a	Section: 8.1
3.	a	Section: 8.1
4.	b	Section: 8.2
5.	b	Section: 8.2
6.	a	Section: 8.2
7.	e	Section: 8.2
8.	a	Section: 8.2
9.	b	Section: 8.2
10.	a	Section: 8.2
11.	c	Section: 8.2
12.	d	Section: 8.2
13.	b	Section: 8.2
14.	b	Section: 8.2
15.	a	Section: 8.2
16.	b	Section: 8.2
17.	*(No answer.)*	Section: 8.2
18.	*(No answer.)*	Section: 8.2
19.	*(No answer.)*	Section: 8.2
20.	a	Section: 8.5
21.	a	Section: 8.6
22.	c	Section: 8.6
23.	a	Section: 8.6
24.	a	Section: 8.6
25.	e	Section: 8.6
26.	b	Section: 8.6
27.	a	Section: 8.6
28.	a	Section: 8.6
29.	e	Section: 8.6
30.	a	Section: 8.6
31.	a	Section: 8.7
32.	a	Section: 8.7
33.	d	Section: 8.7
34.	c	Section: 8.7
35.	c	Section: 8.7
36.	e	Section: 8.7
37.	a	Section: 8.7
38.	a	Section: 8.8
39.	a	Section: 8.8
40.	d	Section: 8.8
41.	b	Section: 8.8
42.	d	Section: 8.8
43.	b	Section: 8.8
44.	d	Section: 8.8
45.	a	Section: 8.8
46.	c	Section: 8.8
47.	b	Section: 8.8
48.	*(No answer.)*	Section: 8.8
49.	a	Section: 8.9
50.	b	Section: 8.9
51.	a	Section: 8.9
52.	a	Section: 8.9
53.	a	Section: 8.9
54.	d	Section: 8.9
55.	a	Section: 8.10
56.	*(No answer.)*	Section: 8.10
57.	a	Section: 8.10
58.	a	Section: 8.10
59.	a	Section: 8.10
60.	e	Section: 8.10
61.	e	Section: 8.10
62.	c	Section: 8.10
63.	b	Section: 8.10
64.	b	Section: 8.10
65.	b	Section: 8.10
66.	c	Section: 8.10
67.	a	Section: 8.10
68.	a	Section: 8.10
69.	b	Section: 8.10
70.	e	Section: 8.10
71.	a	Section: 8.11
72.	b	Section: 8.12
73.	a	Section: 8.13
74.	a	Section: 8.13
75.	c	Section: 8.13
76.	a	Section: 8.13
77.	b	Section: 8.13
78.	a	Section: 8.14
79.	b	Section: 8.14
80.	a	Section: 8.14

Chapter 8B Answers

1.	a	Section: 8.1
2.	a	Section: 8.1
3.	a	Section: 8.1
4.	e	Section: 8.2
5.	c	Section: 8.2
6.	b	Section: 8.2
7.	a	Section: 8.2
8.	d	Section: 8.2
9.	b	Section: 8.2
10.	a	Section: 8.2
11.	c	Section: 8.2
12.	b	Section: 8.2
13.	e	Section: 8.2
14.	c	Section: 8.2
15.	e	Section: 8.2
16.	c	Section: 8.2
17.	*(No answer.)*	Section: 8.2
18.	*(No answer.)*	Section: 8.2
19.	*(No answer.)*	Section: 8.2
20.	a	Section: 8.5
21.	c	Section: 8.6
22.	a	Section: 8.6
23.	d	Section: 8.6
24.	a	Section: 8.6
25.	a	Section: 8.6
26.	b	Section: 8.6
27.	a	Section: 8.6
28.	a	Section: 8.6
29.	c	Section: 8.6
30.	a	Section: 8.6
31.	b	Section: 8.7
32.	d	Section: 8.7
33.	b	Section: 8.7
34.	b	Section: 8.7
35.	d	Section: 8.7
36.	b	Section: 8.7
37.	a	Section: 8.7
38.	a	Section: 8.8
39.	a	Section: 8.8
40.	c	Section: 8.8
41.	d	Section: 8.8
42.	e	Section: 8.8
43.	e	Section: 8.8
44.	a	Section: 8.8
45.	c	Section: 8.8
46.	b	Section: 8.8
47.	c	Section: 8.8
48.	*(No answer.)*	Section: 8.8
49.	b	Section: 8.9
50.	a	Section: 8.9
51.	c	Section: 8.9
52.	a	Section: 8.9
53.	a	Section: 8.9
54.	a	Section: 8.9
55.	a	Section: 8.10
56.	*(No answer.)*	Section: 8.10
57.	a	Section: 8.10
58.	a	Section: 8.10
59.	a	Section: 8.10
60.	b	Section: 8.10
61.	c	Section: 8.10
62.	e	Section: 8.10
63.	e	Section: 8.10
64.	a	Section: 8.10
65.	c	Section: 8.10
66.	e	Section: 8.10
67.	a	Section: 8.10
68.	a	Section: 8.10
69.	b	Section: 8.10
70.	a	Section: 8.10
71.	a	Section: 8.11
72.	b	Section: 8.12
73.	a	Section: 8.13
74.	a	Section: 8.13
75.	c	Section: 8.13
76.	a	Section: 8.13
77.	d	Section: 8.13
78.	c	Section: 8.14
79.	d	Section: 8.14
80.	c	Section: 8.14

1.	a	Section: 9.1
2.	a	Section: 9.2
3.	a	Section: 9.2
4.	a	Section: 9.2
5.	a	Section: 9.2
6.	b	Section: 9.2
7.	b	Section: 9.2
8.	a	Section: 9.2
9.	a	Section: 9.2
10.	b	Section: 9.2
11.	b	Section: 9.3
12.	a	Section: 9.4
13.	b	Section: 9.4
14.	b	Section: 9.4
15.	a	Section: 9.4
16.	a	Section: 9.4
17.	a	Section: 9.4
18.	a	Section: 9.4
19.	a	Section: 9.4
20.	a	Section: 9.4
21.	b	Section: 9.4
22.	a	Section: 9.4
23.	a	Section: 9.4
24.	b	Section: 9.4
25.	a	Section: 9.5
26.	a	Section: 9.5
27.	b	Section: 9.5
28.	c	Section: 9.5
29.	c	Section: 9.5
30.	b	Section: 9.6
31.	a	Section: 9.7
32.	b	Section: 9.7
33.	e	Section: 9.7
34.	c	Section: 9.7
35.	c	Section: 9.7
36.	c	Section: 9.7
37.	c	Section: 9.7
38.	a	Section: 9.7
39.	d	Section: 9.7
40.	e	Section: 9.7
41.	a	Section: 9.7
42.	d	Section: 9.7
43.	c	Section: 9.7
44.	b	Section: 9.10
45.	d	Section: 9.10
46.	d	Section: 9.10
47.	b	Section: 9.10
48.	a	Section: 9.10
49.	a	Section: 9.10
50.	c	Section: 9.10
51.	a	Section: 9.10
52.	*(No answer.)*	Section: 9.10
53.	b	Section: 9.14
54.	b	Section: 9.14
55.	a	Section: 9.14
56.	*(No answer.)*	Section: 9.14
57.	*(No answer.)*	Section: 9.14
58.	a	Section: 9.15
59.	d	Section: 9.15
60.	b	Section: 9.15
61.	d	Section: 9.15
62.	b	Section: 9.15
63.	c	Section: 9.15
64.	b	Section: 9.15
65.	a	Section: 9.15
66.	a	Section: 9.15
67.	b	Section: 9.15
68.	d	Section: 9.15
69.	d	Section: 9.15
70.	a	Section: 9.15
71.	e	Section: 9.15
72.	c	Section: 9.15
73.	a	Section: 9.15
74.	a	Section: 9.15
75.	c	Section: 9.16
76.	d	Section: 9.16
77.	e	Section: 9.16
78.	a	Section: 9.16
79.	*(No answer.)*	Section: 9.16
80.	*(No answer.)*	Section: 9.16

Chapter 9B Answers

1.	a	Section: 9.1
2.	a	Section: 9.2
3.	a	Section: 9.2
4.	a	Section: 9.2
5.	c	Section: 9.2
6.	c	Section: 9.2
7.	a	Section: 9.2
8.	a	Section: 9.2
9.	a	Section: 9.2
10.	a	Section: 9.2
11.	d	Section: 9.3
12.	a	Section: 9.4
13.	a	Section: 9.4
14.	a	Section: 9.4
15.	a	Section: 9.4
16.	c	Section: 9.4
17.	a	Section: 9.4
18.	a	Section: 9.4
19.	a	Section: 9.4
20.	b	Section: 9.4
21.	b	Section: 9.4
22.	a	Section: 9.4
23.	a	Section: 9.4
24.	a	Section: 9.4
25.	a	Section: 9.5
26.	a	Section: 9.5
27.	b	Section: 9.5
28.	a	Section: 9.5
29.	c	Section: 9.5
30.	a	Section: 9.6
31.	a	Section: 9.7
32.	a	Section: 9.7
33.	b	Section: 9.7
34.	a	Section: 9.7
35.	e	Section: 9.7
36.	c	Section: 9.7
37.	a	Section: 9.7
38.	a	Section: 9.7
39.	b	Section: 9.7
40.	a	Section: 9.7
41.	d	Section: 9.7
42.	d	Section: 9.7
43.	b	Section: 9.7
44.	a	Section: 9.10
45.	a	Section: 9.10
46.	c	Section: 9.10
47.	a	Section: 9.10
48.	e	Section: 9.10
49.	e	Section: 9.10
50.	a	Section: 9.10
51.	e	Section: 9.10
52.	*(No answer.)*	Section: 9.10
53.	b	Section: 9.14
54.	c	Section: 9.14
55.	b	Section: 9.14
56.	*(No answer.)*	Section: 9.14
57.	*(No answer.)*	Section: 9.14
58.	b	Section: 9.15
59.	c	Section: 9.15
60.	d	Section: 9.15
61.	e	Section: 9.15
62.	b	Section: 9.15
63.	e	Section: 9.15
64.	b	Section: 9.15
65.	d	Section: 9.15
66.	b	Section: 9.15
67.	b	Section: 9.15
68.	d	Section: 9.15
69.	a	Section: 9.15
70.	c	Section: 9.15
71.	b	Section: 9.15
72.	b	Section: 9.15
73.	a	Section: 9.15
74.	a	Section: 9.15
75.	e	Section: 9.16
76.	e	Section: 9.16
77.	e	Section: 9.16
78.	d	Section: 9.16
79.	*(No answer.)*	Section: 9.16
80.	*(No answer.)*	Section: 9.16

1. b Section: 10.1
2. d Section: 10.1
3. a Section: 10.1
4. c Section: 10.1
5. a Section: 10.1
6. d Section: 10.1
7. c Section: 10.1
8. b Section: 10.2
9. a Section: 10.3
10. e Section: 10.3
11. b Section: 10.3
12. a Section: 10.3
13. a Section: 10.3
14. e Section: 10.3
15. a Section: 10.3
16. e Section: 10.3
17. b Section: 10.3
18. a Section: 10.3
19. b Section: 10.3
20. b Section: 10.3
21. a Section: 10.4
22. b Section: 10.5
23. a Section: 10.5
24. *(No answer.)* Section: 10.7
25. *(No answer.)* Section: 10.7
26. *(No answer.)* Section: 10.7
27. *(No answer.)* Section: 10.7
28. a Section: 10.7
29. a Section: 10.7
30. a Section: 10.7
31. a Section: 10.7
32. c Section: 10.7
33. c Section: 10.7
34. b Section: 10.7
35. b Section: 10.7
36. b Section: 10.7
37. b Section: 10.7
38. c Section: 10.7
39. d Section: 10.7
40. e Section: 10.7
41. c Section: 10.7
42. d Section: 10.7
43. e Section: 10.7
44. e Section: 10.7
45. d Section: 10.7
46. a Section: 10.7
47. b Section: 10.7
48. d Section: 10.7
49. e Section: 10.7
50. d Section: 10.7
51. a Section: 10.10
52. a Section: 10.10
53. a Section: 10.10
54. a Section: 10.10
55. a Section: 10.10
56. a Section: 10.11
57. a Section: 10.12
58. e Section: 10.13
59. e Section: 10.13
60. a Section: 10.13
61. a Section: 10.14
62. c Section: 10.14
63. d Section: 10.16
64. e Section: 10.16
65. d Section: 10.16
66. c Section: 10.16
67. e Section: 10.16
68. *(No answer.)* Section: 10.16
69. a Section: 10.17
70. a Section: 10.17
71. d Section: 10.17
72. c Section: 10.17
73. b Section: 10.17
74. c Section: 10.17
75. a Section: 10.17
76. c Section: 10.17
77. e Section: 10.17
78. a Section: 10.17
79. a Section: 10.17
80. b Section: 10.17

1.	e	Section: 10.1
2.	a	Section: 10.1
3.	b	Section: 10.1
4.	a	Section: 10.1
5.	a	Section: 10.1
6.	e	Section: 10.1
7.	a	Section: 10.1
8.	e	Section: 10.2
9.	a	Section: 10.3
10.	b	Section: 10.3
11.	a	Section: 10.3
12.	b	Section: 10.3
13.	c	Section: 10.3
14.	c	Section: 10.3
15.	d	Section: 10.3
16.	d	Section: 10.3
17.	c	Section: 10.3
18.	c	Section: 10.3
19.	b	Section: 10.3
20.	b	Section: 10.3
21.	a	Section: 10.4
22.	c	Section: 10.5
23.	a	Section: 10.5
24.	*(No answer.)*	Section: 10.7
25.	*(No answer.)*	Section: 10.7
26.	*(No answer.)*	Section: 10.7
27.	*(No answer.)*	Section: 10.7
28.	a	Section: 10.7
29.	a	Section: 10.7
30.	a	Section: 10.7
31.	a	Section: 10.7
32.	a	Section: 10.7
33.	c	Section: 10.7
34.	b	Section: 10.7
35.	c	Section: 10.7
36.	c	Section: 10.7
37.	e	Section: 10.7
38.	e	Section: 10.7
39.	e	Section: 10.7
40.	b	Section: 10.7
41.	c	Section: 10.7
42.	a	Section: 10.7
43.	c	Section: 10.7
44.	b	Section: 10.7
45.	c	Section: 10.7
46.	e	Section: 10.7
47.	e	Section: 10.7
48.	c	Section: 10.7
49.	a	Section: 10.7
50.	c	Section: 10.7
51.	a	Section: 10.10
52.	a	Section: 10.10
53.	a	Section: 10.10
54.	a	Section: 10.10
55.	a	Section: 10.10
56.	a	Section: 10.11
57.	a	Section: 10.12
58.	a	Section: 10.13
59.	b	Section: 10.13
60.	a	Section: 10.13
61.	a	Section: 10.14
62.	e	Section: 10.14
63.	c	Section: 10.16
64.	e	Section: 10.16
65.	a	Section: 10.16
66.	c	Section: 10.16
67.	b	Section: 10.16
68.	*(No answer.)*	Section: 10.16
69.	c	Section: 10.17
70.	e	Section: 10.17
71.	c	Section: 10.17
72.	e	Section: 10.17
73.	e	Section: 10.17
74.	d	Section: 10.17
75.	d	Section: 10.17
76.	b	Section: 10.17
77.	d	Section: 10.17
78.	e	Section: 10.17
79.	b	Section: 10.17
80.	a	Section: 10.17

1.	d	Section: 11.1
2.	d	Section: 11.1
3.	a	Section: 11.2
4.	e	Section: 11.2
5.	b	Section: 11.2
6.	b	Section: 11.2
7.	e	Section: 11.2
8.	d	Section: 11.2
9.	a	Section: 11.2
10.	a	Section: 11.2
11.	a	Section: 11.2
12.	a	Section: 11.2
13.	a	Section: 11.3
14.	a	Section: 11.3
15.	*(No answer.)*	Section: 11.4
16.	e	Section: 11.4
17.	a	Section: 11.4
18.	b	Section: 11.4
19.	a	Section: 11.4
20.	a	Section: 11.4
21.	a	Section: 11.4
22.	a	Section: 11.4
23.	a	Section: 11.4
24.	a	Section: 11.4
25.	c	Section: 11.4
26.	c	Section: 11.4
27.	a	Section: 11.5
28.	b	Section: 11.5
29.	c	Section: 11.5
30.	a	Section: 11.5
31.	e	Section: 11.6
32.	d	Section: 11.8
33.	a	Section: 11.8
34.	a	Section: 11.8
35.	d	Section: 11.8
36.	d	Section: 11.8
37.	a	Section: 11.8
38.	a	Section: 11.8
39.	*(No answer.)*	Section: 11.9
40.	*(No answer.)*	Section: 11.9
41.	e	Section: 11.9
42.	d	Section: 11.9
43.	c	Section: 11.10
44.	b	Section: 11.10
45.	a	Section: 11.10
46.	a	Section: 11.10
47.	*(No answer.)*	Section: 11.11
48.	a	Section: 11.11
49.	a	Section: 11.11
50.	e	Section: 11.12
51.	a	Section: 11.12
52.	a	Section: 11.12
53.	a	Section: 11.12
54.	a	Section: 11.12
55.	b	Section: 11.12
56.	d	Section: 11.12
57.	d	Section: 11.13
58.	d	Section: 11.13
59.	a	Section: 11.13
60.	*(No answer.)*	Section: 11.14
61.	a	Section: 11.14
62.	b	Section: 11.14
63.	d	Section: 11.17
64.	d	Section: 11.17
65.	d	Section: 11.17
66.	e	Section: 11.17
67.	a	Section: 11.17
68.	a	Section: 11.17
69.	c	Section: 11.17
70.	a	Section: 11.17
71.	a	Section: 11.17
72.	a	Section: 11.17
73.	a	Section: 11.17
74.	d	Section: 11.19
75.	a	Section: 11.19
76.	a	Section: 11.19
77.	a	Section: 11.19
78.	a	Section: 11.19
79.	a	Section: 11.19
80.	a	Section: 11.19

Chapter 11B Answers

1.	c	Section: 11.1
2.	c	Section: 11.1
3.	d	Section: 11.2
4.	d	Section: 11.2
5.	d	Section: 11.2
6.	e	Section: 11.2
7.	c	Section: 11.2
8.	e	Section: 11.2
9.	d	Section: 11.2
10.	a	Section: 11.2
11.	a	Section: 11.2
12.	b	Section: 11.2
13.	d	Section: 11.3
14.	a	Section: 11.3
15.	*(No answer.)*	Section: 11.4
16.	c	Section: 11.4
17.	a	Section: 11.4
18.	a	Section: 11.4
19.	a	Section: 11.4
20.	a	Section: 11.4
21.	b	Section: 11.4
22.	a	Section: 11.4
23.	a	Section: 11.4
24.	a	Section: 11.4
25.	c	Section: 11.4
26.	e	Section: 11.4
27.	b	Section: 11.5
28.	a	Section: 11.5
29.	c	Section: 11.5
30.	a	Section: 11.5
31.	d	Section: 11.6
32.	b	Section: 11.8
33.	a	Section: 11.8
34.	a	Section: 11.8
35.	e	Section: 11.8
36.	c	Section: 11.8
37.	e	Section: 11.8
38.	e	Section: 11.8
39.	*(No answer.)*	Section: 11.9
40.	*(No answer.)*	Section: 11.9
41.	a	Section: 11.9
42.	a	Section: 11.9
43.	b	Section: 11.10
44.	b	Section: 11.10
45.	a	Section: 11.10
46.	a	Section: 11.10
47.	*(No answer.)*	Section: 11.11
48.	a	Section: 11.11
49.	c	Section: 11.11
50.	a	Section: 11.12
51.	a	Section: 11.12
52.	a	Section: 11.12
53.	a	Section: 11.12
54.	a	Section: 11.12
55.	d	Section: 11.12
56.	d	Section: 11.12
57.	c	Section: 11.13
58.	b	Section: 11.13
59.	a	Section: 11.13
60.	*(No answer.)*	Section: 11.14
61.	a	Section: 11.14
62.	a	Section: 11.14
63.	b	Section: 11.17
64.	e	Section: 11.17
65.	a	Section: 11.17
66.	a	Section: 11.17
67.	a	Section: 11.17
68.	a	Section: 11.17
69.	d	Section: 11.17
70.	a	Section: 11.17
71.	a	Section: 11.17
72.	a	Section: 11.17
73.	a	Section: 11.17
74.	c	Section: 11.19
75.	a	Section: 11.19
76.	a	Section: 11.19
77.	a	Section: 11.19
78.	a	Section: 11.19
79.	a	Section: 11.19
80.	a	Section: 11.19

1.	d	Section: 12.3
2.	c	Section: 12.4
3.	b	Section: 12.4
4.	c	Section: 12.4
5.	a	Section: 12.5
6.	d	Section: 12.5
7.	a	Section: 12.6
8.	d	Section: 12.6
9.	b	Section: 12.6
10.	b	Section: 12.6
11.	d	Section: 12.7
12.	e	Section: 12.7
13.	c	Section: 12.7
14.	d	Section: 12.7
15.	c	Section: 12.7
16.	a	Section: 12.7
17.	d	Section: 12.7
18.	d	Section: 12.7
19.	a	Section: 12.7
20.	c	Section: 12.8
21.	b	Section: 12.8
22.	a	Section: 12.8
23.	a	Section: 12.10
24.	b	Section: 12.10
25.	a	Section: 12.10
26.	a	Section: 12.10
27.	a	Section: 12.10
28.	a	Section: 12.10
29.	a	Section: 12.10
30.	a	Section: 12.10
31.	a	Section: 12.10
32.	d	Section: 12.10
33.	b	Section: 12.10
34.	a	Section: 12.10
35.	c	Section: 12.10
36.	a	Section: 12.10
37.	a	Section: 12.10
38.	b	Section: 12.10
39.	a	Section: 12.10
40.	*(No answer.)*	Section: 12.10
41.	a	Section: 12.10
42.	e	Section: 12.10
43.	c	Section: 12.11
44.	b	Section: 12.11
45.	c	Section: 12.11
46.	b	Section: 12.11
47.	d	Section: 12.11
48.	b	Section: 12.11
49.	b	Section: 12.12
50.	c	Section: 12.12
51.	d	Section: 12.12
52.	d	Section: 12.12
53.	b	Section: 12.12
54.	b	Section: 12.12
55.	d	Section: 12.12
56.	d	Section: 12.12
57.	d	Section: 12.12
58.	b	Section: 12.12
59.	*(No answer.)*	Section: 12.12
60.	a	Section: 12.12
61.	c	Section: 12.12
62.	c	Section: 12.12
63.	b	Section: 12.12
64.	e	Section: 12.12
65.	a	Section: 12.12
66.	a	Section: 12.12
67.	*(No answer.)*	Section: 12.12
68.	*(No answer.)*	Section: 12.12
69.	c	Section: 12.13
70.	b	Section: 12.13
71.	e	Section: 12.13
72.	d	Section: 12.13
73.	b	Section: 12.13
74.	a	Section: 12.13
75.	e	Section: 12.13
76.	d	Section: 12.13
77.	a	Section: 12.13
78.	c	Section: 12.13
79.	a	Section: 12.13
80.	*(No answer.)*	Section: 12.13

1.	a	Section:	12.3
2.	e	Section:	12.4
3.	a	Section:	12.4
4.	c	Section:	12.4
5.	d	Section:	12.5
6.	a	Section:	12.5
7.	b	Section:	12.6
8.	d	Section:	12.6
9.	c	Section:	12.6
10.	b	Section:	12.6
11.	c	Section:	12.7
12.	c	Section:	12.7
13.	a	Section:	12.7
14.	d	Section:	12.7
15.	c	Section:	12.7
16.	d	Section:	12.7
17.	c	Section:	12.7
18.	d	Section:	12.7
19.	c	Section:	12.7
20.	e	Section:	12.8
21.	a	Section:	12.8
22.	d	Section:	12.8
23.	e	Section:	12.10
24.	b	Section:	12.10
25.	a	Section:	12.10
26.	a	Section:	12.10
27.	a	Section:	12.10
28.	a	Section:	12.10
29.	a	Section:	12.10
30.	a	Section:	12.10
31.	a	Section:	12.10
32.	c	Section:	12.10
33.	c	Section:	12.10
34.	a	Section:	12.10
35.	e	Section:	12.10
36.	a	Section:	12.10
37.	a	Section:	12.10
38.	c	Section:	12.10
39.	a	Section:	12.10
40.	*(No answer.)*	Section:	12.10
41.	e	Section:	12.10
42.	d	Section:	12.10
43.	e	Section:	12.11
44.	d	Section:	12.11
45.	a	Section:	12.11
46.	e	Section:	12.11
47.	c	Section:	12.11
48.	c	Section:	12.11
49.	b	Section:	12.12
50.	e	Section:	12.12
51.	b	Section:	12.12
52.	a	Section:	12.12
53.	a	Section:	12.12
54.	c	Section:	12.12
55.	e	Section:	12.12
56.	d	Section:	12.12
57.	e	Section:	12.12
58.	a	Section:	12.12
59.	*(No answer.)*	Section:	12.12
60.	c	Section:	12.12
61.	b	Section:	12.12
62.	b	Section:	12.12
63.	a	Section:	12.12
64.	d	Section:	12.12
65.	a	Section:	12.12
66.	c	Section:	12.12
67.	*(No answer.)*	Section:	12.12
68.	*(No answer.)*	Section:	12.12
69.	d	Section:	12.13
70.	c	Section:	12.13
71.	c	Section:	12.13
72.	c	Section:	12.13
73.	c	Section:	12.13
74.	c	Section:	12.13
75.	e	Section:	12.13
76.	d	Section:	12.13
77.	*(No answer.)*	Section:	12.13
78.	e	Section:	12.13
79.	a	Section:	12.13
80.	*(No answer.)*	Section:	12.13

Chapter 13A Answers

1.	*(No answer.)*	Section: 13.2
2.	d	Section: 13.2
3.	c	Section: 13.2
4.	b	Section: 13.2
5.	c	Section: 13.2
6.	d	Section: 13.2
7.	e	Section: 13.2
8.	a	Section: 13.2
9.	e	Section: 13.2
10.	d	Section: 13.2
11.	c	Section: 13.2
12.	b	Section: 13.2
13.	d	Section: 13.3
14.	a	Section: 13.3
15.	d	Section: 13.3
16.	a	Section: 13.3
17.	e	Section: 13.3
18.	c	Section: 13.3
19.	a	Section: 13.4
20.	d	Section: 13.4
21.	a	Section: 13.4
22.	b	Section: 13.4
23.	d	Section: 13.7
24.	a	Section: 13.7
25.	*(No answer.)*	Section: 13.7
26.	e	Section: 13.7
27.	a	Section: 13.7
28.	a	Section: 13.7
29.	b	Section: 13.7
30.	c	Section: 13.7
31.	a	Section: 13.7
32.	a	Section: 13.7
33.	a	Section: 13.7
34.	d	Section: 13.7
35.	c	Section: 13.7
36.	a	Section: 13.7
37.	a	Section: 13.7
38.	a	Section: 13.7
39.	a	Section: 13.7
40.	a	Section: 13.7
41.	a	Section: 13.7
42.	a	Section: 13.7
43.	e	Section: 13.7
44.	a	Section: 13.7
45.	c	Section: 13.7
46.	b	Section: 13.7
47.	d	Section: 13.7
48.	d	Section: 13.7
49.	a	Section: 13.7
50.	b	Section: 13.7
51.	c	Section: 13.7
52.	d	Section: 13.7
53.	a	Section: 13.7
54.	b	Section: 13.7
55.	c	Section: 13.7
56.	b	Section: 13.7
57.	b	Section: 13.7
58.	d	Section: 13.7
59.	c	Section: 13.7
60.	b	Section: 13.7
61.	e	Section: 13.7
62.	a	Section: 13.7
63.	c	Section: 13.7
64.	e	Section: 13.7
65.	*(No answer.)*	Section: 13.8
66.	*(No answer.)*	Section: 13.8
67.	d	Section: 13.8
68.	b	Section: 13.8
69.	c	Section: 13.9
70.	b	Section: 13.9
71.	b	Section: 13.9
72.	a	Section: 13.10
73.	d	Section: 13.10
74.	c	Section: 13.10
75.	e	Section: 13.10
76.	b	Section: 13.10
77.	d	Section: 13.10
78.	a	Section: 13.10
79.	b	Section: 13.10
80.	*(No answer.)*	Section: 13.10

Chapter 13B Answers

1.	*(No answer.)*	Section: 13.2
2.	a	Section: 13.2
3.	b	Section: 13.2
4.	e	Section: 13.2
5.	a	Section: 13.2
6.	e	Section: 13.2
7.	e	Section: 13.2
8.	b	Section: 13.2
9.	c	Section: 13.2
10.	d	Section: 13.2
11.	a	Section: 13.2
12.	c	Section: 13.2
13.	c	Section: 13.3
14.	a	Section: 13.3
15.	a	Section: 13.3
16.	a	Section: 13.3
17.	d	Section: 13.3
18.	a	Section: 13.3
19.	a	Section: 13.4
20.	a	Section: 13.4
21.	b	Section: 13.4
22.	a	Section: 13.4
23.	c	Section: 13.7
24.	a	Section: 13.7
25.	*(No answer.)*	Section: 13.7
26.	a	Section: 13.7
27.	a	Section: 13.7
28.	a	Section: 13.7
29.	a	Section: 13.7
30.	c	Section: 13.7
31.	a	Section: 13.7
32.	a	Section: 13.7
33.	a	Section: 13.7
34.	e	Section: 13.7
35.	d	Section: 13.7
36.	e	Section: 13.7
37.	e	Section: 13.7
38.	a	Section: 13.7
39.	a	Section: 13.7
40.	a	Section: 13.7
41.	c	Section: 13.7
42.	e	Section: 13.7
43.	a	Section: 13.7
44.	c	Section: 13.7
45.	b	Section: 13.7
46.	e	Section: 13.7
47.	a	Section: 13.7
48.	a	Section: 13.7
49.	c	Section: 13.7
50.	d	Section: 13.7
51.	b	Section: 13.7
52.	e	Section: 13.7
53.	b	Section: 13.7
54.	d	Section: 13.7
55.	c	Section: 13.7
56.	d	Section: 13.7
57.	e	Section: 13.7
58.	c	Section: 13.7
59.	d	Section: 13.7
60.	c	Section: 13.7
61.	d	Section: 13.7
62.	a	Section: 13.7
63.	a	Section: 13.7
64.	a	Section: 13.7
65.	*(No answer.)*	Section: 13.8
66.	*(No answer.)*	Section: 13.8
67.	d	Section: 13.8
68.	e	Section: 13.8
69.	b	Section: 13.9
70.	a	Section: 13.9
71.	b	Section: 13.9
72.	a	Section: 13.10
73.	d	Section: 13.10
74.	d	Section: 13.10
75.	d	Section: 13.10
76.	e	Section: 13.10
77.	c	Section: 13.10
78.	d	Section: 13.10
79.	a	Section: 13.10
80.	*(No answer.)*	Section: 13.10

1.	b	Section: 14.1	41.	e	Section: 14.7
2.	a	Section: 14.1	42.	a	Section: 14.7
3.	e	Section: 14.2	43.	a	Section: 14.10
4.	e	Section: 14.2	44.	c	Section: 14.10
5.	e	Section: 14.2	45.	a	Section: 14.11
6.	d	Section: 14.2	46.	a	Section: 14.11
7.	c	Section: 14.2	47.	b	Section: 14.11
8.	c	Section: 14.2	48.	a	Section: 14.11
9.	*(No answer.)*	Section: 14.3	49.	a	Section: 14.11
10.	a	Section: 14.4	50.	c	Section: 14.11
11.	d	Section: 14.4	51.	a	Section: 14.11
12.	a	Section: 14.4	52.	a	Section: 14.11
13.	a	Section: 14.4	53.	a	Section: 14.11
14.	a	Section: 14.4	54.	b	Section: 14.11
15.	a	Section: 14.4	55.	b	Section: 14.11
16.	a	Section: 14.4	56.	b	Section: 14.11
17.	c	Section: 14.5	57.	c	Section: 14.11
18.	a	Section: 14.6	58.	a	Section: 14.11
19.	e	Section: 14.6	59.	*(No answer.)*	Section: 14.11
20.	a	Section: 14.6	60.	a	Section: 14.12
21.	b	Section: 14.6	61.	a	Section: 14.12
22.	a	Section: 14.6	62.	e	Section: 14.12
23.	a	Section: 14.6	63.	c	Section: 14.12
24.	e	Section: 14.6	64.	a	Section: 14.12
25.	a	Section: 14.6	65.	a	Section: 14.12
26.	d	Section: 14.7	66.	a	Section: 14.12
27.	c	Section: 14.7	67.	e	Section: 14.12
28.	a	Section: 14.7	68.	a	Section: 14.12
29.	b	Section: 14.7	69.	a	Section: 14.13
30.	a	Section: 14.7	70.	d	Section: 14.13
31.	e	Section: 14.7	71.	c	Section: 14.13
32.	c	Section: 14.7	72.	a	Section: 14.13
33.	b	Section: 14.7	73.	a	Section: 14.13
34.	a	Section: 14.7	74.	d	Section: 14.13
35.	d	Section: 14.7	75.	a	Section: 14.13
36.	c	Section: 14.7	76.	c	Section: 14.13
37.	a	Section: 14.7	77.	e	Section: 14.13
38.	e	Section: 14.7	78.	*(No answer.)*	Section: 14.13
39.	a	Section: 14.7	79.	*(No answer.)*	Section: 14.13
40.	a	Section: 14.7	80.	*(No answer.)*	Section: 14.13

Chapter 14B Answers

1.	c	Section:	14.1
2.	a	Section:	14.1
3.	a	Section:	14.2
4.	b	Section:	14.2
5.	a	Section:	14.2
6.	d	Section:	14.2
7.	a	Section:	14.2
8.	e	Section:	14.2
9.	*(No answer.)*	Section:	14.3
10.	a	Section:	14.4
11.	e	Section:	14.4
12.	a	Section:	14.4
13.	a	Section:	14.4
14.	e	Section:	14.4
15.	a	Section:	14.4
16.	a	Section:	14.4
17.	b	Section:	14.5
18.	c	Section:	14.6
19.	d	Section:	14.6
20.	d	Section:	14.6
21.	e	Section:	14.6
22.	c	Section:	14.6
23.	b	Section:	14.6
24.	e	Section:	14.6
25.	a	Section:	14.6
26.	e	Section:	14.7
27.	e	Section:	14.7
28.	a	Section:	14.7
29.	a	Section:	14.7
30.	a	Section:	14.7
31.	b	Section:	14.7
32.	a	Section:	14.7
33.	c	Section:	14.7
34.	c	Section:	14.7
35.	c	Section:	14.7
36.	c	Section:	14.7
37.	b	Section:	14.7
38.	c	Section:	14.7
39.	b	Section:	14.7
40.	a	Section:	14.7
41.	c	Section:	14.7
42.	b	Section:	14.7
43.	a	Section:	14.10
44.	c	Section:	14.10
45.	a	Section:	14.11
46.	a	Section:	14.11
47.	a	Section:	14.11
48.	c	Section:	14.11
49.	a	Section:	14.11
50.	a	Section:	14.11
51.	a	Section:	14.11
52.	d	Section:	14.11
53.	a	Section:	14.11
54.	a	Section:	14.11
55.	b	Section:	14.11
56.	c	Section:	14.11
57.	c	Section:	14.11
58.	a	Section:	14.11
59.	*(No answer.)*	Section:	14.11
60.	a	Section:	14.12
61.	a	Section:	14.12
62.	e	Section:	14.12
63.	b	Section:	14.12
64.	a	Section:	14.12
65.	a	Section:	14.12
66.	b	Section:	14.12
67.	b	Section:	14.12
68.	a	Section:	14.12
69.	a	Section:	14.13
70.	a	Section:	14.13
71.	e	Section:	14.13
72.	a	Section:	14.13
73.	b	Section:	14.13
74.	c	Section:	14.13
75.	a	Section:	14.13
76.	d	Section:	14.13
77.	a	Section:	14.13
78.	*(No answer.)*	Section:	14.13
79.	*(No answer.)*	Section:	14.13
80.	*(No answer.)*	Section:	14.13

1. e Section: 15.1
2. a Section: 15.1
3. a Section: 15.1
4. c Section: 15.1
5. a Section: 15.1
6. a Section: 15.1
7. a Section: 15.1
8. a Section: 15.1
9. a Section: 15.1
10. a Section: 15.1
11. a Section: 15.1
12. a Section: 15.1
13. c Section: 15.1
14. b Section: 15.1
15. b Section: 15.1
16. a Section: 15.1
17. a Section: 15.1
18. a Section: 15.1
19. a Section: 15.1
20. e Section: 15.1
21. c Section: 15.1
22. e Section: 15.1
23. a Section: 15.3
24. a Section: 15.3
25. a Section: 15.3
26. a Section: 15.3
27. a Section: 15.3
28. c Section: 15.3
29. b Section: 15.3
30. d Section: 15.3
31. b Section: 15.3
32. *(No answer.)* Section: 15.4
33. *(No answer.)* Section: 15.4
34. a Section: 15.5
35. *(No answer.)* Section: 15.6
36. d Section: 15.6
37. a Section: 15.6
38. e Section: 15.6
39. a Section: 15.6
40. a Section: 15.6
41. b Section: 15.6
42. a Section: 15.6
43. a Section: 15.6
44. a Section: 15.6
45. a Section: 15.6
46. c Section: 15.6
47. a Section: 15.6
48. a Section: 15.6
49. a Section: 15.6
50. a Section: 15.6
51. b Section: 15.6
52. a Section: 15.7
53. d Section: 15.7
54. c Section: 15.8
55. e Section: 15.8
56. *(No answer.)* Section: 15.9
57. a Section: 15.9
58. a Section: 15.9
59. a Section: 15.9
60. b Section: 15.9
61. b Section: 15.9
62. c Section: 15.9
63. a Section: 15.9
64. a Section: 15.9
65. a Section: 15.9
66. b Section: 15.9
67. c Section: 15.9
68. a Section: 15.9
69. a Section: 15.9
70. d Section: 15.9
71. d Section: 15.9
72. *(No answer.)* Section: 15.9
73. c Section: 15.11
74. e Section: 15.11
75. b Section: 15.11
76. b Section: 15.11
77. e Section: 15.12
78. d Section: 15.13
79. e Section: 15.13
80. c Section: 15.15

Chapter 15B Answers

1.	b	Section: 15.1
2.	a	Section: 15.1
3.	b	Section: 15.1
4.	a	Section: 15.1
5.	c	Section: 15.1
6.	b	Section: 15.1
7.	e	Section: 15.1
8.	e	Section: 15.1
9.	a	Section: 15.1
10.	a	Section: 15.1
11.	a	Section: 15.1
12.	d	Section: 15.1
13.	a	Section: 15.1
14.	c	Section: 15.1
15.	a	Section: 15.1
16.	a	Section: 15.1
17.	a	Section: 15.1
18.	d	Section: 15.1
19.	c	Section: 15.1
20.	a	Section: 15.1
21.	e	Section: 15.1
22.	b	Section: 15.1
23.	a	Section: 15.3
24.	a	Section: 15.3
25.	a	Section: 15.3
26.	a	Section: 15.3
27.	a	Section: 15.3
28.	a	Section: 15.3
29.	d	Section: 15.3
30.	b	Section: 15.3
31.	a	Section: 15.3
32.	*(No answer.)*	Section: 15.4
33.	*(No answer.)*	Section: 15.4
34.	e	Section: 15.5
35.	*(No answer.)*	Section: 15.6
36.	c	Section: 15.6
37.	b	Section: 15.6
38.	a	Section: 15.6
39.	a	Section: 15.6
40.	a	Section: 15.6
41.	e	Section: 15.6
42.	a	Section: 15.6
43.	a	Section: 15.6
44.	a	Section: 15.6
45.	a	Section: 15.6
46.	c	Section: 15.6
47.	a	Section: 15.6
48.	a	Section: 15.6
49.	e	Section: 15.6
50.	a	Section: 15.6
51.	b	Section: 15.6
52.	b	Section: 15.7
53.	c	Section: 15.7
54.	d	Section: 15.8
55.	c	Section: 15.8
56.	*(No answer.)*	Section: 15.9
57.	a	Section: 15.9
58.	a	Section: 15.9
59.	a	Section: 15.9
60.	c	Section: 15.9
61.	a	Section: 15.9
62.	e	Section: 15.9
63.	a	Section: 15.9
64.	a	Section: 15.9
65.	a	Section: 15.9
66.	b	Section: 15.9
67.	a	Section: 15.9
68.	a	Section: 15.9
69.	a	Section: 15.9
70.	d	Section: 15.9
71.	c	Section: 15.9
72.	*(No answer.)*	Section: 15.9
73.	a	Section: 15.11
74.	d	Section: 15.11
75.	a	Section: 15.11
76.	c	Section: 15.11
77.	b	Section: 15.12
78.	e	Section: 15.13
79.	a	Section: 15.13
80.	d	Section: 15.15

1.	d	Section: 16.3
2.	b	Section: 16.3
3.	e	Section: 16.3
4.	b	Section: 16.3
5.	c	Section: 16.3
6.	e	Section: 16.3
7.	e	Section: 16.3
8.	b	Section: 16.3
9.	e	Section: 16.5
10.	e	Section: 16.5
11.	e	Section: 16.7
12.	d	Section: 16.7
13.	a	Section: 16.8
14.	a	Section: 16.8
15.	d	Section: 16.8
16.	d	Section: 16.8
17.	c	Section: 16.8
18.	c	Section: 16.8
19.	b	Section: 16.8
20.	b	Section: 16.8
21.	e	Section: 16.8
22.	c	Section: 16.9
23.	b	Section: 16.9
24.	d	Section: 16.9
25.	a	Section: 16.10
26.	c	Section: 16.10
27.	a	Section: 16.11
28.	a	Section: 16.11
29.	b	Section: 16.11
30.	d	Section: 16.11
31.	b	Section: 16.11
32.	c	Section: 16.11
33.	a	Section: 16.11
34.	d	Section: 16.11
35.	c	Section: 16.11
36.	b	Section: 16.11
37.	a	Section: 16.11
38.	a	Section: 16.11
39.	a	Section: 16.12
40.	e	Section: 16.12
41.	d	Section: 16.12
42.	a	Section: 16.12
43.	c	Section: 16.12
44.	a	Section: 16.12
45.	a	Section: 16.12
46.	c	Section: 16.12
47.	c	Section: 16.12
48.	a	Section: 16.12
49.	e	Section: 16.12
50.	*(No answer.)*	Section: 6.12
51.	b	Section: 16.13
52.	*(No answer.)*	Section: 16.13
53.	a	Section: 16.14
54.	e	Section: 16.14
55.	a	Section: 16.15
56.	b	Section: 16.15
57.	d	Section: 16.15
58.	a	Section: 16.15
59.	b	Section: 16.15
60.	d	Section: 16.15
61.	c	Section: 16.15
62.	b	Section: 16.15
63.	a	Section: 16.15
64.	b	Section: 16.15
65.	e	Section: 16.15
66.	a	Section: 16.15
67.	d	Section: 16.15
68.	b	Section: 16.15
69.	d	Section: 16.15
70.	b	Section: 16.15
71.	e	Section: 16.15
72.	c	Section: 16.15
73.	a	Section: 16.15
74.	*(No answer.)*	Section: 16.15
75.	e	Section: 16.16
76.	*(No answer.)*	Section: 16.16
77.	*(No answer.)*	Section: 16.16
78.	e	Section: 16.16
79.	a	Section: 16.16
80.	e	Section: 16.16

1.	c	Section:	16.3
2.	a	Section:	16.3
3.	a	Section:	16.3
4.	a	Section:	16.3
5.	c	Section:	16.3
6.	d	Section:	16.3
7.	e	Section:	16.3
8.	d	Section:	16.3
9.	a	Section:	16.5
10.	b	Section:	16.5
11.	c	Section:	16.7
12.	c	Section:	16.7
13.	e	Section:	16.8
14.	a	Section:	16.8
15.	a	Section:	16.8
16.	d	Section:	16.8
17.	c	Section:	16.8
18.	c	Section:	16.8
19.	d	Section:	16.8
20.	a	Section:	16.8
21.	c	Section:	16.8
22.	d	Section:	16.9
23.	b	Section:	16.9
24.	a	Section:	16.9
25.	e	Section:	16.10
26.	d	Section:	16.10
27.	c	Section:	16.11
28.	d	Section:	16.11
29.	b	Section:	16.11
30.	c	Section:	16.11
31.	c	Section:	16.11
32.	a	Section:	16.11
33.	b	Section:	16.11
34.	d	Section:	16.11
35.	b	Section:	16.11
36.	d	Section:	16.11
37.	a	Section:	16.11
38.	a	Section:	16.11
39.	c	Section:	16.12
40.	b	Section:	16.12
41.	a	Section:	16.12
42.	e	Section:	16.12
43.	b	Section:	16.12
44.	c	Section:	16.12
45.	a	Section:	16.12
46.	d	Section:	16.12
47.	c	Section:	16.12
48.	d	Section:	16.12
49.	a	Section:	16.12
50.	*(No answer.)*	Section:	16.12
51.	d	Section:	16.13
52.	*(No answer.)*	Section:	16.13
53.	c	Section:	16.14
54.	a	Section:	16.14
55.	b	Section:	16.15
56.	d	Section:	16.15
57.	c	Section:	16.15
58.	d	Section:	16.15
59.	a	Section:	16.15
60.	d	Section:	16.15
61.	a	Section:	16.15
62.	a	Section:	16.15
63.	a	Section:	16.15
64.	a	Section:	16.15
65.	d	Section:	16.15
66.	d	Section:	16.15
67.	c	Section:	16.15
68.	d	Section:	16.15
69.	b	Section:	16.15
70.	c	Section:	16.15
71.	c	Section:	16.15
72.	c	Section:	16.15
73.	d	Section:	16.15
74.	*(No answer.)*	Section:	16.15
75.	c	Section:	16.16
76.	*(No answer.)*	Section:	16.16
77.	*(No answer.)*	Section:	16.16
78.	b	Section:	16.16
79.	a	Section:	16.16
80.	d	Section:	16.16

1.	a	Section:	17.1
2.	a	Section:	17.1
3.	e	Section:	17.1
4.	a	Section:	17.2
5.	a	Section:	17.2
6.	a	Section:	17.2
7.	b	Section:	17.2
8.	e	Section:	17.2
9.	a	Section:	17.2
10.	a	Section:	17.2
11.	c	Section:	17.2
12.	a	Section:	17.2
13.	a	Section:	17.2
14.	b	Section:	17.4
15.	b	Section:	17.4
16.	c	Section:	17.5
17.	c	Section:	17.5
18.	d	Section:	17.5
19.	d	Section:	17.5
20.	e	Section:	17.5
21.	c	Section:	17.5
22.	d	Section:	17.6
23.	e	Section:	17.6
24.	b	Section:	17.6
25.	d	Section:	17.6
26.	c	Section:	17.6
27.	b	Section:	17.6
28.	d	Section:	17.6
29.	c	Section:	17.6
30.	b	Section:	17.6
31.	*(No answer.)*	Section:	17.6
32.	b	Section:	17.7
33.	a	Section:	17.7
34.	a	Section:	17.7
35.	a	Section:	17.7
36.	d	Section:	17.7
37.	a	Section:	17.7
38.	c	Section:	17.8
39.	a	Section:	17.8
40.	a	Section:	17.8
41.	c	Section:	17.8
42.	e	Section:	17.8
43.	d	Section:	17.8
44.	a	Section:	17.8
45.	b	Section:	17.8
46.	d	Section:	17.8
47.	b	Section:	17.8
48.	*(No answer.)*	Section:	17.8
49.	*(No answer.)*	Section:	17.8
50.	*(No answer.)*	Section:	17.9
51.	e	Section:	17.9
52.	e	Section:	17.9
53.	c	Section:	17.9
54.	b	Section:	17.9
55.	a	Section:	17.9
56.	a	Section:	17.9
57.	a	Section:	17.9
58.	*(No answer.)*	Section:	17.10
59.	e	Section:	17.10
60.	b	Section:	17.10
61.	d	Section:	17.10
62.	e	Section:	17.10
63.	b	Section:	17.10
64.	b	Section:	17.10
65.	d	Section:	17.10
66.	e	Section:	17.10
67.	e	Section:	17.10
68.	b	Section:	17.11
69.	e	Section:	17.11
70.	a	Section:	17.12
71.	c	Section:	17.13
72.	c	Section:	17.13
73.	a	Section:	17.13
74.	d	Section:	17.13
75.	c	Section:	17.15
76.	c	Section:	17.15
77.	b	Section:	17.15
78.	a	Section:	17.15
79.	d	Section:	17.15
80.	e	Section:	17.15

1.	a	Section:	17.1
2.	d	Section:	17.1
3.	a	Section:	17.1
4.	a	Section:	17.2
5.	a	Section:	17.2
6.	a	Section:	17.2
7.	c	Section:	17.2
8.	b	Section:	17.2
9.	a	Section:	17.2
10.	a	Section:	17.2
11.	a	Section:	17.2
12.	a	Section:	17.2
13.	a	Section:	17.2
14.	c	Section:	17.4
15.	e	Section:	17.4
16.	d	Section:	17.5
17.	e	Section:	17.5
18.	b	Section:	17.5
19.	a	Section:	17.5
20.	c	Section:	17.5
21.	d	Section:	17.5
22.	b	Section:	17.6
23.	b	Section:	17.6
24.	b	Section:	17.6
25.	d	Section:	17.6
26.	b	Section:	17.6
27.	d	Section:	17.6
28.	d	Section:	17.6
29.	d	Section:	17.6
30.	c	Section:	17.6
31.	*(No answer.)*	Section:	17.6
32.	a	Section:	17.7
33.	a	Section:	17.7
34.	e	Section:	17.7
35.	a	Section:	17.7
36.	a	Section:	17.7
37.	a	Section:	17.7
38.	b	Section:	17.8
39.	a	Section:	17.8
40.	a	Section:	17.8
41.	c	Section:	17.8
42.	a	Section:	17.8
43.	c	Section:	17.8
44.	a	Section:	17.8
45.	a	Section:	17.8
46.	a	Section:	17.8
47.	a	Section:	17.8
48.	*(No answer.)*	Section:	17.8
49.	*(No answer.)*	Section:	17.8
50.	*(No answer.)*	Section:	17.9
51.	c	Section:	17.9
52.	c	Section:	17.9
53.	c	Section:	17.9
54.	a	Section:	17.9
55.	b	Section:	17.9
56.	d	Section:	17.9
57.	d	Section:	17.9
58.	*(No answer.)*	Section:	17.10
59.	c	Section:	17.10
60.	a	Section:	17.10
61.	b	Section:	17.10
62.	a	Section:	17.10
63.	b	Section:	17.10
64.	b	Section:	17.10
65.	d	Section:	17.10
66.	b	Section:	17.10
67.	c	Section:	17.10
68.	d	Section:	17.11
69.	a	Section:	17.11
70.	a	Section:	17.12
71.	b	Section:	17.13
72.	e	Section:	17.13
73.	b	Section:	17.13
74.	a	Section:	17.13
75.	b	Section:	17.15
76.	e	Section:	17.15
77.	b	Section:	17.15
78.	c	Section:	17.15
79.	d	Section:	17.15
80.	a	Section:	17.15

Chapter 18A Answers

1.	c	Section: 18.1
2.	c	Section: 18.1
3.	d	Section: 18.1
4.	b	Section: 18.3
5.	a	Section: 18.3
6.	b	Section: 18.3
7.	a	Section: 18.3
8.	a	Section: 18.3
9.	a	Section: 18.3
10.	a	Section: 18.3
11.	a	Section: 18.3
12.	a	Section: 18.3
13.	b	Section: 18.3
14.	b	Section: 18.3
15.	a	Section: 18.5
16.	c	Section: 18.5
17.	d	Section: 18.5
18.	d	Section: 18.5
19.	d	Section: 18.5
20.	c	Section: 18.5
21.	e	Section: 18.5
22.	c	Section: 18.5
23.	d	Section: 18.5
24.	b	Section: 18.5
25.	d	Section: 18.5
26.	d	Section: 18.5
27.	*(No answer.)*	Section: 18.5
28.	*(No answer.)*	Section: 18.5
29.	e	Section: 18.5
30.	b	Section: 18.5
31.	c	Section: 18.5
32.	e	Section: 18.5
33.	a	Section: 18.5
34.	b	Section: 18.5
35.	e	Section: 18.5
36.	b	Section: 18.6
37.	d	Section: 18.6
38.	b	Section: 18.6
39.	a	Section: 18.6
40.	c	Section: 18.6
41.	d	Section: 18.6
42.	b	Section: 18.6
43.	e	Section: 18.6
44.	d	Section: 18.6
45.	d	Section: 18.6
46.	a	Section: 18.6
47.	d	Section: 18.6
48.	e	Section: 18.6
49.	e	Section: 18.6
50.	a	Section: 18.6
51.	c	Section: 18.7
52.	d	Section: 18.7
53.	a	Section: 18.7
54.	a	Section: 18.7
55.	c	Section: 18.7
56.	e	Section: 18.7
57.	a	Section: 18.7
58.	c	Section: 18.10
59.	*(No answer.)*	Section: 18.10
60.	a	Section: 18.10
61.	a	Section: 18.10
62.	b	Section: 18.10
63.	b	Section: 18.10
64.	d	Section: 18.10
65.	c	Section: 18.12
66.	b	Section: 18.14
67.	d	Section: 18.14
68.	a	Section: 18.14
69.	e	Section: 18.14
70.	c	Section: 18.14
71.	a	Section: 18.15
72.	a	Section: 18.15
73.	b	Section: 18.15
74.	b	Section: 18.15
75.	c	Section: 18.15
76.	d	Section: 18.15
77.	c	Section: 18.15
78.	*(No answer.)*	Section: 18.15
79.	*(No answer.)*	Section: 18.15
80.	a	Section: 18.16

1.	a	Section: 18.1
2.	d	Section: 18.1
3.	e	Section: 18.1
4.	c	Section: 18.3
5.	a	Section: 18.3
6.	a	Section: 18.3
7.	e	Section: 18.3
8.	a	Section: 18.3
9.	d	Section: 18.3
10.	a	Section: 18.3
11.	a	Section: 18.3
12.	a	Section: 18.3
13.	a	Section: 18.3
14.	b	Section: 18.3
15.	a	Section: 18.5
16.	c	Section: 18.5
17.	d	Section: 18.5
18.	e	Section: 18.5
19.	c	Section: 18.5
20.	e	Section: 18.5
21.	b	Section: 18.5
22.	b	Section: 18.5
23.	e	Section: 18.5
24.	a	Section: 18.5
25.	a	Section: 18.5
26.	d	Section: 18.5
27.	*(No answer.)*	Section: 18.5
28.	*(No answer.)*	Section: 18.5
29.	d	Section: 18.5
30.	a	Section: 18.5
31.	e	Section: 18.5
32.	c	Section: 18.5
33.	a	Section: 18.5
34.	a	Section: 18.5
35.	e	Section: 18.5
36.	d	Section: 18.6
37.	a	Section: 18.6
38.	e	Section: 18.6
39.	a	Section: 18.6
40.	d	Section: 18.6
41.	a	Section: 18.6
42.	d	Section: 18.6
43.	c	Section: 18.6
44.	d	Section: 18.6
45.	e	Section: 18.6
46.	a	Section: 18.6
47.	d	Section: 18.6
48.	e	Section: 18.6
49.	e	Section: 18.6
50.	a	Section: 18.6
51.	e	Section: 18.7
52.	d	Section: 18.7
53.	e	Section: 18.7
54.	c	Section: 18.7
55.	e	Section: 18.7
56.	b	Section: 18.7
57.	e	Section: 18.7
58.	b	Section: 18.10
59.	*(No answer.)*	Section: 18.10
60.	d	Section: 18.10
61.	e	Section: 18.10
62.	b	Section: 18.10
63.	a	Section: 18.10
64.	e	Section: 18.10
65.	c	Section: 18.12
66.	a	Section: 18.14
67.	a	Section: 18.14
68.	d	Section: 18.14
69.	b	Section: 18.14
70.	b	Section: 18.14
71.	b	Section: 18.15
72.	e	Section: 18.15
73.	b	Section: 18.15
74.	a	Section: 18.15
75.	d	Section: 18.15
76.	b	Section: 18.15
77.	e	Section: 18.15
78.	*(No answer.)*	Section: 18.15
79.	*(No answer.)*	Section: 18.15
80.	e	Section: 18.16

1.	d	Section: 19.1
2.	a	Section: 19.1
3.	a	Section: 19.1
4.	b	Section: 19.1
5.	d	Section: 19.1
6.	a	Section: 19.1
7.	a	Section: 19.1
8.	a	Section: 19.1
9.	b	Section: 19.2
10.	a	Section: 19.2
11.	a	Section: 19.2
12.	a	Section: 19.2
13.	d	Section: 19.2
14.	a	Section: 19.2
15.	e	Section: 19.2
16.	*(No answer.)*	Section: 19.2
17.	a	Section: 19.3
18.	a	Section: 19.4
19.	a	Section: 19.4
20.	d	Section: 19.4
21.	b	Section: 19.4
22.	c	Section: 19.4
23.	d	Section: 19.4
24.	c	Section: 19.5
25.	c	Section: 19.6
26.	a	Section: 19.6
27.	a	Section: 19.6
28.	a	Section: 19.6
29.	a	Section: 19.6
30.	a	Section: 19.6
31.	c	Section: 19.6
32.	a	Section: 19.6
33.	e	Section: 19.6
34.	a	Section: 19.6
35.	b	Section: 19.6
36.	a	Section: 19.6
37.	*(No answer.)*	Section: 19.6
38.	*(No answer.)*	Section: 19.6
39.	b	Section: 19.7
40.	d	Section: 19.7
41.	e	Section: 19.7
42.	d	Section: 19.7
43.	a	Section: 19.8
44.	a	Section: 19.8
45.	d	Section: 19.8
46.	a	Section: 19.8
47.	a	Section: 19.8
48.	a	Section: 19.8
49.	b	Section: 19.8
50.	e	Section: 19.9
51.	d	Section: 19.9
52.	c	Section: 19.9
53.	c	Section: 19.9
54.	e	Section: 19.9
55.	a	Section: 19.10
56.	a	Section: 19.10
57.	a	Section: 19.10
58.	e	Section: 19.11
59.	b	Section: 19.11
60.	e	Section: 19.11
61.	a	Section: 19.11
62.	*(No answer.)*	Section: 19.11
63.	a	Section: 19.12
64.	a	Section: 19.12
65.	a	Section: 19.12
66.	e	Section: 19.13
67.	a	Section: 19.13
68.	a	Section: 19.13
69.	b	Section: 19.14
70.	c	Section: 19.14
71.	c	Section: 19.15
72.	a	Section: 19.15
73.	a	Section: 19.16
74.	a	Section: 19.16
75.	a	Section: 19.16
76.	*(No answer.)*	Section: 19.16
77.	a	Section: 19.17
78.	a	Section: 19.17
79.	c	Section: 19.17
80.	d	Section: 19.17

1.	c	Section: 19.1
2.	a	Section: 19.1
3.	a	Section: 19.1
4.	c	Section: 19.1
5.	a	Section: 19.1
6.	a	Section: 19.1
7.	a	Section: 19.1
8.	e	Section: 19.1
9.	c	Section: 19.2
10.	a	Section: 19.2
11.	c	Section: 19.2
12.	a	Section: 19.2
13.	a	Section: 19.2
14.	a	Section: 19.2
15.	e	Section: 19.2
16.	*(No answer.)*	Section: 19.2
17.	a	Section: 19.3
18.	a	Section: 19.4
19.	b	Section: 19.4
20.	d	Section: 19.4
21.	b	Section: 19.4
22.	a	Section: 19.4
23.	a	Section: 19.4
24.	c	Section: 19.5
25.	e	Section: 19.6
26.	c	Section: 19.6
27.	a	Section: 19.6
28.	a	Section: 19.6
29.	a	Section: 19.6
30.	e	Section: 19.6
31.	e	Section: 19.6
32.	a	Section: 19.6
33.	e	Section: 19.6
34.	a	Section: 19.6
35.	e	Section: 19.6
36.	a	Section: 19.6
37.	*(No answer.)*	Section: 19.6
38.	*(No answer.)*	Section: 19.6
39.	e	Section: 19.7
40.	a	Section: 19.7
41.	c	Section: 19.7
42.	e	Section: 19.7
43.	a	Section: 19.8
44.	a	Section: 19.8
45.	b	Section: 19.8
46.	c	Section: 19.8
47.	a	Section: 19.8
48.	e	Section: 19.8
49.	c	Section: 19.8
50.	c	Section: 19.9
51.	b	Section: 19.9
52.	d	Section: 19.9
53.	e	Section: 19.9
54.	a	Section: 19.9
55.	a	Section: 19.10
56.	a	Section: 19.10
57.	a	Section: 19.10
58.	c	Section: 19.11
59.	d	Section: 19.11
60.	d	Section: 19.11
61.	a	Section: 19.11
62.	*(No answer.)*	Section: 19.11
63.	a	Section: 19.12
64.	a	Section: 19.12
65.	b	Section: 19.12
66.	b	Section: 19.13
67.	a	Section: 19.13
68.	a	Section: 19.13
69.	e	Section: 19.14
70.	a	Section: 19.14
71.	b	Section: 19.15
72.	a	Section: 19.15
73.	a	Section: 19.16
74.	a	Section: 19.16
75.	a	Section: 19.16
76.	*(No answer.)*	Section: 19.16
77.	a	Section: 19.17
78.	e	Section: 19.17
79.	a	Section: 19.17
80.	a	Section: 19.17

1.	a	Section: 20.1
2.	a	Section: 20.1
3.	a	Section: 20.1
4.	c	Section: 20.1
5.	a	Section: 20.2
6.	a	Section: 20.3
7.	*(No answer.)*	Section: 20.3
8.	b	Section: 20.3
9.	a	Section: 20.3
10.	a	Section: 20.3
11.	a	Section: 20.3
12.	b	Section: 20.3
13.	d	Section: 20.3
14.	d	Section: 20.3
15.	a	Section: 20.3
16.	b	Section: 20.4
17.	c	Section: 20.4
18.	c	Section: 20.4
19.	c	Section: 20.4
20.	a	Section: 20.4
21.	c	Section: 20.4
22.	c	Section: 20.4
23.	b	Section: 20.4
24.	*(No answer.)*	Section: 20.4
25.	a	Section: 20.5
26.	b	Section: 20.5
27.	e	Section: 20.5
28.	a	Section: 20.6
29.	c	Section: 20.6
30.	*(No answer.)*	Section: 20.6
31.	b	Section: 20.7
32.	a	Section: 20.7
33.	d	Section: 20.7
34.	b	Section: 20.7
35.	c	Section: 20.7
36.	e	Section: 20.7
37.	a	Section: 20.7
38.	b	Section: 20.7
39.	a	Section: 20.7
40.	b	Section: 20.7
41.	a	Section: 20.8
42.	a	Section: 20.8
43.	a	Section: 20.8
44.	a	Section: 20.9
45.	a	Section: 20.9
46.	a	Section: 20.9
47.	b	Section: 20.9
48.	b	Section: 20.9
49.	c	Section: 20.9
50.	c	Section: 20.9
51.	a	Section: 20.10
52.	a	Section: 20.10
53.	e	Section: 20.10
54.	a	Section: 20.10
55.	a	Section: 20.10
56.	a	Section: 20.10
57.	b	Section: 20.10
58.	e	Section: 20.10
59.	a	Section: 20.10
60.	a	Section: 20.10
61.	b	Section: 20.10
62.	e	Section: 20.10
63.	c	Section: 20.10
64.	a	Section: 20.10
65.	a	Section: 20.10
66.	a	Section: 20.10
67.	*(No answer.)*	Section: 20.10
68.	b	Section: 20.11
69.	a	Section: 20.11
70.	*(No answer.)*	Section: 20.11
71.	a	Section: 20.12
72.	a	Section: 20.12
73.	d	Section: 20.12
74.	b	Section: 20.12
75.	d	Section: 20.12
76.	a	Section: 20.12
77.	a	Section: 20.12
78.	a	Section: 20.12
79.	c	Section: 20.12
80.	a	Section: 20.12

1.	a	Section: 20.1
2.	a	Section: 20.1
3.	a	Section: 20.1
4.	e	Section: 20.1
5.	a	Section: 20.2
6.	d	Section: 20.3
7.	*(No answer.)*	Section: 20.3
8.	a	Section: 20.3
9.	a	Section: 20.3
10.	a	Section: 20.3
11.	b	Section: 20.3
12.	a	Section: 20.3
13.	b	Section: 20.3
14.	d	Section: 20.3
15.	b	Section: 20.3
16.	c	Section: 20.4
17.	b	Section: 20.4
18.	a	Section: 20.4
19.	b	Section: 20.4
20.	a	Section: 20.4
21.	b	Section: 20.4
22.	b	Section: 20.4
23.	c	Section: 20.4
24.	*(No answer.)*	Section: 20.4
25.	e	Section: 20.5
26.	e	Section: 20.5
27.	c	Section: 20.5
28.	b	Section: 20.6
29.	a	Section: 20.6
30.	*(No answer.)*	Section: 20.6
31.	d	Section: 20.7
32.	a	Section: 20.7
33.	a	Section: 20.7
34.	d	Section: 20.7
35.	e	Section: 20.7
36.	b	Section: 20.7
37.	e	Section: 20.7
38.	d	Section: 20.7
39.	a	Section: 20.7
40.	a	Section: 20.7
41.	a	Section: 20.8
42.	a	Section: 20.8
43.	c	Section: 20.8
44.	a	Section: 20.9
45.	a	Section: 20.9
46.	a	Section: 20.9
47.	c	Section: 20.9
48.	d	Section: 20.9
49.	e	Section: 20.9
50.	d	Section: 20.9
51.	a	Section: 20.10
52.	a	Section: 20.10
53.	a	Section: 20.10
54.	d	Section: 20.10
55.	a	Section: 20.10
56.	a	Section: 20.10
57.	b	Section: 20.10
58.	b	Section: 20.10
59.	a	Section: 20.10
60.	a	Section: 20.10
61.	c	Section: 20.10
62.	a	Section: 20.10
63.	b	Section: 20.10
64.	c	Section: 20.10
65.	a	Section: 20.10
66.	a	Section: 20.10
67.	*(No answer.)*	Section: 20.10
68.	a	Section: 20.11
69.	a	Section: 20.11
70.	*(No answer.)*	Section: 20.11
71.	a	Section: 20.12
72.	a	Section: 20.12
73.	e	Section: 20.12
74.	a	Section: 20.12
75.	e	Section: 20.12
76.	a	Section: 20.12
77.	d	Section: 20.12
78.	a	Section: 20.12
79.	a	Section: 20.12
80.	a	Section: 20.12

1.	d	Section: 21.1
2.	a	Section: 21.1
3.	b	Section: 21.1
4.	b	Section: 21.1
5.	a	Section: 21.1
6.	a	Section: 21.1
7.	d	Section: 21.1
8.	a	Section: 21.1
9.	d	Section: 21.1
10.	a	Section: 21.1
11.	b	Section: 21.1
12.	d	Section: 21.1
13.	a	Section: 21.1
14.	a	Section: 21.1
15.	a	Section: 21.1
16.	e	Section: 21.1
17.	a	Section: 21.1
18.	a	Section: 21.2
19.	a	Section: 21.2
20.	a	Section: 21.2
21.	a	Section: 21.2
22.	e	Section: 21.2
23.	d	Section: 21.3
24.	b	Section: 21.3
25.	c	Section: 21.3
26.	d	Section: 21.3
27.	e	Section: 21.3
28.	b	Section: 21.3
29.	a	Section: 21.3
30.	c	Section: 21.3
31.	a	Section: 21.3
32.	*(No answer.)*	Section: 21.4
33.	b	Section: 21.4
34.	a	Section: 21.4
35.	d	Section: 21.4
36.	a	Section: 21.4
37.	b	Section: 21.4
38.	e	Section: 21.5
39.	a	Section: 21.5
40.	a	Section: 21.5
41.	a	Section: 21.5
42.	d	Section: 21.5
43.	c	Section: 21.5
44.	b	Section: 21.5
45.	a	Section: 21.5
46.	a	Section: 21.5
47.	a	Section: 21.5
48.	a	Section: 21.5
49.	b	Section: 21.5
50.	a	Section: 21.6
51.	a	Section: 21.6
52.	b	Section: 21.6
53.	a	Section: 21.6
54.	a	Section: 21.6
55.	a	Section: 21.6
56.	e	Section: 21.6
57.	a	Section: 21.6
58.	a	Section: 21.6
59.	a	Section: 21.6
60.	a	Section: 21.6
61.	d	Section: 21.6
62.	a	Section: 21.6
63.	a	Section: 21.6
64.	c	Section: 21.6
65.	*(No answer.)*	Section: 21.6
66.	*(No answer.)*	Section: 21.6
67.	d	Section: 21.7
68.	b	Section: 21.7
69.	c	Section: 21.8
70.	e	Section: 21.8
71.	b	Section: 21.8
72.	*(No answer.)*	Section: 21.8
73.	a	Section: 21.9
74.	d	Section: 21.9
75.	d	Section: 21.9
76.	a	Section: 21.9
77.	*(No answer.)*	Section: 21.9
78.	b	Section: 21.10
79.	a	Section: 21.10
80.	b	Section: 21.10

Chapter 21B Answers

1.	a	Section: 21.1
2.	a	Section: 21.1
3.	e	Section: 21.1
4.	a	Section: 21.1
5.	a	Section: 21.1
6.	a	Section: 21.1
7.	e	Section: 21.1
8.	a	Section: 21.1
9.	b	Section: 21.1
10.	a	Section: 21.1
11.	c	Section: 21.1
12.	e	Section: 21.1
13.	b	Section: 21.1
14.	b	Section: 21.1
15.	a	Section: 21.1
16.	b	Section: 21.1
17.	a	Section: 21.1
18.	c	Section: 21.2
19.	a	Section: 21.2
20.	d	Section: 21.2
21.	e	Section: 21.2
22.	c	Section: 21.2
23.	a	Section: 21.3
24.	c	Section: 21.3
25.	a	Section: 21.3
26.	b	Section: 21.3
27.	c	Section: 21.3
28.	b	Section: 21.3
29.	c	Section: 21.3
30.	a	Section: 21.3
31.	b	Section: 21.3
32.	*(No answer.)*	Section: 21.4
33.	b	Section: 21.4
34.	d	Section: 21.4
35.	c	Section: 21.4
36.	a	Section: 21.4
37.	c	Section: 21.4
38.	b	Section: 21.5
39.	a	Section: 21.5
40.	a	Section: 21.5
41.	a	Section: 21.5
42.	e	Section: 21.5
43.	a	Section: 21.5
44.	a	Section: 21.5
45.	a	Section: 21.5
46.	c	Section: 21.5
47.	a	Section: 21.5
48.	a	Section: 21.5
49.	a	Section: 21.5
50.	b	Section: 21.6
51.	a	Section: 21.6
52.	e	Section: 21.6
53.	a	Section: 21.6
54.	a	Section: 21.6
55.	a	Section: 21.6
56.	a	Section: 21.6
57.	c	Section: 21.6
58.	b	Section: 21.6
59.	a	Section: 21.6
60.	a	Section: 21.6
61.	a	Section: 21.6
62.	a	Section: 21.6
63.	b	Section: 21.6
64.	c	Section: 21.6
65.	*(No answer.)*	Section: 21.6
66.	*(No answer.)*	Section: 21.6
67.	d	Section: 21.7
68.	c	Section: 21.7
69.	d	Section: 21.8
70.	e	Section: 21.8
71.	b	Section: 21.8
72.	*(No answer.)*	Section: 21.8
73.	a	Section: 21.9
74.	e	Section: 21.9
75.	c	Section: 21.9
76.	b	Section: 21.9
77.	*(No answer.)*	Section: 21.9
78.	e	Section: 21.10
79.	a	Section: 21.10
80.	e	Section: 21.10

1.	e	Section: 22.1
2.	c	Section: 22.1
3.	b	Section: 22.1
4.	a	Section: 22.1
5.	b	Section: 22.2
6.	a	Section: 22.3
7.	a	Section: 22.3
8.	a	Section: 22.3
9.	a	Section: 22.3
10.	b	Section: 22.3
11.	c	Section: 22.3
12.	a	Section: 22.3
13.	a	Section: 22.3
14.	c	Section: 22.3
15.	e	Section: 22.3
16.	b	Section: 22.3
17.	a	Section: 22.3
18.	e	Section: 22.3
19.	e	Section: 22.3
20.	c	Section: 22.3
21.	a	Section: 22.3
22.	a	Section: 22.3
23.	a	Section: 22.3
24.	a	Section: 22.3
25.	a	Section: 22.3
26.	b	Section: 22.3
27.	e	Section: 22.3
28.	a	Section: 22.3
29.	*(No answer.)*	Section: 22.3
30.	*(No answer.)*	Section: 22.4
31.	b	Section: 22.4
32.	c	Section: 22.4
33.	a	Section: 22.4
34.	d	Section: 22.4
35.	b	Section: 22.4
36.	b	Section: 22.4
37.	e	Section: 22.4
38.	e	Section: 22.4
39.	a	Section: 22.4
40.	b	Section: 22.4
41.	b	Section: 22.4
42.	b	Section: 22.4
43.	b	Section: 22.4
44.	a	Section: 22.4
45.	b	Section: 22.5
46.	b	Section: 22.6
47.	a	Section: 22.6
48.	*(No answer.)*	Section: 22.6
49.	*(No answer.)*	Section: 22.8
50.	e	Section: 22.8
51.	c	Section: 22.8
52.	c	Section: 22.8
53.	e	Section: 22.8
54.	a	Section: 22.8
55.	a	Section: 22.8
56.	e	Section: 22.8
57.	a	Section: 22.8
58.	d	Section: 22.8
59.	a	Section: 22.8
60.	a	Section: 22.8
61.	e	Section: 22.8
62.	c	Section: 22.8
63.	b	Section: 22.8
64.	e	Section: 22.8
65.	c	Section: 22.8
66.	a	Section: 22.8
67.	b	Section: 22.8
68.	a	Section: 22.8
69.	d	Section: 22.8
70.	b	Section: 22.8
71.	a	Section: 22.8
72.	b	Section: 22.8
73.	a	Section: 22.9
74.	b	Section: 22.9
75.	e	Section: 22.9
76.	b	Section: 22.9
77.	*(No answer.)*	Section: 22.9
78.	e	Section: 22.10
79.	c	Section: 22.10
80.	a	Section: 22.12

1.	c	Section: 22.1
2.	b	Section: 22.1
3.	c	Section: 22.1
4.	a	Section: 22.1
5.	e	Section: 22.2
6.	b	Section: 22.3
7.	a	Section: 22.3
8.	a	Section: 22.3
9.	d	Section: 22.3
10.	c	Section: 22.3
11.	a	Section: 22.3
12.	b	Section: 22.3
13.	b	Section: 22.3
14.	b	Section: 22.3
15.	d	Section: 22.3
16.	b	Section: 22.3
17.	b	Section: 22.3
18.	e	Section: 22.3
19.	e	Section: 22.3
20.	d	Section: 22.3
21.	a	Section: 22.3
22.	a	Section: 22.3
23.	a	Section: 22.3
24.	a	Section: 22.3
25.	a	Section: 22.3
26.	d	Section: 22.3
27.	b	Section: 22.3
28.	a	Section: 22.3
29.	*(No answer.)*	Section: 22.3
30.	*(No answer.)*	Section: 22.4
31.	d	Section: 22.4
32.	d	Section: 22.4
33.	a	Section: 22.4
34.	b	Section: 22.4
35.	d	Section: 22.4
36.	b	Section: 22.4
37.	b	Section: 22.4
38.	b	Section: 22.4
39.	e	Section: 22.4
40.	d	Section: 22.4
41.	c	Section: 22.4
42.	b	Section: 22.4
43.	d	Section: 22.4
44.	a	Section: 22.4
45.	a	Section: 22.5
46.	c	Section: 22.6
47.	a	Section: 22.6
48.	*(No answer.)*	Section: 22.6
49.	*(No answer.)*	Section: 22.8
50.	b	Section: 22.8
51.	c	Section: 22.8
52.	c	Section: 22.8
53.	e	Section: 22.8
54.	a	Section: 22.8
55.	a	Section: 22.8
56.	e	Section: 22.8
57.	a	Section: 22.8
58.	d	Section: 22.8
59.	a	Section: 22.8
60.	a	Section: 22.8
61.	c	Section: 22.8
62.	e	Section: 22.8
63.	d	Section: 22.8
64.	b	Section: 22.8
65.	d	Section: 22.8
66.	e	Section: 22.8
67.	c	Section: 22.8
68.	a	Section: 22.8
69.	e	Section: 22.8
70.	b	Section: 22.8
71.	e	Section: 22.8
72.	a	Section: 22.8
73.	d	Section: 22.9
74.	a	Section: 22.9
75.	d	Section: 22.9
76.	c	Section: 22.9
77.	*(No answer.)*	Section: 22.9
78.	e	Section: 22.10
79.	a	Section: 22.10
80.	c	Section: 22.12